COVARIANT OPERATOR FORMALISM
OF GAUGE THEORIES & QUANTUM GRAVITY

ERRATA

page	line	instead of	should read
26	1↑	$\pi_\alpha = -\psi_\alpha^\dagger$	$\pi_\alpha = -i\psi_\alpha^\dagger$
110	footnote	$: e^A :: e^B := e^{[A^{(+)}, B^{(-)}]}$ $: e^{A+B} :$	$e^A e^B = e^{\frac{1}{2}[A,B]} e^{A+B}$ if $[A, B] = $ c-number
296	Eq. (5.2.4–26)	$\partial_\nu g_{\alpha\tau}$	$\partial_\sigma g_{\alpha\tau}$
314	footnote 4	$M(\hat{x}^\mu, b_\nu)$	$M(\hat{x}^\mu, \bar{c}_\nu)$
316	Eq. (5.5.3–7)	$E_{\mu\nu} = \sum_{X,Y} \cdots$	$E_{\mu\nu} = \kappa \sum_{X,Y} \cdots$
	Eq. (5.5.3–8)	$\kappa \delta_{XY}(\hat{x}^\mu)\tilde{\mathcal{L}}$	$\kappa \delta_{XY}(\hat{x}^\mu)\mathcal{L}$
333	Eq. (5.6.6–11)	$f_{m,2}$	$F_{m,2}$
342	Eq. (5.7.4–3)	$2h_{\nu a}\pi_g^{\mu\nu} + \tilde{g}^{\nu\lambda}\ldots - ih\bar{\psi}\ldots$	$2h_{\nu a}\pi_g^{\mu\nu} - \tilde{g}^{\nu\lambda}\ldots + ih\bar{\psi}\ldots$
	Eq. (5.7.4–4)	$\tilde{g}^{0\nu}\omega_\nu^{ab}$	$-\tilde{g}^{0\nu}\omega_\nu^{ab}$
	Eq. (5.7.4–5)	$-i\tilde{g}^{0\nu}\partial_\nu \bar{t}^{ab}$	$i\tilde{g}^{0\nu}\partial_\nu \bar{t}^{ab}$
	Eq. (5.7.4–6)	$i\tilde{g}^{0\nu}(D_\nu t)^{ab}$	$-i\tilde{g}^{0\nu}(D_\nu t)^{ab}$
	Eq. (5.7.4–7)	$ih\bar{\psi}\gamma^0$	$-ih\bar{\psi}\gamma^0$
350	Eq. (5.8.1–22)	$\tilde{\varphi}_2 \equiv \tilde{\varphi}_{22}$	$\tilde{\varphi}_2 \equiv \tilde{\varphi}_{12}$
355	10↑	$s\varsigma\kappa^{-1}$	$s \underset{\sim}{>} \kappa^{-1}$
365	5↓	the fact the	the fact that the

COVARIANT OPERATOR FORMALISM OF
GAUGE THEORIES AND QUANTUM GRAVITY

World Scientific Lecture Notes in Physics Vol. 27

COVARIANT OPERATOR FORMALISM OF GAUGE THEORIES AND QUANTUM GRAVITY

N. Nakanishi and I. Ojima

Research Institute for Mathematical Sciences
Kyoto University

World Scientific
Singapore • New Jersey • London • Hong Kong

Published by

World Scientific Publishing Co. Pte. Ltd.
P O Box 128, Farrer Road, Singapore 9128
USA office: 687 Hartwell Street, Teaneck, NJ 07666
UK office: 73 Lynton Mead, Totteridge, London N20 8DH

ISBN 9971-50-238-0
     9971-50-239-9 pbk

Printed in Singapore by Utopia Press.

PREFACE

Nowadays, general relativity and gauge theories are well established as the fundamental theories of physics. It is therefore of fundamental importance to construct satisfactory quantum theories of gauge fields and the gravitational field. In this book, we present the covariant canonical operator formalism of abelian and non-abelian gauge theories and of quantum gravity. We emphasize that characteristic features of our presentation are *manifest covariance, non-perturbative nature,* and *non-use of the path-integral approach.*

Although non-covariant gauges are often employed for technical reasons, we believe that it is the right way to respect manifest covariance, because it is the basic ingredient of quantum field theory. Then, as known since the Gupta-Bleuler formalism, use of indefinite metric is indispensable. Although there are still some people who dislike ghosts, we think that, aside from their taste, there are no more reasons than the mathematical hardness in the rigorous treatment. We wish to stress that use of indefinite metric is quite natural as the framework of quantum field theory.

Perturbation theory is quite important and very successful especially in quantum electrodynamics. We of course have no objection to this fact. But it must be noted that perturbative approach is nothing more than an approximation method. The fundamental *formulation* of the theory itself should not be based on the perturbative approach. Such a fundamental problem as color confinement cannot be discussed in the perturbative framework. Especially in quantum gravity, non-perturbative treatment is indispensable because of its perturbative non-renormalizability.

The path-integral approach is currently very fashionable; almost all textbooks on the non-abelian gauge theory are based on the path-integral formalism. It is certainly a very convenient technique for various calculations, but we believe that it should not be regarded as the basic formulation of the theory. Since the path-integral formalism has no such notions as conservation laws, symmetry generators, observables, asymptotic states, etc. in its own framework, the one who works in the path-integral approach must *borrow* basic concepts from the operator formalism. Furthermore, for the theories whose corresponding operator formalism need indefinite metric as in gauge theories and in quantum gravity, one must appeal to analytic continuation (called "Wick rotation") from the outset, thus totally violating the usual probabilistic interpretability of the amplitude.

The purpose of this book is to present the covariant operator formalism in the Heisenberg picture and to discuss some fundamental properties of the theories. We do not intend to discuss practical techniques for calculations, which can be found in many other excellent textbooks. Although we occasionally use such words as perturbation theory, renormalizability, etc. without making explanation, the main stream of the presentation is essentially self-contained.

The theories presented in this book are based on the Lagrangian canonical formalism. Although it is not mathematically rigorous, we believe that the Lagrangian canonical formalism provides us with the standard framework of quantum field theory. Because of extreme hardness of treating indefinite metric rigorously, it is more profitable to formulate the theory heuristically; we rely on the heuristic results unless they are explicitly disproved. Thus the mathematics which we use in the text is of the physicist's standard, though we occasionally refer to the results of the axiomatic field theory. Some mathematical aspects on the indefinite-metric quantum field theory and basic geometrical and cohomological notions are presented in the Appendix without claiming mathematical rigor.

The main contents of this book are based on the work to which either or both of us contributed to a considerable extent. We are unable to refer to all related work; we would like to apologize to the authors whose papers are not properly referred to.

We would like to thank Professor K. K. Phua of the World Scientific Publishing Co. Ltd. for making it possible to publish this book. We are very grateful to Professor T. Muta of Hiroshima University for introducing the World Scientific to us. One of us (N. N.) thanks Dr. M. Abe and the secretaries of the Research Institute for Mathematical Sciences, Kyoto University for typing his manuscript. We both cordially thank Dr. M. Abe for giving helpful instructions on the TEX-typing.

N. Nakanishi
I. Ojima

Kyoto, JAPAN
June 1990

CONTENTS

APPENDIX: MATHEMATICAL SUPPLEMENTS

We here summarize some of notation and convention used in this book.

We employ natural units, $\hbar = c = 1$, where $\hbar = h/2\pi$, h being the Planck constant, and c denotes the light velocity.

We always (except in Sec.2.5) work in the four-dimensional spacetime. As for the general D-dimensional case, remarks are made on footnotes.

We employ Einstein's convention for indices, that is, summation over indices should be understood, unless otherwise indicated, if the same script letter appears twice in a single product.

We use Greek little letters μ, ν, λ, ρ, σ, etc. for spacetime indices; for example, spacetime coordinates are denoted by x^μ ($\mu = 0, 1, 2, 3$). The spatial three-dimensional vectors are denoted by boldfaces.

The Minkowski metric tensor is denoted by $\eta_{\mu\nu}$; $\eta_{00} = -\eta_{11} = -\eta_{22} = -\eta_{33} = 1, \eta_{\mu\nu} = 0$ for $\mu \neq \nu$. The invariant square $\eta_{\mu\nu}x^\mu x^\nu$ is abbreviated as x^2; the scalar product $p_\mu x^\mu$ is abbreviated as px. The totally antisymmetric tensor $\epsilon_{\mu\nu\sigma\tau}$ is defined by $\epsilon_{0123} = 1$ and sign change under any transposition of indices. For gamma matrices, see Sec.1.3.4.

The spacetime differential operator $\partial/\partial x^\mu$ is abbreviated as ∂_μ. The d'Alembertian $\eta^{\mu\nu}\partial_\mu\partial_\nu$ is denoted by $\Box$. When we encounter two or more spacetime points x^μ, y^ν, etc., we write a superscript to the differential operator such as $\partial_\mu{}^x (\equiv \partial/\partial x^\mu)$, $\Box^y (\equiv \eta^{\mu\nu}\partial^2/\partial y^\mu\partial y^\nu)$, etc. The time derivative is often denoted by an upper dot; for example, $\dot{A} \equiv \partial_0 A$, $\ddot{A} \equiv \partial_0{}^2 A$. A middle dot indicates that the preceding defferential operator does not act beyond it; for example, $\partial_\mu AB \cdot C \equiv [\partial_\mu(AB)]C$.

By a symbol $(a \leftrightarrow b)$ we denote the expression which is obtained from the preceding one by interchaging two indices a and b; for example, $A_a B_b - C_{ab} - (a \leftrightarrow b) \equiv A_a B_b - C_{ab} - A_b B_a + C_{ba}$.

The commutator and anticommutator between A and B are denoted by $[A,\ B]$ and by $\{A,\ B\}$, respectively; $[A,\ B]_{\mp} = \{A,\ B\}$ if both A and B are fermionic and $[A,\ B]_{\mp} = [A,\ B]$ otherwise. The equal-time commutator, $[A(x),\ B(y)]$ with $x^0 = y^0$, is abbreviated as $[A(x),\ B(y)]|_0$.

Sometimes we need to take a derivative with respect to a fermionic quantity. In this case, it should always be understood that such differentiation is taken from the *left*, that is, $\frac{\partial A}{\partial \varphi} \equiv (\partial/\partial\varphi)A$. The Euler derivative is denoted by $\delta/\delta\varphi$, that is,

$$\frac{\delta A}{\delta \varphi} \equiv \frac{\partial A}{\partial \varphi} - \partial_\mu \frac{\partial A}{\partial(\partial_\mu \varphi)}.\,.$$

Complex conjugation is denoted by an upper asterisk like A^*; hermitian conjugation is denoted by a dagger like $A^\dagger$. Dirac conjugate of ψ is denoted by $\bar\psi$, but a bar is used also in different senses. The hermitian conjugate term is denoted by h.c. The Fourier transformation is indicated by F.T.

The signature of a real variable α is expressed as $\epsilon(\alpha)$ ($\equiv \alpha/|\alpha|$). The Heaviside step function is denoted by $\theta(\alpha)$ ($= [1 + \epsilon(\alpha)]/2$). An infinitesimal imaginary part is denoted by $\pm i0$; for example,

$$\int d\alpha \, \frac{f(\alpha)}{\alpha + i0} \equiv \lim_{\varepsilon \to +0} \int d\alpha \, \frac{f(\alpha)}{\alpha + i\varepsilon};$$

$\pm i0 x^0$ means $\pm i0$ for $x^0 > 0$ and $\mp i0$ for $x^0 < 0$.

We do not always notationally discriminte a Lie group and its corresponding Lie algebra according to the physicist's convention.

We use the following abbreviated words:

rhs.=right-hand side,

lhs.=left-hand side,

(anti)commutation=commutation or anticommutation,

QED=quantum electrodynamics,

QCD=quantum chromodynamics,

SSB=spontaneous symmetry breaking (or spontaneous breakdown of
       symmetry),

NG=Nambu-Goldstone,

FP=Faddeev-Popov,

BRS=Becchi-Rouet-Stora,

WT=Ward-Takahashi.

The boldfaced words are those which are accessible from the Index.

When we refer to an equation which appears in the same subsection, we omit writing the subsection number, that is, we refer to it by writing the equation number only.

References are indicated like [Hei 29]; The first three Roman letters are those of the first author's family name and the last two numerals are those of the year when the paper was published (If that paper is not yet published, we write "unp" there). In order to distinguish different authors, one extra letter may sometimes be added to the three Roman letters.

CHAPTER ONE

QUANTUM FIELD THEORY
IN THE HEISENBERG PICTURE

Quantum field theory is the fundamental theory of particle physics.
In this chapter, we summarize its general features as the preliminaries
for the succeeding chapters, though it is supposed that the readers are
familiar with quantum field theory.

1.1 GENERAL REMARKS ON QUANTUM FIELD THEORY

In this section, we make some general remarks on quantum field theory.

1.1.1 Basic concepts

A **quantum field**, or simply a **field**, is a fundamental object in quantum field
theory. Mathematically, a quantum field is a finite set of operator-valued generalized functions (distributions or hyperfunctions) of spacetime coordinates x^μ.
When we discuss a field generically, it is denoted by $\varphi(x)$ (or $\varphi_A(x)$ if we discuss
many fields). For particular fields, we use various symbols such as $\phi(x)$, $\psi(x)$, etc.

Quantum field theory is formulated so as to meet the requirement of special
relativity (except for quantum gravity). That is, the theory is invariant under
translations and Lorentz transformations. The Poincaré invariance of the theory
is discussed in some detail in the next subsection.

Quantum fields satisfy certain partial differential equations, which are called
field equations. The field equation should be *local*, that is, every field involved
in it depends only on *one* spacetime point. If a field by itself satisfies linear
differential equations of the same number as that of its components, it is called
free; otherwise it is **interacting**. That is, an interacting field satisfies nonlinear

differential equations, which necessarily involve some products of fields at the *same* spacetime point. Since, however, a field is, apart from its operator nature, not an ordinary function but a *generalized function*, which is not defined point by point, a product of fields at the same spacetime point is *not well-defined* in general. This causes a very difficult problem, called **ultraviolet divergence**. There are several attitudes for coping with this trouble.

1. One gives up field equations, and discusses only the general framework of quantum field theory. This standpoint is called **axiomatic field theory**. This approach, however, makes it impossible to formulate the theory concretely.

2. One first replaces the singular product of fields by a well-defined one by means of point-splitting or some other regularization method. After applying a certain device of removing divergent pieces (called renormalization), one takes a limit to reproduce the original singular product formally. But this approach is successful only in some simple lower-dimensional models.

3. In order to deal with realistic theories concretely, therefore, one must appeal to some approximation method. In perturbation theory, one can obtain an explicit solution in the form of a (non-convergent) series expansion. After making renormalization, one can define the singular product of fields *in each order* of the perturbation series.

We do not wish to discuss this very difficult problem here. Since our primary interest is to develop concrete formalisms of realistic theories without approximation, we simply *postulate* that *the product of fields at the same spacetime point exists uniquely* without specifying how to define it explicitly. Here the uniqueness means that *the ordering of field operators*[1] *in the product is totally irrelevant* apart from an overall statistical signature factor. We emphasize that this statement does not mean to neglect the (anti)commutators of fields at the same spacetime point, but claims that the operator ordering at the same spacetime point is *meaningless*. Only under this understanding, we can reasonably develop the Lagrangian canonical formalism, which is presented in Sec.1.2.

Canonical quantization is carried out by using commutation relations or anticommutation relations according as φ_A is **bosonic** (i.e., obeying Bose statistics)

---

[1]   The word "field operator" is used for emphasizing the operator nature of a field.

or **fermionic** (i.e., obeying Fermi one). Since any two fields at two different point (anti)commute at the equal time, Lorentz invariance implies that

$$[\varphi_A(x), \varphi_B(y)]_{\mp} = 0 \quad \text{for} \quad (x - y)^2 < 0, \tag{1.1.1 - 1}$$

We call Eq. (1) **local (anti)commutativity** or **Einstein causality**, because it means that no action can propagate faster than light.

The operand of field operators is called a **state vector**, or simply a **state**. It is denoted by $| \cdot \rangle$. For two states $|f\rangle$ and $|g\rangle$, their **inner product** is denoted by $\langle g|f\rangle$ and has the property

$$\langle g|f\rangle = \langle f|g\rangle^*. \tag{1.1.1 - 2}$$

We sometimes write $\langle f|$ without considering the inner product. We call $\langle f|$ a **bra-vector** and correspondingly $|f\rangle$ a **ket-vector**.

The totality of states is called a **state-vector space**, and it is usually denoted by $\mathcal{V}$. It is a complex linear space equiped with the inner product. If

$$\langle f|f\rangle > 0 \quad \text{for any} \quad |f\rangle \neq 0, \tag{1.1.1 - 3}$$

then $\mathcal{V}$ is a **Hilbert space** (completion should be understood). In this case, probabilistic interpretation is obvious. But one should note that probabilistic interpretability does *not* necessarily require the validity of Eq. (3) in the whole $\mathcal{V}$. If there are some states in $\mathcal{V}$ which do not satisfy Eq. (3), $\mathcal{V}$ is called an **indefinite-metric Hilbert space**. Its properties are described in detail in the Appendix. In the indefinite-metric Hilbert space, $\langle f|f\rangle$ is called **norm** in abuse of language (**positive norm** if $\langle f|f\rangle > 0$, **zero norm** if $\langle f|f\rangle = 0$, and **negative norm** if $\langle f|f\rangle < 0$). We always assume that $\mathcal{V}$ is **non-degenerate**, that is, if $|f\rangle$ is zero norm then there exists $|g\rangle$ in $\mathcal{V}$ such that $\langle g|f\rangle \neq 0$.

Field operators are represented in $\mathcal{V}$. In general, there are infinitely many inequivalent representations. It is postulated that there exists a unique distinguished state, called the **vacuum**, which should be Poincaré invariant. It is denoted by $|0\rangle$. Any state in $\mathcal{V}$ is essentially constructible by applying field operators on $|0\rangle$ (postulate of **cyclicity**).

Given an operator A, $\langle g|A|f\rangle$ is called **matrix element** of A, though it does not necessarily obey the matrix multiplication law if $\mathcal{V}$ is not a Hilbert space. If

$$\langle g|A^\dagger|f\rangle = \langle f|A|g\rangle^*, \tag{1.1.1 - 4}$$

then $A^\dagger$ is called **hermitian conjugate** of A. If $A^\dagger = A$, A is **hermitian**, and if $A^\dagger = A^{-1}$, A is **unitary**. We do not strictly distinguish hermitian and self-adjoint because we hardly pay attention to the domain in which A is defined.

1.1.2 Poincaré invariance

Quantum field theory is invariant under the **Poincaré group**, which is the totality of translations and Lorentz transformations. Under a Lorentz transformation plus a translation

$$x^\mu \to x'^\mu = L^\mu_{\ \nu} x^\nu + a^\mu \quad \text{(or } x' = Lx + a) \tag{1.1.2 − 1}$$

preserving the Minkowski metric ($\eta_{\mu\nu} L^\mu_{\ \sigma} L^\nu_{\ \tau} = \eta_{\sigma\tau}$), a field $\varphi(x)$ transforms as

$$\varphi(x) \to \varphi'(x') = s(L)\varphi(x), \tag{1.1.2 − 2}$$

where $s(L)$ stands for a finite-dimensional matrix representation[2] of the Lorentz transformation $L \in SO(3,1)$. The above transformation is induced by a unitary operator $U(a,L)$ in the following way:

$$U^{-1}(a,L)\varphi(x)U(a,L) = \varphi'(x) = s(L)\varphi(L^{-1}(x-a)) \tag{1.1.2 − 3}$$

with the composition rule

$$U(a_1, L_1)U(a_2, L_2) = U(L_1 a_2 + a_1, L_1 L_2). \tag{1.1.2 − 4}$$

It is convenient to consider the infinitesimal transformation, for which one can neglect higher order terms. The infinitesimal versions of Eqs. (1) and (2) are

$$x'^\mu = x^\mu + \varepsilon^\mu_{\ \nu} x^\nu + \varepsilon^\mu \tag{1.1.2 − 5}$$

with $\varepsilon^{\mu\nu} = -\varepsilon^{\nu\mu}$ and

$$\varphi'_\alpha(x') = [\delta_\alpha^{\ \beta} - \frac{i}{2}\varepsilon^{\mu\nu}(s_{\mu\nu})_\alpha^{\ \beta}]\varphi_\beta(x) \tag{1.1.2 − 6}$$

---

[2]   For the fields having a half-odd spin, $s(L)$ should be understood as the representation of the universal covering group, $SL(2, \mathbf{C})$, of $SO(3,1)$.

with $s_{\mu\nu} = -s_{\nu\mu}$, respectively. Here, owing to the composition rule $s(L_1)s(L_2) = s(L_1L_2)$, the matrices $s_{\mu\nu}$ satisfy

$$[s_{\mu\nu}, s_{\lambda\rho}] = i(\eta_{\nu\lambda}s_{\mu\rho} - \eta_{\mu\lambda}s_{\nu\rho} + \eta_{\mu\rho}s_{\nu\lambda} - \eta_{\nu\rho}s_{\mu\lambda}). \qquad (1.1.2-7)$$

Likewise, the infinitesimal version of $U(a, L)$ is

$$1 + i\varepsilon^\mu P_\mu - \frac{i}{2}\varepsilon^{\mu\nu}M_{\mu\nu}; \qquad (1.1.2-8)$$

P_μ and $M_{\mu\nu}(= -M_{\nu\mu})$ are called **translation generators** and **Lorentz generators**, respectively, and they are altogether called **Poincaré generators**. Then the infinitesimal form of Eq.(3) becomes

$$[iP_\mu, \varphi_\alpha(x)] = \partial_\mu\varphi_\alpha(x), \qquad (1.1.2-9)$$

$$[iM_{\mu\nu}, \varphi_\alpha(x)] = [(x_\mu\partial_\nu - x_\nu\partial_\mu)\delta_\alpha^{\ \beta} - i(s_{\mu\nu})_\alpha^{\ \beta}]\varphi_\beta(x). \qquad (1.1.2-10)$$

From the composition rule, Eq.(4), we see that the Poincaré generators satisfy the following commutation relations:

$$[P_\mu, P_\nu] = 0, \qquad (1.1.2-11)$$

$$[M_{\mu\nu}, P_\lambda] = i(\eta_{\nu\lambda}P_\mu - \eta_{\mu\lambda}P_\nu), \qquad (1.1.2-12)$$

$$[M_{\mu\nu}, M_{\lambda\rho}] = i(\eta_{\nu\lambda}M_{\mu\rho} - \eta_{\mu\lambda}M_{\nu\rho} + \eta_{\mu\rho}M_{\nu\lambda} - \eta_{\nu\rho}M_{\mu\lambda}). \qquad (1.1.2-13)$$

This algebra is called **Poincaré algebra**. One can confirm the consistency between Eqs. (9),(10) and Eqs. (11)-(13) in the sense of the **Jacobi identity**

$$[A, [B, C]] + [B, [C, A]] + [C, [A, B]] \equiv 0. \qquad (1.1.2-14)$$

The operators which commute with all generators are called **Casimir operators**. The Casimir operators of the Poincaré algebra are P^2 and $W \equiv -w^2$, where we set

$$w^\rho \equiv \frac{1}{2}\epsilon^{\rho\lambda\mu\nu}P_\lambda M_{\mu\nu}, \qquad (1.1.2-15)$$

which commutes with P_σ.

The representations of the Poincaré group are constructed in the following way. Since the translation group is abelian, its irreducible representations are one-dimensional. The representation is specified by the eigenvalue p_μ of P_μ. Hence the Lorentz transformations which change p_μ should be excluded. The totality of Lorentz transformations which leave p_μ invariant is called a **little group** on p_μ. Taking advantage of the technique of induced representations, we have only to construct the irreducible representation of the little group on p_μ. According as p_μ is timelike, lightlike, or spacelike, the corresponding little group is isomorphic to the three-dimensional orthogonal group $SO(3)$, the two-dimensional Euclidean group $E(2)$, or the three-dimensional Lorentz group $SO(2,1)$.

The algebra corresponding to a little group can be found in the following way. Since P_μ takes the value p_μ, we substitute $P_\mu = p_\mu$ into Eq. (12). Then, since the commutator vanishes, the rhs. also must vanish. Thus the generators of the little group are the linear combinations $c^{\mu\nu} M_{\mu\nu}$ which are consistent with

$$c^{\mu\nu}(\eta_{\nu\lambda} p_\mu - \eta_{\mu\lambda} p_\nu) = 0. \qquad (1.1.2 - 16)$$

The commutators satisfied by them are calculated from Eq. (13). For example, if $p_\mu = (p_0 \neq 0, 0, 0, 0)$, the generators of the little group are M_{kl}, and they satisfy the algebra of $SO(3)$; its representation defines spin. The case in which p_μ is lightlike is discussed in Sec.2.2.2.

From the physical reason, it is postulated that $p^2 \geq 0$ and $p_0 \geq 0$. This requirement is called **spectrum condition**. Accordingly, the little group on a spacelike p_μ is unphysical.

1.2 LAGRANGIAN CANONICAL FORMALISM

In this section, we review the general theory of Lagrangian canonical formalism, that is, starting with the Lagrangian density we define canonical variables, their canonical conjugates, and the Hamiltonian, and then set up the canonical commutation relations. We also discuss the symmetry properties of the theory.

1.2.1 Lagrangian density and field equations

We consider a set of fields $\{\varphi_A(x)\}$; they are called **primary fields**. The Lagrangian canonical formalism is based on an **action**

$$I \equiv \int d^4x \mathcal{L}(x), \qquad (1.2.1-1)$$

where I is bosonic, hermitian, and Poincaré invariant. The **Lagrangian density** $\mathcal{L}(x)$ is a local function of primary fields $\varphi_A(x)$, that is , it is constructed from $\varphi_A(x)$ and their derivatives *at the same spacetime point*.[1]

In order for the canonical formalism to be applicable, $\mathcal{L}(x)$ is assumed to contain no second or higher derivatives of $\varphi_A(x)$ and to be at most quadratic with respect to first derivatives of $\varphi_A(x)$. Furthermore, when $\mathcal{L}(x)$ is a polynomial in $\varphi_A(x)$, one always drops a constant term and eliminates linear terms by redefining fields. Then the quadratic part of $\mathcal{L}(x)$ is called **free Lagrangian density**, and the remainder is called **interaction Lagrangian density**. Parameters appearing in the latter are called **coupling constants**.

In Eq.(1), the integration volume is the whole four-dimensional spacetime because any other spacetime region is non-invariant under translations. Hence in considering a variation of I, we need not take account of that of the integration volume. Then the **variational principle** $\delta I = 0$ yields

$$
\begin{aligned}
0 = \delta I &= \int d^4x \delta \mathcal{L}(x) \\
&= \int d^4x \left[\delta\varphi_A \frac{\partial \mathcal{L}}{\partial \varphi_A} + \delta\partial_\mu\varphi_A \frac{\partial \mathcal{L}}{\partial(\partial_\mu\varphi_A)} \right],
\end{aligned} \qquad (1.2.1-2)
$$

---

[1]   If $\mathcal{L}(x)$ is not local, then we have a non-local theory, which is known to have fundamental difficulties

where summation over A should be understood. Using $\delta \partial_\mu \varphi_A = \partial_\mu \delta \varphi_A$ and integrating by parts, we obtain **Euler-Lagrange equations**,

$$\frac{\partial \mathcal{L}}{\partial \varphi_A} - \partial_\mu \frac{\partial \mathcal{L}}{\partial(\partial_\mu \varphi_A)} = 0, \qquad (1.2.1-3)$$

which are called **field equations**.

In the above, an important assumption is that $\delta \varphi_A(x)$ is local and vanishes sufficiently rapidly at infinity in any direction. Hence any term in $\mathcal{L}(x)$ expressible as $\partial_\mu(\,\cdot\,)^\mu$, which is called **total divergence**, can *always* be discarded.[2] It is therefore unreasonable to consider a surface term,[3] which could arise from total divergence owing to the Gauss theorem, in the action I.

1.2.2 Canonical quantization in the ordinary case

In this subsection, we consider canonical quantization in the ordinary case. A more general case is discussed in the next subsection. The most general case is dealt with in Addendum 1.A.

We take φ_A as **canonical variables** and define their **canonical conjugates** by

$$\pi^A \equiv (\partial/\partial \dot{\varphi}_A)\mathcal{L}(\varphi, \partial\varphi, \dot{\varphi}). \qquad (1.2.2-1)$$

Since $\mathcal{L}$ is at most quadratic in $\dot{\varphi}_A \equiv \partial_0 \varphi_A$, π^A is linear in $\dot{\varphi}_B$. We now assume that the simultaneous "linear" equations, Eq.(1), can be solved with respect to $\dot{\varphi}_B$. That is, we assume that it is possible to write

$$\dot{\varphi}_A = \Phi_A(\varphi, \partial\varphi, \pi). \qquad (1.2.2-2)$$

Then, since π^A's are independent, we can set up the **canonical (anti)commutation relations**

$$[\varphi_A(x), \varphi_B(y)]_{\mp,0} = 0, \qquad (1.2.2-3)$$

$$[\pi^A(x), \varphi_B(y)]_{\mp,0} = -i\delta^A{}_B \delta(\mathbf{x} - \mathbf{y}), \qquad (1.2.2-4)$$

$$[\pi^A(x), \pi^B(y)]_{\mp,0} = 0. \qquad (1.2.2-5)$$

---

[2]   A total-divergence term may be meaningful in the "effective" Lagrangian density, which is a non-local function of primary fields. Any quantity F is total divergence in the sense $F = \partial_\mu(\partial^\mu \Box^{-1} F)$ if non-local quantities are admitted.

[3]   It is true that it is often necessary to introduce surface terms, but one should note that such circumstances can arise only after the representation of field operators is considered.

Here we use the following notation:

$$[\varphi_A(x), \varphi_B(y)]_{\mp,0} \equiv \varphi_A(x)\varphi_B(y) - \epsilon_{(AB)}\varphi_B(y)\varphi_A(x) \qquad (1.2.2-6)$$

with $\epsilon_{(AB)} = -1$ if both φ_A and φ_B are fermionic and $\epsilon_{(AB)} = +1$ otherwise (indices in parentheses are irrelevant to the summation convention); the subscript 0 of a bracket means to set $x^0 = y^0$ and $\delta(\mathbf{x} - \mathbf{y})$ is the spatial three-dimensional delta function.

The **Hamiltonian** H is defined by

$$H \equiv \int d\mathbf{x}\, \mathcal{H}(x) \qquad (1.2.2-7)$$

with

$$\mathcal{H}(x) \equiv \dot{\varphi}_B(x)\pi^B(x) - \mathcal{L}(x). \qquad (1.2.2-8)$$

Then the following **Heisenberg equations** hold:

$$[iH, \varphi_A(x)] = \dot{\varphi}_A(x), \qquad (1.2.2-9)$$

$$[iH, \pi^A(x)] = \dot{\pi}^A(x). \qquad (1.2.2-10)$$

Their validity is shown in the following way.

From Eqs.(2)-(5), we have

$$[\dot{\varphi}_A(x), \varphi_B(y)]_{\mp,0} = -i\frac{\partial \Phi_A}{\partial \pi^B}\delta(\mathbf{x} - \mathbf{y}), \qquad (1.2.2-11)$$

$$[\pi^A(x), \dot{\varphi}_B(y)]_{\mp,0} = -i\left(\frac{\partial \Phi_B}{\partial \varphi_A} + \frac{\partial \Phi_B}{\partial(\partial_k \varphi_A)}\partial_k\right)^y \delta(\mathbf{x} - \mathbf{y}), \quad (1.2.2-12)$$

where the superscript y indicates that $\mathbf{y}$ should be used inside the parentheses. Noting Eq.(1), we obtain

$$[\mathcal{L}(x), \varphi_A(y)]_0 = -i\epsilon_{(AB)}\frac{\partial \Phi_B}{\partial \pi^A}\pi^B\delta(\mathbf{x} - \mathbf{y}), \qquad (1.2.2-13)$$

$$[\mathcal{L}(x), \pi^A(y)]_0 = i\left(\frac{\partial \mathcal{L}}{\partial \varphi_A} + \frac{\partial \mathcal{L}}{\partial(\partial_k \varphi_A)}\partial_k\right)^x \delta(\mathbf{x} - \mathbf{y})$$
$$+ i\left(\frac{\partial \Phi_B}{\partial \varphi_A} + \frac{\partial \Phi_B}{\partial(\partial_k \varphi_A)}\partial_k\right)^x \delta(\mathbf{x} - \mathbf{y}) \cdot \pi^B, \quad (1.2.2-14)$$

and then

$$[\mathcal{H}(x), \varphi_A(y)]_0 = -i\dot{\varphi}_A\delta(\mathbf{x} - \mathbf{y}), \qquad (1.2.2-15)$$

$$[\mathcal{H}(x), \pi^A(y)]_0 = -i\left(\frac{\partial \mathcal{L}}{\partial \varphi_A} + \frac{\partial \mathcal{L}}{\partial(\partial_k \varphi_A)}\partial_k\right)^x \delta(\mathbf{x} - \mathbf{y}). \qquad (1.2.2-16)$$

Since the field equation, Eq.(1.2.1–3), is rewritten as

$$\frac{\partial \mathcal{L}}{\partial \varphi_A} - \partial_k \frac{\partial \mathcal{L}}{\partial(\partial_k \varphi_A)} = \dot{\pi}^A, \tag{1.2.2 – 17}$$

Eqs.(15) and (16) yield Eqs.(9) and (10), respectively.

Finally, we demonstrate that the canonical (anti)commutation relations, Eqs.(3)–(5), are invariant under field redefinition. Let $\{\varphi_A\}$ and $\{\varphi'_A\}$ are two systems of canonical variables, each of which is expressible in terms of the other through a set of local relations involving no derivatives:

$$\varphi_A = f_A(\varphi'), \quad \varphi'_A = g_A(\varphi); \tag{1.2.2 – 18}$$

then the canonical quantizations induced by both systems are mutually equivalent, as shown below.

The canonical conjugates are, of course, given by

$$\pi^A \equiv \frac{\partial \mathcal{L}}{\partial \dot{\varphi}_A}, \quad \pi'^A \equiv \frac{\partial \mathcal{L}}{\partial \dot{\varphi}'_A}. \tag{1.2.2 – 19}$$

Since Eq.(18) implies

$$\dot{\varphi}_A = \dot{\varphi}'_B \frac{\partial f_A}{\partial \varphi'_B}, \quad \dot{\varphi}'_A = \dot{\varphi}_B \frac{\partial g_A}{\partial \varphi_B}, \tag{1.2.2 – 20}$$

we have

$$\frac{\partial f_C}{\partial \varphi'_A} \frac{\partial g_B}{\partial \varphi_C} = \frac{\partial g_C}{\partial \varphi_A} \frac{\partial f_B}{\partial \varphi'_C} = \delta^A{}_B, \tag{1.2.2 – 21}$$

and therefore

$$\pi^A = \frac{\partial g_B}{\partial \varphi_A} \pi'^B, \quad \pi'^A = \frac{\partial f_B}{\partial \varphi'_A} \pi^B. \tag{1.2.2 – 22}$$

Setting up the canonical (anti)commutation relations for $\{\varphi_A\}$, we calculate (anti)commutators with respect to $\{\varphi'_A\}$. We then have

$$[\varphi'_A(x), \varphi'_B(y)]_{\mp, 0} = [g_A(\varphi(x)), g_B(\varphi(y))]_{\mp, 0} = 0, \tag{1.2.2 – 23}$$

$$[\pi'^A(x), \varphi'_B(y)]_{\mp, 0} = \frac{\partial f_C}{\partial \varphi'_A}(x) \cdot [\pi^C(x), g_B(\varphi(y))]_{\mp, 0}$$

$$= -i \frac{\partial f_C}{\partial \varphi'_A} \frac{\partial g_B}{\partial \varphi_C} \delta(\mathbf{x} - \mathbf{y})$$

$$= -i \delta^A{}_B \delta(\mathbf{x} - \mathbf{y}), \tag{1.2.2 – 24}$$

$$[\pi'^A(x), \pi'^B(y)]_{\mp, 0} = \frac{\partial f_C}{\partial \varphi'_A}(x) \left[\pi^C(x), \frac{\partial f_D}{\partial \varphi'_B}(y) \right]_{\mp, 0} \pi^D(y) - \epsilon_{(AB)}(A \leftrightarrow B, x \leftrightarrow y)$$

$$= -i \left[\frac{\partial f_C}{\partial \varphi'_A} \frac{\partial^2 f_D}{\partial \varphi_C \partial \varphi'_B} - \epsilon_{(AB)}(A \leftrightarrow B) \right] \pi^D \delta(\mathbf{x} - \mathbf{y}). \tag{1.2.2 – 25}$$

Since

$$\frac{\partial^2 f_D}{\partial \varphi_C \partial \varphi'_B} = \frac{\partial g_E}{\partial \varphi_C} \frac{\partial^2 f_D}{\partial \varphi'_E \partial \varphi'_B}, \qquad (1.2.2-26)$$

Eq.(21) implies

$$\frac{\partial f_C}{\partial \varphi'_A} \frac{\partial^2 f_D}{\partial \varphi_C \partial \varphi'_B} = \frac{\partial^2 f_D}{\partial \varphi'_A \partial \varphi'_B}. \qquad (1.2.2-27)$$

We thus obtain

$$[\pi'^A(x), \pi'^B(y)]_{\mp,0} = 0. \qquad (1.2.2-28)$$

1.2.3 Canonical quantization with multiplier fields

It is rather stringent to assume that Eq.(1.2.2–1) is solvable with respect to $\dot{\varphi}_A$; we often encounter the situation that the rhs. of Eq.(1.2.2–1) does not involve some of $\dot{\varphi}_A$'s, which we denote by $\dot{\varphi}_\alpha$. Here α may run over only some of , but not all of, components of a field. In such a case, the trouble is resolved by considering a new Lagrangian density having the following form:

$$\mathcal{L} = \mathcal{L}_0(\varphi, \partial\varphi, \dot{\varphi}) + F_\alpha(\varphi, \partial\varphi, \dot{\varphi})b^\alpha + G(b), \qquad (1.2.3-1)$$

where $\mathcal{L}_0$ corresponds to the previous Lagrangian density, F_α is linear in $\dot{\varphi}$ and $\det(\partial/\partial\dot{\varphi}_\beta)F_\alpha \neq 0$. Since the new field b^α plays the role of a Lagrange multiplier, it is sometimes called a **multiplier field**. We encounter such Lagrangian densities as Eq.(1) in gauge theories and in quantum gravity.

The field equations are

$$\frac{\partial \mathcal{L}}{\partial \varphi_A} - \partial_\mu \frac{\partial \mathcal{L}}{\partial(\partial_\mu \varphi_A)} = 0, \qquad (1.2.3-2)$$

$$F_\alpha + \frac{\partial G}{\partial b^\alpha} = 0. \qquad (1.2.3-3)$$

Canonical variables are φ_A but *not* b^α. The canonical conjugates are

$$\pi^A \equiv \frac{\partial \mathcal{L}}{\partial \dot{\varphi}_A} = \frac{\partial \mathcal{L}_0}{\partial \dot{\varphi}_A} + \frac{\partial F_\alpha}{\partial \dot{\varphi}_A} b^\alpha. \qquad (1.2.3-4)$$

By assumption, the matrix $((\partial/\partial\dot{\varphi}_\beta)F_\alpha)$ is invertible. It is therefore possible to solve Eq.(4) with respect to $\dot{\varphi}_A (A \neq \alpha)$ and b^α. Note that $\dot{\varphi}_A (A = \alpha)$ is expressible in terms of other variables through Eq.(3). We set up the canonical (anti)commutation relations, Eqs.(1.2.2–3)–(1.2.2–5). In the following, we check the consistency with the Heisenberg equations, Eqs.(1.2.2–9) and (1.2.2–10).

Since from Eq.(4) b^α is expressible as a function of φ, $\partial\varphi$, and π, we can *eliminate* all b^α from $\mathcal{L}$ and the field equations. Then we can write Eq.(1.2.2–2) for *all* A's, and hence proceed in the same way as before. That is, we obtain Eqs.(1.2.2–15) and (1.2.2–16), and therefore the Heisenberg equations. The only difference from the ordinary case is that the Heisenberg equations, Eq.(1.2.2–9), hold not as trivial identities $\dot\varphi_A = \dot\varphi_A$, but those corresponding to $A = \alpha$ are equivalent to field equations, Eq.(3).

1.2.4 Noether theorem

As discussed in Sec.1.1.2, the theory is invariant under the Poincaré group. In general, the theory may have other invariance properties. If I is invariant under certain transformations of primary fields, the theory is said to have a **symmetry** corresponding to those transformations. There are discrete symmetries and continuous ones, but we are interested only in the latter.

A continuous symmetry can be discussed by considering the infinitesimal transformation, which is characterized by an infinitesimal parameter ε. We set

$$\delta^\varepsilon\varphi_A(x) \equiv \varphi'_A(x') - \varphi_A(x), \qquad (1.2.4-1)$$

$$\delta_*^\varepsilon\varphi_A(x) \equiv \varphi'_A(x) - \varphi_A(x), \qquad (1.2.4-2)$$

whence

$$\delta^\varepsilon\varphi_A = \delta_*^\varepsilon\varphi_A + \delta^\varepsilon x^\mu \cdot \partial_\mu\varphi_A, \qquad (1.2.4-3)$$

where $x'^\mu \equiv x^\mu + \delta^\varepsilon x^\mu$. The symmetry considered is called **spacetime symmetry** if $\delta^\varepsilon x^\mu \neq 0$ and **internal symmetry** if vanishing; $\delta^\varepsilon - \delta_*^\varepsilon$ is called the **orbital part** of the symmetry. Poincaré invariance is, of course, a spacetime symmetry.

Although the invariance of I does not necessarily imply

$$\mathcal{L}'(x') = \mathcal{L}(x), \qquad (1.2.4-4)$$

we here restrict our consideration only to this case.[4] Correspondingly, the Jacobian det $\partial_\mu x'^\nu$ is unity, whence $\partial_\mu\delta^\varepsilon x^\mu = 0$. We rewrite Eq.(4) as

$$\delta^\varepsilon\mathcal{L} = \delta_*^\varepsilon\mathcal{L} + \delta^\varepsilon x^\mu \cdot \partial_\mu\mathcal{L} = 0 \qquad (1.2.4-5)$$

---

[4]   General case is discussed in Sec.5.2.4.

with

$$\delta_*^\epsilon \mathcal{L} = \delta_*^\epsilon \varphi_A \cdot \frac{\partial \mathcal{L}}{\partial \varphi_A} + \delta_*^\epsilon \partial_\mu \varphi_A \cdot \frac{\partial \mathcal{L}}{\partial(\partial_\mu \varphi_A)}. \tag{1.2.4-6}$$

Here, from the definition, Eq.(2), ∂_μ commutes with δ_*^ϵ. It is important to note that δ_*^ϵ is a particular case of a variation.[5] Hence, the form invariance of I automatically implies the form invariance of the field equations.

We now define the **Noether current** by

$$\epsilon J^\mu \equiv \delta_*^\epsilon \varphi_A \cdot \frac{\partial \mathcal{L}}{\partial(\partial_\mu \varphi_A)} + \delta^\epsilon x^\mu \cdot \mathcal{L}. \tag{1.2.4-7}$$

It satisfies the conservation law

$$\partial_\mu J^\mu = 0. \tag{1.2.4-8}$$

Indeed, with the help of Eqs.(1.2.3–2),(5), and (6), we have

$$\begin{aligned}
\epsilon \partial_\mu J^\mu &= \delta_*^\epsilon \varphi_A \cdot \partial_\mu \frac{\partial \mathcal{L}}{\partial(\partial_\mu \varphi_A)} + \partial_\mu \delta_*^\epsilon \varphi_A \cdot \frac{\partial \mathcal{L}}{\partial(\partial_\mu \varphi_A)} + \delta^\epsilon x^\mu \cdot \partial_\mu \mathcal{L} \\
&= \delta_*^\epsilon \varphi_A \cdot \frac{\partial \mathcal{L}}{\partial \varphi_A} + \delta_*^\epsilon \partial_\mu \varphi_A \cdot \frac{\partial \mathcal{L}}{\partial(\partial_\mu \varphi_A)} - \delta_*^\epsilon \mathcal{L} \\
&= 0. \tag{1.2.4-9}
\end{aligned}$$

The existence of the conserved current J^μ is known as the **Noether theorem**.

From Eq.(8), we obtain

$$\partial_0 \int d\mathbf{x}\, J^0 = -\int d\mathbf{x}\, \partial_k J^k = 0, \tag{1.2.4-10}$$

that is, the **charge**

$$Q \equiv \int d\mathbf{x}\, J^0(x) \tag{1.2.4-11}$$

is time-independent. Here, however, we have assumed that J^k vanishes at spatial infinity and that the integral in Eq.(11) is convergent.

If Eq.(11) is not convergent, Q is ill-defined as charge operator, that is, we cannot consider eigenstates of Q nor expectation values of Q. But we can still meaningfully consider

---

[5]   We forbid a *non-local* transformation of φ_A, which causes a serious operator-ordering problem in the transformed Lagrangian density.

$$[iQ, F(y)]_{\mp} \equiv i \int d\mathbf{x} \, [J^0(x), F(y)]_{\mp} \qquad (1.2.4-12)$$

for any *local* quantity $F(y)$, because $[J^0(x), F(y)]_{\mp}$ vanishes identically if $|\mathbf{x}|$ is sufficiently large.

We now calculate $[iQ, \varphi_A(y)]_{\mp}$ by means of the canonical (anti)commutation relations. Since Eq.(7) for $\mu = 0$ is

$$\varepsilon J^0 = \delta_*^{\,\varepsilon} \varphi_B \cdot \pi^B + \delta^{\varepsilon} x^0 \cdot \mathcal{L}, \qquad (1.2.4-13)$$

we have

$$
\begin{aligned}
[i\varepsilon J^0(x), \varphi_A(y)]_{\mp 0} =& \delta_*^{\,\varepsilon} \varphi_A \cdot \delta(\mathbf{x} - \mathbf{y}) + i\epsilon_{(AB)} [\delta_*^{\,\varepsilon} \varphi_B(x), \varphi_A(y)]_{\mp 0} \pi^B(x) \\
& + \delta^{\varepsilon} x^0 \cdot \epsilon_{(AB)} \frac{\partial \Phi_B}{\partial \pi^A} \pi^B \delta(\mathbf{x} - \mathbf{y}) \qquad (1.2.4-14)
\end{aligned}
$$

with the aid of Eq.(1.2.2–13). Since $\delta^{\varepsilon} \varphi_B$ can be written in terms of primary fields without using time derivatives, it (anti)commutes with $\varphi_A(y)$ at the equal time. Accordingly, we have

$$
\begin{aligned}
[\delta_*^{\,\varepsilon} \varphi_B(x), \varphi_A(y)]_{\mp 0} &= -\delta^{\varepsilon} x^{\mu} \cdot [\partial_{\mu} \varphi_B(x), \varphi_A(y)]_{\mp 0} \\
&= i\delta^{\varepsilon} x^0 \cdot \frac{\partial \Phi_B}{\partial \pi^A} \delta(\mathbf{x} - \mathbf{y}) \qquad (1.2.4-15)
\end{aligned}
$$

owing to Eqs.(3) and (1.2.2–11). Thus the last two terms of Eq.(14) cancel out. We therefore obtain

$$\varepsilon [iQ, \varphi_A(y)]_{\mp} = \delta_*^{\,\varepsilon} \varphi_A(y), \qquad (1.2.4-16)$$

that is, Q is the **symmetry generator**.

1.2.5 Poincaré generators

Since the time coordinate plays a special role in the canonical quantization, one may feel uneasy about the manifest covariance of the theory. But Poincaré invariance is manifestly guaranteed in the canonical formalism as seen below.

For translations, since

$$\delta_*^{\,\varepsilon} \varphi_A = -\varepsilon^{\mu} \partial_{\mu} \varphi_A, \quad \delta^{\varepsilon} x^{\nu} = \varepsilon^{\mu} \delta^{\nu}{}_{\mu}, \qquad (1.2.5-1)$$

the Noether theorem implies that the **canonical energy-momentum (or stress) tensor**,

$$T^\nu{}_\mu \equiv \partial_\mu\varphi_A \cdot \frac{\partial\mathcal{L}}{\partial(\partial_\nu\varphi_A)} - \delta^\nu{}_\mu\mathcal{L}, \qquad (1.2.5-2)$$

is conserved:

$$\partial_\nu T^\nu{}_\mu = 0. \qquad (1.2.5-3)$$

Hence the translation generators

$$P_\mu \equiv \int d\mathbf{x}\, T^0{}_\mu(x) \qquad (1.2.5-4)$$

satisfy

$$[iP_\mu, \varphi_A(x)] = \partial_\mu\varphi_A(x). \qquad (1.2.5-5)$$

We expect that P_μ's are well-defined. Then they are energy-momentum operators. Especially, P_0 is nothing but the Hamiltonian H.

For Lorentz transformations, since

$$\delta_*^\varepsilon\varphi_A = -(1/2)\varepsilon^{\mu\nu}[(x_\mu\partial_\nu - x_\nu\partial_\mu)\delta_A{}^B - i(s_{\mu\nu})_A{}^B]\varphi_B, \qquad (1.2.5-6)$$
$$\delta^\varepsilon x^\lambda = (1/2)\varepsilon^{\mu\nu}(x_\mu\delta^\lambda{}_\nu - x_\nu\delta^\lambda{}_\mu), \qquad (1.2.5-7)$$

the Noether theorem implies that the **angular-momentum tensor**,

$$\begin{aligned}
\mathcal{M}^\lambda{}_{\mu\nu} &\equiv [(x_\mu\partial_\nu - x_\nu\partial_\mu)\delta_A{}^B - i(s_{\mu\nu})_A{}^B]\varphi_B \cdot \frac{\partial\mathcal{L}}{\partial(\partial_\lambda\varphi_A)} - (x_\nu\delta^\lambda{}_\nu - x_\nu\delta^\lambda{}_\mu)\mathcal{L} \\
&= x_\mu T^\lambda{}_\nu - x_\nu T^\lambda{}_\mu - i(s_{\mu\nu})_A{}^B\varphi_B\frac{\partial\mathcal{L}}{\partial(\partial_\lambda\varphi_A)} \\
&= -\mathcal{M}^\lambda{}_{\nu\mu},
\end{aligned} \qquad (1.2.5-8)$$

is conserved:

$$\partial_\lambda\mathcal{M}^\lambda{}_{\mu\nu} = 0. \qquad (1.2.5-9)$$

Hence the Lorentz generators

$$M_{\mu\nu} \equiv \int d\mathbf{x}\,\mathcal{M}^0{}_{\mu\nu}(x) = -M_{\nu\mu} \qquad (1.2.5-10)$$

satisfy

$$[iM_{\mu\nu}, \varphi_A(x)] = [(x_\mu\partial_\nu - x_\nu\partial_\mu)\delta_A{}^B - i(s_{\mu\nu})_A{}^B]\varphi_A(x). \qquad (1.2.5-11)$$

Thus Eqs.(1.1.2–9) and (1.1.2–10) have been reproduced in the framework of the Lagrangian canonical formalism.

Since $\mathcal{T}_{\mu\nu}$ is not necessarily symmetric, it is convenient to symmetrize it in the following way. First, we set

$$\mathcal{S}^{\lambda}_{\ \mu\nu} \equiv -i(s_{\mu\nu})^{B}_{A}\varphi_{B}\frac{\partial\mathcal{L}}{\partial(\partial_{\lambda}\varphi_{A})} = -\mathcal{S}^{\lambda}_{\ \nu\mu}, \qquad (1.2.5-12)$$

and seek for the quantity $\mathcal{F}_{\lambda\mu\nu}$ satisfying

$$\mathcal{F}_{\lambda\mu\nu} = -\mathcal{F}_{\mu\lambda\nu}, \qquad (1.2.5-13)$$

$$\mathcal{F}_{\lambda\mu\nu} - \mathcal{F}_{\lambda\nu\mu} = \mathcal{S}_{\lambda\mu\nu}. \qquad (1.2.5-14)$$

The solution is

$$\mathcal{F}_{\lambda\mu\nu} \equiv (1/2)(\mathcal{S}_{\lambda\mu\nu} - \mathcal{S}_{\mu\lambda\nu} + \mathcal{S}_{\nu\mu\lambda}). \qquad (1.2.5-15)$$

Then, since Eqs.(9) and (8) imply

$$\begin{aligned} 0 &= \partial^{\lambda}\mathcal{M}_{\lambda\mu\nu} = \delta^{\lambda}_{\mu}\mathcal{T}_{\lambda\nu} - \delta^{\lambda}_{\nu}\mathcal{T}_{\lambda\mu} + \partial^{\lambda}\mathcal{S}_{\lambda\mu\nu} \\ &= \mathcal{T}_{\mu\nu} + \partial^{\lambda}\mathcal{F}_{\lambda\mu\nu} - (\mathcal{T}_{\nu\mu} + \partial^{\lambda}\mathcal{F}_{\lambda\nu\mu}), \end{aligned} \qquad (1.2.5-16)$$

we see

$$\Theta_{\mu\nu} = \Theta_{\nu\mu} \qquad (1.2.5-17)$$

if we set

$$\Theta_{\mu\nu} \equiv \mathcal{T}_{\mu\nu} + \partial^{\lambda}\mathcal{F}_{\lambda\mu\nu}. \qquad (1.2.5-18)$$

Furthermore, Eqs.(3) and (13) imply

$$\partial^{\mu}\Theta_{\mu\nu} = 0. \qquad (1.2.5-19)$$

We call $\Theta_{\mu\nu}$ the **symmetric energy-momentum tensor**. By using $\Theta_{\mu\nu}$, we can rewrite Eqs.(4) and (10) as

$$P_{\mu} = \int d\boldsymbol{x}\,\Theta_{0\mu}(x), \qquad (1.2.5-20)$$

$$M_{\mu\nu} = \int d\boldsymbol{x}\,[x_{\mu}\Theta_{0\nu}(x) - x_{\nu}\Theta_{0\mu}(x)], \qquad (1.2.5-21)$$

respectively.

We now proceed to considering the Poincaré algebra. Since Eq.(5) implies

$$[iP_\mu, F(x)] = \partial_\mu F(x) \qquad (1.2.5-22)$$

for any local operator $F(x)$, we have

$$[P_\mu, P_\nu] = -i \int d\mathbf{x}\, \partial_\mu \Theta_{0\nu} = 0, \qquad (1.2.5-23)$$

where use has been made of Eq.(19) for $\mu = 0$. Likewise, we have

$$[M_{\mu\nu}, P_\lambda] = i \int d\mathbf{x}(x_\mu \partial_\lambda \Theta_{0\nu} - x_\nu \partial_\lambda \Theta_{0\mu}). \qquad (1.2.5-24)$$

For $\lambda = k$, therefore, partial integrations lead us to

$$[M_{\mu\nu}, P_k] = i(\eta_{\nu k} P_\mu - \eta_{\mu k} P_\nu). \qquad (1.2.5-25)$$

For $\lambda = 0$, owing to Eq.(19) we have

$$\begin{aligned}
[M_{\mu\nu}, P_0] =& i \int d\mathbf{x}(-x_\mu \partial^k \Theta_{k\nu} + x_\nu \partial^k \Theta_{k\mu}) \\
=& i \int d\mathbf{x}(\Theta_{\mu\nu} - \eta_{\mu 0}\Theta_{0\nu} - \Theta_{\nu\mu} + \eta_{\nu 0}\Theta_{0\mu}) \\
=& i(\eta_{\nu 0} P_\mu - \eta_{\mu 0} P_\nu). \qquad (1.2.5-26)
\end{aligned}$$

Thus Eqs.(1.1.2–11) and (1.1.2–12) have been reproduced in the framework of the Lagrangian canonical formalism.

Unfortunately, however, a similar method does not work for establishing Eq.(1.1.2–13). But, except for $[M_{0k}, M_{0l}]$, we can reproduce it by direct calculation. Since

$$\mathcal{M}^0{}_{kl} = [(x_k \partial_l - x_l \partial_k)\delta_B^C - i(s_{kl})_B^C]\varphi_C \cdot \pi^B, \qquad (1.2.5-27)$$

the canonical commutation relations yield

$$[\mathcal{M}^0{}_{kl}(x), \pi^A(y)] = i[(x_k \partial_l^x - x_l \partial_k^x)\delta_B^A - i(s_{kl})_B^A]\delta(\mathbf{x} - \mathbf{y}) \cdot \pi^B(x), \quad (1.2.5-28)$$

whence

$$[iM_{kl}, \pi^A] = [(x_k \partial_l - x_l \partial_k)\delta_B^A + i(s_{kl})_B^A]\pi^B. \qquad (1.2.5-29)$$

Because $\mathcal{L}$ is Lorentz scalar, we should have

$$[iM_{\mu\nu}, \mathcal{L}] = (x_\mu \partial_\nu - x_\nu \partial_\mu)\mathcal{L}. \qquad (1.2.5-30)$$

Since

$$\mathcal{M}^0{}_{\lambda\rho} = [(x_\lambda \partial_\rho - x_\rho \partial_\lambda)\delta_A{}^B - i(s_{\rho\lambda})_A{}^B]\varphi_B \cdot \pi^A - (x_\lambda \delta_\rho{}^0 - x_\rho \delta_\lambda{}^0)\mathcal{L}, \qquad (1.2.5-31)$$

we can calculate $[M_{kl}, M_{\lambda\rho}]$ by using Eqs.(11), (29), (30). In this way, after integrating by parts, we obtain

$$
\begin{aligned}
i[M_{kl}, M_{\lambda\rho}] = \int d\mathbf{x} \Big\{ &[(x_\lambda \partial_\rho - x_\rho \partial_\lambda)(x_k \partial_l - x_l \partial_k) \\
&- (x_k \partial_l - x_l \partial_k)(x_\lambda \partial_\rho - x_\rho \partial_\lambda)]\varphi_A \cdot \pi^A \\
&+ \left[(s_{kl})_A{}^B (s_{\lambda\rho})_B{}^C - (s_{\lambda\rho})_A{}^B (s_{kl})_B{}^C\right]\varphi_C \pi^A \\
&+ \left[x_k(\eta_{l\lambda}\delta_\rho{}^0 - \eta_{l\rho}\delta_\lambda{}^0) - x_l(\eta_{k\lambda}\delta_\rho{}^0 - \eta_{k\rho}\delta_\lambda{}^0)\right]\mathcal{L} \Big\}. \qquad (1.2.5-32)
\end{aligned}
$$

Then, with the aid of Eq.(1.1.2–7), it is straightforward to show that

$$i[M_{kl}, M_{\lambda\rho}] = -\eta_{l\lambda} M_{k\rho} + \eta_{k\lambda} M_{l\rho} - \eta_{k\rho} M_{l\lambda} + \eta_{l\rho} M_{k\lambda}. \qquad (1.2.5-33)$$

1.3 FREE FIELD THEORIES

In this section, after summarizing basic properties of the invariant singular functions, we describe some free field theories as the simplest examples of quantum field theory. Since they are elementary, we here present main results only.

1.3.1 Invariant singular functions

A free field, $\varphi(x)$, having a mass $m(\geq 0)$ usually satisfies the **Klein-Gordon equation**

$$(\Box + m^2)\varphi(x) = 0. \tag{1.3.1 - 1}$$

There are two independent c-number Lorentz-invariant solutions to the Klein-Gordon equation:

$$\Delta(x; m^2) \equiv \frac{-i}{(2\pi)^3} \int d^4 p\, \epsilon(p_0)\delta(p^2 - m^2)e^{-\imath p x}, \tag{1.3.1 - 2}$$

$$\Delta^{(1)}(x; m^2) \equiv \frac{1}{(2\pi)^3} \int d^4 p\, \delta(p^2 - m^2)e^{-\imath p x}. \tag{1.3.1 - 3}$$

They are called **invariant delta functions**. They are not ordinary functions but generalized functions, whence they are also called **singular functions**. Both are real; $\Delta(x; m^2)$ is odd while $\Delta^{(1)}(x; m^2)$ is even under the sign change of x^μ. The first one, $\Delta(x; m^2)$, is the commutator function appearing in four-dimensional commutation relations, and correspondingly it vanishes in the spacelike region $x^2 < 0$. Furthermore, it is the solution to the following (singular) Cauchy problem:

$$(\Box + m^2)\Delta(x; m^2) = 0, \tag{1.3.1 - 4}$$

$$\Delta(x; m^2)|_0 = 0, \tag{1.3.1 - 5}$$

$$\partial_0 \Delta(x; m^2)|_0 = -\delta(\boldsymbol{x}), \tag{1.3.1 - 6}$$

where the symbol $|_0$ means to set $x^0 = 0$. On the other hand, $\Delta^{(1)}(x; m^2)$ has no such properties. The explicit expressions[1] for the invariant delta functions are as follows:

$$\Delta(x; m^2) = -\frac{1}{2\pi}\epsilon(x^0)\left[\delta(x^2) - \frac{mJ_1(m\sqrt{x^2})}{2\sqrt{x^2}}\theta(x^2)\right], \qquad (1.3.1-7)$$

$$\Delta^{(1)}(x; m^2) = \frac{m}{4\pi\sqrt{x^2}}N_1(m\sqrt{x^2})\theta(x^2) + \frac{m}{2\pi^2\sqrt{-x^2}}K_1(m\sqrt{-x^2})\theta(-x^2), (1.3.1-8)$$

where J_n and N_n are Bessel functions and K_n is a modified Bessel function. Near the lightcone $x^2 = 0$, they behave like

$$\Delta(x; m^2) = -(2\pi)^{-1}\epsilon(x^0)\left\{\delta(x^2) + m^2\left[-\frac{1}{4} + O(x^2)\right]\theta(x^2)\right\}, \quad (1.3.1-9)$$

$$\Delta^{(1)}(x^2; m^2) = -\frac{1}{2\pi^2}\left\{\mathrm{P}\frac{1}{x^2} - \frac{m^2}{4}\left[\log m^2|x^2| + O(1)\right]\right\}, \qquad (1.3.1-10)$$

where P stands for the Cauchy principal value.

The positive/negative-frequency delta functions $\Delta^{(+)}(x; m^2)$ and $\Delta^{(-)}(x; m^2)$ are defined by

$$\Delta^{(\pm)}(x; m^2) = \mp(2\pi)^{-3}\int d^4p\,\theta(\pm p_0)\delta(p^2 - m^2)e^{-ipx}. \qquad (1.3.1-11)$$

They are related to $\Delta(x; m^2)$ and $\Delta^{(1)}(x; m^2)$ through

$$i\Delta(x; m^2) = \Delta^{(+)}(x; m^2) + \Delta^{(-)}(x; m^2), \qquad (1.3.1-12)$$

$$\Delta^{(1)}(x; m^2) = \Delta^{(+)}(x; m^2) - \Delta^{(-)}(x; m^2), \qquad (1.3.1-13)$$

and they are mutually related through

$$\Delta^{(+)}(-x; m^2) = -\Delta^{(-)}(x; m^2), \qquad (1.3.1-14)$$

$$[\Delta^{(+)}(x; m^2)]^* = -\Delta^{(-)}(x; m^2). \qquad (1.3.1-15)$$

---

[1]   The expressions in the complex D-dimensional spacetime are as follows:

$$\Delta_D(x; m^2) = -\frac{\epsilon(x^0)m^{(D/2)-1}}{2^{D/2}\pi^{(D/2)-1}}(\sqrt{x^2})^{1-(D/2)}J_{1-(D/2)}(m\sqrt{x^2})\theta(x^2),$$

$$\Delta_D^{(1)}(x; m^2) = \frac{m^{(D/2)-1}}{2^{D/2}\pi^{(D/2)-1}}[-(\sqrt{x^2})^{1-(D/2)}N_{1-(D/2)}(m\sqrt{x^2})\theta(x^2)$$

$$+ \frac{2}{\pi}(\sqrt{-x^2})^{1-(D/2)}K_{1-(D/2)}(m\sqrt{-x^2})\theta(-x^2)].$$

Lorentz-invariant solution to the equation

$$(\Box + m^2)G(x) = \delta^4(x) \qquad (1.3.1-16)$$

are called the **Green's functions** (of the Klein-Gordon operator). The following five Green's functions are of interest:

$$\overline{\Delta}(x; m^2) \equiv -(1/2)\epsilon(x^0)\Delta(x; m^2), \qquad (1.3.1-17)$$

$$\Delta_R(x; m^2) \equiv -\theta(x^0)\Delta(x; m^2), \qquad (1.3.1-18)$$

$$\Delta_A(x; m^2) \equiv \theta(-x^0)\Delta(x; m^2), \qquad (1.3.1-19)$$

$$i\Delta_F(x; m^2) \equiv \overline{\Delta}(x; m^2) + (i/2)\Delta^{(1)}(x; m^2), \qquad (1.3.1-20)$$

$$-i\Delta_{\overline{F}}(x; m^2) \equiv \overline{\Delta}(x; m^2) - (i/2)\Delta^{(1)}(x; m^2). \qquad (1.3.1-21)$$

We call $\Delta_R(x; m^2)$ the **retarded Green's function**, $\Delta_A(x; m^2)$ **advanced Green's function**, and $\Delta_F(x; m^2)$ the **causal Green's function** or **Feynman propagator**. The causal nature of $\Delta_F(x; m^2)$ becomes manifest by writing

$$\Delta_F(x; m^2) = \theta(x^0)\Delta^{(+)}(x; m^2) - \theta(-x^0)\Delta^{(-)}(x; m^2). \qquad (1.3.1-22)$$

The momentum representation of some of Green's functions are as follows:

$$\Delta_R(x; m^2) = -(2\pi)^{-4} \int d^4p \, \frac{e^{-ipx}}{p^2 - m^2 + i0p_0}, \qquad (1.3.1-23)$$

$$\Delta_A(x; m^2) = -(2\pi)^{-4} \int d^4p \, \frac{e^{-ipx}}{p^2 - m^2 - i0p_0}, \qquad (1.3.1-24)$$

$$\Delta_F(x; m^2) = i(2\pi)^{-4} \int d^4p \, \frac{e^{-ipx}}{p^2 - m^2 + i0}. \qquad (1.3.1-25)$$

In the massless ($m = 0$) case, the invariant delta functions are particularly called **invariant D functions**, and expressed by using D in place of Δ. From Eqs.(9) and (10), we have

$$D(x) = -(2\pi)^{-1}\epsilon(x^0)\delta(x^2), \qquad (1.3.1-26)$$

$$D^{(1)}(x) = -\frac{1}{2\pi^2}\mathrm{P}\frac{1}{x^2}. \qquad (1.3.1-27)$$

In gauge theories, $D(x)$ is very important; it is called the **Pauli-Jordan D function**. From Eqs.(26) and (27), it is straightforward to see

$$D^{(\pm)}(x) = \mp\frac{1}{4\pi^2} \cdot \frac{1}{x^2 \mp i0x^0}. \qquad (1.3.1-28)$$

As for the Green's functions, we have

$$D_R(x) = (2\pi)^{-1}\theta(x^0)\delta(x^2), \qquad (1.3.1-29)$$

$$D_A(x) = (2\pi)^{-1}\theta(-x^0)\delta(x^2), \qquad (1.3.1-30)$$

$$D_F(x) = -\frac{1}{4\pi^2} \cdot \frac{1}{x^2 - i0}. \qquad (1.3.1-31)$$

1.3.2 Free real scalar field

We consider a free **real** (or **neutral**) **scalar field** $\phi(x)$ having a mass $m(\geq 0)$. Its Lagrangian density is given by

$$\mathcal{L}_S \equiv \frac{1}{2}(\partial^\mu\phi \cdot \partial_\mu\phi - m^2\phi^2). \qquad (1.3.2-1)$$

Its field equation is, of course, the Klein-Gordon equation

$$(\Box + m^2)\phi(x) = 0. \qquad (1.3.2-2)$$

The canonical conjugate of ϕ is $\pi = \dot{\phi}$, and canonical quantization is carried out. Then it is easy to calculate the four-dimensional commutator

$$[\phi(x), \phi(y)] = i\Delta(x - y; m^2). \qquad (1.3.2-3)$$

The canonical energy-momentum tensor is

$$T_{\mu\nu} \equiv \partial_\mu\phi \cdot \partial_\nu\phi - \eta_{\mu\nu}\mathcal{L}_S, \qquad (1.3.2-4)$$

which coincides with the symmetric energy-momentum tensor because ϕ is spinless. Poincaré generators are constructed in a straightforward way. In particular, the Hamiltonian is

$$H \equiv \frac{1}{2}\int d\mathbf{x}\,[\pi^2 + (\boldsymbol{\partial}\varphi)^2 + m^2\phi^2]. \qquad (1.3.2-5)$$

Since Eq.(2) is linear, it is solved by means of Fourier transform. That is, we have the momentum-space representation

$$\phi(x) = (2\pi)^{-3/2}\int d\mathbf{p}\,(2\hat{p}_0)^{-1/2}\left[a(\hat{p})e^{-i\hat{p}x} + a^\dagger(\hat{p})e^{i\hat{p}x}\right], \qquad (1.3.2-6)$$

where $\hat{p}_0 \equiv \sqrt{\boldsymbol{p}^2 + m^2}$ and $\hat{p}_\mu = (\hat{p}_0, p_k)$, so that $\hat{p}^2 = m^2$. The **annihilation operator** $a(\hat{p})$ and the **creation operation** $a^\dagger(\hat{p})$ satisfy the commutation relations

$$[a(\hat{p}), a^\dagger(\hat{q})] = \delta(\boldsymbol{p} - \boldsymbol{q}), \qquad (1.3.2 - 7)$$

$$[a(\hat{p}), a(\hat{q})] = [a^\dagger(\hat{p}), a^\dagger(\hat{q})] = 0. \qquad (1.3.2 - 8)$$

The **vacuum** $|0\rangle$ is defined by

$$a(\hat{p})|0\rangle = 0. \qquad (1.3.2 - 9)$$

The Hamiltonian is rewritten as

$$H = \frac{1}{2} \int d\boldsymbol{p}\, \hat{p}_0 \left[a(\hat{p})a^\dagger(\hat{p}) + a^\dagger(\hat{p})a(\hat{p}) \right]. \qquad (1.3.2 - 10)$$

Unfortunately, $H|0\rangle$ is divergent. Hence we subtract $\langle 0|H|0\rangle$ from H; then the redefined Hamiltonian is

$$H = \int d\boldsymbol{p}\, \hat{p}_0 a^\dagger(\hat{p})a(\hat{p}), \qquad (1.3.2 - 11)$$

which satisfies

$$H|0\rangle = 0, \qquad (1.3.2 - 12)$$

and all other eigenvalues of H are positive definite.

In a free field theory, one can count the absolute number of particles. The **number operator** N is defined by

$$N \equiv \int d\boldsymbol{p}\, a^\dagger(\hat{p})a(\hat{p}), \qquad (1.3.2 - 13)$$

whose eigenvalues are positive semi-definite. The space spanned by all eigenstates of N is (after completion) called the **Fock space**. Fock representation is characterized by the existence of a well-defined (absolute) number operator.

1.3.3 Free complex scalar field

We consider a free **complex** (or **charged**) **scalar field** $\phi(x)$ having a mass m. Its Lagrangian density is given by

$$\mathcal{L}_{\text{CS}} = \partial^\mu \phi^\dagger \cdot \partial_\mu \phi - m^2 \phi^\dagger \phi. \qquad (1.3.3 - 1)$$

If we set $\sqrt{2}\phi = \phi_1 + i\phi_2$, this theory reduces to the theory of two free real scalar fields ϕ_1 and ϕ_2, but it is more convenient not to do so. We regard ϕ and $\phi^\dagger$ as *independent* canonical variables; we therefore have $\pi = \dot{\phi}^\dagger$ and $\pi^\dagger = \dot{\phi}$.

The field equation is, of course,

$$(\Box + m^2)\phi = 0. \qquad (1.3.3-2)$$

The four-dimensional commutation relations are

$$[\phi(x); \phi^\dagger(y)] = i\Delta(x - y; m^2), \qquad (1.3.3-3)$$

$$[\phi(x), \phi(y)] = 0. \qquad (1.3.3-4)$$

The canonical energy-momentum tensor is

$$T_{\mu\nu} = \partial_\mu\phi \cdot \partial_\nu\phi^\dagger + \partial_\mu\phi^\dagger \cdot \partial_\nu\phi - \eta_{\mu\nu}\mathcal{L}_{\mathrm{CS}}. \qquad (1.3.3-5)$$

This theory has an internal symmetry; $\mathcal{L}_{\mathrm{CS}}$ is invariant under the **phase transformation**

$$\phi \rightarrow \phi' = e^{-i\theta}\phi, \qquad \phi^\dagger \rightarrow \phi'^\dagger = e^{i\theta}\phi^\dagger, \qquad (1.3.3-6)$$

where θ is a real parameter. The corresponding Noether current is

$$J^\mu \equiv i(\phi^\dagger\partial^\mu\phi - \phi\partial^\mu\phi^\dagger). \qquad (1.3.3-7)$$

Hence the generator of the phase transformation is

$$Q \equiv i\int d\mathbf{x}(\phi^\dagger\partial_0\phi - \phi\partial_0\phi^\dagger); \qquad (1.3.3-8)$$

indeed

$$[iQ, \phi(x)] = -i\phi(x), \qquad (1.3.3-9)$$

$$[iQ, \phi^\dagger(x)] = i\phi^\dagger(x). \qquad (1.3.3-10)$$

1.3.4 Free Dirac field

The field of spin $1/2$ is called a **Dirac field** or a **spinor field**. Usually, it is a complex field[2] and consists of four components.[3] It obeys Fermi statistics. The free Dirac field $\psi(x)$ satisfies the **Dirac equation**

$$(i\gamma^\mu\partial_\mu - m)\psi = 0, \qquad (1.3.4-1)$$

where the **gamma matrices** γ^μ are 4×4 matrices satisfying

$$\{\gamma^\mu, \gamma^\nu\} = 2\eta^{\mu\nu}, \qquad (1.3.4-2)$$

so that ψ satisfies the Klein-Gordon equation. We employ the convention $\gamma^{0\dagger} = \gamma^0$ and $\gamma^{k\dagger} = -\gamma^k$, whence $\gamma^0\gamma^{\mu\dagger}\gamma^0 = \gamma^\mu$. We set

$$\gamma_5 \equiv i(4!)^{-1}\epsilon_{\mu\nu\sigma\tau}\gamma^\mu\gamma^\nu\gamma^\sigma\gamma^\tau = i\gamma^0\gamma^1\gamma^2\gamma^3, \qquad (1.3.4-3)$$

whence $\{\gamma_5, \gamma^\mu\} = 0$ and $\gamma_5{}^\dagger = \gamma_5$, and

$$\sigma^{\mu\nu} \equiv (1/4)(\gamma^\mu\gamma^\nu - \gamma^\nu\gamma^\mu). \qquad (1.3.4-4)$$

The sixteen matrices

$$\{1, \ \gamma^\mu, \ 2i\sigma^{\mu\nu}, \ \gamma^\mu\gamma_5, \ -i\gamma_5\} \qquad (1.3.4-5)$$

are linearly independent. Hence the representation of Eq.(2) by 4×4 matrices is irreducible.[4]

The Dirac equation is derived from the Lagrangian density

$$\mathcal{L}_\mathrm{D} \equiv \bar{\psi}(i\gamma^\mu\partial_\mu - m)\psi, \qquad (1.3.4-6)$$

where $\bar{\psi} \equiv \psi^\dagger\gamma^0$. Although $\mathcal{L}_\mathrm{D}$ is not hermitian, we can hermitize it by integrating one half of the first term by parts.[5] The canonical variables are ψ_α ($\alpha = 1, 2, 3, 4$)

---

[2]   If it is a real field, it is called a **Majorana field**.

[3]   If massless, it can consist of only two components.

[4]   In D dimensions, the irreducible representation of Eq.(2) is given by $2^{[D/2]} \times 2^{[D/2]}$ matrices, where $[k]$ denotes the largest integer not larger than k. If D is odd, there is no γ_5 For D even, $\gamma_5 = \prod_{\mu=0}^{D-1}\gamma^\mu$ if $D/2$ is odd and $\gamma_5 = i\prod_{\mu=0}^{D-1}\gamma^\mu$ otherwise.

[5]   In the hermitized case, canonical quantization should be carried out by means of the Dirac method (see Addendum 1.A).

only, and their canonical conjugates are $\pi_\alpha = -\psi_\alpha{}^\dagger$ (the minus sign is due to the fermionic nature of ψ). The four-dimensional anticommutation relations are

$$\{\psi_\alpha(x), \bar\psi_\beta(y)\} = iS_{\alpha\beta}(x - y; m), \qquad (1.3.4 - 7)$$

$$\{\psi_\alpha(x), \psi_\beta(y)\} = \{\bar\psi_\alpha(x), \bar\psi_\beta(y)\} = 0, \qquad (1.3.4 - 8)$$

where

$$S(z; m) \equiv (i\gamma^\mu \partial_\mu + m)\Delta(z; m^2). \qquad (1.3.4 - 9)$$

The infinitesimal Lorentz transformation matrix $s_{\mu\nu}$ of the Dirac theory is given by $i\sigma_{\mu\nu}$. Since $\mathcal{L}_\mathrm{D}$ vanishes when the Dirac equation is used, the Poincaré generators simply become

$$P_\mu = i \int d\mathbf{x}\, \psi^\dagger \partial_\mu \psi, \qquad (1.3.4 - 10)$$

$$M_{\mu\nu} = i \int d\mathbf{x}\, \psi^\dagger (x_\mu \partial_\nu - x_\nu \partial_\mu + \sigma_{\mu\nu})\psi. \qquad (1.3.4 - 11)$$

The Dirac theory is invariant under the phase transformation

$$\psi \ \rightarrow\ \psi' = e^{-i\theta}\psi, \qquad \psi^\dagger \ \rightarrow\ \psi'^\dagger = \psi^\dagger e^{i\theta}, \qquad (1.3.4 - 12)$$

and if $m = 0$ under the **chiral transformation**

$$\psi \ \rightarrow\ \psi' = \exp{(-i\theta\gamma_5)}\psi, \qquad \psi^\dagger \ \rightarrow\ \psi'^\dagger = \psi^\dagger \exp{(i\theta\gamma_5)}. \qquad (1.3.4 - 13)$$

The corresponding Noether currents are

$$J^\mu = \bar\psi\gamma^\mu\psi, \qquad (1.3.4 - 14)$$

$$J_5{}^\mu = \bar\psi\gamma^\mu\gamma_5\psi, \qquad (1.3.4 - 15)$$

respectively.

1.4 GENERAL CONSIDERATION IN THE HEISENBERG PICTURE

In Sec.1.2, we have discussed the general framework of the operator formalism of quantum field theory. In this section, we introduce state vectors to represent field operators. Of course, it is natural that the notion of states should have no explicit dependence on spacetime coordinates, as long as the theory contains no external force. Nevertheless, it is often made to describe the theory by transferring the time dependence of field operators to state vectors totally (**Schrödinger picture**) or partially (**interaction picture**, etc.). The original description is, correspondingly, called **Heisenberg picture** in order to discriminate it from other pictures. One should not forget, however, that only the Heisenberg picture is fundamental and that any other picture is derived from it. In this sense, the construction of the theory in the Heisenberg picture is the most important from the theoretical point of view.

1.4.1 Two-point functions

As mentioned in Sec.1.1.1, we postulate the unique existence of a distinguished state, called **vacuum**, which is denoted by $|0\rangle$. It is a state belonging to discrete spectrum, normalized to unity, i.e.,

$$\langle 0|0 \rangle = 1, \qquad\qquad (1.4.1-1)$$

and is Poincaré-invariant, i.e.,

$$P_\mu |0\rangle = 0, \qquad M_{\mu\nu}|0\rangle = 0. \qquad (1.4.1-2)$$

We admit the case in which Eqs.(1) and (2) do not uniquely characterize $|0\rangle$. Its precise characterization is given later.

As shown in Sec.1.3, the four-dimensional (anti)commutator of field operators can be explicitly calculated in free field theories. It is no longer possible to do so if the interaction Lagrangian density is present. It is, however, possible to investigate the structure of the **vacuum expectation value**, $\langle 0| \cdots |0\rangle$, of the four-dimensional (anti)commutator by means of the general postulates of quantum field theory.

For simplicity of description, we consider a (non-free) real scalar field $\phi(x)$. We first investigate

$$W(x,y) \equiv \langle 0|\phi(x)\phi(y)|0\rangle, \qquad (1.4.1-3)$$

which is called the **two-point function**. From Eq.(1.1.2-3) we can write

$$\phi(x) = U(x,1)\phi(0)U^{-1}(x,1). \qquad (1.4.1-4)$$

Furthermore, since P_μ's are mutually commuting, Eq.(1.1.2-8) leads us to

$$U(x,1) = \exp ix^\mu P_\mu \qquad (1.4.1-5)$$

for any finite value of x^μ. Substituting Eqs.(4) and (5) into Eq.(3) and using Eq.(2), we obtain

$$W(x,y) = \langle 0|\phi(0)e^{-i(x-y)P}\phi(0)|0\rangle. \qquad (1.4.1-6)$$

If the state-vector space $\mathcal{V}$ is a Hilbert space, the eigenstates of P_μ form a complete set, that is, we can formally write[1]

$$P_\mu|n\rangle = p_\mu^{(n)}|n\rangle, \qquad (1.4.1-7)$$

$$\sum_n |n\rangle\langle n| = 1, \qquad (1.4.1-8)$$

where $\sum_n$ includes integration over $p_\mu^{(n)}$. Inserting Eq.(8) into Eq.(6), we have

$$W(x,y) = \sum_n \exp\left[-i(x-y)p^{(n)}\right] |\langle n|\phi(0)|0\rangle|^2. \qquad (1.4.1-9)$$

Because of the spectrum condition $(p^{(n)})^2 \geq 0$ with $p_0^{(n)} \geq 0$, we may write Eq.(9) as

$$W(x,y) = \int d^4p\, e^{-i(x-y)p}(2\pi)^{-3}\theta(p_0)\rho(p^2), \qquad (1.4.1-10)$$

where

$$\rho(p^2) = \int_0^\infty ds\, \delta(s-p^2)\rho(s) \geq 0. \qquad (1.4.1-11)$$

Using Eq.(1.3.1-11), we obtain

$$W(x,y) = \int_0^\infty ds\, \Delta^{(+)}(x-y;s)\rho(s). \qquad (1.4.1-12)$$

---

[1]   Strictly speaking, we should consider wave-packet states rather than non-normalizable eigenstates of P_μ.

This formula is called the **spectral representation** of $W(x, y)$, and $\rho(s)$ is called a **spectral function**.

From Eq.(12), we immediately obtain

$$\langle 0|[\phi(x), \phi(y)]|0\rangle = i \int_0^\infty ds\, \Delta(x - y; s)\rho(s). \qquad (1.4.1 - 13)$$

The **time-ordered product** is defined by

$$\mathrm{T}\phi(x)\phi(y) \equiv \theta(x^0 - y^0)\phi(x)\phi(y) + \theta(y^0 - x^0)\phi(y)\phi(x), \qquad (1.4.1 - 14)$$

for which

$$\langle 0|\mathrm{T}\phi(x)\phi(y)|0\rangle = \int_0^\infty ds\, \Delta_F(x - y; s)\rho(s). \qquad (1.4.1 - 15)$$

We differentiate Eq.(13) by x^0 and set $x^0 = y^0$. Then, owing to the canonical commutation relation[2] and Eq.(1.3.1-6), we have

$$-i\delta(\mathbf{x} - \mathbf{y}) = -i \int_0^\infty ds\, \delta(\mathbf{x} - \mathbf{y})\rho(s), \qquad (1.4.1 - 16)$$

namely,

$$\int_0^\infty ds\, \rho(s) = 1. \qquad (1.4.1 - 17)$$

If particle contents of the theory are known, the spectrum condition can be made more precise. If there are only one kind of scalar particles having a physical mass m_r, which is generally different from m, then we have $p^2 = m_r^2$ for one-particle states and $p^2 \geq (2m_r)^2$ for many-particle states. In this case, we can write

$$\rho(s) = Z\delta(s - m_r^2) + \sigma(s)\theta(s - 4m_r^2) \qquad (1.4.1 - 18)$$

with $Z \geq 0$ and $\sigma(s) \geq 0$. Substituting Eq.(18) into Eq.(17), we find

$$1 - Z = \int_{4m_r^2}^\infty ds\, \sigma(s), \qquad (1.4.1 - 19)$$

whence

$$0 \leq Z \leq 1. \qquad (1.4.1 - 20)$$

---

[2]   We assume that the interaction Lagrangian density does not contain $\dot{\phi}$.

If $Z = 1$, then $\sigma(s) \equiv 0$, that is, we have

$$\langle 0|\phi(x)\phi(y)|0\rangle = \Delta^{(+)}(x - y; m_r{}^2), \qquad (1.4.1 - 21)$$

whence

$$\langle 0|(\Box + m_r{}^2)\phi(x) \cdot (\Box + m_r{}^2)\phi(y)|0\rangle = 0. \qquad (1.4.1 - 22)$$

The metric positivity, Eq.(1.1.1-3), therefore, implies

$$(\Box + m_r{}^2)\phi(x)|0\rangle = 0. \qquad (1.4.1 - 23)$$

Then the separating property of vacuum (see Corollary A.2-6 in Appendix A.2) implies

$$(\Box + m_r{}^2)\phi(x) = 0. \qquad (1.4.1 - 24)$$

Thus $\phi(x)$ is a free field, that is, as long as $\phi(x)$ is non-free, we have $Z < 1$.

Since the coefficient of the discrete spectrum is $Z < 1$, it is convenient to re-normalize it to unity. We therefore consider

$$\phi^{(r)}(x) \equiv Z^{-1/2}\phi(x), \qquad (1.4.1 - 25)$$

and call it a **renormalized field**. Correspondingly, Z is called a (wave-function) **renormalization constant**.

In the above discussion, we have assumed that the canonical commutation relations are consistent with the general principles of the theory including the metric positivity of $\mathcal{V}$. But such an optimistic standpoint may cause troubles. Consider a current

$$j^\mu \equiv \bar{\psi}\gamma^\mu\psi. \qquad (1.4.1 - 26)$$

If we naively apply the canonical anticommutation relations, we find

$$[j^0(x), j^k(y)]_0 = 0. \qquad (1.4.1 - 27)$$

On the other hand, the spectral representation consistent with $\partial_\mu j^\mu = 0$ is

$$\langle 0|[j^\mu(x), j^\nu(y)]|0\rangle = i \int_0^\infty ds\, \pi(s)(-s\eta^{\mu\nu} - \partial^\mu\partial^\nu)\Delta(x - y; s), \qquad (1.4.1 - 28)$$

where the metric positivity implies

$$\pi(s) \geq 0 \quad (\pi(s) \not\equiv 0). \qquad (1.4.1 - 29)$$

Hence we have

$$\langle 0|[j^0(x), j^k(y)]_0|0\rangle = i \int_0^\infty ds\, \pi(s) \partial^k \delta(\boldsymbol{x} - \boldsymbol{y})$$
$$\neq 0 \qquad\qquad (1.4.1-30)$$

in contradiction with Eq.(27)[Got 55, Sch 59].

The widely accepted resolution of the difficulty is to distrust the canonical result, Eq.(27). Indeed, if one calculates $[j^0, j^k]_0$ by point-splitting, one finds that it does not vanish at the coinciding limit. Such a non-canonical term as Eq.(30) is generally called a **Schwinger term** [Sch 59]. However, the Schwinger term is a very pathological concept; it violates the Jacobi identity [Joh 66] and its expression is generally dependent on the method of calculation. Furthermore, no physical result follows from the Schwinger term. In Sec.5.8.3, a possible resolution which respects the canonical result is proposed: The appearance of the Schwinger term is due to the negligence of gravity. Only in the two-dimensional spacetime, the Schwinger term is significant [cf. Eq.(2.5.3-15)] because it is finite and because there is no gravity.

1.4.2 Asymptotic fields

In particle-physics experiments, what are really observed are not fields but particles. In the scattering experiments, two particles collide with each other and after complicated interactions some particles are observed. Both before and after the collision, particles are so distantly located from each other that they can be regarded as free.[3] That is, in the remote past and in the remote future compared with the interaction time, we encounter only free particles. It is thus quite important to describe free particle nature in the asymptotic regions $x^0 \to \pm\infty$.

As a simple example, we consider a scalar field $\phi^{(r)} = Z^{-1/2}\phi$ satisfying a (renormalized) field equation

$$(\Box + m_r{}^2)\phi^{(r)}(x) = J(x), \qquad\qquad (1.4.2-1)$$

where we assume that $J(x)$ has *no discrete spectrum* on mass shell. We can integrate Eq.(1) as

$$\phi^{(r)}(x) = \phi^{\text{in}}(x) + \int d^4 y\, \Delta_R(x - y; m_r{}^2) J(y), \qquad\qquad (1.4.2-2)$$

---

[3]   If there are massless particles, special care is required.

$$\phi^{(r)}(x) = \phi^{\text{out}}(x) + \int d^4y\, \Delta_A(x - y; m_r{}^2)J(y), \qquad (1.4.2 - 3)$$

where

$$(\Box + m_r{}^2)\phi^{\text{as}}(x) = 0 \qquad (1.4.2 - 4)$$

with $\phi^{\text{as}} = \phi^{\text{in}}$ or ϕ^{out}. The integral equations, Eqs.(2) and (3), are called **Yang-Feldman equations** [Yan 50].

From the definitions, Eqs.(1.3.1-18) and (1.3.1-19), of Δ_R and Δ_A, we can infer that

$$\phi^{(r)}(x) \rightarrow \phi^{\text{in}}(x) \quad \text{as} \quad x^0 \rightarrow -\infty, \qquad (1.4.2 - 5)$$

$$\phi^{(r)}(x) \rightarrow \phi^{\text{out}}(x) \quad \text{as} \quad x^0 \rightarrow +\infty. \qquad (1.4.2 - 6)$$

In this sense, $\phi^{\text{in}}(x)$ and $\phi^{\text{out}}(x)$ are called an **in-field** and an **out-field**, respectively. They are altogether called **asymptotic fields** and denoted by $\phi^{\text{as}}(x)$.

From Eq.(1.3.1-18) we have

$$\partial_0 \Delta_R(x - y; m_r{}^2) = -\theta(x^0 - y^0)\partial_0\Delta(x - y; m_r{}^2) - \delta(x^0 - y^0)\Delta(x - y; m_r{}^2),$$
$$(1.4.2 - 7)$$

but the last term vanishes identically. Hence we can infer

$$\partial_0\phi^{(r)}(x) \rightarrow \partial_0\phi^{\text{in}}(x) \quad \text{as} \quad x^0 \rightarrow -\infty. \qquad (1.4.2 - 8)$$

It should be noted, however, that the same thing is no longer true for $\partial_0{}^2\phi^{(r)}$ because we then encounter a non-vanishing term

$$-\delta(x^0 - y^0)\partial_0\Delta(x - y; m_r{}^2) = \delta^4(x - y). \qquad (1.4.2 - 9)$$

The equal-time commutators concerning ϕ^{in} and $\partial_0\phi^{\text{in}}$ are the same as those concerning $\phi^{(r)}$ and $\partial_0\phi^{(r)}$ because the second term of Eq.(2) essentially vanishes as $x^0 \rightarrow -\infty$. Combining this result with Eq.(4), we obtain

$$[\phi^{\text{in}}(x), \phi^{\text{in}}(y)] = i\Delta(x - y; m_r{}^2). \qquad (1.4.2 - 10)$$

Likewise for ϕ^{out}. Thus ϕ^{as} is a free field. But we must remember that ϕ^{as} is a *non-local* field in the Heisenberg picture.

Although the above reasoning clarifies what the asymptotic field is, the mathematical meaning of Eqs.(5) and (6) is not clear. The asymptotic equality between

$\phi^{(r)}(x)$ and $\phi^{as}(x)$ can be defined in the following sense [Leh 55]: For any two states $|f\rangle$ and $|g\rangle \in \mathcal{V}$,

$$\langle f|\phi^{(r)}(x) - \phi^{as}(x)|g\rangle \ \rightarrow \ 0 \quad \text{as} \quad x^0 \ \rightarrow \ \mp\infty. \qquad (1.4.2-11)$$

Such convergence as above is called **weak convergence** in mathematics, in order to discriminate it from the norm convergence in the mathematical sense (**strong convergence**). It should be noted that weak convergence is meaningful even if $\mathcal{V}$ is an indefinite-metric Hilbert space. Weak convergence is really "weak" because the weak limit of a product of two operators is *not*, in general, equal to the product of the weak limits of those operators. Indeed, a product of primary fields may have an asymptotic field different from any primary-field's asymptotic fields. In general, if the two-point function $\langle 0|\Phi_1(x)\Phi_2(y)|0\rangle$ for *local* operators $\Phi_1(x)$ and $\Phi_2(y)$ has a discrete spectrum, then there exist asymptotic fields $\Phi_1{}^{as}$ and $\Phi_2{}^{as}$ such that

$$\langle 0|\Phi_1{}^{as}(x)\Phi_2{}^{as}(y)|0\rangle = \text{ discrete spectrum of } \langle 0|\Phi_1(x)\Phi_2(y)|0\rangle. \qquad (1.4.2-12)$$

Since $[\Phi_1{}^{as}(x), \Phi_2{}^{as}(y)]_{\mp}$ is a c-number [Gre 62, Rob 62], it is equal to the discrete spectrum of $\langle 0| [\Phi_1(x), \Phi_2(y)]_{\mp} |0\rangle$.

The (anti)commutator between a symmetry generator Q and a local operator $\Phi(x)$ is not spoiled by taking weak limit, provided that the symmetry is not broken. That is, if

$$[iQ, \Phi(x)]_{\mp} = \Psi(x), \qquad (1.4.2-13)$$

then we have

$$[iQ, \Phi^{as}(x)]_{\mp} = \Psi^{as}(x). \qquad (1.4.2-14)$$

In particular, from the commutators with the Poincaré generators, we obtain the covariance of asymptotic fields under the Poincaré algebra. From Eq.(14) we see that Q is expressible as an integral over a quadratic function of asymptotic fields.

1.4.3 Asymptotic states and asymptotic completeness

Since asymptotic fields are free fields, we can construct their Fock representation on the basis of the vacuum $|0\rangle$. Since the vacuum of the Fock space is unique, this fact uniquely characterizes $|0\rangle$ even when its Poincaré invariance cannot do so.

The state-vectors of the above Fock space are called **asymptotic states**. We call an in-field asymptotic state an **in-state** and an out-field one an **out-state**. The Fock space spanned by all in-states is denoted by $\mathcal{V}^{\text{in}}$; $\mathcal{V}^{\text{out}}$ is similarly defined. In the axiomatic field theory [Haa 58, Haa 59, Rue 62], it is rigorously proved that the (wave-packet) asymptotic states exist as strong limits of some states in $\mathcal{V}$, provided that $\mathcal{V}$ is a Hilbert space and that there exists a finite gap between the discrete spectrum and the continuous one. Therefore, both $\mathcal{V}^{\text{in}}$ and $\mathcal{V}^{\text{out}}$ are subspaces of $\mathcal{V}$.

We now introduce a very crucial postulate, called **asymptotic completeness**: We postulate that $\mathcal{V}^{\text{in}} = \mathcal{V}$. Then the PCT theorem (see Appendix A.2) implies $\mathcal{V}^{\text{out}} = \mathcal{V}$. We thus have

$$\mathcal{V}^{\text{in}} = \mathcal{V} = \mathcal{V}^{\text{out}}. \qquad (1.4.3-1)$$

The asymptotic completeness is important because *it is the only known general principle which uniquely determines the representation space of field operators.* We may say that the asymptotic completeness is a certain minimality requirement because any other representation space must be *larger*, that is, it contains non-asymptotic extra states in addition to all asymptotic states. The choice of such extra states is not only necessarily model-dependent but also often *"author-dependent"* (see Sec.2.5). This fact may lead authors to futile controversy. Furthermore, extra states are usually devoid of particle contents, whence it is quite questionable in what way they are observed, unless one can invent a mechanism by which they become unobservable.

Under the asymptotic completeness, any operator is (at least formally) expandable into a series of normal products of asymptotic fields, where **normal product** is a product of free fields such that any annihilation operator lies on the right of any creation operator. This is because we can always choose expansion coefficients in such a way that all matrix elements of that operator in terms of asymptotic states coincide with those of that series. In this sense, under the asymptotic completeness, *any operator is expressible in terms of asymptotic fields.*

Since $\mathcal{V}^{\text{in}} = \mathcal{V}^{\text{out}}$, the transformation between the in-states $|\text{in}\rangle$ and the out-states $|\text{out}\rangle$ is a unitary transformation S, called **S-matrix**, because it is symbolically written as

$$S = \{\langle\text{out}|\text{in}\rangle\}. \qquad (1.4.3-2)$$

More precisely, S should be called **S-operator**; it is defined by

$$\varphi^{\mathrm{out}}(x) = S^{-1}\varphi^{\mathrm{in}}(x)S, \qquad\qquad (1.4.3-3)$$

with $S^{-1} = S^{\dagger}$ and

$$S|0\rangle = |0\rangle. \qquad\qquad (1.4.3-4)$$

The S-operator S commutes with any symmetry generator Q because of the time independence of Q. Especially, S commutes with the Poincaré generators. Since one-particle states are uniquely specified by the eigenvalue of P_{μ} and spin component, the one-particle restriction of S is a unit operator. The expression for Q in terms of φ^{out} has the same form as that in term of φ^{in}, because if $Q = F(\varphi^{\mathrm{in}})$ then

$$\begin{aligned}
F(\varphi^{\mathrm{out}}) = F(S^{-1}\varphi^{\mathrm{in}}S) &= S^{-1}F(\varphi^{\mathrm{in}})S \\
&= S^{-1}QS = Q.
\end{aligned} \qquad\qquad (1.4.3-5)$$

Hence we may write $Q = F(\varphi^{\mathrm{as}})$.

1.4.4 Reduction formula

The S-operator introduced in Sec.1.4.3 is the concept of practical importance because the transition probabilities of particle reactions are essentially given by the absolute squares of its matrix elements. It is therefore very important to express S in terms of field operators in a more direct way. The most standard way of doing this is called **Lehmann-Symanzik-Zimmermann** (or **LSZ**) **formalism** [Leh 55, Leh 57].

For simplicity of description, we consider only a real scalar field $\phi(x)$. The Klein-Gordon operator $\Box + m_r^2$ is denoted by K. We intoduce a complete set of orthonormal positive-frequency solutions, $f_k(x)$, to the Klein-Gordon equation:

$$Kf_k = 0, \qquad\qquad (1.4.4-1)$$

$$i\int d\mathbf{x}\,[f_k^*(x)\partial_0 f_l(x) - \partial_0 f_k^*(x)\cdot f_l(x)] = \delta_{kl}, \qquad\qquad (1.4.4-2)$$

$$\sum_k f_k(x)f_k^*(y) = \Delta^{(+)}(x-y;m_r^2). \qquad\qquad (1.4.4-3)$$

We then define

$$\phi_k(x^0) \equiv i \int d\mathbf{x} \, [\phi(x) \partial_0 f_k(x) - \partial_0 \phi(x) \cdot f_k(x)], \qquad (1.4.4-4)$$

and introduce the asymptotic-field creation operators $\phi^{in}{}_k{}^\dagger$ and $\phi^{out}{}_k{}^\dagger$ as the weak $x^0 \to \mp\infty$ limits of $\phi_k(x^0)$; of course, $\phi^{as}{}_k{}^\dagger$ is independent of time.

The **time-ordered product** of field operators is defined by

$$T\phi(x_1) \cdots \phi(x_n) \equiv \sum_\sigma \left[\prod_{j=1}^{n-1} \theta(x_{\sigma(j)} - x_{\sigma(j+1)}) \right] \phi(x_{\sigma(1)}) \cdots \phi(x_{\sigma(n)}), \quad (1.4.4-5)$$

where σ stands for a permutation of $\{1, 2, \cdots, n\}$ and the summation runs over all permutations. The vacuum expectation value of Eq.(5) is called a τ-**function**:

$$\tau(x_1, \cdots, x_n) \equiv \langle 0|T\phi(x_1) \cdots \phi(x_n)|0 \rangle. \qquad (1.4.4-6)$$

We first prove that

$$\langle 0|T\phi(x_1) \cdots \phi(x_n) \cdot \phi^{in}{}_k{}^\dagger|0 \rangle = i \int d^4y \, K^y \tau(x_1, \cdots, x_n, y) \cdot f_k(y). \qquad (1.4.4-7)$$

We substitute

$$\phi^{in}{}_k{}^\dagger = i \lim_{y^0 \to -\infty} \int d\mathbf{y} \, [\phi(y) \partial_0 f_k(y) - \partial_0 \phi(y) \cdot f_k(y)] \qquad (1.4.4-8)$$

into the lhs. of Eq.(7). Since $x_j{}^0 > y^0$ for any j, $\phi(y)$ can be included under the T symbol. Then, by using

$$\langle 0|\phi^{out}{}_k{}^\dagger T\phi(x_1) \cdots \phi(x_n)|0 \rangle = 0, \qquad (1.4.4-9)$$

we can rewrite the integral as follows:

$$\lim_{y^0 \to -\infty} \int d\mathbf{y} \, [\cdots] = - \int_{-\infty}^{+\infty} d^4y \, \partial_0{}^y [\cdots]. \qquad (1.4.4-10)$$

We thus obtain

$$\langle 0|T\phi(x_1) \cdots \phi(x_n) \cdot \phi^{in}{}_k{}^\dagger|0 \rangle$$
$$= -i \int d^4y \, [\langle 0|T\phi(x_1) \cdots \phi(x_n)\phi(y)|0 \rangle (\partial_0)^2 f_k(y)$$
$$- (\partial_0{}^y)^2 \langle 0|T\phi(x_1) \cdots \phi(x_n)\phi(y)|0 \rangle \cdot f_k(y)], \quad (1.4.4-11)$$

from which Eq.(7) immediately follows by using Eq.(1) and by integrating by parts.

If no indices coincide with each other, the above reasoning can be easily generalized to proving the formula

$$\langle 0| \prod_{j=1}^{s} \phi^{\text{out}}_{l_j} \, \text{T}\phi(x_1)\cdots\phi(x_n) \cdot \prod_{i=1}^{r} \phi^{\text{in}}_{k_i}{}^{\dagger} |0\rangle$$

$$= i\int d^4y\, K^y \langle 0| \prod_{j=1}^{s} \phi^{\text{out}}_{l_j} \, \text{T}\phi(x_1)\cdots\phi(x_n)\phi(y) \prod_{i=1}^{r-1} \phi^{\text{in}}_{k_i}{}^{\dagger} |0\rangle \cdot f_{k_r}(y)$$

$$= i\int d^4y\, K^y \langle 0| \prod_{j=1}^{s-1} \phi^{\text{out}}_{l_j} \, \text{T}\phi(y)\phi(x_1)\cdots\phi(x_n) \prod_{i=1}^{r} \phi^{\text{in}}_{k_i}{}^{\dagger} |0\rangle \cdot f_{l_s}^*(y). (1.4.4-12)$$

This formula is called **reduction formula**. By using Eq.(12) repeatedly, we can express

$$\langle 0| \prod_{j} \phi^{\text{in}}_{l_j} \cdot S \prod_{i} \phi^{\text{in}}_{k_i}{}^{\dagger} |0\rangle = \langle 0| \prod_{j} \phi^{\text{out}}_{l_j} \cdot \prod_{i} \phi^{\text{in}}_{k_i}{}^{\dagger} |0\rangle \qquad (1.4.4-13)$$

in terms of τ-functions.

Taking account of the possible coincidence of indices, we can write the result in a simple operator form:

$$S =: \exp\Big(\int d^4y\, \phi^{\text{in}}(y) K^y \frac{\delta}{\delta J(y)}\Big) : \langle 0|\text{T} \exp i\int d^4x\, J(x)\phi(x)|0\rangle \big|_{J=0},$$
$$(1.4.4-14)$$

where the pair of double dots stands for normal product and $J(x)$ is a c-number source function.

For further discussions on asymptotic fields and asymptotic states, especially in the indefinite-metric cases, see Appendix $A.4$.

1.5 SPONTANEOUS BREAKDOWN OF SYMMETRY

Spontaneous breakdown of symmetry is the breakdown of symmetry at the level of the representation of field operators. It is a very interesting phenomenon which cannot occur in a system of finite degrees of freedom. It was originally found in solid-state physics, and then by analogy it was brought into quantum field theory by Nambu and Jona-Lasinio [Nam 61]. The central subject in the spontaneous breakdown of symmetry is the Goldstone theorem claiming the existence of a massless mode [Gol 61].

1.5.1 Goldstone theorem

A symmetry generator Q, of course, commutes with P_μ, but the vacuum $|0\rangle$ need not be a simultaneous eigenstate of Q and P_μ in quantum field theory. That is, we may have

$$Q|0\rangle \neq |0\rangle\langle 0|Q|0\rangle. \qquad (1.5.1-1)$$

Although Eq.(1) most clearly characterizes the spontaneous breakdown of the symmetry generated by Q, it is problematic to write Eq.(1) explicitly because $Q|0\rangle$ does not exist at least in the positive-metric case. Indeed, if $Q|0\rangle$ were well-defined, we would have

$$\langle 0|Q \cdot Q|0\rangle = \int d\mathbf{x} \int d\mathbf{y}\, \langle 0|J^0(x)J^0(y)|0\rangle$$
$$= \int d\mathbf{x} \int d\mathbf{y}\, f(x-y) = \infty \qquad (1.5.1-2)$$

because of translational invariance and metric positivity.

Instead of Eq.(1), the spontaneous breakdown of symmetry is characterized by the **Goldstone commutator**

$$\langle 0|[iQ, \chi(x)]_\mp |0\rangle \neq 0 \qquad (1.5.1-3)$$

for some operator $\chi(x)$. Indeed, if $Q|0\rangle = |0\rangle\langle 0|Q|0\rangle$, the lhs. of Eq.(3) would vanish; that is, if $Q|0\rangle$ is sensible, Eq.(3) implies Eq.(1).

The simplest example of Eq.(3) is found in the free massless scalar field theory. This theory is invariant under the transformation $\phi \rightarrow \phi' = \phi + \alpha$, α

being a c-number constant. The corresponding generator is given by

$$Q = \int d\mathbf{x}\, \partial_0 \phi(x). \qquad (1.5.1-4)$$

Since

$$[iQ, \phi(x)] = 1, \qquad (1.5.1-5)$$

this symmetry is spontaneously broken. Here, the fact that $\phi(x)$ is massless is crucial.

In general, the following theorem, which is called **Goldstone theorem**, holds: *If Eq.(3) holds, then $\chi(x)$ contains a massless discrete spectrum.* Here, as remarked in Sec.1.2.4, Eq.(3) should be understood as

$$\int d\mathbf{x}\, \langle 0|[J^0(x), \chi(y)]_{\mp}|0\rangle = c \neq 0, \qquad (1.5.1-6)$$

where c is a constant because of translational invariance. Proof goes as follows [Gol 62]. (Proof can be made more rigorously [Kas 66, Eza 67, Ree 68].)

Because of translational invariance, we can write

$$\langle 0|[J^0(x), \chi(y)]_{\mp}|0\rangle = \int d^4p\, f(p)e^{-ip(x-y)}. \qquad (1.5.1-7)$$

Combining Eq.(7) with Eq.(6), we have

$$c = (2\pi)^3 \int d^4p\, f(p)\delta(\mathbf{p})e^{-ip(x-y)}, \qquad (1.5.1-8)$$

whose Fourier transform is

$$c\delta^4(p) = (2\pi)^3 f(p)\delta(\mathbf{p}). \qquad (1.5.1-9)$$

Since $c \neq 0$, $f(p)$ must become proportional to $\delta(p_0)$ when $\mathbf{p} = 0$, that is, it contains a *one-dimensional* δ-like singularity. This fact implies that $f(p)$ has a massless discrete spectrum in its spectral representation.

This massless spectrum is called **Goldstone mode**. In the covariant field theory, it must represent a particle, which is called **Goldstone boson** or **Nambu-Goldstone (NG) boson** (or **NG fermion** if χ is fermionic). For the internal symmetry, Q is Lorentz invariant, whence χ must be spinless, that is, the NG boson is spinless.

It is quite instructive to provide an alternative proof of the Goldstone theorem by means of asymptotic fields [Ume 65, Sen 67]. Let $\{\varphi^{as}{}_A(x)\}$ be the totality of independent asymptotic fields. Since $\varphi^{as}{}_A$ satisfies a free-field equation, $[iQ, \varphi^{as}{}_A]_{\mp}$ also does, whence it must be linear with respect to asymptotic fields; that is,

$$[iQ, \varphi^{as}{}_A(x)]_{\mp} = \alpha_A{}^B \varphi^{as}{}_B(x) + \beta_A, \qquad (1.5.1-10)$$

where $\alpha_A{}^B$ is a differential operator with c-number constant coefficients, β_A being a c-number constant. Of course, $\alpha_A{}^B = 0$ if the quantum numbers of $\varphi^{as}{}_A$ and $\varphi^{as}{}_B$ are not common.

Since

$$\langle 0|\varphi^{as}{}_A(x)|0\rangle = 0 \qquad (1.5.1-11)$$

for any $\varphi^{as}{}_A$, Eq.(10) yields

$$\langle 0|[iQ, \varphi^{as}{}_A(x)]_{\mp}|0\rangle = \beta_A. \qquad (1.5.1-12)$$

Comparing Eq.(12) with Eq.(3), we find that a necessary and sufficient condition for the spontaneous breakdown of symmetry is that we do not have $\beta_A = 0$ for all A's. Let $\beta_A = c \neq 0$ for some A, and $\chi^{as}(x)$ be the corresponding asymptotic field:

$$\langle 0|[iQ, \chi^{as}(x)]_{\mp}|0\rangle = c \neq 0. \qquad (1.5.1-13)$$

Since χ^{as} is a free field, it satisfies a linear differential equation $\tilde{K}\chi^{as} = 0$. Then Eq.(13) implies that $\tilde{K}c = 0$, that is, the differential operator $\tilde{K}$ cannot have a constant term. Thus χ^{as} must contain a massless mode.

Since any (anti)commutator of asymptotic fields is a c-number (see Appendix $A.4$), Eq.(10) shows that Q consists of quadratic terms and linear ones of asymptotic fields. From the above result, we see that the existence of a linear term in Q characterizes the spontaneous breakdown of symmetry.

Mathematical discussions on the spontaneous breakdown of symmetry in the indefinite-metric cases are given in Appendix $A.3$.

1.5.2 Goldstone model

Whether or not the spontaneous breakdown of symmetry occurs is often dependent on the values of the parameters involved in the theory.

We consider a complex scalar field having a ϕ^4 interaction. Its Lagrangian density is given by

$$\mathcal{L}_G \equiv \partial^\mu \phi^\dagger \cdot \partial_\mu \phi - u\phi^\dagger \phi - (\lambda/4)(\phi^\dagger \phi)^2. \qquad (1.5.2-1)$$

Here we assume $\lambda > 0$ because if $\lambda < 0$ there is no vacuum, but we do not prescribe the sign of the mass term. The field equation which follows from Eq.(1) is

$$(\Box + u)\phi + (\lambda/2)(\phi^\dagger \phi)\phi = 0. \qquad (1.5.2-2)$$

If ϕ is expanded in powers of its asymptotic field, the zeroth order term is

$$\langle 0|\phi(x)|0\rangle \equiv v/\sqrt{2}. \qquad (1.5.2-3)$$

Hence, in the zeroth-order approximation, Eq.(2) yields

$$(4u + \lambda|v|^2)v = 0. \qquad (1.5.2-4)$$

Since $\lambda > 0$, the solution to Eq.(4) is $v = 0$ only, if $u > 0$. This solution is consistent with the phase invariance of $\mathcal{L}_G$, that is, the phase symmetry is not spontaneously broken. On the other hand, if $u < 0$, then Eq.(4) has infinitely many solutions

$$|v| = \sqrt{-4u/\lambda} \qquad (1.5.2-5)$$

in addition to $v = 0$. The corresponding potential energy becomes

$$\begin{aligned} E_0 &\equiv u|v/\sqrt{2}|^2 + (\lambda/4)|v/\sqrt{2}|^4 \\ &= -u^2/\lambda < 0. \end{aligned} \qquad (1.5.2-6)$$

That is, it is lower than the energy corresponding to $v = 0$. Thus the vacuum should realize Eq.(3) with Eq.(5). Since

$$[iQ, \phi(x)] = -i\phi(x) \qquad (1.5.2-7)$$

for the phase-transformation generator Q [see Eq.(1.3.3-8)], Eq.(3) with Eq.(5) shows that this symmetry is spontaneously broken. The theory given by $\mathcal{L}_G$ with $u < 0$ is called the **Goldstone model** [Gol 61], which is a very instructive example of spontaneous breakdown.

Although v is, in general, complex, we can make it real positive by redefining ϕ by the phase factor of v^{-1}. Hence, without loss of generality, we may assume $v > 0$. It is convenient to write

$$\sqrt{2}\phi(x) = v + \varphi(x) + i\chi(x), \qquad (1.5.2-8)$$

where both real scalar fields $\varphi(x)$ and $\chi(x)$ have a vanishing vacuum expectation value. Substituting Eq.(8) into Eq.(1), we have

$$\mathcal{L}_{G} = \frac{1}{2}(\partial^{\mu}\varphi \cdot \partial_{\mu}\varphi - M^{2}\varphi^{2}) + \frac{1}{2}\partial^{\mu}\chi \cdot \partial_{\mu}\chi$$
$$- \frac{1}{2}gM\varphi(\varphi^{2} + \chi^{2}) - \frac{1}{8}g^{2}(\varphi^{2} + \chi^{2})^{2} - E_{0}, \qquad (1.5.2-9)$$

where

$$M^{2} \equiv -2u > 0, \qquad (1.5.2-10)$$

$$g \equiv \sqrt{\lambda/2} > 0, \qquad (1.5.2-11)$$

whence

$$v = M/g. \qquad (1.5.2-12)$$

The above zeroth-order approximation will be good if g is very small. In this case, φ is a massive scalar field having a mass M, while χ is massless. Of course, the mass of φ receives quantum correction, but χ remains exactly massless because of the Goldstone theorem. The asymptotic field, χ^{as}, of χ satisfies

$$[\chi^{\text{as}}(x), \chi^{\text{as}}(y)] = iD(x - y). \qquad (1.5.2-13)$$

The generator Q can be expressed as

$$Q = Z^{-1/2}v \int d\mathbf{x}\, \partial_{0}\chi^{\text{as}}(x), \qquad (1.5.2-14)$$

where Z denotes the renormalization constant of χ.

In the above, the representation of the field operator has been constructed on the basis of a particular choice of the phase of v. For various choices of it , we have physically equivalent but unitarily inequivalent representations. There is no need for considering all irreducible representations simultaneously so as to recover the invariance under the phase transformation.

1.5.3 Nambu-Jona-Lasinio model

In the Goldstone model, the NG boson $\chi(x)$ is a part of a primary field $\phi(x)$. But, in general, $\chi(x)$ may be a composite field. Such a simple example is the **Nambu-Jona-Lasinio model** [Nam 61].

The Lagrangian density of this model is given by

$$\mathcal{L}_{\mathrm{NJL}} \equiv i\bar{\psi}\gamma^\mu \partial_\mu \psi + g[(\bar{\psi}\psi)^2 - (\bar{\psi}\gamma_5\psi)^2]. \qquad (1.5.3-1)$$

It is invariant under the chiral transformation

$$\psi \rightarrow \psi' = \exp(-i\theta\gamma_5)\psi, \qquad \bar{\psi} \rightarrow \bar{\psi}' = \bar{\psi}\exp(-i\theta\gamma_5), \qquad (1.5.3-2)$$

as is easily confirmed by using $\gamma_5{}^2 = 1$ and $\exp(-i\theta\gamma_5) = \cos\theta - i\gamma_5\sin\theta$. If this invariance is not broken, ψ must be massless. But, according to the analysis based on the self-consistent self-energy equation (though it suffers, unfortunately, from serious divergence difficulty), it is possible for ψ to acquire a non-zero mass if the coupling constant g is negative and sufficiently large.

The chiral transformation, Eq.(2), is generated by

$$Q_5 \equiv \int d\mathbf{x}\, \psi^\dagger \gamma_5 \psi. \qquad (1.5.3-3)$$

Indeed, the canonical anticommutation relation yields

$$[iQ_5, \psi] = -i\gamma_5\psi. \qquad (1.5.3-4)$$

Hence, for the **pseudoscalar density**

$$\rho \equiv -i\bar{\psi}\gamma_5\psi, \qquad (1.5.3-5)$$

we have

$$[iQ_5, \rho] = -2\bar{\psi}\psi. \qquad (1.5.3-6)$$

Therefore, if

$$\langle 0|\bar{\psi}(x)\psi(x)|0\rangle \neq 0, \qquad (1.5.3-7)$$

then the chiral invariance is spontaneously broken, and $\rho(x)$ contains the Goldstone mode. That is, the NG boson is a composite particle of ψ and $\bar{\psi}$.

Addendum 1.A DIRAC METHOD OF QUANTIZATION

We describe the **Dirac method of quantization** for a singular system and show that it is independent of the choice of canonical variables.

For simplicity of description, we consider a system of finite degrees of bosonic freedom only. Canonical variables q_i $(i = 1, \ldots, n)$ are functions of time τ, and the Lagrangian L is a function of q_i and $\dot{q}_i \equiv dq_i/d\tau$. One wishes to define the canonical conjugates p_i by $\partial L/\partial \dot{q}_i$, but, in general, the simultaneous equations

$$p_i = \partial L/\partial \dot{q}_i \tag{1.A − 1}$$

are solvable with respect to $\dot{q}_i$ *only partially*. That is, Eq.(1) may yield some relations involving no $\dot{q}_i$. Such relations are called **constraints**, and the system having constraints is called a **singular system**. Dirac [Dir 64] presented a general method for quantizing such a singular system.

As mentioned above, we suppose that Eq.(1) implies the existence of indepedent constraints

$$\phi_\alpha = 0, \tag{1.A − 2}$$

where ϕ_α $(\alpha = 1, \ldots, r \leq n)$ are functions of q_i and p_i. We consider a variation of the Hamiltonian $p_i \dot{q}_i − L$:

$$\delta(p_i \dot{q}_i − L) = \dot{q}_i \delta p_i − \frac{\partial L}{\partial q_i} \delta q_i + (p_i − \frac{\partial L}{\partial \dot{q}_i}) \delta \dot{q}_i. \tag{1.A − 3}$$

If Eq.(1) is used, Eq.(3) shows that the Hamiltonian is expressible in terms of q_i and p_i. Of course, this statement is valid under the validity of the constraints, Eq.(2). Hence we should consider a generalized Hamiltonian

$$H \equiv p_i \dot{q}_i − L + v_\alpha \phi_\alpha, \tag{1.A − 4}$$

where v_α $(\alpha = 1, \ldots, r)$ are undetermined functions.

The time development of a quantity χ is given by

$$\dot{\chi} = (\chi, \, H)_{\mathrm{P}}, \tag{1.A − 5}$$

where $(A, \, B)_{\mathrm{P}}$ is the **Poisson bracket** defined by

$$(A, \, B)_{\mathrm{P}} \equiv \frac{\partial A}{\partial q_i} \frac{\partial B}{\partial p_i} − \frac{\partial A}{\partial p_i} \frac{\partial B}{\partial q_i}. \tag{1.A − 6}$$

Since the constraints must hold for any τ, consistency requires to have

$$(\phi_\alpha,\ H)_P = 0, \qquad\qquad (1.A-7)$$

Some of Eq.(7) may be satisfied by choosing v_α appropriately, but the remainder may be new conditions, which must be regarded as constraints. We should therefore consider Eq.(7) for these new constraints. Repeating this procedure until we no longer obtain new constraints, we obtain a set of independent constraints Eq.(2) with $\alpha = 1, \ldots, s$ $(r \le s \le 2n)$.

If $(\phi_\alpha,\ \chi)_P$ $(\alpha = 1, \ldots, s)$ are all written as linear combinations of constraints, χ is called of **first class**. Otherwise, χ is of **second class**. Let $\mathcal{A}$ be the $s \times s$ matrix formed by $(\phi_\alpha,\ \phi_\beta)_P$. If $\det\mathcal{A} = 0$, certain combinations ϕ' of ϕ_α are first-class constraints. When quantized, these first-class constraints are interpreted as conditions restricting state vectors in such forms as $\phi'|f\rangle = 0$, because there exist some quantities χ such that $(\phi',\ \chi)_P \ne 0$. When applied to quantum field theory, however, such conditions are generally inconsistent with the existence of the vacuum (see Sec.2.2.3). We therefore do not wish to have any first-class constraints.

Since the first-class constraints are nothing but the generators of gauge transformations (i.e., canonical transformations which leave H and all constraints invariant), they become of second class by adding gauge-fixing constraints. Indeed, if we start with the Lagrangian modified by gauge fixing in such a way that no gauge invariance remains, then we encounter no first-class constraints. In the following, therefore, we assume that all constraints are of second class.

In this case, the matrix $\mathcal{A}$ is invertible, whence s is even because $\mathcal{A}$ is antisymmetric. We then define the **Dirac bracket** of two quantities A and B by

$$(A,\ B)_D \equiv (A,\ B)_P - (A,\ \phi_\alpha)_P(\mathcal{A}^{-1})_{\alpha\beta}(\phi_\beta,\ B)_P. \qquad (1.A-8)$$

The Dirac bracket has the same properties as those (antisymmetry, Leibniz rule, and Jacobi identity) of the Poisson bracket. Furthermore, *the Dirac bracket between ϕ_α and any quantity χ vanishes*:

$$(\phi_\alpha,\ \chi)_D = 0. \qquad\qquad (1.A-9)$$

Quantization is carried out by replacing the Dirac bracket by $-i$ times a commutator:

$$(A,\ B)_D \ \rightarrow \ -i[A,\ B]. \qquad\qquad (1.A-10)$$

Here, because of Eq.(9), *we can regard Eq.(2) as operator identities*. We no longer need to worry about the existence of constraints. The Dirac method of quantization is thus quite useful for quantizing a singular system.

Now, there arises a question: Is the Dirac bracket independent of the choice of canonical variables? It is well known that the Poisson bracket is invariant under the canonical transformation. In the following, we show that the Dirac bracket is also invariant under the canonical transformation [Kug unp.c].

The Dirac bracket can easily be expressed in terms of the **Lagrange bracket**, which is defined as follows. Let z_j $(j = 1, \ldots, 2n)$ be functions of q_ι and p_ι such that q_ι and p_ι are expressible in terms of z_j. Then the Lagrange bracket between z_j and z_k is

$$(z_j, \; z_k)_{\mathrm{L}} \equiv \frac{\partial q_\iota}{\partial z_j} \frac{\partial p_\iota}{\partial z_k} - \frac{\partial q_\iota}{\partial z_k} \frac{\partial p_\iota}{\partial z_j}. \qquad (1.A-11)$$

The $2n \times 2n$ matrix formed by $(z_j, \; z_k)_{\mathrm{L}}$ is the inverse matrix of the one formed by the corresponding Poisson brackets:

$$(z_j, \; z_k)_{\mathrm{L}}(z_j, \; z_\ell)_{\mathrm{P}} = \delta_{k\ell}. \qquad (1.A-12)$$

Thus, as is well known, the Lagrange bracket is invariant under the canonical transformation.

We choose $z_j = \phi_\alpha$ for $j = 2n - s + \alpha$ and $z_j = z_a$ for $j = a = 1, 2, \ldots, 2n - s$. Then, with the aid of Eqs.(9), (8) and (12), we have

$$\begin{aligned}
(z_a, \; z_b)_{\mathrm{L}}(z_a, \; z_c)_{\mathrm{D}} &= (z_a, \; z_b)_{\mathrm{L}}(z_a, \; z_c)_{\mathrm{D}} + (\phi_\alpha, \; z_b)_{\mathrm{L}}(\phi_\alpha, \; z_c)_{\mathrm{D}} \\
&= (z_j, \; z_b)_{\mathrm{L}}(z_j, \; z_c)_{\mathrm{D}} \\
&= \delta_{bc}. \qquad (1.A-13)
\end{aligned}$$

Thus the $(2n - s) \times (2n - s)$ matrix formed by the Dirac brackets $(z_a, \; z_c)_{\mathrm{D}}$ is the inverse matrix of the restricted one of the Lagrange brackets $(z_a, \; z_b)_{\mathrm{L}}$. It follows from this fact that *the Dirac bracket is invariant under the canonical transformation*.

Finally, we present two important examples of the invariance of the Dirac brackets [Kug unp.c]. The action is not altered by adding total divergence. In the present formulation, this amounts to considering a change of L by $\dot{F}$, where F is a function of q_ι only (i.e., involving no $\dot{q}_\iota$). In this case, the canonical variables

p_i change by $\partial F/\partial \dot{q}_i$. Thus the Poisson brackets remain invariant, and so are the Dirac brackets.

The second example is the situation in which multipliers are present as discussed in Sec.1.2.3. Corresponding to Eq.(1.2.3-1), we consider the Lagrangian

$$L = L_0(q, \dot{q}) + F_\alpha(q, \dot{q})\, b_\alpha + G(b). \qquad (1.A-14)$$

Here L_0 involves no $\dot{q}_\alpha$,[1] F_α is linear in $\dot{q}_A$ and $\det \partial F_\alpha/\partial \dot{q}_\beta \neq 0$. In the Dirac quantization, not only q_a but also b_α are taken as canonical variables. The canonical conjugates of q_α and b_α are

$$p_\alpha = \frac{\partial L}{\partial \dot{q}_\alpha} = \frac{\partial F_\beta}{\partial \dot{q}_\alpha} b_\beta, \qquad (1.A-15)$$

$$\pi_\alpha = \frac{\partial L}{\partial \dot{b}_\alpha} = 0. \qquad (1.A-16)$$

They are constraints because $\partial F_\beta/\partial \dot{q}_\alpha$ involves no $\dot{q}_A$. Hence we write

$$\phi_{1\alpha} \equiv p_\alpha - \frac{\partial F_\beta}{\partial \dot{q}_\alpha} b_\beta, \qquad \phi_{2\alpha} \equiv \pi_\alpha. \qquad (1.A-17)$$

Then, since

$$(\phi_{2\alpha},\ \phi_{2\beta})_{\mathrm{P}} = 0, \qquad (1.A-18)$$

$$(\phi_{1\alpha},\ \phi_{2\beta})_{\mathrm{P}} = -\partial F_\beta/\partial \dot{q}_\alpha, \qquad (1.A-19)$$

we have

$$\det \mathcal{A} = (\det \partial F_\beta/\partial \dot{q}_\alpha)^2 \neq 0, \qquad (1.A-20)$$

that is, $\mathcal{A}$ is invertible. Thus the constraints $\phi_{1\alpha}$ and $\phi_{2\alpha}$ are of second class.

Because $(\partial F_\beta/\partial \dot{q}_\alpha)$ is invertible, $\{q_A, b_\alpha, p_B, \pi_\beta\}$ is expressible in terms of $\{q_A, p_B, \phi_{1\alpha}, \phi_{2\beta}\}$, which we adopt as z_j. Then Eq.(13) implies that

$$(q_A,\ q_B)_{\mathrm{D}} = (p_A,\ p_B)_{\mathrm{D}} = 0, \qquad (1.A-21)$$

$$(q_A,\ p_B)_{\mathrm{D}} = \delta_{AB}. \qquad (1.A-22)$$

We thus obtain the canonical commutation relations which are exactly the same as those in which canonical variables are q_A only but not b_α. The validity of this result is crucially due to the fact that L involves no $\dot{b}_\alpha$, that is, it is not correct to regard a primary variable b_α as non-canonical if L involves $\dot{b}_\alpha$.

---

[1] $\{q_\alpha\}$ is a subset of $\{q_A\}$ and has a one-to-one correspondence with $\{b_\alpha\}$.

CHAPTER TWO

ABELIAN GAUGE THEORY

It is most well-established how to describe the electromagnetic interaction. It is mediated by the electromagnetic field, which couples with every charged field in a gauge-invariant way. The electromagnetic field is a prototype of the abelian gauge field, and therefore we regard both almost as synonym throughout this chapter, that is, we first intoduce A_μ as the electromagnetic field and then regard it as a generic abelian gauge field without noticing so explicitly.

2.1 UNQUANTIZED ELECTROMAGNETIC FIELD

In the present section, we describe the theory of the classical electromagnetic field, that is, all quantities considered in this section are c-numbers.

2.1.1 Maxwell theory

The electric field E_j and the magnetic field H_k are put together into a four-dimensional antisymmetric tensor $F_{\mu\nu} = -F_{\nu\mu}$ by the relations $F_{j0} = E_j$ and $\epsilon_{ijk}F^{ij} = H_k$. The **electromagnetic field strength** $F_{\mu\nu}$ satisfies the (vacuum) **Maxwell equations**

$$\partial_\mu(\epsilon^{\mu\nu\sigma\tau}F_{\sigma\tau}) = 0, \qquad (2.1.1-1)$$

$$\partial_\mu F^{\mu\nu} = -ej^\nu, \qquad (2.1.1-2)$$

where e is the electromagnetic coupling constant and the electric current j_μ satisfies the conservation law

$$\partial_\mu j^\mu = 0. \qquad (2.1.1-3)$$

From Eq.(1), there exists A_μ such that

$$F_{\mu\nu} = \partial_\mu A_\nu - \partial_\nu A_\mu; \qquad (2.1.1-4)$$

A_μ is the **electromagnetic potential**. Then Eq.(2) becomes

$$\Box A_\mu - \partial_\mu \partial^\nu A_\nu = -ej_\mu. \qquad (2.1.1-5)$$

Let Λ be an arbitrary non-singular function. Then, evidently, Eq.(4) and therefore Eq.(5) are invariant under the **local gauge transformation**

$$A_\mu \rightarrow A'_\mu = A_\mu + \partial_\mu \Lambda. \qquad (2.1.1-6)$$

If we use such Λ as $\Box \Lambda = -\partial^\mu A_\mu$, then the transformed electromagnetic potential (omitting a prime) satisfies

$$\partial^\mu A_\mu = 0, \qquad (2.1.1-7)$$

which is called the **Lorentz condition**. Although it has no physical significance, it is very convenient to set it up. Then Eq.(5) reduces to

$$\Box A_\mu = -ej_\mu. \qquad (2.1.1-8)$$

Of course, Eq.(8) is still invariant under the gauge transformation, Eq.(6), with Λ satisfying $\Box \Lambda = 0$.

2.1.2 Coupling with charged fields

In the classical Maxwell theory, the electromagnetic field couples with the **electric current** j^μ determined by actual charge distribution. In the first-quantized theory, however, j^μ is expressed in terms of **charged fields**, which are complex fields and therefore non-classical quantities, though yet c-numbers. The theory of charged fields is invariant under a phase transformation, and the Noether theorem implies the existence of the conserved current j^μ (see Sec.2.1.3).

First, we consider the free Dirac field (see Sec.1.3.4):

$$(i\gamma^\mu \partial_\mu - m)\psi = 0. \qquad (2.1.2-1)$$

It is a general rule for introducing the electromagnetic field that the momentum p_μ is replaced by $p_\mu + qeA_\mu$, where qe denotes the electric charge which is carried by the charged field. Owing to the quantum-theory identification rule "$p_\mu = i\partial_\mu$", we replace Eq.(1) by

$$[i\gamma^\mu(\partial_\mu - iqeA_\mu) - m]\psi = 0. \tag{2.1.2 - 2}$$

Its conjugate equation is

$$\bar{\psi}[-i\gamma^\mu(\overleftarrow{\partial}_\mu + iqeA_\mu) - m] = 0. \tag{2.1.2 - 3}$$

From Eqs.(2) and (3), we find that the electric current,

$$j_D^\mu \equiv q\bar{\psi}\gamma^\mu\psi, \tag{2.1.2 - 4}$$

of the Dirac field ψ satisfies the conservation law

$$\partial_\mu j_D^\mu = 0. \tag{2.1.2 - 5}$$

Likewise, we consider a complex scalar field ϕ, which satisfies

$$[(\partial^\mu - iqeA^\mu)(\partial_\mu - iqeA_\mu) + m^2]\phi = 0, \tag{2.1.2 - 6}$$

where qe denotes the electric charge of ϕ. Its electric current j_S^μ is defined by

$$j_S^\mu \equiv iq(\phi^\dagger\partial^\mu\phi - \partial^\mu\phi^\dagger \cdot \phi) + 2q^2 eA^\mu\phi^\dagger\phi, \tag{2.1.2 - 7}$$

which satisfies the conservation law

$$\partial_\mu j_S^\mu = 0. \tag{2.1.2 - 8}$$

If charged fields directly interact with each other, the electric current of each field is no longer conserved, but their total sum defines the conserved electric current j^μ.

It is very important to note that charged fields couple with the electromagnetic field through A_μ but not $F_{\mu\nu}$. This fact implies that what is fundamental in quantum theory is *not* the gauge-invariant $F_{\mu\nu}$ but the gauge-variant A_μ. Indeed, the **Aharonov-Bohm effect** [Aha 59] shows that there is a physically observable consequence of Eq.(2) which cannot be described in terms of $F_{\mu\nu}$ alone. If A_μ can be written as $\partial_\mu\Phi$ outside a cylindrical region D and if the continuation of Φ into D becomes singular, $F_{\mu\nu}$ vanishes outside D but A_μ cannot be gauged away by Eq.(2.1.1-6), because Λ must be nonsingular in the *whole* spacetime. Then two coherent electron beams which pass through the right and left sides of D exhibit the interference depending on Φ (more precisely, magnetic flux inside D). The real existence of the Aharonov-Bohm effect has been experimentally demonstrated in a very clear-cut way [Ton 86].

Thus A_μ should be regarded as the primary entity. Hereafter *we refer to A_μ as the* **electromagnetic field**.

2.1.3 Lagrangian densities

It is very important that the electromagnetic theory is derivable from the action principle.

The classical Lagrangian density of A_μ is given by

$$\mathcal{L}_{\text{EM}} = -\frac{1}{4} F^{\mu\nu} F_{\mu\nu}, \qquad (2.1.3-1)$$

where $F_{\mu\nu}$ is *defined* by Eq.(2.1.1-4) as a quantity derivable from A_μ. In the so-called **first-order formalism**, one takes both A_μ and $F_{\mu\nu}$ as primary fields and derives Eq.(2.1.1-4) as a consequence of the action principle by adopting a Lagrangian density

$$\mathcal{L}'_{\text{EM}} = +\frac{1}{4} F^{\mu\nu} F_{\mu\nu} - \frac{1}{2} F^{\mu\nu} (\partial_\mu A_\nu - \partial_\nu A_\mu). \qquad (2.1.3-2)$$

But, throughout this book, we do not employ the first-order formalism because there is no compelling reason for regarding $F_{\mu\nu}$ as primary.

Apart from non-electromagnetic interaction terms, the Lagrangian densities for charged fields are as follows: For a Dirac field ψ, we have

$$\mathcal{L}_{\text{D}} = \bar{\psi}(i\gamma^\mu D_\mu - m)\psi; \qquad (2.1.3-3)$$

for a complex scalar field ϕ, we set

$$\mathcal{L}_{\text{S}} = (D^\mu \phi)^\dagger D_\mu \phi - m^2 \phi^\dagger \phi. \qquad (2.1.3-4)$$

In the above, D_μ denotes **covariant differentiation** defined by

$$D_\mu \equiv \partial_\mu - iqeA_\mu. \qquad (2.1.3-5)$$

Evidently, $\mathcal{L}_{\mathrm{EM}}$ is invariant under the local gauge transformation, Eq.(2.1.1-6), while $\mathcal{L}_{\mathrm{D}}$ is invariant under it if ψ simultaneously transforms as

$$\psi \ \rightarrow \ \psi' = e^{iqe\Lambda}\psi, \qquad\qquad (2.1.3-6)$$

because then

$$(D_\mu\psi)' = e^{iqe\Lambda}D_\mu\psi. \qquad\qquad (2.1.3-7)$$

Likewise for $\mathcal{L}_{\mathrm{S}}$.

In general, the derivative which transforms just like the primary field under the local gauge transformation is called **covariant derivative**. In the **abelian gauge theory**, covariant differentiation is only of the form given in Eq.(5), though the value of q depends on the charged field.[1] Covariant derivative has interesting geometrical interpretation, which is discussed in Sec.3.2.1.

The invariance under the local gauge transformation requires to replace each simple differentiation ∂_μ (except in $\mathcal{L}_{\mathrm{EM}}$) by covariant differentiation D_μ, but it cannot forbid the appearance of such a term (**Pauli term**) as

$$i\bar{\psi}\sigma^{\mu\nu}\psi F_{\mu\nu}. \qquad\qquad (2.1.3-8)$$

It is customary to impose renormalizability on the choice of interaction Lagrangian densities in order to exclude the Pauli term in the four-dimensional theory. If all A_μ's except those in $\mathcal{L}_{\mathrm{EM}}$ are only those which are involved in D_μ, the interaction is called **minimal**. In this case, of course, the Pauli term is excluded.

Let $\mathcal{L}$ be the total Lagrangian density, which may involve non-electromagnetic interactions. Let $\{\varphi_\alpha\}$ be a set of complex fields, and suppose that $\mathcal{L}$ is invariant under the phase transformation

$$\varphi_\alpha \ \rightarrow \ \varphi'_\alpha = e^{iq_\alpha e\theta}\varphi_\alpha, \qquad\qquad (2.1.3-9)$$

where θ is a constant, then, as discussed in Sec.1.2.4, the Noether theorem implies that the current,

$$-ej^\mu \equiv \sum_\alpha \left[iq_\alpha e\varphi_\alpha \frac{\partial\mathcal{L}}{\partial(\partial_\mu\varphi_\alpha)} - iq_\alpha e\varphi_\alpha^\dagger \frac{\partial\mathcal{L}}{\partial(\partial_\mu\varphi_\alpha^\dagger)} \right], \qquad (2.1.3-10)$$

---

[1]   For the electromagnetic theory, q is always an integral multiple of 1/3, but its reason is not yet known.

is conserved:

$$\partial_\mu j^\mu = 0. \tag{2.1.3 - 11}$$

Of course, for $\mathcal{L} = \mathcal{L}_D$ and for $\mathcal{L} = \mathcal{L}_S$, Eq.(10) reduces to Eq.(2.1.2-4) and to Eq.(2.1.2-7), respectively.

In the electromagnetic theory, $\mathcal{L}$ is invariant under the local gauge transformation

$$A_\mu \ \rightarrow \ A'_\mu = A_\mu + \partial_\mu \Lambda, \tag{2.1.3 - 12}$$

$$\varphi_\alpha \ \rightarrow \ \varphi'_\alpha = e^{iq_\alpha e \Lambda} \varphi_\alpha. \tag{2.1.3 - 13}$$

Since Eq.(9) can be regarded as a special case, Λ =const, of the local gauge transformation, the phase transformation is also called the **global gauge transformation**, or the **gauge transformation of the first kind**, while calling the local one the **gauge transformation of the second kind**.

If the electromagnetic interaction is minimal, $\partial_\mu \varphi_\alpha$ and $\partial_\mu \varphi_\alpha^\dagger$ are always accompanied with $(-iq_\alpha \varphi_\alpha)eA_\mu$ and $(+iq_\alpha \varphi_\alpha^\dagger)eA_\mu$, respectively. Hence Eq.(10) reduces to

$$ej^\mu = (\partial/\partial A_\mu)(\mathcal{L} - \mathcal{L}_{\text{EM}}). \tag{2.1.3 - 14}$$

2.2 GENERAL REMARKS ON COVARIANT QUANTIZATION

Before entering the presentation of the quantized theory of the electromagnetic field, we discuss some general aspects of it without particularizing its framework in this section.

The electromagnetic field cannot be properly quantized without fixing gauge because the differential operator $\eta_{\mu\nu}\Box - \partial_\mu\partial_\nu$, which should govern the two-point function of A_μ, is not invertible. We do not consider quantizations in non-covariant gauges[1] but concentrate our attention to the covariant gauges.

It is shown in Sec.2.2.2 that A_μ must be quantized with indefinite metric. We describe the indefinite-metric quantum field theory in the Appendix in detail.

2.2.1 Incompatibility between classical equations and quantum principles

It is characteristic of gauge theories that classical equations must be modified at quantization if one wishes to maintain manifest covariance. We here present the essence of the proof of the proposition that *the free Maxwell equation,*

$$\Box A_\mu - \partial_\mu\partial^\nu A_\nu = 0, \qquad (2.2.1-1)$$

is incompatible with manifest covariance if A_μ is a nontrivial quantized vector field. Its rigorous proof can be made in the Wightman framework of quantum field theory [Str 67].

We consider the two-point function

$$W_{\mu\nu}(x,y) \equiv \langle 0|A_\mu(x)A_\nu(y)|0\rangle. \qquad (2.2.1-2)$$

From the manifest covariance under the Poincaré group, it can be represented as

$$W_{\mu\nu}(x,y) = \eta_{\mu\nu}D_1(x-y) + \partial_\mu\partial_\nu D_2(x-y), \qquad (2.2.1-3)$$

---

[1]   It seems impossible to formulate the quantized theory of A_μ in the axial, light-cone, and temporal gauges in a satisfactory way. Contrary to the naive expectation, indefinite metric is indispensable in those gauges [Nak 83c], unless one accepts the violation of translational invariance or the use of a limiting procedure.

where D_1 and D_2 are some Lorentz-invariant functions. By using the fact that A_μ satisfies Eq.(1), we have

$$(\eta_{\mu\nu}\Box - \partial_\mu\partial_\nu)D_1(x) = 0. \qquad (2.2.1-4)$$

Contracting μ and ν, we find

$$\Box D_1(x) = 0, \qquad (2.2.1-5)$$

whence Eq.(4) reduces to

$$\partial_\mu\partial_\nu D_1(x) = 0. \qquad (2.2.1-6)$$

Since the Lorentz-invariant solution of Eq.(6) is a constant c only, Eq.(3) becomes

$$W_{\mu\nu}(x,y) = \eta_{\mu\nu}c + \partial_\mu\partial_\nu D_2(x-y). \qquad (2.2.1-7)$$

Accordingly, Eqs.(2.1.1-4) and (2) yield

$$\langle 0|F_{\mu\nu}(x)F_{\lambda\rho}(y)|0\rangle = 0. \qquad (2.2.1-8)$$

Since $F_{\mu\nu}$ is a physical field, the state

$$\int d^4x\, f^{\mu\nu}(x)F_{\mu\nu}(x)|0\rangle, \qquad (2.2.1-9)$$

where $f^{\mu\nu}(x)$ is an arbitrary real test function, should belong to a positive-metric Hilbert space. Hence Eq.(8) implies that Eq.(9) vanishes identically. Since $f^{\mu\nu}(x)$ is arbitrary, we have

$$F_{\mu\nu}(x)|0\rangle = 0. \qquad (2.2.1-10)$$

Owing to the separating property of vacuum (see Corollary A.2-6 in Appendix A.2), Eq.(10) is equivalent to

$$F_{\mu\nu}(x) = 0, \qquad (2.2.1-11)$$

that is, A_μ is trivial $(= \partial_\mu\Phi)$.

It is evident that the above difficulty cannot be resolved by *adding* the Lorentz condition, Eq.(2.1.1-7). It is necessary to *modify* the Maxwell equation, Eq.(1), itself. Thus the classical equation cannot survive in quantum theory as a *local* field equation.

2.2.2 Indispensability of indefinite metric

If A_μ is covariantly quantized, all four components of it must become operators.
Owing to the Minkowski character, A_0 should be quantized with indefinite metric. If A_μ is massive, it is possible to separate the negative-norm part covariantly
because of the fact that the corresponding little group $SO(3)$ is compact. Since,
however, the electromagnetic field A_μ is massless, it is impossible to avoid negative
norm without violating manifest covariance, owing to the fact that the corresponding little group $E(2)$ is noncompact. In this subsection, we explicitly demonstrate
the impossibility of covariantly quantizing A_μ in the positive-metric Hilbert space
without reference to any field equations [Mat 74].

Let P_μ and $M_{\mu\nu}$ be Poincaré generators. We set

$$J_i \equiv \epsilon_{ijk}M^{jk}, \tag{2.2.2 − 1}$$

$$K_j \equiv M_{0j}; \tag{2.2.2 − 2}$$

then from the commutation relation, Eq.(1.1.2-13), we find[2]

$$[J_i, J_j] = i\epsilon_{ijk}J_k, \tag{2.2.2 − 3}$$

$$[K_i, K_j] = -i\epsilon_{ijk}J_k, \tag{2.2.2 − 4}$$

$$[J_i, K_j] = i\epsilon_{ijk}K_k. \tag{2.2.2 − 5}$$

Hence with

$$M_i \equiv \frac{1}{2}(J_i + iK_i), \tag{2.2.2 − 6}$$

$$N_j \equiv \frac{1}{2}(J_j - iK_j), \tag{2.2.2 − 7}$$

we have

$$[M_i, N_j] = 0, \tag{2.2.2 − 8}$$

$$[M_i, M_j] = i\epsilon_{ijk}M_k, \tag{2.2.2 − 9}$$

$$[N_i, N_j] = i\epsilon_{ijk}N_k. \tag{2.2.2 − 10}$$

---

[2]   For three-dimensional quantities such as J_i and K_j, we use δ_{ij} as metric rather than
$\eta_{ij} = -\delta_{ij}$, without using superscripts.

We further set

$$M_\pm \equiv M_1 \pm iM_2, \qquad N_\pm \equiv N_1 \pm iN_2; \qquad (2.2.2-11)$$

then

$$[M_\pm, M_3] = \mp M_\pm, \qquad [M_+, M_-] = 2M_3, \qquad (2.2.2-12)$$

$$[N_\pm, N_3] = \mp N_\pm, \qquad [N_+, N_-] = 2N_3. \qquad (2.2.2-13)$$

Now, in order to construct the little-group algebra on a lightlike vector $p_\mu = (p_0, 0, 0, p_0)$, we *formally* set $P_\mu = p_\mu$ in the commutation relation

$$[M_{\mu\nu}, P_\lambda] = i\eta_{\nu\lambda}P_\mu - i\eta_{\mu\lambda}P_\nu, \qquad (2.2.2-14)$$

as discussed in Sec.1.1.2. We then find

$$\text{``}[J_1, p_\lambda]\text{''} = -i\eta_{2\lambda}p_0, \qquad (2.2.2-15)$$

$$\text{``}[J_2, p_\lambda]\text{''} = i\eta_{1\lambda}p_0, \qquad (2.2.2-16)$$

$$\text{``}[\dot{J}_3, p_\lambda]\text{''} = 0; \qquad (2.2.2-17)$$

$$\text{``}[K_1, p_\lambda]\text{''} = i\eta_{1\lambda}p_0, \qquad (2.2.2-18)$$

$$\text{``}[K_2, p_\lambda]\text{''} = i\eta_{2\lambda}p_0, \qquad (2.2.2-19)$$

$$\text{``}[K_3, p_\lambda]\text{''} = i(\eta_{3\lambda} - \eta_{0\lambda})p_0. \qquad (2.2.2-20)$$

Thus what are consistent with $P_\mu = p_\mu$ are $J_3, J_1 + K_2$, and $J_2 - K_1$, that is, they are the generators of the little group $E(2)$:

$$[J_3, L_1] = iL_2, \qquad (2.2.2-21)$$

$$[J_3, L_2] = -iL_1, \qquad (2.2.2-22)$$

$$[L_1, L_2] = 0, \qquad (2.2.2-23)$$

where

$$J_3 \equiv M_3 + N_3, \qquad (2.2.2-24)$$

$$L_1 \equiv J_1 + K_2 = M_- + N_+, \qquad (2.2.2-25)$$

$$L_2 \equiv J_2 - K_1 = i(M_- - N_+). \qquad (2.2.2-26)$$

Our next task is to calculate how these generators act on the helicity eigenstates $u(\rho, \sigma)$, where $\rho = \pm$ and $\sigma = \pm$ according as

$$M_3 u(\pm, \sigma) = \pm \frac{1}{2} u(\pm, \sigma), \qquad (2.2.2-27)$$

$$N_3 u(\rho, \pm) = \pm \frac{1}{2} u(\rho, \pm). \qquad (2.2.2-28)$$

Since M_+ and M_- are raising and lowering operators, respectively, we of course have

$$M_\pm u(\pm, \sigma) = 0, \qquad (2.2.2-29)$$

$$M_\pm u(\mp, \sigma) = u(\pm, \sigma). \qquad (2.2.2-30)$$

Likewise for $N_\pm$ and σ.

For simplicity of notation, we write $u^{(\alpha)} \equiv u(\rho, \sigma)$ with $\alpha = 1$ for $++$, $\alpha = 2$ for $--$, $\alpha = 3$ for $-+$, and $\alpha = 4$ for $+-$. Let

$$J_3 u^{(\alpha)} = \sum_\beta u^{(\beta)} (J_3)_{\beta\alpha}. \qquad (2.2.2-31)$$

Then the non-vanishing elements of the matrix $(J_3)_{\beta\alpha}$ are

$$(J_3)_{11} = -(J_3)_{22} = 1 \qquad (2.2.2-32)$$

only. Likewise, the non-vanishing elements of the matrices $(L_1)_{\beta\alpha}$ and $(L_2)_{\beta\alpha}$ are

$$(L_1)_{14} = (L_1)_{24} = (L_1)_{31} = (L_1)_{32} = 1, \qquad (2.2.2-33)$$

$$-(L_2)_{14} = (L_2)_{24} = (L_2)_{31} = -(L_2)_{32} = i, \qquad (2.2.2-34)$$

respectively.

Let e^μ be the unit vectors in the directions of the coordinate axes, and set

$$u^{(\alpha)} = u^{(\alpha)}_\mu e^\mu. \qquad (2.2.2-35)$$

Since

$$(J_3)_{\beta 3} = (L_1)_{\beta 3} = (L_2)_{\beta 3} = 0 \qquad \text{for any } \beta, \qquad (2.2.2-36)$$

$u^{(3)}$ is invariant under the little group, whence p_μ can be identified with $u^{(3)}_\mu$. Since $u^{(1)}$ and $u^{(2)}$ are eigenvectors of J_3 with non-vanishing eigenvalues, they are orthogonal to $u^{(3)}$. The remaining vector $u^{(4)}$ does not arise under the action of J_3, L_1, and L_2 on $u^{(j)}$ ($j = 1, 2, 3$), that is, $u^{(1)}$, $u^{(2)}$, and $u^{(3)}$ span an invariant subspace.

We consider a massless vector field $A_\mu(x)$, whose Fourier transform $a_\mu(p)$ is a function of a lightlike vector p_ν.[3] Lorentz transformation $\Lambda_\mu{}^{\mu'}$ is represented by a (pseudo-)unitary operator and the vacuum $|0\rangle$ is invariant under it; then

$$\langle 0 | a_\mu(p) a_\nu^\dagger(p) | 0 \rangle = \langle 0 | a'_\mu(p) a'_\nu{}^\dagger(p) | 0 \rangle, \qquad (2.2.2 - 37)$$

where use has been made of $a_\nu(-p) = a_\nu^\dagger(p)$. In the above,

$$a'_\mu(p) = \Lambda_\mu{}^{\mu'} a_{\mu'}(\Lambda^{-1} p), \qquad (2.2.2 - 38)$$

but if Λ belongs to the little group on p, then $\Lambda^{-1} p = p$.

We substitute the expansion

$$a_\mu(p) = \sum_\alpha \hat{a}_\alpha u_\mu^{(\alpha)}(p) \qquad (2.2.2 - 39)$$

into Eq.(37); then

$$\sum_{\alpha\beta} \langle 0 | \hat{a}_\alpha \hat{a}_\beta^\dagger | 0 \rangle u_\mu^{(\alpha)} u_\nu^{(\beta)*}$$
$$= \sum_{\alpha'\beta'} \langle 0 | \hat{a}_{\alpha'} \hat{a}_{\beta'}^\dagger | 0 \rangle \Lambda_\mu{}^{\mu'} \Lambda_\nu{}^{\nu'} u_{\mu'}^{(\alpha')} u_{\nu'}^{(\beta')*}$$
$$= \sum_{\alpha'\beta'\alpha\beta} \langle 0 | \hat{a}_{\alpha'} \hat{a}_{\beta'}^\dagger | 0 \rangle u_\mu^{(\alpha)} u_\nu^{(\beta)*} \mathcal{D}_{\alpha\alpha'} \mathcal{D}_{\beta\beta'}{}^*, \qquad (2.2.2 - 40)$$

where

$$\Lambda u^{(\alpha)} = \sum_\beta u^{(\beta)} \mathcal{D}_{\beta\alpha}. \qquad (2.2.2 - 41)$$

Thus, setting

$$\mathcal{M}_{\alpha\beta} \equiv \langle 0 | \hat{a}_\alpha \hat{a}_\beta^\dagger | 0 \rangle, \qquad (2.2.2 - 42)$$

we have

$$\mathcal{M}_{\alpha\beta} = \sum_{\alpha'\beta'} \mathcal{M}_{\alpha'\beta'} \mathcal{D}_{\alpha\alpha'} \mathcal{D}_{\beta\beta'}{}^*, \qquad (2.2.2 - 43)$$

---

[3]   Here, we essentially consider the asymptotic field (see Sec.2.3.6).

or

$$\mathcal{M} = \mathcal{D}\mathcal{M}\mathcal{D}^\dagger \qquad (2.2.2-44)$$

in the matrix notation.

We consider an infinitesimal Lorentz transformation

$$\mathcal{D} = 1 + i(\varepsilon_1 J_3 + \varepsilon_2 L_1 + \varepsilon_3 L_2). \qquad (2.2.2-45)$$

Then Eq.(44) becomes

$$J_3\mathcal{M} = \mathcal{M}J_3^\dagger, \quad L_1\mathcal{M} = \mathcal{M}L_1^\dagger, \quad L_2\mathcal{M} = \mathcal{M}L_2^\dagger. \qquad (2.2.2-46)$$

The first condition implies that $\mathcal{M}$ is of the form $\begin{pmatrix} D & 0 \\ 0 & C \end{pmatrix}$, where D is a diagonal 2×2 matrix, C being an arbitrary 2×2 matrix. The second condition then leads us to

$$\mathcal{M} = \begin{pmatrix} f & 0 & 0 & 0 \\ 0 & f & 0 & 0 \\ 0 & 0 & 2g & f \\ 0 & 0 & f & 0 \end{pmatrix} \qquad (2.2.2-47)$$

with f and g arbitrary; Eq.(47) satisfies also the third condition. The hermiticity implies that f and g are real.

Now, the eigenvalues of the matrix, Eq.(47), are

$$f, \; f, \; g \pm \sqrt{g^2 + f^2}. \qquad (2.2.2-48)$$

As long as $f \neq 0$, *at least one of the eigenvalues is negative.* (If $f = 0$ then $\langle 0 | a_\mu a_\nu^\dagger | 0 \rangle$ is proportional to $p_\mu p_\nu$, that is, it is trivial.) Thus we conclude that A_μ cannot be quantized in the positive-metric Hilbert space.

2.2.3 General remarks on subsidiary conditions

As proved in Sec.2.2.2, the manifestly covariant theory of a massless vector field is necessarily quantized with indefinite metric. On the other hand, the positivity of metric is vital to the probabilistic interpretability of quantum theory. In order to have a physically sensible theory, we introduce a **subsidiary condition**, which is a condition defining the **physical subspace** of the indefinite-metric Hilbert space. If we can show that the physical subspace is independent of time and free of negative-norm states, we have a probabilistically interpretable theory. Indeed, as shown in Appendix $A.2$, the physical S-matrix, which is defined by restricting the total S-matrix to the physical subspace, is truly unitary (modulo zero norm).

Since a subsidiary condition is a condition defining a *subspace*, it must be *linear* with respect to states. Nevertheless, because of the difficulty in finding an appropriate subsidiary condition, nonlinear subsidiary conditions were often proposed in the past. Let G be a certain operator; then the totality of the states $|f\rangle$ satisfying

$$\langle f|G|f \rangle = 0 \qquad\qquad (2.2.3-1)$$

does *not* constitute a subspace.[4] If Eq.(1) is replaced by

$$\langle f'|G|f \rangle = 0, \qquad\qquad (2.2.3-2)$$

it cannot be regarded as a condition defining a set of states, because to define $|f\rangle$ we must specify $|f'\rangle$ but to define $|f'\rangle$ we must specify $|f\rangle$.[5]

A subsidiary condition is generally of the following form: The physical subspace $\mathcal{V}_{\text{phys}}$ is the totality of the states $|f\rangle$ (called **physical states**) satisfying

$$G|f\rangle = 0. \qquad\qquad (2.2.3-3)$$

It is indispensable to require that the vacuum $|0\rangle$ be contained in $\mathcal{V}_{\text{phys}}$, that is, we should have

$$G|0\rangle = 0. \qquad\qquad (2.2.3-4)$$

Then G cannot be a local operator, because if so then the separating property of vacuum (see Corollary $A.2$-6 in Appendix $A.2$) implies that $G = 0$. Thus a subsidiary condition should be a nonlocal one.

Since G is a nontrivial operator, there must exist an operator φ which does not commute with G. If G is a spacetime-dependent operator, we often encounter the following situation;

$$[G,\, \varphi] = c + r, \qquad\qquad (2.2.3-5)$$

where c is a *non-zero* c-number and r is either zero or such a q-number that there exists a physical state $|f\rangle$ satisfying

$$\langle f|r|f \rangle = 0 \qquad\qquad (2.2.3-6)$$

---

[4]   For example, the totality of lightlike vectors is not a subspace of the Minkowski space.

[5]   If $|f'\rangle$ is arbitrary, then Eq.(2) reduces to Eq.(3) below.

and $\langle f|f \rangle \neq 0$. Furthermore, if G is *hermitian*, then we encounter a contradiction because

$$
\begin{aligned}
0 &= (\langle f|G\rangle\varphi|f\rangle - \langle f|\varphi(G|f\rangle) \\
&= \langle f|[G,\ \varphi]|f\rangle = c\langle f|f\rangle \neq 0.
\end{aligned}
\qquad (2.2.3-7)
$$

Thus G should be *non-hermitian*.

Nevertheless, subsidiary conditions based on a spacetime-dependent hermitian G have been so far proposed many times. One can evade the above contradiction if one admits a *continuous* set of states. For example, let x and p be a coordinate and its conjugate momentum; they satisfy $[x,\ p] = i$. If one considered its expectation value in a definite eigenstate of p, then one would encounter a contradiction similar to Eq.(7). But such a procedure is not admissible in the case of continuous spectrum; one should consider two generically different states and set up the normalization of δ-function type. Thus in order to adopt a hermitian G, one must introduce a state-vector space having a continuously infinite number of independent vectors — a *nonseparable* space.

A nonseparable space is, however, a too large space. In such a space, an infinitesimal symmetry transformation is *not* necessarily represented by an infinitesimal change of a state vector, whence the corresponding symmetry generator may not exist. Furthermore, when we define the quantum-theoretical probability, inner product of states must be supplemented by some kind of integration weight, but we have no general principle for determining it.

As discussed in Appendix $A.2$, we usually formulate quantum field theory in a separable space by smearing field operators by means of rapidly decreasing test functions. It is a very fundamental postulate that we *can* always consider such wave-packet states. Without this postulate we no longer have a healthy framework of quantum field theory. One might say that it is sufficient to have such wave-packet states in the physical subspace. But it is impossible to construct desirable wave-packet states without solving the subsidiary condition explicitly.

We therefore restrict our consideration to a separable space. Then it is generally expected that G is *either* a spacetime-dependent, non-hermitian, nonlocal operator *or* a spacetime-independent one (such as a symmetry generator). If G is a symmetry generator, Eq.(4) shows that this symmetry must be *unbroken* spontaneously.

2.3 COVARIANT QUANTUM THEORY OF THE ELECTROMAGNETIC FIELD

The covariant operator formalism of the electromagnetic field was correctly formulated first by Gupta [Gup 50] and completed by Bleuler [Ble 50]. The original one was given in the Feynman (or Fermi) gauge, but it is more convenient to consider a set of covariant gauges specified by a gauge parameter α. For this purpose, one introduces an auxiliary scalar field, called B-field. Although the introduction of the B-field is not necessarily indispensable in the electromagnetic theory, it becomes very crucial in later developments. The B-field formalism was originally proposed by Nakanishi [Nak 66] and by Lautrup [Lau 67] independently.[1]

2.3.1 B-field formalism

In order to quantize a gauge theory, it is necessary to introduce **gauge fixing**. For making gauge fixing covariantly, we introduce an auxiliary hermitian scalar field, $B(x)$, called the **B-field**, and take the following Lagrangian density as the proper part of the electromagnetic field:

$$\mathcal{L}_{\text{EM}} = -\frac{1}{4}F^{\mu\nu}F_{\mu\nu} + B\partial^\mu A_\mu + \frac{1}{2}\alpha B^2, \qquad (2.3.1-1)$$

where α is a real number, called a **gauge parameter**. Two special choices are of particular interest: $\alpha = 1$, **Feynman** (or **Fermi**) **gauge** and $\alpha = 0$, **Landau gauge**. For $\alpha \neq 0$, we can rewrite Eq.(1) as

$$\mathcal{L}_{\text{EM}} = -\frac{1}{4}F^{\mu\nu}F_{\mu\nu} - \frac{1}{2}\alpha^{-1}(\partial^\mu A_\mu)^2 + \frac{1}{2}\alpha^{-1}(\partial^\mu A_\mu + \alpha B)^2. \qquad (2.3.1-2)$$

---

[1]   Prior to their work, a similar auxiliary field was introduced by some authors [Uti 59, Sch 63], but not in the framework of quantum field theory After its proposal, many authors [Luk 67, Got 67, Nak 67, Yok 68] investigated the B-field formalism from various standpoints. Systematic presentations also were made [Nak 72b, Nak 75a]. The present section is based on [Nak 75a].

In some of recent literature, the B-field formalism is attributed to Stueckelberg, unfortunately without confirming what his formalism is. As stated at the last paragraph of Sec 2.4.1, the Stueckelberg formalism is quite different from the B-field formalism both in its aim and in its construction. One of the authors (N.N.) would like to thank R.Stora for correspondence on this matter, which clarified why the confusion had arisen in the literature.

Since the last term decouples, we can eliminate B. But we do not do so because the Landau-gauge case is the most important and because the explicit use of B makes the theory more transparent.

We add the Lagrangian density of charged fields (see Sec.2.1.3) to Eq.(1). Then the field equations for the electromagnetic field are

$$\partial_\mu F^{\mu\nu} - \partial^\nu B = -ej^\nu, \qquad (2.3.1-3)$$

$$\partial^\mu A_\mu + \alpha B = 0. \qquad (2.3.1-4)$$

Compared with Eq.(2.1.1-2), Eq.(3) contains an extra term $-\partial^\nu B$. As discussed in Sec.2.2.1, such a modification of the Maxwell equation is indispensable in quantized theory. We call Eq.(3) the **quantum Maxwell equation**. Furthermore, as seen from Eq.(4), the Lorentz condition $\partial^\mu A_\mu = 0$ holds as an operator equation only in the Landau-gauge case.

Multiplying Eq.(3) by ∂_ν, we obtain

$$\Box B = 0, \qquad (2.3.1-5)$$

that is, B is a *free* massless field, though A_μ is interacting. This is a very important characteristic of the abelian gauge theory with linear covariant gauge fixing.

2.3.2 Canonical quantization

Canonical quantization can be consistently carried out by taking A_μ but *not* B and charged fields as canonical variables. We restrict our description to the electromagnetic-field part only.

The canonical conjugate, π^μ, of A_μ is defined by

$$\pi^\mu \equiv (\partial/\partial\dot{A}_\mu)\mathcal{L}_{\text{EM}}$$
$$= F^{0\mu} + \eta^{0\mu}B. \qquad (2.3.2-1)$$

We set up the canonical commutation relations

$$[A_\mu(x), \ A_\nu(y)]_0 = 0, \qquad (2.3.2-2)$$

$$[\pi^\mu(x), \ A_\nu(y)]_0 = -i\delta^\mu{}_\nu\delta(\boldsymbol{x} - \boldsymbol{y}), \qquad (2.3.2-3)$$

$$[\pi^\mu(x), \ \pi^\nu(y)]_0 = 0. \qquad (2.3.2-4)$$

As seen from Eq.(1), $\dot{A}_k$ and B are directly expressible in terms of π^μ, but $\dot{A}_0$ and $\dot{B}$ are not; they are expressible in terms of others by solving Eq.(2.3.1-4) and the zeroth component of Eq.(2.3.1-3):

$$\dot{A}_0 = -\partial^k A_k - \alpha B, \qquad\qquad (2.3.2-5)$$

$$\dot{B} = -\partial^k F_{0k} + ej_0. \qquad\qquad (2.3.2-6)$$

Here j_0 commutes with A_μ and π^μ at the equal time. From Eqs.(2)–(6) we obtain

$$[A_\mu(x),\ \dot{A}_\nu(y)]_0 = -i\left[\eta_{\mu\nu} - (1-\alpha)\delta^0{}_\mu \delta^0{}_\nu\right]\delta(\boldsymbol{x}-\boldsymbol{y}), \qquad (2.3.2-7)$$

$$[A_\mu(x),\ B(y)]_0 = i\eta_{\mu 0}\delta(\boldsymbol{x}-\boldsymbol{y}), \qquad\qquad (2.3.2-8)$$

$$[A_\mu(x),\ \dot{B}(y)]_0 = -[\dot{A}_\mu(x),\ B(y)]_0 = -i\delta^k{}_\mu \partial_k \delta(\boldsymbol{x}-\boldsymbol{y}), \qquad (2.3.2-9)$$

$$[B(x),\ B(y)]_0 = 0, \qquad\qquad (2.3.2-10)$$

$$[B(x),\ \dot{B}(y)]_0 = 0. \qquad\qquad (2.3.2-11)$$

A charged field, say ψ, commutes with A_ν, $\dot{A}_\mu$, and B, but owing to Eq.(6), we have

$$[\psi(x),\ \dot{B}(y)]_0 = e[\psi(x),\ j_0(y)]_0 = e\psi(x)\delta(\boldsymbol{x}-\boldsymbol{y}), \qquad (2.3.2-12)$$

where we have set $q = +1$ (*hereafter we consider this case only*).

2.3.3 Properties of the B-field and subsidiary condition

Since B satisfies a free-field equation, Eq.(2.3.1-5), a non-local current

$$J_\mu(z,y) \equiv \partial_\mu^z D(y-z) \cdot B(z) - D(y-z)\partial_\mu B(z) \qquad (2.3.3-1)$$

is conserved ($\partial_\mu^z J^\mu = 0$). Hence we obtain an integral representation for $B(y)$:

$$B(y) = \int d\boldsymbol{z}\,[\partial_0^z D(y-z) \cdot B(z) - D(y-z)\partial_0 B(z)]. \qquad (2.3.3-2)$$

Indeed, since the rhs. of Eq.(2) is independent of z^0, the lhs. is obtained by setting $z^0 = y^0$.

Owing to the z^0 independence of Eq.(2), we can calculate four-dimensional commutation relations $[\Phi(x),\ B(y)]$ by applying Eq.(2) with $z^0 = x^0$, and by using equal-time commutation relations. In this way, we obtain

$$[B(x),\ B(y)] = 0, \qquad\qquad (2.3.3-3)$$

$$[A_\mu(x),\ B(y)] = -i\partial_\mu D(x-y), \qquad\qquad (2.3.3-4)$$

$$[\psi(x),\ B(y)] = e\psi(x)D(x-y). \qquad\qquad (2.3.3-5)$$

From Eq.(4) we have

$$[F_{\mu\nu}(x),\ B(y)] = 0, \qquad\qquad (2.3.3-6)$$

whence

$$[j_\mu(x),\ B(y)] = 0 \qquad\qquad (2.3.3-7)$$

because of Eq.(2.3.1-3).

The above results suggest a remarkable similarity between $[\Phi(x),\ B(y)]$ and the infinitesimal local gauge transformation. Let $\Phi(x)$ be any local quantity which transforms as

$$\Phi'(x) - \Phi(x) = \pounds(\Phi)\varepsilon(x) \qquad\qquad (2.3.3-8)$$

under an infinitesimal local gauge transformation characterized by $\varepsilon(x)$ at the classical level, where $\pounds(\Phi)$ is a differential operator acting on $\varepsilon(x)$ [e.g., $\pounds(A_\mu) = \partial_\mu$, $\pounds(\psi) = ie\psi$]. Then, in general, we have

$$[\Phi(x),\ B(y)] = -i\pounds(\Phi)^x D(x-y). \qquad\qquad (2.3.3-9)$$

Thus $B(y)$ plays a role of a kind of "generator" of local gauge transformations. Later (in Sec.5.6.4), we shall see that such a formula as Eq.(9) is beautifully extended to quantum gravity.

In a way similar to Eq.(2), we can define the positive/negative frequency part of B by

$$B^{(\pm)}(x) \equiv -i\int dz\,[\partial_0^z D^{(\pm)}(x-z) \cdot B(z) - D^{(\pm)}(x-z)\partial_0 B(z)]. \quad (2.3.3-10)$$

We then set up **Gupta subsidiary condition**

$$B^{(+)}(x)|f\rangle = 0, \qquad\qquad (2.3.3-11)$$

that is, the physical subspace $\mathcal{V}_{\text{phys}}$ is the totality of the states $|f\rangle$ (physical states) satisfying Eq.(11). Because

$$[iP_0,\ B^{(+)}(x)] = \partial_0 B^{(+)}(x), \qquad (2.3.3-12)$$

$\mathcal{V}_{\text{phys}}$ is time-independent, and indeed it is Poincaré-invariant. The fact that Eq.(11) excludes negative-norm states is shown in Sec.2.3.6.

From Eq.(2.3.1-3), we have

$$\langle f|\partial_\mu F^{\mu\nu} + ej^\nu|g\rangle = 0 \qquad (2.3.3-13)$$

for any $|f\rangle$, $|g\rangle \in \mathcal{V}_{\text{phys}}$ because $B^{(-)} = \left(B^{(+)}\right)^\dagger$. Thus the classical Maxwell equation holds in the physical subspace. Likewise, from Eq.(2.3.1-4), we have

$$\langle f|\partial^\mu A_\mu|g\rangle = 0. \qquad (2.3.3-14)$$

On the other hand, it is interesting to note that owing to Eq.(2.3.2-6) the charge operator[2]

$$Q \equiv e \int d\mathbf{x}\, j^0(x) \qquad (2.3.3-15)$$

is equivalent to

$$\int d\mathbf{x}\, \partial_0 B(x) \qquad (2.3.3-16)$$

as the generator of phase transformations, because $\int d\mathbf{x}\, \partial_k F^{k0}$ commutes with any local operator. This integral is *not zero* but *ill-defined* (and hence Eq.(16) is also so) as an operator.

The theory is invariant under the *c-number* gauge transformation

$$A_\mu \ \rightarrow \ A'_\mu = A_\mu + \partial_\mu \Lambda, \qquad (2.3.3-17)$$

$$B \ \rightarrow \ B' = B, \qquad (2.3.3-18)$$

$$\psi \ \rightarrow \ \psi' = e^{ie\Lambda}\psi, \qquad (2.3.3-19)$$

where Λ is a c-number function satisfying

$$\Box \Lambda = 0. \qquad (2.3.3-20)$$

---

[2]   More precisely, we should use the renormalized current (see Sec.2.3.7).

Hence we have the corresponding generator

$$Q_\Lambda \equiv -\int d\boldsymbol{x}\,(\Lambda\partial_0 B - \partial_0\Lambda \cdot B). \qquad (2.3.3-21)$$

Indeed, Eq.(9) yields

$$[iQ_\Lambda,\ \Phi(x)] = \pounds(\Phi)\Lambda(x) \qquad (2.3.3-22)$$

by using an integral representation of Λ similar to Eq.(2). Since

$$\langle\,0\,|[iQ_\Lambda,\ A_\mu(x)]|\,0\,\rangle = \partial_\mu\Lambda(x), \qquad (2.3.3-23)$$

Q_Λ is *always* spontaneously broken unless Λ is a constant. Then the Goldstone theorem (Sec.1.5.1) implies that A_μ contains a massless spectrum [Fer 71, Bra 74].

2.3.4 Free electromagnetic field

In order to obtain some more insights into the electromagnetic field, we consider the free-field case by setting $j_\mu = 0$ *only in this subsection*.

First, the canonical energy-momentum tensor is given by

$$\begin{aligned}
\mathcal{T}_{\mathrm{EM}}{}^\mu{}_\nu &\equiv \partial_\nu A_\lambda \frac{\partial\mathcal{L}_{\mathrm{EM}}}{\partial(\partial_\mu A_\lambda)} - \delta^\mu{}_\nu \mathcal{L}_{\mathrm{EM}} \\
&= -F^{\mu\lambda}\partial_\nu A_\lambda + B\partial_\nu A^\mu - \delta^\mu{}_\nu \mathcal{L}_{\mathrm{EM}}.
\end{aligned} \qquad (2.3.4-1)$$

The symmetric energy-momentum tensor,

$$\begin{aligned}
\Theta_{\mathrm{EM}\mu\nu} &\equiv \frac{1}{4}\eta_{\mu\nu}F^{\lambda\rho}F_{\lambda\rho} - F_\mu{}^\lambda F_{\nu\lambda} - A_\mu\partial_\nu B \\
&\quad - A_\nu\partial_\mu B + \eta_{\mu\nu}\left(\partial^\lambda B \cdot A_\lambda - \frac{1}{2}\alpha B^2\right),
\end{aligned} \qquad (2.3.4-2)$$

is obtained by using Eq.(2.3.1-3) with $j_\mu = 0$, i.e.,

$$\partial_\mu F^{\mu\nu} - \partial^\nu B = 0, \qquad (2.3.4-3)$$

and adding total divergence (see Sec.1.2.5).

From Eq.(2), the Hamiltonian is given by

$$H_{\mathrm{EM}} = \int d\boldsymbol{x}\left(\frac{1}{4}F^{kl}F_{kl} - \frac{1}{2}F^{0k}F_{0k} - \dot{B}A_0 + \partial^k B \cdot A_k - \frac{1}{2}\alpha B^2\right). \qquad (2.3.4-4)$$

Of course, the first and second terms are the magnetic and electric energies, respectively.

Now, substituting Eq.(2.3.1-4) into Eq.(3), we obtain

$$\Box A_\mu - (1-\alpha)\partial_\mu B = 0, \qquad (2.3.4-5)$$

that is, even in the free case, A_μ does not satisfy the d'Alembert equation except for $\alpha = 1$. But, because of Eq.(2.3.1-5), Eq.(5) yields

$$\Box^2 A_\mu = 0. \qquad (2.3.4-6)$$

That is a **dipole-ghost** equation.

In order to construct an integral representation for A_μ, we need a dipole-ghost invariant D function, which is denoted by $E(x)$. It is defined by

$$\begin{aligned}
E(x) &\equiv -(\partial/\partial m^2)\Delta(x; m^2)|_{m^2=0} \\
&= -i(2\pi)^{-3} \int d^4p\, \epsilon(p_0)\delta'(p^2)e^{-ipx} \\
&= -(8\pi)^{-1}\epsilon(x^0)\theta(x^2). \qquad (2.3.4-7)
\end{aligned}$$

It satisfies

$$\Box E(x) = D(x), \qquad (2.3.4-8)$$

$$E(x)|_0 = \partial_0 E(x)|_0 = \partial_0{}^2 E(x)|_0 = 0, \qquad (2.3.4-9)$$

$$\partial_0{}^3 E(x)|_0 = -\delta(\mathbf{x}). \qquad (2.3.4-10)$$

By using $E(x)$, we have an integral representation for A_μ:

$$\begin{aligned}
A_\nu(y) = \int d\mathbf{z}\, [\partial_0{}^z D(y-z) \cdot A_\nu(z) - D(y-z)\partial_0 A_\nu(z) \\
+ \partial_0{}^z E(y-z)\cdot \Box A_\nu(z) - E(y-z)\partial_0\Box A_\nu(z)], (2.3.4-11)
\end{aligned}$$

in which $\Box A_\nu$ may be replaced by $(1-\alpha)\partial_\nu B$. Setting $z_0 = x_0$ in Eq.(11), we can calculate the four-dimensional commutation relation; we obtain

$$[A_\mu(x),\ A_\nu(y)] = -i\eta_{\mu\nu}D(x-y) + i(1-\alpha)\partial_\mu\partial_\nu E(x-y). \qquad (2.3.4-12)$$

As $iD(x)$ is divided into the positive-frequency part $D^{(+)}(x)$ and the negative-frequency part $D^{(-)}(x)$, we divide $iE(x)$ into $E^{(+)}(x)$ and $E^{(-)}(x)$:

$$iE(x) = E^{(+)}(x) + E^{(-)}(x), \quad E^{(+)}(x) = -E^{(-)}(-x), \quad (2.3.4-13)$$

$$\Box E^{(\pm)}(x) = D^{(\pm)}(x), \quad (2.3.4-14)$$

$$\begin{aligned}
E^{(\pm)}(x) &\equiv -(\partial/\partial m^2)\Delta^{(\pm)}(x; m^2)|_{m^2=0} \\
&= \pm(2\pi)^{-3} \int d^4p\, \theta(\pm p_0)\delta'(p^2)e^{-ipx} \\
&= \mp(4\pi)^{-2}\left[\log(-x^2 \pm i0x^0) + \text{const}\right]. \quad (2.3.4-15)
\end{aligned}$$

Here $E^{(\pm)}(x)$ contains an infinite constant, but we may replace it by a finite one without introducing any trouble.[3]

The positive/negative frequency part of A_μ is defined by

$$\begin{aligned}
A_\mu^{(\pm)}(x) &\equiv -i \int dz \Big\{ \partial_0^z D^{(\pm)}(x-z) \cdot A_\mu(z) - D^{(\pm)}(x-z)\partial_0 A_\mu(z) \\
&\quad + (1-\alpha)\left[\partial_0^z E^{(\pm)}(x-z) \cdot \partial_\mu B(z) - E^{(\pm)}(x-z)\partial_0\partial_\mu B(z)\right] \Big\}. \quad (2.3.4-16)
\end{aligned}$$

Note that the unspecified constant in $E^{(\pm)}$ does not yield any ambiguity in the definition of $A_\mu^{(\pm)}$ because $\partial_0 \int dz\, \partial_k B(z)$ and $\int dz\, \partial_0^2 B(z) = -\int dz\, \partial_k \partial^k B(z)$ may be set equal to zero.

2.3.5 Vacuum expectation values

We return to the interacting case in which j_μ is a non-trivial operator. In this subsection, we investigate various two-point functions.

We, of course, postulate the existence of a unique vacuum $|0\rangle$. As discussed in Sec.1.4.1, we have $\langle 0|0\rangle = 1$ and $P_\mu|0\rangle = M_{\mu\nu}|0\rangle = 0$. We further require $|0\rangle$ to belong to $\mathcal{V}_{\text{phys}}$, that is,

$$B^{(+)}(x)|0\rangle = 0, \quad (2.3.5-1)$$

whence

$$\langle 0|B(x)|0\rangle = 0. \quad (2.3.5-2)$$

---

[3]   We encounter a similar infrared divergence in the two-dimensional free massless scalar theory, but in that case the problem is more serious (see Sec.2.5.2).

Since there is no constant Lorentz vector, we should have

$$\langle\, 0\,|A_\mu(x)|\,0\,\rangle = 0. \qquad (2.3.5-3)$$

For any local operator $\Phi(x)$, Eq.(1) yields

$$\langle\, 0\,|\Phi(x)B(y)|\,0\,\rangle = \langle\, 0\,|[\Phi(x),\ B^{(-)}(y)]|\,0\,\rangle. \qquad (2.3.5-4)$$

Hence, from Eqs.(2.3.3-3), (2.3.3-4), and (2.3.3-7) together with Eq.(2.3.3-10), we obtain

$$\langle\, 0\,|B(x)B(y)|\,0\,\rangle = 0, \qquad (2.3.5-5)$$

$$\langle\, 0\,|A_\mu(x)B(y)|\,0\,\rangle = -\partial_\mu D^{(+)}(x-y), \qquad (2.3.5-6)$$

$$\langle\, 0\,|j_\mu(x)B(y)|\,0\,\rangle = 0. \qquad (2.3.5-7)$$

Since $j_\mu(x)$ commutes with $B(y)$, $j_\mu(x)|\,0\,\rangle$ is a physical state. Accordingly, the intermediate states which are to be inserted into $\langle\, 0\,|j_\mu(x)j_\nu(y)|\,0\,\rangle$ can be restricted to the positive-norm states only (see Appendix $A.2$). Then, as discussed in Sec.1.4.1, we have a spectral representation

$$e^2\langle\, 0\,|j_\mu(x)j_\nu(y)|\,0\,\rangle = \int_0^\infty ds\,\pi(s)(-s\eta_{\mu\nu}-\partial_\mu\partial_\nu)\Delta^{(+)}(x-y;s), \qquad (2.3.5-8)$$

where use has been made of $\partial^\mu j_\mu = 0$. Here

$$e^2\sum_\lambda\langle\, 0\,|j_\mu(0)|p,\lambda\rangle\langle p,\lambda|j_\nu(0)|\,0\,\rangle = (2\pi)^{-3}\theta(p_0)(-p^2\eta_{\mu\nu}+p_\mu p_\nu)\pi(p^2). \qquad (2.3.5-9)$$

Considering $\mu = \nu = 0$ component of Eq.(9), we have an important result

$$\pi(s) \geq 0. \qquad (2.3.5-10)$$

Now, the quantum Maxwell equation, Eq.(2.3.1-3), together with Eq.(2.3.1-4) is rewritten as

$$\Box A_\mu = -ej_\mu + (1-\alpha)\partial_\mu B. \qquad (2.3.5-11)$$

Hence, owing to Eqs.(5) and (7), we find

$$\Box^x\Box^y\langle\, 0\,|A_\mu(x)A_\nu(y)|\,0\,\rangle = e^2\langle\, 0\,|j_\mu(x)j_\nu(y)|\,0\,\rangle. \qquad (2.3.5-12)$$

Substituting Eq.(8) into Eq.(12) and integrating, we obtain

$$\langle\, 0\,|A_\mu(x)A_\nu(y)|\,0\,\rangle = (a\eta_{\mu\nu} + b\partial_\mu\partial_\nu)D^{(+)}(x-y) + (c\eta_{\mu\nu} + d\partial_\mu\partial_\nu)E^{(+)}(x-y)$$
$$+ \int_{+0}^{\infty} ds\, s^{-2}\pi(s)(-s\eta_{\mu\nu} - \partial_\mu\partial_u)\Delta^{(+)}(x-y;s). \quad (2.3.5-13)$$

The integration constants a, b, c, and d are determined in the following way. From Eqs.(2.3.1-4) and (6), we find

$$c = 0, \quad a + d = -\alpha. \quad (2.3.5-14)$$

Next, we construct the spectral representation for $\langle\,0\,|[A_\mu(x),\ A_\nu(y)]|\,0\,\rangle$ from Eq.(13) and make use of the equal-time commutation relation, Eq.(2.3.2-7), for $\mu \neq 0$, $\nu \neq 0$; then

$$-i\eta_{kl}\delta(\mathbf{x} - \mathbf{y}) = i(a\eta_{kl} + b\partial_k\partial_l)\delta(\mathbf{x} - \mathbf{y})$$
$$+ i\int_{+0}^{\infty} ds\, s^{-2}\pi(s)(-s\eta_{kl} - \partial_k\partial_l)\delta(\mathbf{x} - \mathbf{y}). \quad (2.3.5-15)$$

Hence we find

$$-a = Z_3 \equiv 1 - \int_{+0}^{\infty} ds\, s^{-1}\pi(s), \quad (2.3.5-16)$$

$$b = Z_3 K \equiv \int_{+0}^{\infty} ds\, s^{-2}\pi(s). \quad (2.3.5-17)$$

Substituting Eqs.(14), (16), and (17) into Eq.(13), we finally obtain

$$\langle\, 0\,|A_\mu(x)A_\nu(y)|\,0\,\rangle$$
$$= Z_3(-\eta_{\mu\nu} + K\partial_\mu\partial_\nu)D^{(+)}(x-y) + (Z_3 - \alpha)\partial_\mu\partial_\nu E^{(+)}(x-y)$$
$$+ \int_{+0}^{\infty} ds\, s^{-2}\pi(s)(-s\eta_{\mu\nu} - \partial_\mu\partial_\nu)\Delta^{(+)}(x-y;s), \quad (2.3.5-18)$$

and

$$\langle\, 0\,|[A_\mu(x),\ A_\nu(y)]|\,0\,\rangle$$
$$= iZ_3\left[(-\eta_{\mu\nu} + K\partial_\mu\partial_\nu)D(x-y) + (1 - Z_3^{-1}\alpha)\partial_\mu\partial_\nu E(x-y)\right]$$
$$+ i\int_{+0}^{\infty} ds\, s^{-2}\pi(s)(-s\eta_{\mu\nu} - \partial_\mu\partial_\nu)\Delta(x-y;s). \quad (2.3.5-19)$$

The constant Z_3 is the **renormalization constant** of the photon wave-function. As seen from Eq.(2.3.3-4), the transverse part of $A_\nu(y)|0\rangle$ is a physical state, whence we postulate that $Z_3 > 0$. Thus from Eqs.(16) and (17) with Eq.(10), we obtain

$$0 < Z_3 \leq 1, \qquad (2.3.5-20)$$

$$K \geq 0, \qquad (2.3.5-21)$$

where the equality holds only in the free-field case. The term proportional to K is called the **Källén term** [Käl 52]. It can be removed by considering $A_\mu - \frac{1}{2}Z_3 K \partial_\mu B$ in place of A_μ.

2.3.6 Asymptotic fields

As done in Sec.1.4.2, we introduce asymptotic fields. There is some subtlety concerning charged fields related to infrared problem, which is discussed in Addendum 2.A. We here consider the electromagnetic-field part only. We set

$$Z_3^{-1/2} \dot{A}_\mu - \frac{1}{2}KZ_3^{1/2}\partial_\mu B \; \to \; A^{\text{as}}_\mu, \qquad (2.3.6-1)$$

$$Z_3^{1/2} B \; = \; B^{\text{as}}, \qquad (2.3.6-2)$$

where "as" represents "in" (as $x^0 \to -\infty$) or "out" (as $x^0 \to +\infty$). Then from Eqs.(2.3.5-19), (2.3.3-4), and (2.3.3-3), we have

$$[A^{\text{as}}_\mu(x), \; A^{\text{as}}_\nu(y)] = -i\eta_{\mu\nu}D(x-y) + i(1-\alpha_r)\partial_\mu\partial_\nu E(x-y) \qquad (2.3.6-3)$$

with

$$\alpha_r \equiv Z_3^{-1}\alpha, \qquad (2.3.6-4)$$

$$[A^{\text{as}}_\mu(x), \; B^{\text{as}}(y)] = -i\partial_\mu D(x-y), \qquad (2.3.6-5)$$

$$[B^{\text{as}}(x), \; B^{\text{as}}(y)] = 0. \qquad (2.3.6-6)$$

Furthermore, A^{as}_μ and B^{as} satisfy

$$\Box A^{\text{as}}_\mu - (1-\alpha_r)\partial_\mu B^{\text{as}} = 0, \qquad (2.3.6-7)$$

$$\partial^\mu A^{\text{as}}_\mu + \alpha_r B^{\text{as}} = 0, \qquad (2.3.6-8)$$

$$\Box B^{\text{as}} = 0. \qquad (2.3.6-9)$$

That is, $A_\mu{}^{\text{as}}$ and B^{as} are the same as those of the free electromagnetic theory with replacement of α by α_r. Hence we can use the results of Sec.2.3.4.

It is easier to analyze free fields in momentum space, but because of the presence of dipole ghost the usual three-dimensional Fourier transform becomes impossible. Instead, we rewrite Eq.(2.3.4-16) as

$$A_\mu^{\text{as}(+)}(x) = (2\pi)^{-3/2} \int d^4p \, a_\mu(p) e^{-\imath px} \qquad (2.3.6-10)$$

with

$$a_\mu(p) \equiv -\,i(2\pi)^{-3/2}\theta(p_0) \int dz \left\{ \delta(p^2)e^{\imath pz} \left[ip_0 A_\mu^{\text{as}}(z) - \partial_0 A_\mu^{\text{as}}(z)\right] \right.$$
$$\left. + \delta'(p^2)e^{\imath pz}(1 - \alpha_r) \left[ip_0 \partial_\mu B^{\text{as}}(z) - \partial_0 \partial_\mu B^{\text{as}}(z)\right] \right\}. \qquad (2.3.6-11)$$

Note that the support of $a_\mu(p)$ is $p^2 = 0$ only, as it should be. Likewise we have

$$B^{\text{as}(+)}(x) = (2\pi)^{-3/2} \int d^4p \, b(p) e^{-\imath px} \qquad (2.3.6-12)$$

with

$$b(p) \equiv -i(2\pi)^{-3/2}\theta(p_0)\delta(p^2) \int dz \, e^{\imath pz} \left[ip_0 B^{\text{as}}(z) - \partial_0 B^{\text{as}}(z)\right]. \qquad (2.3.6-13)$$

It is evident from the definition, Eq.(13), that

$$p^2 b(p) = 0. \qquad (2.3.6-14)$$

By using $p^2\delta'(p^2) = -\delta(p^2)$, it is straightforward to show that

$$p^2 a_\mu(p) - i(1 - \alpha_r)p_\mu b(p) = 0. \qquad (2.3.6-15)$$

It needs a little manipulation to calculate $p^\mu a_\mu(p)$. We rewrite $p^\mu e^{\imath pz}$ by $-i\partial^\mu e^{\imath pz}$, integrate by parts with paying special care for $\mu = 0$, and then make use of Eqs.(7)–(9); in this way, we find

$$p^\mu a_\mu(p) - i\alpha_r b(p) = 0. \qquad (2.3.6-16)$$

Of course, Eqs.(15), (16), and (14) are the four-dimensional Fourier transforms of the positive-frequency parts of Eqs.(7), (8), and (9), respectively, as it should be.

We can also show that $a_\mu(p)$ and $b(p)$ satisfy the following commutation relations by using Eqs.(3), (5), and (6):

$$[a_\mu(p),\ a_\nu{}^\dagger(q)]$$
$$= \theta(p_0)\left[-\eta_{\mu\nu}\delta(p^2) - (1 - \alpha_r)p_\mu p_\nu \delta'(p^2)\right]\delta^4(p - q), \quad (2.3.6 - 17)$$
$$[a_\mu(p),\ b^\dagger(q)] = \theta(p_0)ip_\mu\delta(p^2)\delta^4(p - q), \quad (2.3.6 - 18)$$
$$[b(p),\ b^\dagger(q)] = 0. \quad (2.3.6 - 19)$$

The subsidiary condition, Eq.(2.3.3-11), is transcribed into

$$b(p)|f\rangle = 0. \quad (2.3.6 - 20)$$

Now, we consider one-photon states. Because of Eq.(16), there are four independent one-photon states having p_μ $(p_0 > 0)$. For simplicity, we take a Lorentz frame $p_1 = p_2 = 0$. Then we choose[4]

$$|p,\ j\rangle_T \equiv a_j{}^\dagger(p)|0\rangle \quad (j = 1, 2), \quad (2.3.6 - 21)$$
$$|p\rangle_L \equiv a_3{}^\dagger(p)|0\rangle, \quad (2.3.6 - 22)$$
$$|p\rangle_S \equiv b^\dagger(p)|0\rangle \quad (2.3.6 - 23)$$

as independent one-photon states. Here $|p,\ j\rangle_T$ $(j = 1, 2)$ are **transverse photon states**, $|p\rangle_L$ a **longitudinal photon state**, and $|p\rangle_S$ a **scalar photon state**. From Eqs.(14)–(19), we obtain

$$p^2|p,\ j\rangle_T = p^2|p\rangle_S = 0, \quad (2.3.6 - 24)$$
$$p^2|p\rangle_L = -i(1 - \alpha_r)p_3|p\rangle_S; \quad (2.3.6 - 25)$$
$$_T\langle q,\ i\,|\,p,\ j\rangle_T = \delta_{ij}\delta(p^2)\delta^4(p - q), \quad (2.3.6 - 26)$$
$$_L\langle q\,|\,p\rangle_L = \left[\delta(p^2) - (1 - \alpha_r)p_3{}^2\delta'(p^2)\right]\delta^4(p - q), \quad (2.3.6 - 27)$$
$$_T\langle q,\ j\,|\,p\rangle_L = {}_T\langle q,\ j\,|\,p\rangle_S = {}_S\langle q\,|\,p\rangle_S = 0, \quad (2.3.6 - 28)$$
$$_S\langle q\,|\,p\rangle_L = -ip_3\delta(p^2)\delta^4(p - q); \quad (2.3.6 - 29)$$
$$b(q)|p,\ j\rangle_T = b(q)|p\rangle_S = 0, \quad (2.3.6 - 30)$$
$$b(q)|p\rangle_L = -ip_3\delta(p^2)\delta(p - q) \neq 0. \quad (2.3.6 - 31)$$

---

[4]   Here (a_1, a_2), a_3, and b correspond to $(\hat{a}_1, \hat{a}_2)$, $\hat{a}_3$, and $\hat{a}_4$, respectively, of Eq.(2.2.2-39).

Thus, the transverse photons $|p, j\rangle_\mathrm{T}$ satisfy $p^2 = 0$, have positive norm, belong to $\mathcal{V}_\mathrm{phys}$, and are orthogonal to both $|p\rangle_\mathrm{L}$ and $|p\rangle_\mathrm{S}$; the longitudinal photon $|p\rangle_\mathrm{L}$ does not satisfy $p^2 = 0$ precisely unless $\alpha_r = 1$, does not necessarily have positive norm, does not belong to $\mathcal{V}_\mathrm{phys}$, and is not orthogonal to $|p\rangle_\mathrm{S}$; the scalar photon $|p\rangle_\mathrm{S}$ satisfies $p^2 = 0$, has zero norm, belongs to $\mathcal{V}_\mathrm{phys}$, but is not orthogonal to $|p\rangle_\mathrm{L}$.

In general, the physical subspace $\mathcal{V}_\mathrm{phys}$ is spanned by basis states which contain no longitudinal photons, whence its metric is positive semi-definite. Zero-norm states of $\mathcal{V}_\mathrm{phys}$ are those which contain at least one scalar photon; their totality is denoted by $\mathcal{V}_0$. Then, according to the general theorem (see Theorem A.2-2 in Appendix A.2), the physical S-matrix S_phys defined in the quotient space $\mathcal{V}_\mathrm{phys}/\mathcal{V}_0$ is unitary in the genuine sense. As far as S_phys is concerned, we may use any one of positive-metric subspaces of $\mathcal{V}_\mathrm{phys}$, but it is a subspace *non*-invariant under Poincaré group.

2.3.7 Renormalized theory

Quantum electrodynamics (QED) is a theory of the interacting system of the electromagnetic field and the electron field. It starts with the Lagrangian density

$$
\begin{aligned}
\mathcal{L} = & \mathcal{L}_\mathrm{EM} + \mathcal{L}_\mathrm{D} \\
= & -\frac{1}{4} F^{\mu\nu} F_{\mu\nu} + B\partial^\mu A_\mu + \frac{1}{2}\alpha B^2 \\
& + e\bar{\psi}\gamma^\mu\psi A_\mu + \bar{\psi}(i\gamma^\mu\partial_\mu - m)\psi.
\end{aligned} \qquad (2.3.7-1)
$$

Its perturbative approach is extremely successful. Although it suffers from ultraviolet divergences, all of them can be absorbed into renormalization constants. After replacing the renormalized mass m_r and the renormalized coupling constant e_r by the observed mass m_{ob} and the observed coupling constant e_{ob}, respectively, we no longer encounter ultraviolet divergences as far as perturbation expansion is concerned.

It is worth emphasizing that the renormalization procedure is necessary to be performed even if we do not require the removal of ultraviolet divergences. We here discuss non-perturbative aspects of renormalization.

Four **renormalization constants** δm, Z_1, Z_2, and Z_3 are introduced,[5] where we know $Z_3 > 0$ from Sec.2.3.5 and assume $Z_2 > 0$. We then define the following renormalized quantities:

$$A^{(r)}_\mu \equiv Z_3^{-1/2} A_\mu, \tag{2.3.7 - 2}$$

$$B^{(r)} \equiv Z_3^{1/2} B, \tag{2.3.7 - 3}$$

$$\psi^{(r)} \equiv Z_2^{-1/2} \psi, \tag{2.3.7 - 4}$$

$$m_r \equiv m + \delta m, \tag{2.3.7 - 5}$$

$$e_r \equiv Z_1^{-1} Z_2 Z_3^{1/2} e, \tag{2.3.7 - 6}$$

$$\alpha_r \equiv Z_3^{-1} \alpha. \tag{2.3.7 - 7}$$

We express $\mathcal{L}$ in terms of those renormalized quantities:

$$\mathcal{L} = -\frac{1}{4} Z_3 F^{(r)\mu\nu} F^{(r)}_{\mu\nu} + B^{(r)} \partial^\mu A^{(r)}_\mu + \frac{1}{2} \alpha_r B^{(r)2}$$
$$+ Z_1 e_r \bar{\psi}^{(r)} \gamma^\mu \psi^{(r)} A^{(r)}_\mu + Z_2 \bar{\psi}^{(r)} (i\gamma^\mu \partial_\mu - m_r + \delta m) \psi^{(r)}. \tag{2.3.7 - 8}$$

Then the **renormalized free Lagrangian density** is

$$\mathcal{L}_0^{(r)} = -\frac{1}{4} F^{(r)\mu\nu} F^{(r)}_{\mu\nu} + B^{(r)} \partial^\mu A^{(r)}_\mu + \frac{1}{2} \alpha_r B^{(r)2}$$
$$+ \bar{\psi}^{(r)} (i\gamma^\mu \partial_\mu - m_r) \psi^{(r)}, \tag{2.3.7 - 9}$$

and $\mathcal{L}_I^{(r)} \equiv \mathcal{L} - \mathcal{L}_0^{(r)}$ is the **renormalized interaction Lagrangian density**. The renormalized perturbation series is constructed in this way, but we do not consider it here.

In the Heisenberg picture, the **renormalized field equations** are as follows:

$$\Box A^{(r)}_\mu - (1 - \alpha_r) \partial_\mu B^{(r)} = -e_r j^{(r)}_\mu, \tag{2.3.7 - 10}$$

$$\partial^\mu A^{(r)}_\mu + \alpha_r B^{(r)} = 0, \tag{2.3.7 - 11}$$

$$\Box B^{(r)} = 0, \tag{2.3.7 - 12}$$

where the **renormalized current** is defined by

$$j^{(r)}_\mu \equiv Z_1 Z_3^{-1} \bar{\psi}^{(r)} \gamma_\mu \psi^{(r)} - e_r^{-1} (Z_3^{-1} - 1) \partial_\mu B^{(r)}, \tag{2.3.7 - 13}$$

---

[5]   Here, of course, the **Ward identity** $Z_1 = Z_2$ holds, but it is not relevant to our discussion.

which satisfies

$$\partial^\mu j^{(r)}_\mu = 0. \qquad (2.3.7-14)$$

One should pay attention to the presence of the second term on the rhs. of Eq.(13) [Nak 75a]; it cancels the massless spectrum involved in the first term.

The commutation relations between renormalized fields have the same form as those between unrenormalized ones:

$$[B^{(r)}(x),\ B^{(r)}(y)] = 0, \qquad (2.3.7-15)$$

$$[A^{(r)}_\mu(x),\ B^{(r)}(y)] = -i\partial_\mu D(x-y), \qquad (2.3.7-16)$$

$$[j^{(r)}_\mu(x),\ B^{(r)}(y)] = 0, \qquad (2.3.7-17)$$

$$[\psi^{(r)}(x),\ B^{(r)}(y)] = e_r\psi^{(r)}(x)D(x-y). \qquad (2.3.7-18)$$

2.3.8 Gaugeon formalism

As seen in Eq.(2.3.7-7), the gauge parameter α must be renormalized; that is, except for the Landau gauge, the renormalized theory has a gauge different from that of the unrenormalized one. If one insists on having $\alpha_r = \alpha \neq 0$, thcn manifest covariance is necessarily violated [Nak 75a].

Since the gauge parameter is not a physical quantity, it is desirable that one can freely adjust its value. It is, however, impossible to change the value of α within the framework presented so far. In order to realized such a *q-number* gauge transformation, one must introduce a set of new degrees of freedom, called a **gaugeon**. The gaugeon formalism, proposed by Yokoyama [Yok 74a], starts with the Lagrangian density which consists of the Landau-gauge electromagnetic-field one and of the gaugeon one:

$$\mathcal{L}_{\text{EMG}} = -\frac{1}{4}F^{\mu\nu}F_{\mu\nu} - \partial^\mu B \cdot A_\mu - \partial^\mu G \cdot \partial_\mu C - \frac{1}{2}\epsilon(C+\gamma B)^2, \qquad (2.3.8-1)$$

where $\epsilon = \pm 1$ and γ is a real parameter. They are related to α through

$$\alpha = -\epsilon\gamma^2. \qquad (2.3.8-2)$$

The new scalar field G is called the **gaugeon field**, and C is its accompanying scalar field.

We consider the following q-number gauge transformation:

$$A_\mu' = A_\mu - (\gamma - \gamma')\partial_\mu G, \quad B' = B, \qquad (2.3.8-3)$$

$$G' = G, \quad C' = C + (\gamma - \gamma')B, \qquad (2.3.8-4)$$

$$\psi' = e^{-\imath e(\gamma - \gamma')G}\psi. \qquad (2.3.8-5)$$

Then $\mathcal{L}_D$ remains invariant and $\mathcal{L}_{EMG}$ is transformed into an expression which has exactly the same form as in Eq.(1) but is expressed in terms of primed quantities ($\epsilon' = \epsilon$). Thus the above transformation can be regarded as a gauge transformation from $\alpha = -\epsilon\gamma^2$ to $\alpha' = -\epsilon\gamma'^2$ (the sign of α cannot be changed).

The field equations which follow from Eq.(1) together with the charged-field Lagrangian density are as follows:

$$\partial_\mu F^{\mu\nu} - \partial^\nu B = -ej^\nu, \qquad (2.3.8-6)$$

$$\partial^\mu A_\mu = \epsilon\gamma(C + \gamma B), \qquad (2.3.8-7)$$

$$\Box C = 0, \qquad (2.3.8-8)$$

$$\Box G = \epsilon(C + \gamma B). \qquad (2.3.8-9)$$

Since $\Box(C + \gamma B) = 0$, Eq.(9) shows that the gaugeon field G is a dipole ghost.

As seen from Eq.(1), for $\gamma = 0$ the gaugeon part decouples from the rest. In this case, therefore, the theory is equivalent to the usual B-field formalism in the Landau gauge. The Lagrangian density for other values of γ can be obtained from that for $\gamma = 0$ by a q-number transformation, whence the theory for $\gamma \neq 0$ is also equivalent to the B-field formalism in the *Landau* gauge. Thus we do *not* obtain a theory which is equivalent to the B-field formalism in a non-Landau gauge as long as we start with Eq.(1).

In spite of this fact, it is quite interesting that *the gaugeon formalism precisely reproduces the Green's function in the α gauge*. Although Eq.(7) differs from Eq.(2.3.1-4) by a term proportional to C, it is inessential because C commutes with any of A_μ, B, and ψ in all times.

2.4 MASSIVE VECTOR FIELD

Gauge invariance is violated by the introduction of a mass term. This fact, however, does not necessarily imply that a gauge field is always massless. It is quite
an important discovery that if the global symmetry corresponding to gauge invariance is spontaneously broken, then the gauge field acquires a non-zero mass. This
mechanism is known as the Higgs mechanism, on the basis of which the electroweak
theory was capable of describing the short-range weak interactions.

In this section, after presenting various formalisms for a genuine massive
vector field, we discuss how the Higgs mechanism is covariantly described in the
B-field formalism.

2.4.1 Proca formalism

There are a number of formalisms describing a massive vector field.

The **Proca formalism** [Pro 36] is the most time-honored formalism for a
vector field. Let U_μ be a massive vector field and set

$$F_{\mu\nu} = \partial_\mu U_\nu - \partial_\nu U_\mu. \qquad (2.4.1-1)$$

The (free) Lagrangian density in the Proca formalism is given by

$$\mathcal{L}_{\mathrm{Pr}} = -\frac{1}{4}F^{\mu\nu}F_{\mu\nu} + \frac{1}{2}m^2 U^\mu U_\mu. \qquad (2.4.1-2)$$

The vector field U_μ couples with matter fields through a conserved current j_μ,

$$\partial^\mu j_\mu = 0, \qquad (2.4.1-3)$$

as in the case of the electromagnetic field.

From Eq.(2), the field equation is

$$\Box U_\mu - \partial_\mu \partial^\nu U_\nu + m^2 U_\mu = -e j_\mu. \qquad (2.4.1-4)$$

Applying ∂^μ to it, we obtain

$$\partial^\mu U_\mu = 0 \qquad (2.4.1-5)$$

because of $m \neq 0$, that is, the Lorentz condition is automatically obtained as an operator relation. Owing to this crucial fact, U_μ is quantized *without using indefinite metric*.[1] Substitution of Eq.(5) into Eq.(4) yields

$$(\Box + m^2)U_\mu = -ej_\mu. \qquad (2.4.1-6)$$

Canonical quantization is carried out by adopting U_1, U_2, U_3 as canonical variables. Their canonical conjugates are

$$\pi^k \equiv (\partial/\partial\dot{U}_k)\mathcal{L}_{\mathrm{Pr}} = F^{k0}. \qquad (2.4.1-7)$$

On the other hand, from the zeroth component of Eq.(4) and Eq.(5), we have

$$U_0 = m^{-2}(-\partial_k\pi^k - ej_0), \qquad (2.4.1-8)$$
$$\dot{U}_0 = -\partial^k U_k. \qquad (2.4.1-9)$$

After setting up the canonical commutation relations, therefore, we can calculate commutators involving U_0 and/or $\dot{U}_0$. It is important to note that U_0 does *not* commutes with U_k at the equal time because of Eq.(8).

In the free-field case ($j_\mu = 0$), we can easily calculate the four-dimensional commutation relation, which is given by

$$[U_\mu(x),\ U_\nu(y)] = i(-\eta_{\mu\nu} - m^{-2}\partial_\mu\partial_\nu)\Delta(x - y; m^2). \qquad (2.4.1-10)$$

Likewise, we have

$$\langle\, 0\,|U_\mu(x)U_\nu(y)|\,0\,\rangle = (-\eta_{\mu\nu} - m^{-2}\partial_\mu\partial_\nu)\Delta^{(+)}(x - y; m^2). \qquad (2.4.1-11)$$

However, owing to the non-commutativity between $\theta(x^0 - y^0)$ and ∂_0, we find that

$$\langle\, 0\,|TU_\mu(x)U_\nu(y)|\,0\,\rangle = (-\eta_{\mu\nu} - m^{-2}\partial_\mu\partial_\nu)\Delta_{\mathrm{F}}(x - y; m^2) - im^{-2}\delta_\mu{}^0\delta_\nu{}^0\delta^4(x - y),$$
$$(2.4.1-12)$$

that is, it is *not* Lorentz covariant and does *not* satisfy the Lorentz condition.

Although the positivity of state-space metric is achieved in the Proca formalism, it brings us such pathologies as above. Furthermore, as seen from Eq.(12), the high-energy asymptotic behavior is very bad, violating the usual power-counting rule based on the canonical dimension.

---

[1]   Remember that the little group on a timelike vector is compact.

Since the Proca formalism fully makes use of the non-vanishing mass, we cannot consider the massless limit.

There is another time-honored formalism called the **Stueckelberg formalism** [Stu 38], which is a massive extension of the Feynman-gauge electrodynamics. In addition to a massive vector field A_μ, one introduces a *massive* scalar field B; they have the *same* mass m.[2] The free Lagrangian density is given by

$$\mathcal{L}_{\mathrm{St}} = -\frac{1}{2}\partial^\mu A^\nu \cdot \partial_\mu A_\nu + \frac{1}{2}m^2 A^\mu A_\mu + \frac{1}{2}\partial^\mu B \cdot \partial_\mu B - \frac{1}{2}m^2 B^2. \qquad (2.4.1-13)$$

The theory is quantized with indefinite metric, and therefore one must set up a subsidiary condition. The vector field corresponding to the Proca field is given by $U_\mu = A_\mu + m^{-1}\partial_\mu B$, but it does not satisfy the Lorentz condition. Because what couples with j_μ is not A_μ but U_μ, the pathologies encountered in the Proca formalism is not essentially resolved in the Stueckelberg formalism.

2.4.2 B-field formalism of a massive vector field

A masive (neutral) vector field is most naturally described by the B-field formalism [Nak 72a, Gho 72]. This theory is free of the pathologies encountered in the Proca formalism and has a smooth massless limit.

The (free) Lagrangian density for a massive vector field A_μ is obtained from that for the electromagnetic field by simply adding a mass term, that is,

$$\mathcal{L}_V = -\frac{1}{4}F^{\mu\nu}F_{\mu\nu} + \frac{1}{2}m^2 A^\mu A_\mu + B\partial^\mu A_\mu + \frac{1}{2}\alpha B^2. \qquad (2.4.2-1)$$

Coupling with matter fields is assumed to be the same as in the case of the electromagnetic field.

Field equations are as follows:

$$\partial_\mu F^{\mu\nu} + m^2 A^\nu - \partial^\nu B = -ej^\nu. \qquad (2.4.2-2)$$

$$\partial^\mu A_\mu + \alpha B = 0. \qquad (2.4.2-3)$$

Applying ∂_ν on Eq.(2) and using $\partial^\mu j_\mu = 0$ and Eq.(3), we obtain

$$(\Box + \alpha m^2)B = 0. \qquad (2.4.2-4)$$

---

[2]   It is non-trivial to extend the Stueckelberg formalism to the unequal-mass one [Fuj 64].

Thus B is a free field, but its mass depends on α. For $\alpha < 0$, it becomes imaginary,[3] whence we have to restrict ourselves to considering $\alpha \geq 0$ only. Since a massive vector theory is not a gauge theory, we need not take acount of the gauge parameter seriously.

Canonical quantization is carried out in the same way as in the case of the electeomagetic field. The four-dimensional commutators involving B are modified into the following form:

$$[B(x),\ B(y)] = -im^2 \Delta(x - y; \alpha m^2), \qquad (2.4.2-5)$$

$$[A_\mu(x),\ B(y)] = -i\partial_\mu \Delta(x - y; \alpha m^2), \qquad (2.4.2-6)$$

$$[j_\mu(x),\ B(y)] = 0. \qquad (2.4.2-7)$$

It is important to note that B is no longer a zero-norm field but a negative-norm one.

The subsidiary condition

$$B^{(+)}(x)|f\rangle = 0 \qquad (2.4.2-8)$$

is, of course, set up as in the massless case.

As done in Sec.2.3.5, we calculate $\langle 0\,|\,[A_\mu(x),\ A_\nu(y)]\,|\,0\rangle$ by using

$$(\Box + m^2)A_\mu = -ej_\mu + (1 - \alpha)\partial_\mu B \qquad (2.4.2-9)$$

and

$$e^2\langle 0\,|\,[j_\mu(x),\ j_\nu(y)]\,|\,0\rangle$$
$$= i\int_0^\infty ds\,\pi(s)\,(-s\eta_{\mu\nu} - \partial_\mu\partial_\nu)\,\Delta(x - y; s) \qquad (2.4.2-10)$$

together with Eqs.(5) and (7). We obtain

$$\langle 0\,|\,[A_\mu(x),\ A_\nu(y)]\,|\,0\rangle = im^{-2}\partial_\mu\partial_\nu \Delta(x - y; \alpha m^2)$$
$$+ i\int_{+0}^\infty ds\,\rho(s)\,(-\eta_{\mu\nu} - s^{-1}\partial_\mu\partial_\nu)\,\Delta(x - y;\ s) \qquad (2.4.2-11)$$

with

$$\rho(s) \equiv s(s - m^2)^{-2}\pi(s) \geq 0, \qquad (2.4.2-12)$$

[3]   An imaginary-mass field cannot be satisfactorily quantized [Nak 72b].

where we have *assumed* that $\pi(s)$ involves no massless discrete spectrum. The first term of the rhs. of Eq.(11), which is of negative norm, plays the crucial role in resolving the pathologies encounterd in the Proca formalism, that is, in T-product, its corresponding term cancels not only the extra term of Eq.(2.4.1-12) but also the bad high-energy behavior.

From the equal-time commutation relation, we obtain

$$
\begin{aligned}
-i\eta_{k\ell}\delta(\mathbf{x}-\mathbf{y}) =& im^{-2}\partial_k\partial_\ell\delta(\mathbf{x}-\mathbf{y}) \\
& + i\int_0^\infty ds\,\rho(s)\left(-\eta_{k\ell}-s^{-1}\partial_k\partial_\ell\right)\delta(\mathbf{x}-\mathbf{y}), \quad (2.4.2-13)
\end{aligned}
$$

whence

$$
\int_0^\infty ds\,\rho(s) = 1, \qquad\qquad\qquad (2.4.2-14)
$$

$$
\int_0^\infty ds\,\frac{\rho(s)}{s} = \frac{1}{m^2}. \qquad\qquad (2.4.2-15)
$$

Thus the bare (i.e., unrenormalized) mass squared of a vector field is a harmonic average of the mass-squared distribution.

Let m_r be the physical (i.e., renormalized) mass of A_μ; then

$$
\rho(s) = Z\delta(s-m_r{}^2) + \theta(s-s_0)\sigma(s) \qquad\qquad (2.4.2-16)
$$

with $s_0 > 0$; Eq.(12) implies that $Z > 0$ and $\sigma(s) \geq 0$. Substituting Eq.(16) into Eq.(14) and (15), we obtain

$$
0 < Z = 1 - \int_{s_0}^\infty ds\,\sigma(s) \leq 1, \qquad\qquad (2.4.2-17)
$$

$$
\frac{1}{m^2} = \frac{Z}{m_r{}^2} + \int_{s_0}^\infty \frac{\sigma(s)}{s}. \qquad\qquad (2.4.2-18)
$$

The latter sum rule is called the **Johnson theorem** [Joh 61a]. From this formula, we see that if $m \to 0$ then $m_r \to 0$.[4] Thus *the abelian gauge field cannot acquire a non-zero physical mass.* This result holds *if j_μ contains no massless discrete spectrum when $m \neq 0$,* as assumed in writing Eq.(11). We need not assume Eq.(16) explicitly: Since the rhs. of Eq.(15) goes to infinity as $m \to 0$, $\rho(s)$ must develop a massless discrete spectrum in the limit.

---

[4]   The original formulation of the Johnson theorem was made in the Proca formalism; therefore it was not legitimate to take $m \to 0$.

The above theorem is quite important because it proves the masslessness of the abelian gauge field. One can evade it only when j_μ contains a massless discrete spectrum, as is realized in the Higgs mechanism (see Sec.2.4.4).

We define the asymptotic fields $A^{\mathrm{as}}{}_\mu$ and B^{as} by

$$Z^{-1/2} A_\mu \;\rightarrow\; A^{\mathrm{as}}{}_\mu, \qquad\qquad (2.4.2-19)$$

$$Z^{1/2} k B = B^{\mathrm{as}}, \qquad\qquad (2.4.2-20)$$

where

$$k \equiv Z^{-1} m_r{}^2/m^2 = 1 + Z^{-1} m_r{}^2 \int_{s_0}^\infty ds\,\sigma(s)/s \geq 1. \qquad (2.4.2-21)$$

Then we have

$$[A^{\mathrm{as}}{}_\mu(x),\ A^{\mathrm{as}}{}_\nu(y)] = i\left(-\eta_{\mu\nu} - m_r{}^{-2}\partial_\mu\partial_\nu\right)\Delta(x-y;m_r{}^2)$$
$$+ ikm_r{}^{-2}\partial_\mu\partial_\nu\Delta(x-y;\alpha_r m_r{}^2), \qquad (2.4.2-22)$$

$$[A^{\mathrm{as}}{}_\mu(x),\ B^{\mathrm{as}}(y)] = -ik\partial_\mu\Delta(x-y;\alpha_r m_r{}^2), \qquad (2.4.2-23)$$

$$[B^{\mathrm{as}}(x),\ B^{\mathrm{as}}(y)] = -ikm_r{}^2\Delta(x-y;\alpha_r m_r{}^2), \qquad (2.4.2-24)$$

and

$$\left(\Box + m_r{}^2\right) A^{\mathrm{as}}{}_\mu - (1-\alpha_r)\partial_\mu B^{\mathrm{as}} = 0, \qquad (2.4.2-25)$$

$$\partial^\mu A^{\mathrm{as}}{}_\mu + \alpha_r B^{\mathrm{as}} = 0, \qquad (2.4.2-26)$$

$$\left(\Box + \alpha_r m_r{}^2\right) B^{\mathrm{as}} = 0, \qquad (2.4.2-27)$$

where

$$\alpha_r \equiv Z^{-1} k^{-1}\alpha = (m^2/m_r{}^2)\alpha. \qquad (2.4.2-28)$$

It is easily confirmed that as $m_r \rightarrow 0$ the above results reproduce the corresponding ones for the electromagnetic field (see Sec.2.3.6, but take account of the Källén term).

We define

$$U^{\mathrm{as}}{}_\mu \equiv A^{\mathrm{as}}{}_\mu - m_r{}^{-2}\partial_\mu B^{\mathrm{as}}. \qquad (2.4.2-29)$$

Then $U^{\mathrm{as}}{}_\mu$ satisfies

$$[U^{\mathrm{as}}{}_\mu(x),\ U^{\mathrm{as}}{}_\nu(y)] = i\left(-\eta_{\mu\nu} - m_r{}^{-2}\partial_\mu\partial_\nu\right)\Delta(x-y;m_r{}^2), \quad (2.4.2-30)$$

$$[U^{\mathrm{as}}{}_\mu(x),\ B^{\mathrm{as}}(y)] = 0, \qquad (2.4.2-31)$$

$$\left(\Box + m_r{}^2\right) U^{\mathrm{as}}{}_\mu = 0, \qquad (2.4.2-32)$$

$$\partial^\mu U^{\mathrm{as}}{}_\mu = 0, \qquad (2.4.2-33)$$

that is, it is physical and behaves like a Proca field. On the other hand, owing to Eq.(24), the negative-norm B^{as} is unphysical. Thus the physical subspace has positive-definite metric.

Finally, we remark that an extension of the gaugeon formalism (Sec.2.3.8) to a massive vector field can be made in a satisfactory way [Yok 74b]. In this formalism, any $\Phi = B,\ C,\ G$ satisfies

$$(\Box + m^2)\Box\Phi = 0. \qquad (2.4.2 - 34)$$

In particular, the mass of B is independent of α. Thus the gaugeon formalism of a massive vector field is rather different from the B-field formalism with $\alpha \neq 0$.

2.4.3 Higgs model

As mentioned in Sec.1.5.1, the Goldstone theorem states that if a continuous symmetry is spontaneously broken, there appears the NG particle which is necessarily exactly massless and, if it is an internal symmetry, spinless. Since, however, no massless spinless particles are observed in nature, it is quite difficult to apply the idea of spontaneous breakdown to the description of the real world. Fortunately, one can evade the Goldstone theorem if spontaneous breakdown occurs in gauge theory. Furthermore, in this case, the gauge field becomes massive "by eating the NG boson".[5] This mechanism is called the **Higgs mechanism** [And 63, Hig 64a, Hig 64b, Eng 64, Gur 64]. The situation is most clearly demonstrated in the **Higgs model** [Hig 66, Lee 72].

The Higgs model is nothing but the Goldstone model (see Sec.1.5.2) in which the complex scalar field ϕ couples with an abelian gauge field A_μ in the minimal way. That is, the Higgs-model Lagrangian density is given by

$$\mathcal{L}_{Hi} = -\frac{1}{4}F^{\mu\nu}F_{\mu\nu} + (\partial_\mu + ieA^\mu)\phi^\dagger \cdot (\partial_\mu - ieA_\mu)\phi - u\phi^\dagger\phi - \frac{1}{4}\lambda(\phi^\dagger\phi)^2 \quad (2.4.3 - 1)$$

with $u < 0$ and $\lambda > 0$. As in the Goldstone model, the phase invariance is spontaneously broken by

$$\langle 0 | \phi | 0 \rangle = v/\sqrt{2}, \qquad (2.4.3 - 2)$$

where

$$v = (-4u/\lambda)^{1/2} > 0. \qquad (2.4.3 - 3)$$

---

[5]   This phrase is quite misleading See later discussion.

We set

$$\sqrt{2}\phi(x) = v + \varphi(x) + i\chi(x), \qquad (2.4.3-4)$$

where φ and χ are real scalar fields having a vanishing vacuum expectation value. We rewrite Eq.(1) as

$$
\begin{aligned}
\mathcal{L}_{\text{Hi}} = &-\frac{1}{4}F^{\mu\nu}F_{\mu\nu} + \frac{1}{2}M^2(A^\mu - M^{-1}\partial^\mu\chi)(A_\mu - M^{-1}\partial_\mu\chi) \\
&+ \frac{1}{2}(\partial^\mu\varphi \cdot \partial_\mu\varphi - \mu^2\varphi^2) + eA^\mu(\chi\partial_\mu\varphi - \varphi\partial_\mu\chi) \\
&+ eMA^\mu A_\mu\varphi + \frac{1}{2}e^2 A^\mu A_\mu(\varphi^2 + \chi^2) - \frac{1}{2}g\mu\varphi(\varphi^2 + \chi^2) \\
&- \frac{1}{8}g^2(\varphi^2 + \chi^2)^2 + \text{const}, \qquad (2.4.3-5)
\end{aligned}
$$

where

$$M \equiv ev > 0, \qquad (2.4.3-6)$$

$$\mu^2 \equiv -2u > 0, \qquad (2.4.3-7)$$

$$g \equiv \sqrt{\lambda/2} > 0. \qquad (2.4.3-8)$$

As seem in Eq.(5), the kinetic term of the NG boson χ disappears if $A_\mu - M^{-1}\partial_\mu\chi$ is regarded as a new vector field, and the latter acquires a "bare" mass M. Interaction terms can be made small by making e and g small while v large.

The above semi-classical discussion can be made more precise in the fully quantized covariant formalism, as shown by Nakanishi [Nak 73a]. We start with the Lagrangian density

$$\mathcal{L}_{\text{H}} = \mathcal{L}_{\text{Hi}} + B\partial^\mu A_\mu + \frac{1}{2}\alpha B^2. \qquad (2.4.3-9)$$

Canonical quantization is carried out, and as in Sec.2.3.3 we have

$$[B(x),\ B(y)] = 0, \qquad (2.4.3-10)$$

$$[A_\mu(x),\ B(y)] = -i\partial_\mu D(x-y), \qquad (2.4.3-11)$$

$$[\phi(x),\ B(y)] = e\phi(x)D(x-y), \qquad (2.4.3-12)$$

and set up the subsidiary condition

$$B^{(+)}(x)|f\rangle = 0. \qquad (2.4.3-13)$$

Substituting Eq.(4) into Eq.(12), we find

$$[\varphi(x),\ B(y)] = ie\chi(x)D(x-y), \tag{2.4.3-14}$$

$$[\chi(x),\ B(y)] = -ie[v+\varphi(x)]D(x-y). \tag{2.4.3-15}$$

Accordingly, we have

$$\langle\,0\,|\,[\varphi(x),\ B(y)]\,|\,0\,\rangle = 0, \tag{2.4.3-16}$$

$$\langle\,0\,|\,[\chi(x),\ B(y)]\,|\,0\,\rangle = -iMD(x-y). \tag{2.4.3-17}$$

Eq.(17) is an extremely important result; it shows that the phase invariance is spontaneously broken because its generator is given by $\int d\mathbf{y}\,\partial_0 B(y)$ [see Eq.(38) below], that $\chi(x)$ is the NG field, that it is massless, and most importantly that it is *unphysical* (see below). It should be noted that χ is *not* eaten by the gauge field but becomes unobservable because of the subsidiary condition.

Since the transverse part of A_μ does not contain a massless discrete spectrum, reasoning similar to Sec.2.3.5 yields

$$\langle\,0\,|\,[A_\mu(x),\ A_\nu(y)]\,|\,0\,\rangle = iL\partial_\mu\partial_\nu D(x-y) - i\alpha\partial_\mu\partial_\nu E(x-y)$$
$$+ i\int_{+0}^{\infty} ds\,\rho(s)\left(-\eta_{\mu\nu} - s^{-1}\partial_\mu\partial_\nu\right)\Delta(x-y;s) \tag{2.4.3-18}$$

with

$$\int_{+0}^{\infty} ds\,\rho(s) = 1, \tag{2.4.3-19}$$

$$L \equiv \int_{+0}^{\infty} ds\,\rho(s)/s > 0. \tag{2.4.3-20}$$

From Eqs.(16) and (17) together with Eq.(2.3.1-4), Lorentz covariance, and a canonical commutation relation, we have

$$\langle\,0\,|\,[A_\mu(x),\ \varphi(y)]\,|\,0\,\rangle = 0, \tag{2.4.3-21}$$

$$\langle\,0\,|\,[A_\mu(x),\ \chi(y)]\,|\,0\,\rangle = i\alpha M\partial_\mu E(x-y). \tag{2.4.3-22}$$

Since A_μ acquires a non-zero mass,[6] we denote its physical mass by M_r. Then we can write

$$\rho(s) = Z\delta\left(s-M_r{}^2\right) + \theta\left(s-s_0\right)\sigma(s), \tag{2.4.3-23}$$

---

[6]   This fact is consistent with Eq.(2.3.3-23); indeed , since $A_\mu - M^{-1}\partial_\mu\chi$ commutes with Q_Λ, the NG boson of Q_Λ is nothing but χ [Nak 80d].

with $s_0 > 0$, whence

$$L = \frac{Z}{M_r{}^2} + \int_{s_0}^{\infty} ds \, \frac{\sigma(s)}{s}. \qquad (2.4.3-24)$$

In contrast to the case of a genuine massive vector field, L is *not* equal to M^{-2} by the Z factor of χ (see below).

Now, we introduce asymptotic fields. For simplicity, we set $A_\mu \to \hat{A}^{\mathrm{as}}{}_\mu$, $B = \hat{B}^{\mathrm{as}}$, $\chi \to \hat{\chi}^{\mathrm{as}}$, and $\varphi \to \hat{\varphi}^{\mathrm{as}}$ without making wave-function renormalization. From the above results, we have

$$[\hat{A}^{\mathrm{as}}{}_\mu(x), \, \hat{A}^{\mathrm{as}}{}_\nu(y)] = iZ(-\eta_{\mu\nu} - M_r{}^{-2}\partial_\mu\partial_\nu)\Delta(x - y; M_r{}^2)$$
$$+ iL\partial_\mu\partial_\nu D(x - y) - i\alpha\partial_\mu\partial_\nu E(x - y), \qquad (2.4.3-25)$$

$$[\hat{A}^{\mathrm{as}}{}_\mu(x), \, \hat{B}^{\mathrm{as}}(y)] = -i\partial_\mu D(x - y), \qquad (2.4.3-26)$$

$$[\hat{A}^{\mathrm{as}}{}_\mu(x), \, \hat{\chi}^{\mathrm{as}}(y)] = i\alpha M \partial_\mu E(x - y), \qquad (2.4.3-27)$$

$$[\hat{B}^{\mathrm{as}}(x), \, \hat{B}^{\mathrm{as}}(y)] = 0, \qquad (2.4.3-28)$$

$$[\hat{\chi}^{\mathrm{as}}(x), \, \hat{B}^{\mathrm{as}}(y)] = -iMD(x - y). \qquad (2.4.3-29)$$

Because $\hat{\varphi}^{\mathrm{as}}$ has a mass generically equal to neither 0 nor M_r, it commutes with any other asymptotic fields. We assume that the set $\{\hat{A}^{\mathrm{as}}{}_\mu, \, \hat{B}^{\mathrm{as}}, \, \hat{\chi}^{\mathrm{as}}, \, \hat{\varphi}^{\mathrm{as}}\}$ is *complete*.

Since $\hat{A}^{\mathrm{as}}{}_\mu$ is an admixture of massive and massless fields, we extract a massive vector field $\hat{U}^{\mathrm{as}}{}_\mu$ from it by setting

$$\hat{U}^{\mathrm{as}}{}_\mu \equiv \hat{A}^{\mathrm{as}}{}_\mu - a\partial_\mu\hat{\chi}^{\mathrm{as}} - b\partial_\mu\hat{B}^{\mathrm{as}}, \qquad (2.4.3-30)$$

where a and b are constants to be determined. We require that $\hat{U}^{\mathrm{as}}{}_\mu$ be purely massive, whence

$$[\hat{U}^{\mathrm{as}}{}_\mu(x), \, \hat{B}^{\mathrm{as}}(y)] = 0, \qquad (2.4.3-31)$$

$$[\hat{U}^{\mathrm{as}}{}_\mu(x), \, \hat{\chi}^{\mathrm{as}}(y)] = 0. \qquad (2.4.3-32)$$

From Eq.(31), we find $a = M^{-1}$. On the other hand, Eq.(32) implies that

$$[\hat{\chi}^{\mathrm{as}}(x), \, \hat{\chi}^{\mathrm{as}}(y)] = ibM^2 D(x - y) + i\alpha M^2 E(x - y). \qquad (2.4.3-33)$$

We then have

$$[\hat{U}^{\mathrm{as}}{}_\mu(x), \, \hat{U}^{\mathrm{as}}{}_\nu(y)] = [\hat{U}^{\mathrm{as}}{}_\mu(x), \, \hat{A}^{\mathrm{as}}{}_\nu(y)]$$
$$= iZ(-\eta_{\mu\nu} - M_r{}^{-2}\partial_\mu\partial_\nu)\Delta(x - y; M_r{}^2) + i(L - b)\partial_\mu\partial_\nu D(x - y). \qquad (2.4.3-34)$$

Since the massless spectrum should vanish, we find $b = L$. Thus Eq.(30) becomes

$$\hat{U}^{\mathrm{as}}{}_\mu = \hat{A}^{\mathrm{as}}{}_\mu - M^{-1}\partial_\mu \hat{\chi}^{\mathrm{as}} - L\partial_\mu \hat{B}^{\mathrm{as}}. \qquad (2.4.3-35)$$

Since $\Box\hat{\chi}^{\mathrm{as}} + \alpha M \hat{B}^{\mathrm{as}}$ commutes with all asymptotic fields, it must vanish because of asymptotic completeness. We thus have

$$\Box\hat{\chi}^{\mathrm{as}} = -\alpha M \hat{B}^{\mathrm{as}}, \qquad (2.4.3-36)$$

that is, the NG boson $\hat{\chi}^{\mathrm{as}}$ is a dipole ghost for $\alpha \neq 0$ [Nak 75b]. From Eqs.(35) and (36), we have $\partial^\mu \hat{U}^{\mathrm{as}}{}_\mu = 0$.

From Eq.(13) and the above consideration, it is evident that $\hat{U}^{\mathrm{as}}{}_\mu$, $\hat{B}^{\mathrm{as}}$, and $\hat{\varphi}^{\mathrm{as}}$ are physical, while the NG boson $\hat{\chi}^{\mathrm{as}}$ is unphysical. The physical subspace has positive semi-definite metric as in the case of the electromagnetic field. Since $\hat{B}^{\mathrm{as}}$ is of zero norm, observable particles are all massive. Thus it is beautifully explained in the B-field formalism how the Higgs mechanism works without contradicting the Goldstone theorem.

It should be noted that all commutators between the asymptotic fields can be very simply reproduced essentially (i.e., apart from renormalization) by identifying the free part of $\mathcal{L}_{\mathrm{H}}$ with the Lagrangian density for asymptotic fields [Nis 73]. We call this method the **free-Lagrangian method**, which is applied to quantum gravity in Sec.5.8.1.

Finally, we make a remark about the charge operator

$$Q \equiv e \int d\mathbf{x}\, j^0(x). \qquad (2.4.3-37)$$

Since (the transverse part of) A_μ becomes massive, the surface integral $\int d\mathbf{x}\, \partial_k F^{k0}$ vanishes, whence

$$Q = \int d\mathbf{x}\, \partial_0 B(x). \qquad (2.4.3-38)$$

As discussed in Sec.1.5.1, in general, when symmetry is spontaneously broken, the corresponding charge operator Q is not well-defined. Indeed, if Q is a well-defined operator, $Q|0\rangle$ is a translationally invariant state linearly independent of $|0\rangle$, contradicting the Borchers theorem (see Appendix A.2). One should note, however, that the Borchers theorem *cannot* be proved in the *indefinite-metric* Hilbert space. Hence, in the Higgs model, there is a possibility that Q is a well-defined operator. In Sec.2.5.2, we persent an example of a well-defined charge operator whose symmetry is spontaneously broken. Further discussions on the spontaneous symmetry breaking are made in Appendix $A.3$.

2.4.4 Pre-Higgs model

The model which is obtained from the Higgs model by simply adding a mass term of the gauge field is called the **pre-Higgs model**. The reasons why this model is interesting are as follows:

1. The gauge field acquires a non-zero physical mass in the Higgs model. How is the Johnson theorem stated in Sec.2.4.2 evaded?
2. The NG boson is physical in the pre-Higgs model. Then why can it become unphysical in the Higgs limit? [This is strange because "physical" and "unphysical" are characterized by equality and by inequality, respectively, and the limit of an equality should be an equality.]

The Lagrangian density of the pre-Higgs model is

$$\mathcal{L}_{\mathrm{PH}} = \mathcal{L}_{\mathrm{H}} + \frac{1}{2}m^2 A^\mu A_\mu, \qquad (2.4.4-1)$$

where $\mathcal{L}_{\mathrm{H}}$ is given by Eq.(2.4.3-9). Field equations, Eqs.(2.4.2-2)-(2.4.2-4), and commutation relations, Eqs.(2.4.2-5)-(2.4.2-7), for a massive vector theory remain unchanged because they are independent of representation. The subsidiary condition is the same as before.

On the other hand, the vacuum expectation values of commutators become as follows:

$$\begin{aligned}
\langle 0 | [A_\mu(x),\ A_\nu(y)] | 0 \rangle \\
= &-i(m^{-2} - L)\partial_\mu\partial_\nu D(x-y) + im^{-2}\partial_\mu\partial_\nu\Delta(x-y;\ \alpha m^2) \\
&+ i\int_{+0}^{\infty} ds\, \rho(s)(-\eta_{\mu\nu} - s^{-1}\partial_\mu\partial_\nu)\Delta(x-y;\ s) \qquad (2.4.4-2)^7
\end{aligned}$$

with L defined by Eq.(2.4.3-20),

$$\langle 0 | [A_\mu(x),\ B(y)] | 0 \rangle = -i\partial_\mu\Delta(x-y;\ \alpha m^2), \qquad (2.4.4-3)$$

$$\langle 0 | [A_\mu(x),\ \chi(y)] | 0 \rangle = im^{-2}M\partial_\mu[D(x-y) - \Delta(x-y;\ \alpha m^2)] \qquad (2.4.4-4)$$

---

[7]   We can obtain this result directly without using $\langle 0 | [j_\mu,\ j_\nu] | 0 \rangle$.

with M defined by Eq.(2.4.3-6),

$$\langle\, 0\,|\,[\chi(x),\, B(y)]\,|\,0\,\rangle = -iM\Delta(x-y;\, \alpha m^2), \qquad (2.4.4-5)$$

$$\langle\, 0\,|\,[B(x),\, B(y)]\,|\,0\,\rangle = -im^2\Delta(x-y;\, \alpha m^2), \qquad (2.4.4-6)$$

etc. From Eqs.(2.4.2-9),(2), (3), and (6), we obtain

$$e^2\langle\, 0\,|\,[j_\mu(x),\, j_\nu(y)]\,|\,0\,\rangle = -im^2(1-m^2L)\partial_\mu\partial_\nu D(x-y)$$
$$+ i\int_{+0}^{\infty} ds\,(s-m^2)^2\rho(s)(-\eta_{\mu\nu}-s^{-1}\partial_\mu\partial_\nu)\Delta(x-y;\, s).\, (2.4.4-7)$$

Since j_μ is physical [see Eq.(2.4.2-7)], its spectral function is non-negative; that is, $\rho(s)\ge 0$ and

$$1 - m^2 L > 0, \qquad (2.4.4-8)$$

where the equality case is excluded by later consideration.

It is important to note that j_μ *contains a massless discrete spectrum as long as $m\ne 0$*. This is the very reason why the Johnson theorem is evaded in the Higgs model [Nak 73a]. The existence of a physical massless spectrum is due to the fact that the NG boson is physical in the pre-Higgs model, as seen later.

Because of the existence of a physical massless boson, the existence of stable massive particles is quite questionable. But, for simplicity, we here assume that all of A_μ, B, χ, and φ have their asymptotic field as in the case of the Higgs model. From Eqs.(2)-(6), we have

$$[\hat{A}^{\text{as}}_\mu(x),\, \hat{A}^{\text{as}}_\nu(y)] = -iZ(-\eta_{\mu\nu}-M_r^{-2}\partial_\mu\partial_\nu)\Delta(x-y;\, M_r^2)$$
$$- i(m^{-2}-L)\partial_\mu\partial_\nu D(x-y) + im^{-2}\partial_\mu\partial_\nu\Delta(x-y;\, \alpha m^2), \quad (2.4.4-9)$$

$$[\hat{A}^{\text{as}}_\mu(x),\, \hat{B}^{\text{as}}(y)] = -i\partial_\mu\Delta(x-y;\, \alpha m^2), \qquad (2.4.4-10)$$

$$[\hat{A}^{\text{as}}_\mu(x),\, \hat{\chi}^{\text{as}}(y)] = -im^{-2}M\partial_\mu\,[D(x-y)-\Delta(x-y;\, \alpha m^2)], \quad (2.4.4-11)$$

$$[\hat{\chi}^{\text{as}}(x),\, \hat{B}^{\text{as}}(y)] = -iM\Delta(x-y;\, \alpha m^2), \qquad (2.4.4-12)$$

$$[\hat{B}^{\text{as}}(x),\, \hat{B}^{\text{as}}(y)] = -im^2\Delta(x-y;\, \alpha m^2). \qquad (2.4.4-13)$$

The asymptotic field $\hat{\varphi}^{\text{as}}$ commutes with all others because it has a different mass.

We define the Proca field $\hat{U}^{\text{as}}_\mu$ by

$$\hat{U}^{\text{as}}_\mu \equiv \hat{A}^{\text{as}}_\mu - M^{-1}(1-m^2L)\partial_\mu\hat{\chi}^{\text{as}} - L\partial_\mu\hat{B}^{\text{as}}, \qquad (2.4.4-14)$$

so as to satisfy Eqs.(2.4.3-31) and (2.4.3-32) and also $(\Box + M_r{}^2)\hat{U}^{\text{as}}{}_\mu = 0$. Then from the consistency we should have $1 - m^2L \neq 0$ and

$$
\begin{aligned}
[\hat{\chi}^{\text{as}}(x),\ \hat{\chi}^{\text{as}}(y)] =& im^{-2}M^2(1 - m^2L)^{-1}D(x - y) \\
&- im^{-2}M^2\Delta(x - y;\ \alpha m^2).
\end{aligned} \tag{2.4.4 - 15}
$$

Hence $\Box\hat{\chi}^{\text{as}}$ cannot vanish. Instead, we can show that $\Box\hat{\chi}^{\text{as}} + \alpha M \hat{B}^{\text{as}}$ commutes with all asymptotic fields, and therefore we must have

$$
\Box\hat{\chi}^{\text{as}} + \alpha M \hat{B}^{\text{as}} = 0 \tag{2.4.4 - 16}
$$

because of asymptotic completeness.

The NG boson is defined by

$$
\hat{X}^{\text{as}} \equiv mM^{-1}\hat{\chi}^{\text{as}} - m^{-1}\hat{B}^{\text{as}}. \tag{2.4.4 - 17}
$$

Indeed, it is straightforward to show that

$$
\Box\hat{X}^{\text{as}} = 0, \tag{2.4.4 - 18}
$$

$$
[\hat{X}^{\text{as}}(x),\ \hat{X}^{\text{as}}(y)] = i(1 - m^2L)^{-1}D(x - y), \tag{2.4.4 - 19}
$$

and $\hat{X}^{\text{as}}$ commutes with $\hat{U}^{\text{as}}{}_\mu$, $\hat{B}^{\text{as}}$, and $\hat{\varphi}^{\text{as}}$; therefore $\hat{X}^{\text{as}}$ is *physical*. Furthermore, the generator

$$
Q \equiv e \int d\mathbf{x}\, j^0(x) = - \int d\mathbf{x}\, [m^2 \hat{A}^{\text{as}0}(x) - \partial^0 \hat{B}^{\text{as}}(x)], \tag{2.4.4 - 20}
$$

which commutes with $\hat{U}^{\text{as}}{}_\mu$, $\hat{B}^{\text{as}}$, and $\hat{\varphi}^{\text{as}}$, has a non-vanishing commutator with $\hat{X}^{\text{as}}$:

$$
[iQ,\ \hat{X}^{\text{as}}(x)] = -m. \tag{2.4.4 - 21}
$$

Now, we make an important observation. As $m \to 0$, the NG boson $\hat{X}^{\text{as}}$ has no limit. In the Higgs limit, the NG boson is $\hat{\chi}^{\text{as}}$. Thus *there arises rearrangement of the NG boson as $m \to 0$*. This is the reason why the observability of the NG boson changes [Kon 76].

Finally, the physical subspace is constructed by $\hat{U}^{\text{as}}{}_\mu$, $\hat{X}^{\text{as}}$, and $\hat{\varphi}^{\text{as}}$, and it has positive-definite metric [see Eq.(8)].

2.4.5 Chiral gauge model

In the Higgs model, spontaneous breakdown of symmetry is caused by a primary field ϕ, but the existence of such a field is not a prerequisite for the Higgs mechanism. The Higgs mechanism having no primary agency of spontaneous breakdown is called the **dynamical Higgs mechanism**. An example of the dynamical Higgs mechanism is provided by the **chiral gauge model**, which is obtained from the Nambu-Jona-Lasinio model (see Sec.1.5.3) by adding a chiral gauge field. Both gauge field and Dirac field, which are primarily massless, become massive in this model [Freu 70]. It is, however, incompatible with asymptotic completeness that all asymptotic fields are massive and cannot transform under chiral transformation [Aur 72]. This dilemma is resolved by the existence of the unphysical NG boson, as in the case of the Higgs model [Nak 73b].

The Lagrangian density of the chiral gauge model is

$$\mathcal{L}_{CG} \equiv -\frac{1}{4}F^{\mu\nu}F_{\mu\nu} + B\partial^\mu A_\mu + \frac{1}{2}\alpha B^2 - e\bar{\psi}\gamma^\mu\gamma_5\psi A_\mu + i\bar{\psi}\gamma^\mu\partial_\mu\psi + g\left[(\bar{\psi}\psi)^2 - (\bar{\psi}\gamma_5\psi)^2\right]. \tag{2.4.5-1}$$

It is invariant under the chiral transformation

$$\psi \rightarrow \psi' = \exp(-i\theta\gamma_5)\psi, \tag{2.4.5-2}$$

and the corresponding chiral current is

$$j_5{}^\mu \equiv \bar{\psi}\gamma^\mu\gamma_5\psi. \tag{2.4.5-3}$$

Owing to the conservation law $\partial_\mu j_5{}^\mu = 0$, B satisfies $\Box B = 0$, and the subsidiary condition, Eq.(2.4.3-13), is set up.

As in the case of the electromagnetic field, it is easy to show that

$$[\psi(x),\ B(y)] = -e\gamma_5\psi(x)D(x-y), \tag{2.4.5-4}$$

$$[\bar{\psi}(x),\ B(y)] = -e\bar{\psi}(x)\gamma_5 D(x-y). \tag{2.4.5-5}$$

For the pseudoscalar density [see Eq.(1.5.3-5)]

$$\rho(x) \equiv -i\bar{\psi}(x)\gamma_5\psi(x), \tag{2.4.5-6}$$

we have

$$[\rho(x),\ B(y)] = 2ie\bar{\psi}(x)\psi(x)D(x-y). \qquad (2.4.5-7)$$

The spontaneous breakdown of chiral invariance is characterized by

$$\langle\,0\,|\,\bar{\psi}(x)\psi(x)\,|\,0\,\rangle = -M_0/2g \neq 0, \qquad (2.4.5-8)$$

where a real constant M_0 turns out to be the zeroth approximation to the physical mass of the Dirac field, as is seen from the term $g(\bar{\psi}\psi)^2$ in Eq.(1). From Eqs.(7) and (8), we obtain

$$\langle\,0\,|\,[\rho(x),\ B(y)]\,|\,0\,\rangle = -iMD(x-y) \qquad (2.4.5-9)$$

with

$$M \equiv eM_0/g. \qquad (2.4.5-10)$$

Eq.(9) is the most crucial one, which just corresponds to Eq.(2.4.3-17) of the Higgs model. The pseudoscalar density $\rho(x)$ must contain an unphysical massless NG boson. If we denote its asymptotic field by $\hat{\chi}^{\text{as}}$, then everything concerning the asymptotic fields $\hat{A}^{\text{as}}{}_{\mu}$, $\hat{B}^{\text{as}}$, and $\hat{\chi}^{\text{as}}$ goes exactly in the same way as in the Higgs model.[8] Hence, we do not repeat further description here.

One can also add a mass term to the chiral gauge field. The discussion is completely parallel to the case of the pre-Higgs model [Kon 76, SuzI 85].

---

[8]   But we note that the free-Lagrangian method does not work in this model because $\hat{\chi}^{\text{as}}$ is a composite field.

2.5 TWO-DIMENSIONAL MODELS*

It is quite instructive to consider two-dimensional models, though not realistic, because of their simplicity. For example, we can test the validity of postulates of quantum field theory in exactly solvable models. Of course, since two-dimensional spacetime is too much special, we must always be very careful about how general the qualitative consequences of two-dimensional models are.

Two-dimensional models have been studied by quite a large number of authors from various standpoints, and their conclusions are often mutually contradictory in such ways that the reasons are not necessarily attributable to errors. The moral which is drawn from this embarrassing situation is that one must clearly state one's principles on which one's analysis is based. It is necessary to exclude prejudiced assumptions even if they are time-honored and/or used by many authors.

Our analysis made here is based on the following principles. We start with a Lagrangian density and carry out canonical quantization in a way similar to the four-dimensional case as much as possible. Our most crucial postulate is to respect *asymptotic completeness*, because we always postulate it in our analysis of gauge theories. In most work, the representation is determined by unconsciously regarding some quantities introduced for technical convenience as fundamental ones. This is the most essential one of the reasons why many diverse conclusions have been claimed so far.

2.5.1 Two-dimensional D functions

The two-dimensional spacetime is quite peculiar. There is no *a priori* distinction between time and space in the two-dimensional Minkowski space; the distinction between them is afterwards brought by imposing the positivity of energy. The spacelike region consists of two disconnected domains only in the two-dimensional case. The two-dimensional Lorentz group is abelian; we cannot speak of little group. Spin has no intrinsic meaning; we introduce scalar, spinor, vector, etc. only by analogy from higher dimensions.

---

*   This section is devoted to a specialized topic. To read it is unnecessary for understanding the following chapters.

In this subsection, as preliminaries, we present the definitions of two-dimensional D functions and their properties [Nak 77a]. The notation used is as follows: $x^\mu = (x^0,\ x^1)$, $x^2 = (x^0)^2 - (x^1)^2$, $\Box = \partial_0{}^2 - \partial_1{}^2$; $\eta_{00} = -\eta_{11} = 1$, $\eta_{01} = \eta_{10} = 0$, $\epsilon_{00} = \epsilon_{11} = 0$, $\epsilon_{01} = -\epsilon_{10} = 1$; hence $\epsilon_{\mu\nu}\epsilon_{\sigma\tau} = \eta_{\mu\tau}\eta_{\nu\sigma} - \eta_{\mu\sigma}\eta_{\nu\tau}$, $\epsilon^{\mu\nu}\epsilon_{\nu\tau} = \delta^\mu{}_\tau$.

In the massless case, since nothing discriminates time and space (before taking account of the positivity of energy), there are two odd invariant solutions to the d'Alembert equation:

$$\Box D_2(x) = \Box \tilde{D}_2(x) = 0, \qquad (2.5.1-1)$$

$$D_2(x) \equiv (2\pi i)^{-1} \int d^2p\, \epsilon(p_0)\delta(p^2)e^{-ipx}$$

$$= -\frac{1}{2}\epsilon(x^0)\theta(x^2), \qquad (2.5.1-2)$$

$$\tilde{D}_2(x) \equiv (2\pi i)^{-1} \int d^2p\, \epsilon(p_1)\delta(p^2)e^{-ipx}$$

$$= -\frac{1}{2}\epsilon(x^1)\theta(-x^2). \qquad (2.5.1-3)$$

They are mutually related through

$$\tilde{D}_2(x^0,\ x^1) = -D_2(-x^1,\ -x^0), \qquad (2.5.1-4)$$

$$\partial_0\tilde{D}_2 = \partial_1 D_2, \quad \partial_1\tilde{D}_2 = \partial_0 D_2. \qquad (2.5.1-5)$$

Integrating Eq.(5) over x^1, we have

$$D_2(x) = \int_{-\infty}^{x^1} dz^1\, \partial_0\tilde{D}_2(x^0,\ z^1), \qquad (2.5.1-6)$$

$$\tilde{D}_2(x) = \int_{-\infty}^{x^1} dz^1\, \partial_0 D_2(x^0,\ z^1) + \frac{1}{2}. \qquad (2.5.1-7)$$

Note the appearance of an additional constant $1/2$ in Eq.(7).

In order to define the positive-frequency part, we need $\theta(p_0)\delta(p^2)$, but, unfortunately, it is *not well-defined* in two dimensions because $\int d^2p\, \theta(p_0)\delta(p^2)$ is infrared divergent. This fact causes a number of troubles in the two-dimensional massless quantum field theory, but at the same time it is a very challenging problem to overcome this infrared difficulty.

We here employ Klaiber's definition of $D_2{}^{(\pm)}(x)$ [Kla 68]:

$$D_2{}^{(\pm)}(x) \equiv \pm(2\pi)^{-1} \int_{-\infty}^{+\infty} dp_1\, (2\hat{p}_0)^{-1} \left[e^{\mp i\hat{p}x} - \theta(\kappa - \hat{p}_0) \right], \qquad (2.5.1-8)$$

where $\hat{p}_0 = |p_1|$, $\hat{p}_\mu = (\hat{p}_0, \ p_1)$, and κ is a positive constant. Although the subtraction term in Eq.(8) makes $D_2^{(\pm)}(x)$ κ-dependent, it does not violate the Lorentz invariance of $D_2^{(\pm)}(x)$. Carrying out the p_1 integration, we obtain

$$
\begin{aligned}
D_2^{(\pm)}(x) &= \mp(4\pi)^{-1}\log(-\mu^2 x^2 \pm i0x^0) \\
&= \mp(4\pi)^{-1}\left[\log\mu(x^0 - x^1 \mp i0) + \log\mu(x^0 + x^1 \mp i0) + i\pi\right], \quad (2.5.1-9)
\end{aligned}
$$

where $\mu \equiv \kappa e^\gamma$, γ being Euler's constant. One should note the similarity between Eqs.(9) and (2.3.4–15). In the present case, we must write the additional constant explicitly. Although the positivity of the Fourier transform of $D_2^{(+)}(x)$ is lost, other properties are all right:

$$D_2^{(-)}(x) = -D_2^{(+)}(-x) = -\left[D_2^{(+)}(x)\right]^*, \qquad (2.5.1-10)$$

$$\Box D_2^{(\pm)}(x) = 0, \qquad (2.5.1-11)$$

$$iD_2(x) = D_2^{(+)}(x) + D_2^{(-)}(x). \qquad (2.5.1-12)$$

We can likewise define $\tilde{D}_2^{(\pm)}(x)$. But it must be emphasized that *the interchangeability of time and space breaks down at this stage because of the positivity of energy.* We have

$$
\begin{aligned}
\tilde{D}_2^{(\pm)}(x) &\equiv \pm(2\pi)^{-1}\int_{-\infty}^{+\infty} dp_1\,(2p_1)^{-1}\left[e^{\mp i\hat{p}x} - \theta(\kappa - \hat{p}_0)\right] \\
&= \pm\frac{1}{4\pi}\log\frac{x^0 - x^1 \mp i0}{x^0 + x^1 \mp i0} \quad (2.5.1-13)
\end{aligned}
$$

independently of κ. Although Eqs.(10)–(12) with tildes hold, $\tilde{D}_2^{(\pm)}(x)$ is *not* Lorentz invariant; indeed

$$(x_0\partial_1 - x_1\partial_0)\tilde{D}_2^{(\pm)}(x) = \mp(2\pi)^{-1}. \qquad (2.5.1-14)$$

The spatial asymptotic behaviors of $D_2^{(\pm)}(x)$ and $\tilde{D}_2^{(\pm)}(x)$ are very important. As $x^1 \to \pm\infty$, $D_2^{(+)}(x)$ is logarithmically divergent, while

$$\tilde{D}_2^{(+)}(x) \ \to \ \mp i/4. \qquad (2.5.1-15)$$

It is the crucial problem how to avoid the troubles caused by the non-existence of the $x^1 \to \pm\infty$ limit of $D_2^{(+)}(x)$ in constructing the theory of a massless scalar field. Since

$$\partial_0 D_2^{(\pm)}(x) = \partial_1 \tilde{D}_2^{(\pm)}(x) = \mp(2\pi)^{-1}\frac{x^0}{x^2 \mp i0x^0}, \qquad (2.5.1-16)$$

$$\partial_1 D_2^{(\pm)}(x) = \partial_0 \tilde{D}_2^{(\pm)}(x) = \pm(2\pi)^{-1}\frac{x^1}{x^2 \mp i0x^0}, \qquad (2.5.1-17)$$

we see that as $|x^1| \to \infty$

$$\partial_0 D_2^{(+)}(x) = \partial_1 \tilde{D}_2^{(+)}(x) \sim (x^1)^{-2}, \qquad (2.5.1-18)$$

$$\partial_1 D_2^{(+)}(x) = \partial_0 \tilde{D}_2^{(+)}(x) \sim (x^1)^{-1}. \qquad (2.5.1-19)$$

In what follows, the very good asymptotic behavior of $\partial_0 D_2^{(+)}(x)$ plays a crucial role.

From Eqs.(16) and (15), we have

$$\tilde{D}_2^{(\pm)}(x) = \int_{-\infty}^{x^1} dz^1 \, \partial_0 D_2^{(\pm)}(x^0, \, z^1) + \frac{i}{4}, \qquad (2.5.1-20)$$

$$\int_{-\infty}^{+\infty} dx^1 \, \partial_0 D_2^{(\pm)}(x) = -\frac{i}{2}. \qquad (2.5.1-21)$$

We cannot, however, integrate Eq.(17) because $D_2^{(\pm)}(x)$ diverges as $x^1 \to \pm\infty$, but if we define the whole spatial integral by

$$\int_{-\infty}^{+\infty} dx^1 \, (\, \cdots \,) \equiv \lim_{L \to \infty} \int_{-L}^{+L} dx^1 \, (\, \cdots \,), \qquad (2.5.1-22)$$

then we have

$$\int_{-\infty}^{+\infty} dx^1 \, \partial_0 \tilde{D}_2^{(\pm)}(x) = 0. \qquad (2.5.1-23)$$

Hereafter, *we always employ Eq.(22)*.

By using the explicit expressions, we can show that [Nak 77a]

$$i \int_{-\infty}^{+\infty} dz^1 [D_2^{(\pm)}(x-z)\partial_0{}^z D_2^{(\pm)}(z-y) - \partial_0{}^z D_2^{(\pm)}(x-z) \cdot D_2^{(\pm)}(z-y)]$$
$$= D_2^{(\pm)}(x-y), \qquad (2.5.1-24)$$

$$i \int_{-\infty}^{+\infty} dz^1 [D_2^{(\pm)}(x-z)\partial_0{}^z D_2^{(\mp)}(z-y) - \partial_0{}^z D_2^{(\pm)}(x-z) \cdot D_2^{(\mp)}(z-y)] = 0, \qquad (2.5.1-25)$$

$$i \int_{-\infty}^{+\infty} dz^1 [D_2^{(\pm)}(x-z)\partial_0{}^z \tilde{D}_2^{(\pm)}(z-y) - \partial_0{}^z D_2^{(\pm)}(x-z) \cdot \tilde{D}_2^{(\pm)}(z-y)]$$
$$= \tilde{D}_2^{(\pm)}(x-y), \qquad (2.5.1-26)$$

$$i \int_{-\infty}^{+\infty} dz^1 [D_2^{(\pm)}(x-z)\partial_0{}^{'}\tilde{D}_2^{(\mp)}(z-y) - \partial_0{}^z D_2^{(\pm)}(x-z) \cdot \tilde{D}_2^{(\mp)}(z-y)] = 0. \qquad (2.5.1-27)$$

The last two hold only in the sense of Eq.(22).

For a massive case, there arises no special matter. We have

$$\Delta_2(x;\ m^2) \equiv (2\pi i)^{-1} \int d^2p\, \epsilon(p_0)\delta(p^2 - m^2)e^{-ipx}$$

$$= -\frac{1}{2}\epsilon(x^0)J_0(m\sqrt{x^2})\theta(x^2), \qquad (2.5.1-28)$$

$$\Delta_2^{(\pm)}(x;\ m^2) = \pm(2\pi)^{-1} \int d^2p\, \theta(\pm p_0)\delta(p^2 - m^2)e^{-ipx}$$

$$= \pm(2\pi)^{-1}K_0(m\sqrt{-x^2 \pm i0x^0}), \qquad (2.5.1-29)$$

where J_0 and K_0 denote a Bessel function and a modified Bessel function, respectively.

For a massless dipole ghost, we have

$$E_2(x) \equiv (2\pi i)^{-1} \int d^2p\, \epsilon(p_0)\delta'(p^2)e^{-ipx}$$

$$= -(1/8)\epsilon(x^0)x^2\theta(x^2), \qquad (2.5.1-30)$$

$$E_2^{(\pm)}(x) \equiv \mp(16\pi)^{-1}x^2 \left[\log(-\mu^2x^2 \pm i0x^0) - 2\right] \qquad (2.5.1-31)$$

in such a way that

$$iE_2(x) = E_2^{(+)}(x) + E_2^{(-)}(x), \qquad (2.5.1-32)$$

$$\Box E_2^{(\pm)}(x) = D_2^{(\pm)}(x). \qquad (2.5.1-33)$$

2.5.2 Two-dimensional massless scalar field

The two-dimensional free massless scalar field plays very important roles in solving several two-dimensional models. At first sight, one may suppose that the two-dimensional free massless scalar field would be the simplest object in quantum field theory, but it is quite a subtle one, owing to the severe infrared divergence and owing to the introduction of the conjugate field.

The Lagrangian density of the free massless scalar field ϕ is, of course, given by

$$\mathcal{L}_\phi = \frac{1}{2}\partial^\mu\phi \cdot \partial_\mu\phi. \qquad (2.5.2-1)$$

Then its field equation is

$$\Box\phi = 0, \qquad (2.5.2-2)$$

and canonical quantization yields

$$[\phi(x), \ \phi(y)] = iD_2(x - y). \qquad (2.5.2-3)$$

Now, the first hurdle is how to define the positive/negative frequency part $\phi^{(\pm)}(x)$. The naive definition involves infrared divergence. Something fundamental must therefore be violated. One may violate the mathematical theory of distributions by restricting test functions artificially [Wig 67]. One may violate manifest translational invariance by introducing such a cutoff term as in Eq.(2.5.1-8) in the definition of $\phi^{(\pm)}$ [Kla 68]. Or, one may violate the positivity of state-vector space [Nak 77a]. In comparison with the four-dimensional gauge theory, the last choice seems to be most natural, whence hereafter we follow this line of thought. In this case, we of course encounter another hurdle; we must find a satisfactory subsidiary condition (see Sec.2.2.3).

We thus define

$$\phi^{(\pm)}(x) \equiv i \int_{-\infty}^{+\infty} dz^1 \left[D_2^{(\pm)}(x - z)\partial_0\phi(z) - \partial_0^{\ z} D_2^{(\pm)}(x - z) \cdot \phi(z) \right]. \quad (2.5.2-4)$$

Owing to the subtraction term ·in Eq.(2.5.1-8), $\phi^{(\pm)}(x)$ contains an infrared mode, but we define the vacuum by[1]

$$\phi^{(+)}(x)\,|\,0\,\rangle = 0, \quad \langle\,0\,|\,0\,\rangle = 1. \qquad (2.5.2-5)$$

With the aid of Eqs.(2.5.1–24) and (2.5.1–25), we can show that

$$[\phi^{(\pm)}(x), \ \phi^{(\mp)}(y)] = D_2^{(\pm)}(x - y), \qquad (2.5.2-6)$$

$$[\phi^{(\pm)}(x), \ \phi^{(\pm)}(y)] = 0, \qquad (2.5.2-7)$$

whence

$$\langle\,0\,|\,\phi(x)\phi(y)\,|\,0\,\rangle = D_2^{(+)}(x - y). \qquad (2.5.2-8)$$

As $|x^1 - y^1| \to \infty$, Eq.(8) behaves like $\log|x^1 - y^1|$. Thus, unlike the usual theory, we *cannot* suppose that $\phi(x) \to 0$ as $|x^1| \to \infty$. We may, however, postulate that

$$|x^1|^{-\epsilon}\phi(x) \to 0 \quad \text{as} \quad |x^1| \to \infty \qquad (2.5.2-9)$$

---

[1]   Note that if $\phi^{(\pm)}$ were defined in a translationally non-invariant way, we could not define the vacuum as in Eq.(5).

for any $\varepsilon > 0$. Furthermore, we can assume, without contradicting Eq.(8), that

$$\partial_0 \phi(x) \sim |x^1|^{-2}, \qquad (2.5.2 - 10)$$

$$\partial_1 \phi(x) \sim |x^1|^{-1} \qquad (2.5.2 - 11)$$

as $|x^1| \to \infty$. The good asymptotic behavior of $\partial_0 \phi(x)$ is quite important. The convergence of the integral in Eq.(4) is guaranteed by Eqs.(9) and (10).

The **number operator** is defined by

$$N \equiv i \int_{-\infty}^{+\infty} dx^1 \left[\phi^{(-)}(x) \partial_0 \phi^{(+)}(x) - \partial_0 \phi^{(-)}(x) \cdot \phi^{(+)}(x) \right]. \qquad (2.5.2 - 12)$$

Of course, we have $N^\dagger = N$, $\partial_\mu N = 0$, $N|0\rangle = 0$, and

$$\left[iN, \ \phi^{(\pm)}(x) \right] = \mp i \phi^{(\pm)}(x), \qquad (2.5.2 - 13)$$

whence

$$N \prod_{j=1}^{n} \phi^{(-)}(x^{(j)}) |0\rangle = n \prod_{j=1}^{n} \phi^{(-)}(x^{(j)}) |0\rangle. \qquad (2.5.2 - 14)$$

Thus the Fock vacuum $|0\rangle$ is uniquely characterized by $n = 0$, as it should be. Any state can be arbitrarily well approximated by a polynomial state $R(\phi^{(-)})|0\rangle$.

The (operator) **irreducibility** of $\phi(x)$ is of paramount importance in our theory,[2] that is, we can *prove* the following property:[3]

$$[A, \ \phi^{(\pm)}(x)] = 0 \quad \text{for any} \quad x^\mu \qquad (2.5.2 - 15)$$

if and only if A is a c-number.

Only the converse needs a proof. If A commutes with $\phi^{(\pm)}(x)$, then it does with N. Accordingly, $N(A|0\rangle) = A(N|0\rangle) = 0$, whence the uniquiness of $|0\rangle$ implies $A|0\rangle = \lambda|0\rangle$, where λ is a c-number. Then $AR(\phi^{(-)})|0\rangle = R(\phi^{(-)})A|0\rangle = \lambda R(\phi^{(-)})|0\rangle$, that is, $A = \lambda$.

Now, we proceed to introducing the **conjugate field** $\tilde{\phi}(x)$, which satisfies

$$\partial_\mu \phi + \epsilon_{\mu\nu} \partial^\nu \tilde{\phi} = 0, \quad \text{i.e.,} \quad \partial_0 \tilde{\phi} = \partial_1 \phi \ \ \& \ \ \partial_1 \tilde{\phi} = \partial_0 \phi. \qquad (2.5.2 - 16)$$

---

[2]   Indeed, much confusion has arisen owing to the negligence or misunderstanding of irreducibility

[3]   The usual expression "$[A, \ \phi(x)] = 0$ for any x^μ" is weaker than Eq.(15) because of the noncommutativity of integrations. One of the author (N.N.) would like to thank A. Nakamura for pointing out the inadequacy of $[A, \ \phi(x)] = 0$ in the present model.

Thus $\tilde{\phi}(x)$ is *determined* by $\phi(x)$ apart from an additional constant operator. If translational covariance of $\tilde{\phi}(x)$ is required, we have

$$\tilde{\phi}(x) \equiv \int_{-\infty}^{x^1} dz^1 \, \partial_0 \phi(x^0, z^1) \qquad (2.5.2-17)$$

uniquely except for the addition of an arbitrary function of symmetry generators. Note that Eq.(17) is *convergent* owing to Eq.(10).

From Eq.(16) we have

$$\Box \tilde{\phi}(x) = 0. \qquad (2.5.2-18)$$

Furthermore, with the aid of Eqs.(2.5.1-7) and (2.5.1-6), we obtain[4]

$$[\phi(x), \, \tilde{\phi}(y)] = i\tilde{D}_2(x-y) + i/2, \qquad (2.5.2-19)$$

$$[\tilde{\phi}(x), \, \tilde{\phi}(y)] = iD_2(x-y). \qquad (2.5.2-20)$$

Because of the symmetry between $\phi(x)$ and $\tilde{\phi}(x)$, it might be tempting to treat them on the equal footing. Such a standpoint, however, leads us to inconsistency unless the theory is totally altered. In our standpoint, $\tilde{\phi}(x)$ is *not* a fundamental quantity, and its properties are *derived* from those of $\phi(x)$, but *not vice versa*.

If we define $\tilde{\phi}^{(\pm)}(x)$ by replacing ϕ by $\phi^{(\pm)}$ in Eq.(17), then $[\tilde{\phi}^{(+)}(x), \, \tilde{\phi}^{(-)}(y)]$ is divergent. This is the second hurdle. In order to avoid it, we define $\tilde{\phi}^{(\pm)}$ by

$$\tilde{\phi}^{(\pm)}(x) \equiv i \int_{-\infty}^{+\infty} dz^1 \, \left[D_2^{(\pm)}(x-z)\partial_0\tilde{\phi}(z) - \partial_0{}^z D_2^{(\pm)}(x-z) \cdot \tilde{\phi}(z) \right].$$
$$(2.5.2-21)$$

The difference between the above two definitions is essentially the order of integrations. In the correct ordering, the asymmetric integration ($-\infty$ to x^1) precedes the symmetric one ($-\infty$ to $+\infty$).

In general, since the contributions from infinity are not necessarily vanishing, one must be extremely careful about the order of limits [Nak 80a] especially because such quantities as $\phi^{(\pm)}$ and $\tilde{\phi}^{(\pm)}$ already contain limiting process.

---

[4]   One might feel it more aesthetic to define $\tilde{\phi}(x)$ by

$$\frac{1}{2} \int_{-\infty}^{+\infty} dz^1 \, \epsilon(x^1 - z^1)\partial_0\phi(x^0, z^1),$$

because then the additional term $i/2$ in Eq.(19) disappears. But this term plays an important role in the next subsection.

With the aid of Eqs.(2.5.1-24)-(2.5.1-27) and (2.5.1-21), we derive from Eqs. (19) and (20) that

$$[\phi^{(\pm)}(x), \; \tilde{\phi}^{(\mp)}(y)] = \tilde{D}_2^{(\pm)}(x-y) + i/8, \qquad (2.5.2-22)$$

$$[\phi^{(\pm)}(x), \; \tilde{\phi}^{(\pm)}(y)] = i/8, \qquad (2.5.2-23)$$

$$[\tilde{\phi}^{(\pm)}(x), \; \tilde{\phi}^{(\mp)}(y)] = D_2^{(\pm)}(x-y), \qquad (2.5.2-24)$$

$$[\tilde{\phi}^{(\pm)}(x), \; \tilde{\phi}^{(\pm)}(y)] = 0. \qquad (2.5.2-25)$$

Now, we introduce important conserved charges. Since $\partial_\mu \phi$ is a conserved current, we see that

$$\Phi \equiv \int_{-\infty}^{+\infty} dx^1 \, \partial_0 \phi(x) \qquad (2.5.2-26)$$

is a well-defined constant operators. Likewise, we introduce well-defined constant operators

$$\Phi^{(\pm)} \equiv \int_{-\infty}^{+\infty} dx^1 \, \partial_0 \phi^{(\pm)}(x) = [\Phi^{(\mp)}]^\dagger; \qquad (2.5.2-27)$$

of course,

$$\Phi = \Phi^{(+)} + \Phi^{(-)}. \qquad (2.5.2-28)$$

It is easily seen that $\Phi^{(\pm)}$ commutes with $\phi^{(\pm)}(x)$, $\tilde{\phi}^{(\pm)}(x)$, $\tilde{\phi}^{(\mp)}(x)$, $\Phi^{(\pm)}$, and $\Phi^{(\mp)}$, but

$$\left[i\Phi^{(\pm)}, \; \phi^{(\mp)}(x) \right] = 1/2. \qquad (2.5.2-29)$$

Since $\partial_\mu \tilde{\phi}$ is also a conserved current, it might be tempting to introduce

$$\tilde{\Phi} \equiv \int_{-\infty}^{+\infty} dx^1 \, \partial_0 \tilde{\phi}(x) = \int_{-\infty}^{+\infty} dx^1 \, \partial_1 \phi(x). \qquad (2.5.2-30)$$

Because of irreducibility [see Eq.(15)], however, $\tilde{\Phi}$ must equal $\langle 0 | \tilde{\Phi} | 0 \rangle = 0$ if it is well-defined. Thus we do not consider $\tilde{\Phi}$.[5]

Since $[\tilde{\phi}^{(+)}(x), \; \phi^{(+)}(y)] \neq 0$, it is impossible to have $\tilde{\phi}^{(+)}(x) | 0 \rangle = 0$. Indeed, we have

$$\tilde{\phi}^{(+)}(x) | 0 \rangle = (1/4)\Phi^{(-)} | 0 \rangle, \qquad (2.5.2-31)$$

---

[5]   At this point, a number of authors [Had 79a, Had 79b, Arat 81, Suz 82a, Suz 82b] prefer to regard $\tilde{\Phi}$ as a sensible operator on the basis of symmetrical treatment of $\phi(x)$ and $\tilde{\phi}(x)$ at the sacrifice of irreducibility.

because both sides are one-particle states and their difference is annihilated by $\phi^{(+)}(y)$. By using the commutation relations and Eq.(31), we find

$$\langle 0 \mid \phi(x)\tilde{\phi}(y) \mid 0 \rangle = \tilde{D}_2^{(+)}(x-y) + i/4, \qquad (2.5.2-32)$$

$$\langle 0 \mid \tilde{\phi}(x)\tilde{\phi}(y) \mid 0 \rangle = D_2^{(+)}(x-y). \qquad (2.5.2-33)$$

We now discuss the Lorentz transformation property. Since the energy-momentum tensor is

$$T_{\mu\nu} \equiv \partial_\mu \phi \cdot \partial_\nu \phi - \eta_{\mu\nu}\mathcal{L}_\phi, \qquad (2.5.2-34)$$

Poincaré generators are

$$P_0 \equiv \frac{1}{2} \int_{-\infty}^{+\infty} dx^1 \, [(\partial_0 \phi)^2 + (\partial_1 \phi)^2], \qquad (2.5.2-35)$$

$$P_1 \equiv \int_{-\infty}^{+\infty} dx^1 \, \partial_0 \phi \cdot \partial_1 \phi, \qquad (2.5.2-36)$$

$$M_{01} \equiv \int_{-\infty}^{+\infty} dx^1 \, \left\{ x_0 \partial_0 \phi \cdot \partial_1 \phi - \frac{1}{2} x_1 \, [(\partial_0 \phi)^2 + (\partial_1 \phi)^2] \right\}. \quad (2.5.2-37)$$

We can then show that [Nak 77e]

$$[iP_\mu, \, \phi^{(\pm)}(x)] = \partial_\mu \phi^{(\pm)}(x), \qquad (2.5.2-38)$$

$$[iM_{01}, \, \phi^{(\pm)}(x)] = (x_0\partial_1 - x_1\partial_0)\phi^{(\pm)}(x), \qquad (2.5.2-39)$$

$$[iP_\mu, \, \tilde{\phi}^{(\pm)}(x)] = \partial_\mu \tilde{\phi}^{(\pm)}(x), \qquad (2.5.2-40)$$

$$[iM_{01}, \, \tilde{\phi}^{(\pm)}(x)] = (x_0\partial_1 - x_1\partial_0)\tilde{\phi}^{(\pm)}(x) \pm i\pi^{-1}\Phi^{(\pm)}. \quad (2.5.2-41)$$

In deriving Eq.(41), use has been made of

$$\lim_{x^1 \to -\infty} x^1 \partial_1 \phi^{(\pm)}(x) = \mp i\pi^{-1}\Phi^{(\pm)}, \qquad (2.5.2-42)$$

which follows from irreducibility. It is noteworthy that $\tilde{\phi}(x)$ is *not a Lorentz scalar*. Indeed, because of Eqs.(32) and (2.5.1-14), *it is impossible that both $\phi(x)$ and $\tilde{\phi}(x)$ are Lorentz scalars.* [6]

We note that the Poincaré invariance of $\Phi^{(\pm)}$ follows from Eqs.(38) and (39).

---

[6] This fact is a strong evidence against the standpoint that both $\phi(x)$ and $\tilde{\phi}(x)$ are regarded as fundamental quantities on the same level, because then Lorentz invariance would have to be regarded as spontaneously broken without having the NG mode.

We must set up a subsidiary condition because indefinite metric has been introduced to avoid infrared difficulty. We define the physical subspace $\mathcal{V}_{\text{phys}}$ by the totality of the states $|f\rangle$ satisfying

$$\Phi^{(+)}|f\rangle = 0. \qquad (2.5.2-43)$$

Evidently, $\mathcal{V}_{\text{phys}}$ is Poincaré invariant. The positive semi-definiteness of its metric can be proved in the following way [Nak 77a, Nak 80a].

Without loss of generality, we may restrict ourselves to considering the one-particle sector only. Let $\{f_n(x)\}$ be a set of one-particle wave functions satisfying the d'Alembert equation:

$$f_n(x) \equiv \int dp_1 \, (2\hat{p}_0)^{-1} \left[\tilde{f}_n(p_1) e^{-i\hat{p}x} - \tilde{f}_n(0)\theta(\kappa - \hat{p}_0) \right], \qquad (2.5.2-44)$$

where $\hat{p}_0 = |p_1|$, $\hat{p}_\mu = (\hat{p}_0, p_1)$. We suppose that this set is complete in the sense of

$$\sum_n \tilde{f}_n(p_1)\tilde{f}_n^*(q_1) = \delta(p_1 - q_1), \qquad (2.5.2-45)$$

that is,

$$\sum_n \left[f_n(x)\partial_0 f_n^*(y) - \partial_0 f_n(x) \cdot f_n^*(y) \right] = 2\pi i D_2^{(+)}(x-y). \qquad (2.5.2-46)$$

An arbitrary one-particle state $|f\rangle$ is expressed as

$$|f\rangle = \left\{ \sum_n c_n \int_{-\infty}^{+\infty} dx^1 \left[\phi^{(-)}(x)\partial_0 f_n(x) - \partial_0 \phi^{(-)}(x) \cdot f_n(x) \right] + c\Phi^{(-)} \right\} |0\rangle \qquad (2.5.2-47)$$

with some coefficients c_n and c. From Eqs.(29) and (44), we then have

$$\Phi^{(+)}|f\rangle = -(\pi/2)\sum_n c_n \tilde{f}_n(0)|0\rangle, \qquad (2.5.2-48)$$

whence $|f\rangle$ satisfies the subsidiary condition, Eq.(43), if and only if

$$\sum_n c_n \tilde{f}_n(0) = 0. \qquad (2.5.2-49)$$

On the other hand, the norm of $|f\rangle$ is

$$\langle f|f\rangle = \pi \int dp_1 \, |p_1|^{-1} \left[\left| \sum_n c_n \tilde{f}_n(p_1) \right|^2 - \left| \sum_n c_n \tilde{f}_n(0) \right|^2 \theta(\kappa - |p_1|) \right]$$
$$- \pi \text{Re} \left[c^* \sum_n c_n \tilde{f}_n(0) \right]. \qquad (2.5.2-50)$$

Accordingly, if $|f\rangle$ is a physical state, then we see that

$$\langle f | f \rangle = \pi \int dp_1 \, |p_1|^{-1} | \sum_n c_n \tilde{f}_n(p_1)|^2 \geq 0. \qquad (2.5.2-51)$$

Thus $\mathcal{V}_{\text{phys}}$ is positive semi-definite.

Obviously, $\Phi^{(-)}|0\rangle$ is an only zero-norm one-particle physical state. In general, $\Phi^{(-)}\mathcal{V}_{\text{phys}}$ is the zero-norm subspace of $\mathcal{V}_{\text{phys}}$. Thus the structure of the physical subspace is quite similar to that in the gauge theory.

Because of Eq.(29), the symmetry induced by $\Phi^{(\pm)}$ is spontaneously broken, and the corresponding NG boson is $\phi(x)$ itself. Nevertheless, $\Phi^{(\pm)}$ must be a well-defined operator because otherwise the subsidiary condition, Eq.(43), would become meaningless. Then, as discussed at the end of Sec.2.4.3, translationally invariant states $(|0\rangle$ and $\Phi^{(-)}|0\rangle)$ are no longer unique.[7]

2.5.3 Two-dimensional massless Dirac field

Although there is no spinor representation in two dimensions, we can construct a two-dimensional analogue of the Dirac field. It is a two-component field $\psi = \begin{pmatrix} \psi_1 \\ \psi_2 \end{pmatrix}$, and the γ-matrices are defined by

$$\gamma^0 = \begin{pmatrix} 0 & 1 \\ 1 & 0 \end{pmatrix}, \quad \gamma^1 = \begin{pmatrix} 0 & 1 \\ -1 & 0 \end{pmatrix}, \quad \gamma^5 \equiv \gamma^0 \gamma^1 = \begin{pmatrix} -1 & 0 \\ 0 & 1 \end{pmatrix}, \qquad (2.5.3-1)$$

where

$$\gamma^\mu \gamma^5 = \epsilon^{\mu\nu} \gamma_\nu. \qquad (2.5.3-2)$$

The massless Dirac equation is

$$\gamma^\mu \partial_\mu \psi = 0, \quad \text{i.e.,} \quad (\partial_0 + \gamma^5 \partial_1)\psi = 0. \qquad (2.5.3-3)$$

Thus both components decouple.

The anticommutation relations are

$$\{\psi_\alpha(x), \psi_\beta(y)\} = 0, \qquad (2.5.3-4)$$

$$\{\psi_\alpha(x), \bar{\psi}_\beta(y)\} = -(\gamma^\mu)_{\alpha\beta} \partial_\mu D(x-y)$$
$$= \begin{pmatrix} 0 & \delta(x^0 + x^1 - y^0 - y^1) \\ \delta(x^0 - x^1 - y^0 + y^1) & 0 \end{pmatrix} \quad (2.5.3-5)$$

---

[7] Indeed, we can explicitly show [Nak 80a] that there exists a state $|\Omega\rangle$ which cannot be approximated by a state of the form $R(\phi)\Phi^{(-)}|0\rangle$, where $R(\phi)$ is any quasi-local operator made of ϕ; the existence of $|\Omega\rangle$ invalidates the proof of the Borchers theorem

with $\bar{\psi} \equiv \psi^\dagger \gamma^0$. Owing to the presence of ∂_μ, the two-dimensional Dirac theory is free of infrared difficulty. Accordingly, we can safely use the momentum representation

$$\psi(x) = (1/\sqrt{2\pi}) \int dp_1 \left[a(p_1)e^{-\imath \hat{p}x} + b^\dagger(p_1)e^{\imath \hat{p}x} \right] u(p_1) \qquad (2.5.3-6)$$

with

$$\{ a(p_1),\ a^\dagger(q_1) \} = \{ b(p_1),\ b^\dagger(q_1) \} = \delta(p_1 - q_1), \qquad (2.5.3-7)$$

$$u(p_1) \equiv \begin{pmatrix} (\hat{p}_0 + p_1)/2\hat{p}_0 \\ (\hat{p}_0 - p_1)/2\hat{p}_0 \end{pmatrix} = \begin{pmatrix} \theta(p_1) \\ \theta(-p_1) \end{pmatrix}. \qquad (2.5.3-8)$$

The vector current

$$j^\mu(x) \equiv\ :\bar{\psi}(x)\gamma^\mu\psi(x): \qquad (2.5.3-9)$$

is conserved, where colons stand for normal ordering. Because of Eq.(2), the axial-vector current $j_5{}^\mu(x)$ is not independent of $j_\nu(x)$:

$$j_5{}^\mu(x) \equiv\ :\bar{\psi}(x)\gamma^\mu\gamma^5\psi(x): := \epsilon^{\mu\nu}j_\nu(x). \qquad (2.5.3-10)$$

From their conservation laws, we therefore see

$$\Box j_\mu = 0. \qquad (2.5.3-11)$$

Indeed, by explicit calculation [Kla 68], we find

$$j_\mu(x) = \frac{i}{\sqrt{2\pi}} \int \frac{dk_1}{\sqrt{2\hat{k}_0}} \hat{k}_\mu \left[c(k_1)e^{-\imath \hat{k}x} - c^\dagger(k_1)e^{\imath \hat{k}x} \right], \qquad (2.5.3-12)$$

where

$$c(k_1) \equiv -(i/\sqrt{\hat{k}_0}) \int dp_1 \{ \theta(k_1 p_1) \left[a^\dagger(p_1)a(p_1 + k_1) - b^\dagger(p_1)b(p_1 + k_1) \right]$$
$$+ \theta(p_1(k_1 - p_1)) a(p_1)b(k_1 - p_1) \}. \qquad (2.5.3-13)$$

Since $c(k_1)$ satisfies

$$[c(k_1),\ c^\dagger(p_1)] = \delta(k_1 - p_1) \qquad (2.5.3-14)$$

and annihilates the vacuum, we can write[8]

$$j_\mu = -(1/\sqrt{\pi})\partial_\mu\phi \qquad (2.5.3-15)$$

---

[8]   Therefore, the canonical commutation relation of ϕ implies the existence of the Schwinger term (see Sec.1.4.1):

$$[j_0(x),\ j_1(y)]_0 = i\pi^{-1}\delta'(x^1 - y^1).$$

in terms of a free massless scalar field ϕ. The negative-norm infrared mode of ϕ does not contribute to j_μ because of the presence of ∂_μ. Thus apart from the infrared mode, ϕ is expressible in terms of ψ.

It is important that the converse is also possible in some sense. The **fermion-boson equivalence** is an interesting property of two-dimensional models, first found in the massive Thirring model by Coleman [Col 75] and Mandelstam [Man 75]. For the free massless Dirac field, we consider the quantity defined by

$$\hat{\psi}(x) \equiv: \exp i\sqrt{\pi}[\phi(x) - \gamma^5 \tilde{\phi}(x)] : u, \qquad (2.5.3 - 16)$$

where colons indicate to arrange the product in the order $\tilde{\phi}^{(-)}$, $\phi^{(-)}$, $\phi^{(+)}$, $\tilde{\phi}^{(+)}$, and $u \equiv \begin{pmatrix} u_1 \\ u_2 \end{pmatrix}$ with u_α being a complex number such that

$$|u_\alpha|^2 = \mu/2\pi, \qquad (2.5.3 - 17)$$

where μ is the infrared cufoff parameter ($\mu = \kappa e^\gamma$) introduced in Eq.(2.5.1-9). It is straightforward to see that[9]

$$\gamma^\mu \partial_\mu \hat{\psi}(x) = 0, \qquad (2.5.3 - 18)$$

$$\hat{\psi}_\alpha(x)\hat{\psi}_\beta(y) = \exp\left[-M_{\alpha\beta}(x - y)\right] \cdot \hat{\psi}_\beta(y)\hat{\psi}_\alpha(x), \qquad (2.5.3 - 19)[10]$$

$$\hat{\psi}_\alpha(x)\hat{\psi}_\beta^{\,\dagger}(y) = \exp\left[+M_{\alpha\beta}(x - y)\right] \cdot \hat{\psi}_\beta^\dagger(y)\hat{\psi}_\alpha(x), \qquad (2.5.3 - 20)$$

where

$$M_{\alpha\beta}(z) \equiv i\pi\left\{ \left[1 + (-1)^{\alpha+\beta}\right] D_2(z) - (-1)^\alpha \left[\tilde{D}_2(z) - 1/2\right]\right.$$
$$\left. - (-1)^\beta \left[\tilde{D}_2(z) + 1/2\right] \right\}. \qquad (2.5.3 - 21)$$

Since

$$M_{\alpha\beta}(z) = i\pi\left[(-1)^\alpha\theta(z^1) - (-1)^\beta\theta(-z^1)\right] \quad \text{for} \quad z^2 < 0, \qquad (2.5.3 - 22)$$

$\hat{\psi}$ obeys Fermi statistics. Here the existence of the extra term $i/2$ in Eq.(2.5.2-19) is crucial to ensure Fermi statistics for $\alpha \neq \beta$.

---

[9] $e^A e^B = e^{[A, B]}e^B e^A$ if $[A, B]$ =c-number.

[10] Hereafter in this subsection, no summation convention is used for α, β.

Next, we show that $\hat{\psi}$ and $\hat{\psi}^\dagger$ satisfy the canonical anticommutation relations [Tan 76]. From Eq.(16) we have[11]

$$\hat{\psi}_\alpha(x)\hat{\psi}_\beta(y) = u_\alpha u_\beta \exp\left[-M^{(+)}_{\alpha\beta}(x-y)\right] \cdot \Psi^{(+)}_{\alpha\beta}(x,\,y), \quad (2.5.3-23)$$

$$\hat{\psi}_\alpha(x)\hat{\psi}^\dagger_\beta(y) = u_\alpha u_\beta^* \exp\left[+M^{(+)}_{\alpha\beta}(x-y)\right] \cdot \Psi^{(-)}_{\alpha\beta}(x,\,y), \quad (2.5.3-24)$$

where

$$\begin{aligned}
M^{(+)}_{\alpha\beta} &\equiv \pi\{\left[1 + (-1)^{\alpha+\beta}\right] D_2^{(+)}(z) - (-1)^\alpha \left[\tilde{D}_2^{(+)}(z) - i/4\right] \\
&\quad - (-1)^\beta \left[\tilde{D}_2^{(+)}(z) + i/4\right]\} \\
&= -(1/4)\{\left[1 + (-1)^\alpha\right]\left[1 + (-1)^\beta\right] \log\mu(z^0 - z^1 - i0) \\
&\quad + \left[1 - (-1)^\alpha\right]\left[1 - (-1)^\beta\right] \log\mu(z^0 + z^1 - i0) \\
&\quad + i\pi\left[1 - (-1)^\alpha + (-1)^\beta + (-1)^{\alpha+\beta}\right]\}, \quad (2.5.3-25)
\end{aligned}$$

$$\Psi^{(\pm)}_{\alpha\beta}(x,\,y) \equiv\; :\exp i\sqrt{\pi}\left[\phi(x) \pm \phi(y) - (-1)^\alpha \tilde{\phi}(x) \mp (-1)^\beta \tilde{\phi}(y)\right]:\,. \quad (2.5.3-26)$$

The reversed-order products $\hat{\psi}_\beta(y)\hat{\psi}_\alpha(x)$ and $\hat{\psi}^\dagger_\beta(y)\hat{\psi}_\alpha(x)$ are also expressed as in Eq.(23) and as in Eq.(24), respectively, by replacing $M^{(+)}_{\alpha\beta}(x-y)$ by $M^{(+)}_{\beta\alpha}(y-x)$. Since

$$\exp\left[\mp M^{(+)}_{\alpha\beta}(0,\,x^1-y^1)\right] = -\exp\left[\mp M^{(+)}_{\beta\alpha}(0,\,y^1-x^1)\right] \quad \text{for} \quad \alpha \neq \beta, \quad (2.5.3-27)$$

we have

$$\{\psi_\alpha(x),\,\psi_\beta(y)\}_0 = \{\psi_\alpha(x),\psi_\beta^\dagger(y)\}_0 = 0 \quad \text{for} \quad \alpha \neq \beta. \quad (2.5.3-28)$$

For $\alpha = \beta$, we find

$$\exp\left[\mp M^{(+)}_{\alpha\alpha}(0,\,x^1-y^1)\right] = \pm i[-(-1)^\alpha \mu(x^1-y^1) - i0]^{\pm 1}. \quad (2.5.3-29)$$

Since Eq.(29) is continuous at $x^1 = y^1$ in the upper sign case, we obtain

$$\{\psi_\alpha(x),\,\psi_\alpha(y)\}_0 = 0. \quad (2.5.3-30)$$

In the lower sign case, Eq.(29) yields

$$\begin{aligned}
\{\psi_\alpha(x),\,\psi_\alpha^\dagger(y)\}_0 &= \frac{i}{2\pi}\left[\frac{1}{(-1)^\alpha(x^1-y^1)+i0} + \frac{1}{(-1)^\alpha(y^1-x^1)+i0}\right] \\
&\quad \cdot \left[\Psi^{(-)}_{\alpha\alpha}(x,\,y)\right]_0 \\
&= \delta(x^1 - y^1). \quad (2.5.3-31)
\end{aligned}$$

---

[11] $:e^A: \cdot :e^B: := e^{[A^{(+)},B^{(-)}]} :e^{A+B}:.$

Thus the canonical anticommutation relations are established.

We calculate the current according to Johnson's definition [Joh 61b]. With ε^ν infinitesimal, we consider

$$[\hat{\psi}^\dagger(x+\varepsilon)\gamma^0\gamma_\mu]_\alpha\hat{\psi}_\alpha(x)$$
$$=(-1)^{(\alpha-1)\mu}|u_\alpha|^2\exp M^{(+)}_{\alpha\alpha}(\varepsilon)\cdot\Psi^{(-)}_{\alpha\alpha}(x,\ x+\varepsilon)$$
$$=(-1)^{(\alpha-1)\mu}(2\pi)^{-1}(-i)[\varepsilon^0-(-1)^\alpha\varepsilon^1-i0]^{-1}$$
$$\cdot[1-i\sqrt{\pi}\varepsilon^\nu\partial_\nu\phi(x)+(-1)^\alpha i\sqrt{\pi}\varepsilon^\nu\partial_\nu\tilde{\phi}(x)+O(\varepsilon^2)].\ (2.5.3-32)$$

Taking $\varepsilon^2\neq 0$ and setting $\tilde{\varepsilon}^0=\varepsilon^1$, $\tilde{\varepsilon}^1=\varepsilon^0$, we define the current $\hat{\jmath}_\mu(x)$ by

$$\hat{\jmath}_\mu(x)\equiv\frac{1}{2}\lim_{\varepsilon\to 0}[\hat{\jmath}_\mu(x;\ \varepsilon)+\hat{\jmath}_\mu(x;\ \tilde{\varepsilon})]\qquad(2.5.3-33)$$

with

$$\hat{\jmath}_\mu(x;\ \varepsilon)\equiv\frac{1}{2}\sum_{\alpha=1}^{2}\left\{[\hat{\psi}^\dagger(x+\varepsilon)\gamma^0\gamma_\mu]_\alpha\hat{\psi}_\alpha(x)-\hat{\psi}_\alpha(x)[\hat{\psi}^\dagger(x-\varepsilon)\gamma^0\gamma_\mu]_\alpha\right\}.$$
$$(2.5.3-34)$$

Here, from Eq.(32) we have

$$\hat{\jmath}_\mu(x;\ \varepsilon)=(2\pi)^{-1}(\varepsilon^2)^{-1}\sum_{\alpha=1}^{2}(-1)^{(\alpha-1)\mu}[\varepsilon^0+(-1)^\alpha\varepsilon^1]$$
$$\cdot(-\sqrt{\pi})\varepsilon^\nu[\partial_\nu\phi(x)+(-1)^\alpha\epsilon_{\nu\sigma}\partial^\sigma\phi(x)+O(\varepsilon^2)].\ (2.5.3-35)$$

Then we find

$$\hat{\jmath}_\mu(x)=-(1/\sqrt{\pi})\partial_\mu\phi(x),\qquad(2.5.3-36)$$

just as in Eq.(15).

From the above analysis, it is tempting to identify $\hat{\psi}$ with ψ. There are, however, several differences between them.

1. While ψ is a local field, $\hat{\psi}$ is a *non-local* one. Locality plays an important role in symmetry properties; for instance, the ψ theory has a (fermionic) symmetry generator

$$Q_u\equiv\int_{-\infty}^{+\infty}dx^1\,(\psi^\dagger u+u^\dagger\psi),\qquad(2.5.3-37)$$

but the corresponding generator does *not* exist in the ϕ theory because of the non-locality of $\hat{\psi}$.

2. Because of the Pauli principle, we understand the non-existence of $(\psi_\alpha)^{-1}$, but it is easy to write down $(\hat{\psi}_\alpha)^{-1}$ explicitly (in the form of anti-normal ordered product). In the latter case, the consistency with the Pauli principle can be assured only by the non-associativity of singular products.

3. $\hat{\psi}$ contains a contribution from the negative-norm infrared mode of ϕ. Moreover, it explicitly depends on the infrared cutoff parameter μ in its normalization.

4. We, of course, have $\langle 0 | \psi | 0 \rangle = 0$, but

$$\langle 0 | \hat{\psi}(x) | 0 \rangle = u \neq 0. \qquad (2.5.3-38)$$

In the ψ theory the phase invariance is not broken, but in the ϕ theory the corresponding transformation $\phi \rightarrow \phi' = \phi + \text{const}$ is necessarily spontaneously broken.

5. Under the Lorentz transformation, ψ transforms as

$$[iM_{01}, \psi(x)] = (x_0\partial_1 - x_1\partial_0 - \frac{1}{2}\gamma^5)\psi(x) \qquad (2.5.3-39)$$

as usual, while $\hat{\psi}$ transforms as

$$\begin{aligned}[iM_{01}, \hat{\psi}(x)] =&(x_0\partial_1 - x_1\partial_0)\hat{\psi}(x) \\ &+ (1/\sqrt{\pi})\gamma^5[\hat{\psi}(x)\Phi^{(+)} - \Phi^{(-)}\hat{\psi}(x)], \qquad (2.5.3-40)\end{aligned}$$

as is shown by using Eqs.(2.5.2-39) and (2.5.2-41) [Nak 77e]. If $\hat{\psi}$ transformed like Eq.(39), we would encounter a contradiction with Eq.(38).

6. ψ has *two* independent (field) degrees of freedom, but ϕ or $\hat{\psi}$ has only one. In other words, ψ is, so to speak, "charged", while ϕ or $\hat{\psi}$ is "neutral". Accordingly, if $\psi = \hat{\psi}$ held strictly, then, expressing ϕ in terms of ψ_2, one could express ψ_1 in terms of ψ_2, in contradiction to the independence of ψ_1 and ψ_2.

If one dare to construct a quantity having all properties of ψ from the ϕ theory, it is inevitable to introduce the so-called "spurion operators", which are constant operators *inexpressible in terms of ϕ*, thus *violating the irreducibility of the operator algebra.*

Finally, we briefly mention about the **Thirring model** [Thi 58, Joh 61b]. Its Lagrangian density is given by

$$\mathcal{L}_{\text{Th}} = i\bar{\psi}\gamma^\mu\partial_\mu\psi + g\bar{\psi}\gamma^\mu\psi \cdot \bar{\psi}\gamma_\mu\psi. \qquad (2.5.3-41)$$

This model is known to be exactly solvable. A complete set of the Wightman functions was constructed by Klaiber [Kla 68]. His construction is based on the free massless Dirac field. The Thirring field ψ, however, has *no* discrete spectrum.[12] What has a discrete spectrum is the current j_μ, which can be expressed as $\text{const}\partial_\mu\phi$. Thus *the asymptotic field is a massless scalar field* ϕ. From the requirement of asymptotic completeness, therefore, we should construct ψ out of ϕ. This can be done explicitly [Nak 77b, Nak 77c]; we have

$$\psi(x) = Z^{1/2} : \exp\left[ia\phi(x) - ib\gamma^5\tilde\phi(x)\right] : u \qquad (2.5.3-42)$$

where Z is a (divergent) renormalization constant, and parameters a and b are determined by

$$ab = \pi, \quad -a + b = (g/2\pi)(a + b) \qquad (2.5.3-43)$$

for $|g| < 2\pi$.[13]

Since, as mentioned above,

$$j_\mu = -(2\pi)^{-1}(a + b)\partial_\mu\phi, \qquad (2.5.3-44)$$

the generator of phase transformations is proportional to Φ defined by Eq.(2.5.2-26). Accordingly, it is *spontaneously broken* in contradiction to the **Coleman theorem** [Col 73], which states that there is no spontaneous breakdown of continuous symmetry in two dimensions. This theorem is, however, based on the impossibility of realizing the two-dimensional free massless scalar field in the *positive-metric* Hilbert space. *It is no longer valid in the indefinite-metric Hilbert space*, as emphasized by Nakanishi [Nak 77b].

Finally, we note that it is possible to extend the above consideration to the $U(N)$-symmetric Thirring model [Mit 78].

2.5.4 Schwinger model

The **Schwinger model** is the two-dimensional massless quantum electrodynamics. It is exactly solvable and physically equivalent to the two-dimensional free massive vector field theory, as found by Schwinger [Sch 62b]. As in the Thirring model, the Dirac field ψ has no discrete spectrum,[14] that is, ψ is *confined*. The

---

[12]   The two-point function behaves like $\gamma^\mu p_\mu(p^2 + i0)^{-1+\alpha}$ with $\alpha \equiv g^2/(4\pi^2 - g^2)$ near $p^2 = 0$.

[13]   For $|g| > 2\pi$, we need a negative-norm scalar field.

[14]   The two-point function behaves like $\gamma^\mu p_\mu(p^2 + i0)^{-5/4}$ near $p^2 = 0$

physical part of the gauge field A_μ becomes *massive*; indeed, one can describe this phenomenon as the dynamical Higgs mechanism [Ito 75, Nak 75b].

An operator solution to the Schwinger model was constructed first by Lowenstein and Swieca [Low 71]. But their solution is constructed by making use of a free massless Dirac field.[15] Since ψ has no discrete spectrum, the Lowenstein-Swieca solution does not meet the requirement of asymptotic completeness. In what follows, we present an operator solution expressed in terms of asymptotic fields only, constructed by Nakanishi [Nak 77b, Nak 77c].[16] This solution exists for any linear covariant gauge in contrast with the Lowenstein-Swieca one which exists only in the Landau gauge.

The Lagrangian density of the Schwinger model is

$$\mathcal{L}_{\text{Schw}} = -\frac{1}{4}F^{\mu\nu}F_{\mu\nu} + B\partial^\mu A_\mu + \frac{1}{2}\alpha B^2 + \bar\psi\gamma^\mu(i\partial_\mu + eA_\mu)\psi \qquad (2.5.4-1)$$

with $F_{\mu\nu} \equiv \partial_\mu A_\nu - \partial_\nu A_\mu$, just as in quantum electrodynamics. The field equations are

$$\gamma^\mu(i\partial_\mu + eA_\mu)\psi = 0, \qquad (2.5.4-2)$$

$$\partial_\mu F^{\mu\nu} - \partial^\nu B = -ej^\nu, \qquad (2.5.4-3)$$

$$\partial^\mu A_\mu + \alpha B = 0. \qquad (2.5.4-4)$$

As in Sec.2.3.1, the current conservation implies

$$\Box B = 0. \qquad (2.5.4-5)$$

Canonical quantization can be carried out in the same way as in Sec.2.3.2 ($\pi^0 = B$, $\pi^1 = F_{01}$).

The asymptotic fields are as follows. Since it is known (see [Sch 62b]) that A_μ has a discrete spectrum at

$$M^2 \equiv e^2/\pi, \qquad (2.5.4-6)$$

there should exist a massive vector asymptotic field U_μ. Since B satisfies a free-field equation, Eq.(5), B itself must be an asymptotic field. Furthermore, since B

---

[15]   Then, one is led to the existence of spurions, which implies the so-called "θ-vacua". But closer investigation shows that they are not truly vacua [Nak 77g, Abe 78a, Abe 78b, Nakw 84a, Nakw 84b].

[16]   Subsequently, various operator solutions have been proposed.

is a zero-norm field, there must exist another massless asymptotic field, which we denote by X. Those three asymptotic fields satisfy the following field equations and commutation relations:

$$(\Box + M^2)U_\mu = 0, \qquad \partial_\mu U^\mu = 0, \qquad (2.5.4-7)$$

$$\Box B = 0, \quad \Box X = -\alpha M B; \qquad (2.5.4-8)$$

$$[U_\mu(x),\, U_\nu(y)] = i\left(-\eta_{\mu\nu} - M^{-2}\partial_\mu\partial_\nu\right)\Delta_2(x-y;\, M^2), \qquad (2.5.4-9)$$

$$[X(x),\, B(y)] = -iM D_2(x-y), \qquad (2.5.4-10)$$

$$[X(x),\, X(y)] = iD_2(x-y) + i\alpha M^2 E_2(x-y), \qquad (2.5.4-11)$$

and the other commutators vanish. Although Δ_2, D_2, and E_2 are the two-dimensional ones defined in Sec.2.5.1, one should note that the properties of these asymptotic fields are the same as those in the Higgs model discussed in Sec.2.4.3.

The vector field U_μ is equivalent to a (pseudo)scalar field $\tilde{U}$, defined by

$$\tilde{U} \equiv M^{-1}\epsilon^{\mu\nu}\partial_\mu U_\nu. \qquad (2.5.4-12)$$

Indeed, because of Eq.(7), we have

$$U_\mu = -M^{-1}\epsilon_{\mu\nu}\partial^\nu\tilde{U}, \qquad (2.5.4-13)$$

whence

$$\partial_\mu\tilde{U} = -M\epsilon_{\mu\nu}U^\nu. \qquad (2.5.4-14)$$

It is easy to see that

$$(\Box + M^2)\tilde{U} = 0, \qquad (2.5.4-15)$$

$$[\tilde{U}(x),\, \tilde{U}(y)] = i\Delta_2(x-y;\, M^2). \qquad (2.5.4-16)$$

All the above properties can be derived from a Lagrangian density

$$\mathcal{L}^{\mathrm{as}} \equiv \frac{1}{2}\partial^\mu\tilde{U}\cdot\partial_\mu\tilde{U} - \frac{1}{2}M^2\tilde{U}^2 - M^{-1}\partial^\mu X\cdot\partial_\mu B$$
$$- \frac{1}{2}M^{-2}\partial^\mu B\cdot\partial_\mu B + \frac{1}{2}\alpha B^2. \qquad (2.5.4-17)$$

From Eq.(17), we can easily write down the expressions for the Poincaré generators P_μ and M_{01}.

Since B satisfies the d'Alembert equation, we can define its conjugate field by

$$\tilde{B}(x) \equiv \int_{-\infty}^{x^1} dz^1\, \partial_0 B(x^0,\, z^1) \qquad (2.5.4-18)$$

according to Eq.(2.5.2-17). On the other hand, since X is a dipole-ghost field for $\alpha \neq 0$, its conjugate field is definable only in the Landau gauge.[17]

The positive/negative frequency part of X is defined by

$$X^{(\pm)}(x) \equiv i \int_{-\infty}^{+\infty} dz^1 \Big[D_2^{(\pm)}(x-z)\partial_0 X(z) - \partial_0^{\,z} D_2^{(\pm)}(x-z) \cdot X(z)$$
$$+ E_2^{(\pm)}(x-z)\partial_0 \Box X(x) - \partial_0^{\,z} E_2^{(\pm)}(x-z) \cdot \Box X(z) \Big]. \quad (2.5.4-19)$$

We can then show that

$$[\, X^{(\pm)}(x),\, B^{(\mp)}(y)\,] = -M D_2^{(\pm)}(x-y), \qquad (2.5.4-20)$$

$$[\, X^{(\pm)}(x),\, B^{(\pm)}(y)\,] = 0, \qquad (2.5.4-21)$$

$$[\, X^{(\pm)}(x),\, \tilde{B}^{(\mp)}(y)\,] = -M \left[\tilde{D}_2^{(\pm)}(x-y) + i/8 \right], \qquad (2.5.4-22)$$

$$[\, X^{(\pm)}(x),\, \tilde{B}^{(\pm)}(y)\,] = -iM/8, \qquad (2.5.4-23)$$

$$[\, X^{(\pm)}(x),\, X^{(\mp)}(y)\,] = D_2^{(\pm)}(x-y) + \alpha M^2 E_2^{(\pm)}(x-y), \quad (2.5.4-24)$$

$$[\, X^{(\pm)}(x),\, X^{(\pm)}(y)\,] = 0, \qquad (2.5.4-25)$$

While $B^{(+)}$, $B^{(-)}$, $\tilde{B}^{(+)}$ and $\tilde{B}^{(-)}$ commute with each other. Of course, $\tilde{U}^{(\pm)}$ commutes with anyone other than $\tilde{U}^{(\mp)}$.

As in Sec.2.5.2, we introduce

$$Q \equiv \int_{-\infty}^{+\infty} dx^1\, \partial_0 B(x) = Q^{(+)} + Q^{(-)}, \qquad (2.5.4-26)$$

$$Q^{(\pm)} \equiv \int_{-\infty}^{+\infty} dx^1\, \partial_0 B^{(\pm)}(x) = [Q^{(\mp)}]^\dagger. \qquad (2.5.4-27)$$

The above commutation relations imply that $Q^{(\pm)}$ commutes with all but $X^{(\mp)}$;

$$[\, iQ^{(\pm)},\, X^{(\mp)}(x)\,] = -M/2. \qquad (2.5.4-28)$$

---

[17]   This is the reason why the Lowenstein-Swieca solution, which needs $\tilde{X}$, exists only in the Landau gauge.

The vacuum $|0\rangle$ is defined by

$$\tilde{U}^{(+)}(x)|0\rangle = B^{(+)}(x)|0\rangle = X^{(+)}(x)|0\rangle = 0, \quad \langle 0|0\rangle = 1. \qquad (2.5.4-29)$$

As in Eq.(2.5.2-31), we have

$$\tilde{B}^{(+)}(x)|0\rangle = (1/4)Q^{(-)}|0\rangle. \qquad (2.5.4-30)$$

Now, we present the operator solution to the Schwinger model which meets the requirement of asymptotic completeness:

$$\psi(x) \equiv Z^{1/2} : \exp\left\{i\sqrt{\pi}\left[\gamma^5\tilde{U}(x) + M^{-1}\gamma^5\tilde{B}(x) + X(x)\right]\right\} : u, \quad (2.5.4-31)$$
$$A_\mu(x) \equiv U_\mu(x) + M^{-2}\partial_\mu B(x) + M^{-1}\partial_\mu X(x), \qquad (2.5.4-32)$$

where u is the constant two-component quantity defined in Sec.2.5.3 ($|u_\alpha|^2 = \mu/2\pi$, $\mu = e^\gamma\kappa$) and the renormalization constant Z is given by $Z \equiv \sqrt{M/2\kappa}$. Note the similarity between Eq.(32) and Eq.(2.4.3-35) of Higgs model.

The quantities corresponding to Eq.(2.5.3-21) and to Eq.(2.5.3-25) are

$$M_{\alpha\beta}(z) \equiv i\pi\left\{(-1)^{\alpha+\beta}\Delta_2(z; M^2) + D_2(z) + \alpha M^2 E_2(z)\right.$$
$$\left. - (-1)^\alpha\left[\tilde{D}_2(z) - 1/2\right] - (-1)^\beta\left[\tilde{D}_2(z) + 1/2\right]\right\}, \quad (2.5.4-33)$$

$$M^{(+)}{}_{\alpha\beta}(z) \equiv \pi\left\{(-1)^{\alpha+\beta}\Delta_2^{(+)}(z; M^2) + D_2^{(+)}(z) + \alpha M^2 E_2^{(+)}(z)\right.$$
$$\left. - (-1)^\alpha\left[\tilde{D}_2^{(+)}(z) - i/4\right] - (-1)^\beta\left[\tilde{D}_2^{(+)}(z) + i/4\right]\right\}, \quad (2.5.4-34)$$

respectively. From Eq.(33) we see that ψ and $\psi^\dagger$ satisfy local anticommutativity.

By direct computation, we obtain

$$[\psi(x), B(y)] = \sqrt{\pi}M\psi(x)D_2(x - y), \qquad (2.5.4-35)$$

$$[A_\mu(x), \psi(y)] = \sqrt{\pi}\left\{M^{-1}\epsilon_{\mu\nu}\partial^\nu[\Delta_2(x - y; M^2) - D_2(x - y)]\gamma^5\right.$$
$$\left. - \alpha M\partial_\mu E_2(x - y)\right\}\psi(y), \qquad (2.5.4-36)$$

$$[A_\mu(x), A_\nu(y)] = i\left(-\eta_{\mu\nu} - M^{-2}\partial_\mu\partial_\nu\right)\Delta_2(x - y; M^2)$$
$$+ iM^{-2}\partial_\mu\partial_\nu D_2(x - y) - i\alpha\partial_\mu\partial_\nu E_2(x - y), \quad (2.5.4-37)$$

$$[A_\mu(x), B(y)] = -i\partial_\mu D_2(x - y), \qquad (2.5.4-38)$$

and $[B, B] = 0$. Thus those commutators vanish for $(x - y)^2 < 0$, that is, local commutativity is satisfied.

Since the Schwinger model is a gauge theory, the definition of the point-split current should contain a phase factor of a line integral over A_μ so as to be gauge invariant [Sch 51], that is,

$$j_\mu(x; \varepsilon) \equiv \sum_{\alpha=1}^{2} \frac{1}{2} \Big\{ [\bar\psi(x + \varepsilon)\gamma_\mu]_\alpha \psi_\alpha(x) \exp \Big[ie \int_x^{x+\varepsilon} dz^\lambda A_\lambda(z) \Big]$$
$$- \psi_\alpha(x)[\bar\psi(x - \varepsilon)\gamma_\mu]_\alpha \exp \Big[-ie \int_{x-\varepsilon}^x dz^\lambda A_\lambda(z) \Big] \Big\}. \quad (2.5.4 - 39)$$

As in Sec.2.5.3, we calculate the $\varepsilon^\lambda \to 0$ limit of $j_\mu(x; \varepsilon)$, which turns out to be independent of the direction of ε^λ, with the aid of

$$\Delta_2^{(+)}(\varepsilon; M^2) \cong -(4\pi)^{-1} [\log(-\varepsilon^2 + i0\varepsilon^0) + 2(\gamma + \log M/2)]. \quad (2.5.4 - 40)$$

We then obtain

$$ej_\mu(x) \equiv e \lim_{\varepsilon \to 0} j_\mu(x; \varepsilon) = M^2 U_\mu(x) + \partial_\mu B(x)$$
$$= -M \partial_\mu X(x) + M^2 A_\mu(x). \quad (2.5.4 - 41)$$

From Eqs.(31) and (32), it is easy to show that the field equation, Eq.(2), is satisfied in the sense of

$$-i\gamma^\mu \partial_\mu \psi(x) = e\gamma^\mu : A_\mu(x)\psi(x) :, \quad (2.5.4 - 42)$$

whose rhs. may be rewritten as

$$e \cdot \frac{1}{2} \lim_{\varepsilon \to 0} \gamma^\mu [A_\mu(x + \varepsilon)\psi(x) + \psi(x)A_\mu(x - \varepsilon)]. \quad (2.5.4 - 43)$$

From Eqs.(32) and (41), we see that the field equations, Eqs.(3) and (4), are satisfied.

It can be shown in the same way as in Sec.2.5.3 that the canonical anticommutation relations for ψ and $\psi^\dagger$ are satisfied. The other canonical commutation relations follow from Eqs.(35)–(38). Thus the set $\{\psi(x), A_\mu(x), B(x)\}$ constructed above is certainly the solution to the Schwinger model.

Since $F_{01} = -M\tilde{U}$ is massive, the charge $e \int dx^1 j_0(x)$ equals to Q defined by Eq.(26). Hence

$$\langle 0 | [iQ, X] | 0 \rangle = -M \neq 0 \quad (2.5.4 - 44)$$

implies that invariance under phase transformations is *spontaneously broken* and that $X(x)$ is the NG boson.

The local gauge transformation is generated by

$$Q_\Lambda \equiv - \int_{-\infty}^{+\infty} dx^1 \left[\Lambda(x)\partial_0 B(x) - \partial_0 \Lambda \cdot B(x) \right], \qquad (2.5.4 - 45)$$

where Λ satisfies $\Box \Lambda = 0$. Under Q_Λ, only X transforms into $X + M\Lambda$. Hence ψ, A_μ, and B correctly transform under the local gauge transformation [see Eqs.(2.3.3-17)–(2.3.3-19)]. Of course, $j_\mu(x)$ is gauge invariant.

The Lagrangian density, Eq.(1), is invariant under the chiral transformation $\psi \rightarrow \psi' = e^{-i\gamma^5 \theta}\psi$, but the corresponding current $\epsilon^{\mu\nu} j_\nu(x)$ is not conserved. The conserved chiral current, which is gauge non-invariant but chiral-gauge invariant, is given by

$$j_5{}^\mu(x) \equiv \epsilon^{\mu\nu} j_\nu(x) - (M/\sqrt{\pi})\epsilon^{\mu\nu} A_\nu(x). \qquad (2.5.4 - 46)$$

thus $\partial_\mu(\epsilon^{\mu\nu} j_\nu)$ equals $-(e/\pi)F_{01}$, which is called **chiral anomaly**. Since Eq.(46) is rewritten as

$$j_5{}^\mu(x) = -(1/\sqrt{\pi})\epsilon^{\mu\nu}\partial_\nu X(x), \qquad (2.5.4 - 47)$$

the chiral charge

$$Q_5 \equiv e \int dx^1 j_5{}^0(x) = M \int_{-\infty}^{+\infty} dx^1 \partial_1 X(x) \qquad (2.5.4 - 48)$$

commutes with all asymptotic fields. Thus Q_5 is *not* a sensible quantity. The non-existence of Q_5 is consistent with the violation of chiral symmetry by anomaly.

The Poincaré generators P_μ and M_{01} of the Schwinger model can be easily constructed from $\mathcal{L}^{as}$. We then have

$$[iP_\mu, \psi(x)] = \partial_\mu \psi(x), \qquad (2.5.4 - 49)$$

$$[iM_{01}, \psi(x)] = (x_0 \partial_1 - x_1 \partial_0)\psi(x)$$
$$- (1/\sqrt{\pi})M^{-1}\gamma^5 \left[\psi(x)Q^{(+)} - Q^{(-)}\psi(x) \right] \qquad (2.5.4 - 50)$$

as in Eq.(2.5.3-40). The covariance of the field equations holds in the sense of the above transformation.

The physical states $|f\rangle$ are defined by the subsidiary condition

$$B^{(+)}(x)|f\rangle = 0 \qquad (2.5.4 - 51)$$

as in the ordinary quantum electrodynamics. Then we automatically have $Q^{(+)} | f \rangle = 0$. The physical subspace $\mathcal{V}_{\text{phys}}$ is Poincaré invariant and has positive semi-definite metric. Any Fock state containing $X^{(-)}$ is unphysical. The zero-norm subspace of $\mathcal{V}_{\text{phys}}$ is $B^{(-)}\mathcal{V}_{\text{phys}}$. Thus the gauge-invariant algebra is physically equivalent to a massive vector or (pseudo)scalar field theory.

The $(n + m)$-point function is given by

$$
\langle 0 | \prod_{j=1}^{n} \psi_{\alpha_j}(x^{(j)}) \prod_{k=1}^{m} \psi_{\beta_k}{}^{\dagger}(y^{(k)}) | 0 \rangle
$$

$$
= Z^{(n+m)/2} \prod_{j=1}^{n} u_{\alpha_j} \prod_{k=1}^{m} u_{\beta_k}{}^{*} \exp\left[- \sum_{1 \leq j < k \leq n} M^{(+)}{}_{\alpha_j \alpha_k}(x^{(j)} - x^{(k)})\right.
$$

$$
\left. - \sum_{1 \leq j < k \leq m} M^{(+)}{}_{\beta_j \beta_k}(y^{(j)} - y^{(k)}) + \sum_{j=1}^{n}\sum_{k=1}^{m} M^{(+)}{}_{\alpha_j \alpha_k}(x^{(j)} - y^{(k)})\right], \quad (2.5.4 - 52)
$$

where $M^{(+)}{}_{\alpha\beta}(z)$ is defined by Eq.(34). It is remarkable that *it does not vanish for any values of n and m*. This fact is the manifestation of the spontaneous breakdown of the phase invariance. Also, cluster property is violated; this fact corresponds to the confinement of ψ.

Finally, we remark that the above formalism can be extended to the **chiral Schwinger model** [Miy 87] and to the **generalized chiral Schwinger model** [Miy 88], in which A_μ couples with ψ in a left-right asymmetric way. In those models, although the anomaly violates gauge invariance, it is shown that the trouble is overcome by adding a **Wess-Zumino term**, which involves a new scalar field θ, to the Lagrangian density. The operator solution is constructed quite satisfactorily. The non-free component of ψ is confined in contrast to the result in the unitary-gauge formulation, which violates the canonical nature of the theory [Miy 89].

2.5.5 Pre-Schwinger model

The **pre-Schwinger model** is the Schwinger model in which A_μ has a bare mass m. It is exactly slvable [Thi 64, Dub 67, Low 71]. The operator solution which meets the requirement of asymptotic completeness is easily obtained by extending the consideration made in Sec.2.5.4 [Nak 77f].

The Lagrangian density of the pre-Schwinger model is obtained by adding

$\frac{1}{2}m^2 A^\mu A_\mu$ to Eq.(2.5.4-1). Its structure of asymptotic fields is essentially the same as in the pre-Higgs model (see Sec.2.4.4). That is, asymptotic fields are a massive vector field U_μ having a mass squared

$$M_r{}^2 \equiv m^2 + e^2/\pi, \tag{2.5.5 - 1}$$

the B field having a mass squared αm^2, which has negative norm, and a massless scalar field ϕ described in Sec.2.5.2.

The operator solution to the pre-Schwinger model is then given by

$$\psi(x) \equiv Z^{1/2} : \exp\left[i\frac{e}{M_r}\gamma^5\tilde{U}(x) + i\frac{e}{m^2}B(x)\right.$$
$$\left. + i\sqrt{\pi}\frac{M_r}{m}\phi(x) - i\sqrt{\pi}\frac{m}{M_r}\gamma^5\tilde{\phi}(x)\right] : u, \tag{2.5.5 - 2}$$

$$A_\mu(x) \equiv U_\mu(x) + m^{-2}\partial_\mu B(x) + (e/\sqrt{\pi}mM_r)\partial_\mu\phi(x), \tag{2.5.5 - 3}$$

where $\tilde{U} \equiv M_r{}^{-1}\epsilon^{\mu\nu}\partial_\mu U_\nu$, $|u_\alpha|^2 = \mu/2\pi$ ($\mu = e^\gamma\kappa$), and

$$Z \equiv \exp\left(-\frac{e^2}{2\pi M_r{}^2}\log\frac{M_r}{2\kappa} + \frac{e^2}{2\pi m^2}\log\frac{\sqrt{\alpha}m}{2\kappa}\right). \tag{2.5.5 - 4}$$

The current j_μ, which is defined as in the Schwinger model, is

$$ej_\mu(x) = \frac{e^2}{\pi}U_\mu(x) - \frac{em}{\sqrt{\pi}M_r}\partial_\mu\phi(x). \tag{2.5.5 - 5}$$

Hence the charge is

$$Q \equiv e\int_{-\infty}^{+\infty} dx^1\, j_0(x) = -\frac{em}{\sqrt{\pi}M_r}\Phi, \tag{2.5.5 - 6}$$

where Φ is defined by Eq.(2.5.2-26). Since

$$\langle 0\,|\,[\,iQ,\ \phi(x)\,]\,|\,0\,\rangle = -em/\sqrt{\pi}M_r \neq 0, \tag{2.5.5 - 7}$$

the phase invariance is spontaneously broken and the corresponding NG boson is $\phi(x)$.

Lorentz transformation property is similar to the case of the Schwinger model. As for the subsidiary condition, we need $Q^{(+)}|f\rangle = 0$ in addition to $B^{(+)}|f\rangle = 0$.

The pre-Schwinger model tends to the Thirring model (see Sec.2.5.3) as $m \rightarrow \infty$ and $e \rightarrow \infty$ in such a way that e/m remains finite.

What is interesting is the $m \rightarrow 0$ limit. In the same way as in the Higgs limit of the pre-Higgs model (see Sec.2.4.4), the rearrangement of the NG boson arises. Indeed, setting

$$X \equiv M_r m^{-1} \phi + (e/\sqrt{\pi}) m^{-2} B, \qquad (2.5.5-8)$$

which satisfies

$$\Box X = -\alpha(e/\sqrt{\pi}) B, \qquad (2.5.5-9)$$

$$[X(x),\ B(y)] = -i(e/\sqrt{\pi}) \Delta_2(x-y;\ \alpha m^2), \qquad (2.5.5-10)$$

$$[X(x),\ X(y)] = im^{-2} \left[M_r{}^2 D_2(x-y) - (e^2/\pi) \Delta_2(x-y;\ \alpha m^2) \right], \quad (2.5.5-11)$$

we can reproduce the operator solution of the Schwinger model, except for the Z factor,[18] as $m \rightarrow 0$.

2.5.6 Schroer model

The vector-coupling theory of the spinor-scalar system is known to be exactly solvable in any dimensions, but only in the two-dimensional case it is free of ultraviolet-divergence difficulty. The **Schroer model** [Schr 63] is the two-dimensional vector-coupling theory of a massive spinor and a massless scalar.

The Lagrangian density of the Schroer model is

$$\mathcal{L}_{\text{Schr}} = \frac{1}{2} \partial^\mu \phi \cdot \partial_\mu \phi + \bar{\psi}(i\gamma^\mu \partial_\mu - m)\psi + e\bar{\psi}\gamma^\mu \psi \partial_\mu \phi. \qquad (2.5.6-1)$$

The field equations are, therefore,

$$(i\gamma^\mu \partial_\mu - m)\psi = -e\gamma^\mu \psi \partial_\mu \phi, \qquad (2.5.6-2)$$

$$\Box \phi = 0 \qquad (2.5.6-3)$$

because of the current conservation.

---

[18]   The calculation of Z involves a limiting procedure, which is not commutable with $m \rightarrow 0$.

Let $\psi^{(0)}$ be an auxiliary free massive spinor field satisfying

$$(i\gamma^\mu \partial_\mu - m)\psi^{(0)} = 0; \qquad (2.5.6-4)$$

then an operator solution is given by

$$\psi(x) \equiv Z^{1/2} : e^{ie\phi(x)} : \psi^{(0)}(x). \qquad (2.5.6-5)$$

By using Eq.(5), one can easily calculate the two-point function $\langle 0 | T\psi(x)\bar{\psi}(y) | 0 \rangle$, and finds that it has no discrete spectrum as long as $e \neq 0$.[19] More generally, one can prove that there is no massive discrete spectrum in this model by using the *positivity* of the Hilbert-space metric. On the basis of this result, Schroer [Schr 63] introduced the notion of **"infraparticle"**, which is a particle which interacts with a *massless* field and loses its discrete spectrum owing to infrared radiation. It was conjectured that a similar thing would take place in quantum electrodynamics (see Addendum 2.A).

We should note, however, that the confinement of a spinor field is quite a common phenomenon in such two-dimensional models as the Thirring model, the Schwinger model, the pre-Schwinger model, etc. Here the free massless scalar field ϕ, which must be quantized with indefinite metric, plays an important role. Schroer's analysis, in which ψ and $\psi^{(0)}$ cannot be represented in a common Hilbert space, should be reexamined from the point of view of *indefinite metric* [Nak 77d]. Although ψ itself has no discrete spectrum, a composite field[20] $e^{-ie\phi}\psi$ does have a discrete spectrum, that is, $\psi^{(0)}$ is an asymptotic field of a composite field. Thus the operator solution expressed in terms of asymptotic fields is exactly the same as Eq.(5), but we must note here that ψ and $\psi^{(0)}$ are represented in a *common indefinite-metric* Hilbert space in contrast with Schroer's treatment.

We define the current $j_\mu(x)$ as in Sec.2.5.3; then we find

$$j_\mu(x) = j^{(0)}_\mu(x) - (e/2\pi)\partial_\mu \phi(x), \qquad (2.5.6-6)$$

where $j^{(0)}_\mu$ denotes the free current of $\psi^{(0)}$. Hence the commutation relations with the generator

$$Q \equiv e \int dx^1 j_0(x) \qquad (2.5.6-7)$$

---

[19]   It behaves like $(\gamma^\mu p_\mu + m)(p^2 - m^2 + i0)^{-1+\alpha}$ with $\alpha \equiv e^2/2\pi$ near $p^2 = m^2$.

[20]   This product is of formal nature as in the Lagrangian density.

are as follows:

$$[iQ, \ \phi(x)] = -e^2/2\pi, \qquad\qquad (2.5.6-8)$$

$$[iQ, \ \psi^{(0)}(x)] = -ie\psi^{(0)}(x), \qquad\qquad (2.5.6-9)$$

$$[iQ, \ \psi(x)] = -ie\left(1 + \frac{e^2}{2\pi}\right)\psi(x). \qquad (2.5.6-10)$$

As is seen from Eq.(8), the phase invariance is spontaneously broken and $\phi(x)$ is the corresponding NG boson. Furthermore, Eqs.(9) and (10) show that ψ transforms differently from $\psi^{(0)}$. This is the reason why $\psi^{(0)}$ is not the asymptotic field of ψ itself.

We can thus conclude that the confinement of ψ is understood as a result of spontaneous breakdown of the phase invariance. We need not introduce such a special notion as "infraparticle" to understand the Schroer model.

Addendum 2.A   ON THE EXISTENCE OF A CHARGED ASYMPTOTIC FIELD

Because of the masslessness of the electromagnetic field A_μ, soft photons in the cloud of a charged particle cause an infrared problem. According to the renormalization group analysis in quantum electrodynamics, which is supported by the exact result calculated in the Bloch-Nordsieck model, the electron propagator behaves like

$$(p^2 - m_r{}^2 + i0)^\beta (\gamma^\mu p_\mu - m_r + i0)^{-1} \qquad (2.A-1)$$

with[1]

$$\beta \equiv e_r{}^2 (\alpha_r - 3)/8\pi^2 \qquad (2.A-2)$$

near $p^2 = m_r{}^2$ [Bog 59]. The above result suggests that the electron's two-point function does *not* have a discrete spectrum at $p^2 = m_r{}^2$ except in the **Yennie gauge** $\alpha_r = 3$. A similar phenomenon is encountered in the exactly solvable Schroer model (see Sec.2.5.6). In general, a particle (called "infraparticle" by Schroer) which interacts with a massless field seems to lose its discrete spectrum. If this is true, we cannot define a charged asymptotic field.

The infrared problem is most seriously taken into account in the coherent-state formalism [Chu 65, Kul 70]. But it does not resolve the trouble of the non-existence of a charged asymptotic field. Moreover, we encounter inconsistency between the subsidiary condition Eq.(2.3.3-11) and the infrared-coherence condition. Although Zwanziger [Zwa 76] abandons the former, his standpoint is not acceptable from the viewpoint of our formalism [Oji 79c].

On the other hand, from the rigorous approach, it can be proved that A_μ *does have its asymptotic field* [Buc 75, Buc 77]. Hence the photon's energy-momentum $P^{\mathrm{ph}}{}_\mu$ can be constructed. One then defines the charged-field energy-momentum by $P^{\mathrm{ch}}{}_\mu \equiv P_\mu - P^{\mathrm{ph}}{}_\mu$ and characterizes the charged asymptotic state by an isolated spectrum of $(P^{\mathrm{ch}})^2$ [Mor 83].

We here take a naiver approach. As discussed in Sec.2.5.6, the absence of a discrete spectrum of ψ in the Schroer model is due to the spontaneous breakdown of the phase invariance. It does not therefore support the absence of the electron's

---

[1]   The expression for β will not receive higher-order corrections [Yen 61]. It is confirmed explicitly that there is no fourth-order correction [Sol 56].

discrete spectrum. It is instructive, however, to remember that a composite operator $e^{-ie\phi}\psi$ does have a discrete spectrum in the Schroer model. We show that a similar thing takes place in quantum electrodynamics if we employ the gaugeon formalism discussed in Sec.2.3.8.

To our electromagnetic field Lagrangian density Eq.(2.3.1-1), we add a gaugeon part:

$$\mathcal{L}_{\mathrm{EMG}} = -\frac{1}{4}F^{\mu\nu}F_{\mu\nu} + B\partial^{\mu}A_{\mu} + \frac{1}{2}\alpha B^2 - \partial^{\mu}G \cdot \partial_{\mu}C - \frac{1}{2}\epsilon C^2. \qquad (2.A-3)$$

By a q-number gauge transformation[2]

$$\psi' = e^{ie\lambda G}\psi, \qquad A'_{\mu} = A_{\mu} + \lambda\partial_{\mu}G,$$
$$B' = B, \quad G' = G, \quad C' = C - \lambda B \qquad (2.A-4)$$

with $\alpha - \epsilon\lambda^2 = 3$, the expressions for the Green's functions are transformed into those in the Yennie gauge, in which β vanishes. We may thus infer that *a composite operator $e^{ie\lambda G}\psi$ has a discrete spectrum*. In this way, we can claim the existence of the charged asymptotic field in the ordinary sense.

[2]   All quantities below are renormalized ones.

CHAPTER THREE

NON-ABELIAN GAUGE THEORIES
AND BRS TRANSFORMATION

To formulate a covariant operator formalism of non-abelian gauge theories, it is crucial to introduce the notion of BRS symmetry. In this chapter, we present it as a quantized version of local gauge invariance.

3.1 HISTORICAL BACKGROUNDS: DIFFERENCE BETWEEN ABELIAN AND NON-ABELIAN GAUGE THEORIES

Nowadays, it has been well established that the elementary-particle physics is described by non-abelian gauge theories. **Quantum chromodynamics (QCD)**, which is an (unbroken) $SU(3)$ non-abelian gauge theory, is a theory of strong interactions; hadrons are believed to be colorless composite particles of a **quark** and an antiquark or of three quarks bound by exchanging gauge bosons called **gluons**. The **electroweak theory**, which is a broken $SU(2) \times U(1)$ non-abelian gauge theory, describes electromagnetic and weak interactions in a unified way. By combining QCD and the electroweak theory, we have the **standard theory**, which is in excellent agreement with recent high-energy experiments. Thus it is an extremely important problem to formulate the non-abelian gauge theory in a satisfactory way. In this and next chapters, we present its covariant operator formalism and discuss its general framework.

In Chapter 2, we have seen that a covariant operator formalism of abelian gauge theory can be neatly formulated with the aid of B-field. What is most crucial in that case is the *free* d'Alembert equation, Eq.(2.3.1-5), of B-field,

$$\Box B = 0, \qquad\qquad (3.1-1)$$

which follows, irrespective of the presence of interactions, from the conservation law, Eq.(2.1.3-11), of the electric current j_μ,

$$0 = e\partial_\mu j^\mu = \Box B. \tag{3.1 - 2}$$

This enables us to decompose B-field into positive and negative frequency (or annihilation and creation operator) parts , $B^{(+)}$ and $B^{(-)}$, owing to which the subsidiary condition, Eq.(2.3.3-11), specifying physical states

$$B^{(+)}(x)|\text{phys}\rangle = 0 \tag{3.1 - 3}$$

can be set up in a consistent manner. [Recall here that a field operator $\varphi(x)$ can be decomposed into positive and negative frequency parts in a relativistic-invariant sense only when it has a timelike and/or lightlike momentum support, which implies that $\varphi(x)$ is a *(generalized) free field* having a c-number commutator according to the Greenberg-Robinson theorem [Dell 61, Gre 62, Rob 62] (see Appendix $A.4$).]

In spite of many attempts to extend this approach to the non-abelian case, none of them were successful, because the B-fields in the non-abelian gauge theories fail to satisfy such a free equation as Eq.(1). This is just an inevitable consequence of the non-linear self-interaction of gauge fields due to non-abelian nature of the theory, as will be seen in the next section. One of the most serious defects affecting the approaches of this sort can be found in the breakdown of unitarity of the (physical) S-matrix as pointed out first by Feynman [Fey 63]: Contrary to the abelian cases, the contributions from the unphysical longitudinal and scalar modes to intermediate states do not exactly cancel out owing to the self-coupling of non-abelian gauge field (and likewise for the gravitational field). Feynman [Fey 63] and later DeWitt [DeW 67] found in the context of perturbation theory that this violation of unitarity can be rephrased as the missing contributions of a pair of massless scalar (vector, in the case of gravitational field) fermions to closed loops in Feynman diagrams. Subsequently, a clear-cut explanation for the appearance of these fermions with strange statistics was given by Faddeev and Popov [Fad 67, Fad 69] from the viewpoint of path-integral formalism, and, since then, these "fictitious" particles have come to be called **Faddeev-Popov (FP) ghosts**.

These circumstances clearly show that we should not stick to pursuing superficial formal analogies, but should find the essential meaning of Eqs.(1) and (3) in such a form that the extension to the general cases including non-abelian gauge theories becomes transparent. For this purpose, it is very important to note the role of B-field as a *generator of the local gauge transformation* mentioned in Eq.(2.3.3-9). Namely, the subsidiary condition Eq.(3) represents essentially the requirement that local gauge invariance lost through the quantization procedure should be preserved in the physical subspace specified by Eq.(3). Such essence of local gauge invariance at the quantum level is inherited by Green's functions in the form of the well-known Ward-Takahashi identities following from Eq.(1). In view of the indispensability of FP ghosts in the non-abelian cases, the corresponding Ward-Takahashi identities should necessarily involve these unphysical ghosts. Such identities were found by Slavnov [Sla 72] and Taylor [Tay 71], and hence, sometimes are called **Slavnov-Taylor identities**. It is remarkable that these complicated Slavnov-Taylor identities were shown to follow from just one special type of global symmetry transformation involving FP ghost in an essential way; it is **Becchi-Rouet-Stora (BRS) transformation** (or sometimes called **BRST transformation**) found by Becchi, Rouet and Stora [Bec 76a] (and independently by Tyutin [Tyu unp]). This transformation can properly be viewed as a quantized version of local gauge transformation, on the basis of which the covariant operator formalism of non-abelian gauge fields and general relativity will be formulated in the rest of this book. The notion of BRS transformation has turned out to have very universal importance and play crucial roles in any type of gauge theories. To be explicit, however, we explain it first in the most familiar case of non-abelian gauge theory of Yang-Mills type starting from the very beginning.

3.2 LOCAL GAUGE INVARIANCE IN CLASSICAL YANG-MILLS THEORY

At first sight, FP ghosts may look awkward artifacts introduced from outside just for quantization at the sacrifice of geometric beauty of gauge theory at the classical level. As will become clear through the following developments, however, this is not the case; they play a fundamental role in the BRS transformation, which is crucial for the whole theory of quantized non-abelian gauge fields. The natural and intrinsic essence of BRS transformation and of FP ghosts can be revealed by clarifying their geometric meaning as well as their relation with the constraint conditions following from local gauge invariance. For this reason, we present first in this section the non-abelian gauge theory at the classical level, paying attention to its geometric aspects. While the abelian gauge theory described in Chapter 2 is characterized as the theory with local $U(1)$ symmetry, a non-abelian gauge theory is obtained by replacing the abelian $U(1)$ group with a non-abelian Lie group G in the following way.

3.2.1 Local gauge transformations and covariant derivatives

Let G and $LieG$ be a finite dimensional Lie group and its associated Lie algebra, respectively. As usual, we denote the Lie bracket of $LieG$ by $[\,\cdot\,,\,\cdot\,]$ which defines a $LieG$-valued bilinear mapping on $LieG$ satisfying

$$[X,\ Y] = -[Y,\ X], \qquad\qquad \text{(antisymmetry)} \quad (3.2.1-1)$$

$$[[X,Y],Z] + [[Y,Z],X] + [[Z,X],Y] = 0. \text{ (Jacobi identity)} \quad (3.2.1-2)$$

When we choose a basis $\{X_1,\cdots,X_n\}$ of $LieG$, its structure is characterized by the structure constants $f^c{}_{ab}$,

$$[X_a,\ X_b] = f^c{}_{ab}X_c, \qquad\qquad (3.2.1-3)$$

which satisfy the relations

$$f^c{}_{ab} = -f^c{}_{ba}, \qquad\qquad (3.2.1-4)$$

$$f^d{}_{ab}f^e{}_{dc} + f^d{}_{bc}f^e{}_{da} + f^d{}_{ca}f^e{}_{db} = 0, \qquad\qquad (3.2.1-5)$$

corresponding to Eqs.(1) and (2). This Jacobi identity, Eq.(2) or (5), guarantees that the linear map ad defined on $LieG$ by

$$(ad(X))(Y) \equiv [X, \ Y] \quad \text{for } Y \in LieG \qquad (3.2.1-6)$$

gives the **adjoint representation** of Lie algebra $LieG$, representing $LieG$ in itself regarded as a *vector space* :

$$ad([X, \ Y]) = [ad(X), \ ad(Y)]. \qquad (3.2.1-7)$$

This can be easily checked as follows :

$$
\begin{aligned}
ad([X, \ Y])(Z) &= [[X, \ Y], \ Z] \\
&= [X, \ [Y, \ Z]] - [Y, \ [X, \ Z]] \\
&= ad(X)\Big((ad(Y))(Z)\Big) - ad(Y)\Big((ad(X))(Z)\Big) \\
&= \Big[ad(X), \ ad(Y)\Big](Z). \qquad (3.2.1-8)
\end{aligned}
$$

In view of Eq.(3), the matrix elements of this representation in the above basis $\{X_a\}_{1 \leq a \leq n}$ are given by the structure constants f^c_{ab} :

$$ad(X_a)^c{}_b = f^c_{ab}. \qquad (3.2.1-9)$$

Now a symmetry of the theory under the group G is extended to the local gauge invariance (or, gauge invariance of the second kind) by introducing a **nonabelian gauge field** $A_\mu{}^a$ associated with G, which is usually called a **Yang-Mills field** if G is a *compact* group of an internal symmetry. From the viewpoint of differential geometry, it is nothing but a **connection 1-form** on the spacetime M taking its value in the Lie algebra $LieG$:[1]

$$\omega_x = g\mathcal{A}_\mu(x)dx^\mu = X_a g A_\mu{}^a(x)dx^\mu, \qquad (3.2.1-10)$$

where g is the coupling constant intervening between geometric context (connection 1-form ω) and physical one (Yang-Mills field $A_\mu{}^a$). If the theory involves

---

[1]   As for some basic differential-geometric notions, such as differential forms, exterior differential, and so on, see the brief explanation given in Appendix A.1. To avoid the unnecessary mathematical technicalities, we refrain from presenting the notion of connection in its general form based upon the principal bundles, just restricting the case of spacetime M being flat Minkowskian.

a matter field $\varphi^\alpha(x)$ transforming under G according to a representation ρ, the local gauge invariance is attained by replacing the differential operator ∂_μ and the exterior differential $d = dx^\mu \partial_\mu$ acting on φ^α, respectively, with the **covariant differentiation** ∇_μ and the **covariant exterior differential** $\tilde{\nabla}$ given by

$$
\begin{aligned}
\nabla_\mu &= \partial_\mu + \bar\rho(g\mathcal{A}_\mu) = \partial_\mu + g\bar\rho(X_a)A_\mu{}^a \\
&= \partial_\mu - ig\tau_a A_\mu{}^a,
\end{aligned}
\tag{3.2.1 − 11}
$$

$$
\begin{aligned}
\tilde{\nabla} &= d + \bar\rho(\omega) = dx^\mu(\partial_\mu + \bar\rho(g\mathcal{A}_\mu)) \\
&= dx^\mu \nabla_\mu.
\end{aligned}
\tag{3.2.1 − 12}
$$

Here $\bar\rho$ is the representation of Lie algebra $LieG$ derived from from ρ (differential representation),

$$
\rho(\exp[\alpha^a X_a]) = \exp[\alpha^a \bar\rho(X_a)],
\tag{3.2.1 − 13}
$$

where exp is the exponential mapping from (a neighborhood of 0 in) $LieG$ to (a neighborhood of the unit e in) G. $\tau_a \equiv i\bar\rho(X_a)$ are generators of $LieG$ in the "hermitian" matrix form (if ρ is a *unitary* representation) which is commonly used in physics :

$$
[\tau_a, \tau_b] = if^c{}_{ab}\tau_c.
\tag{3.2.1 − 14}
$$

The field strength, $\mathcal{F}_{\mu\nu} \equiv X_a F_{\mu\nu}{}^a$, defined by

$$
\mathcal{F}_{\mu\nu} \equiv \partial_\mu \mathcal{A}_\nu - \partial_\nu \mathcal{A}_\mu + g[\mathcal{A}_\mu, \ \mathcal{A}_\nu],
\tag{3.2.1 − 15a}
$$

$$
\begin{aligned}
F_{\mu\nu}{}^a &\equiv \partial_\mu A_\nu{}^a - \partial_\nu A_\mu{}^a + gf^a{}_{bc}A_\mu{}^b A_\nu{}^c \\
&= -F_{\nu\mu}{}^a,
\end{aligned}
\tag{3.2.1 − 15b}
$$

just corresponds to the **curvature 2-form** Ω defined as the covariant exterior derivative of the connection 1-form ω:

$$
\begin{aligned}
\Omega &= \tilde{\nabla}\omega = d\omega + \frac{1}{2}[\omega, \ \omega] \\
&= \frac{1}{2}gX_a(\partial_\mu A_\nu{}^a - \partial_\nu A_\mu{}^a + gf^a{}_{bc}A_\mu{}^b A_\nu{}^c)dx^\mu \wedge dx^\nu \\
&= \frac{1}{2}gX_a F_{\mu\nu}{}^a dx^\mu \wedge dx^\nu = \frac{1}{2}g\mathcal{F}_{\mu\nu}dx^\mu \wedge dx^\nu.
\end{aligned}
\tag{3.2.1 − 16}
$$

This quantity Ω satisfies the relations,

$$
\tilde{\nabla}^2\varphi = \bar\rho(\Omega)\varphi,
\tag{3.2.1 − 17}
$$

$$
\tilde{\nabla}\Omega = 0,
\tag{3.2.1 − 18}
$$

which are called, respectively, **Ricci identity** and **Bianchi identity**. These relations are equivalently written as

$$([\nabla_\mu, \nabla_\nu]\varphi)^\alpha(x) = gF_{\mu\nu}{}^a \bar{\rho}(X_a)^\alpha{}_\beta\varphi^\beta(x), \qquad (3.2.1-19)$$

$$D_\lambda F_{\mu\nu}{}^a + D_\mu F_{\nu\lambda}{}^a + D_\nu F_{\lambda\mu}{}^a = 0, \qquad (3.2.1-20)$$

where D_μ denotes the covariant differentiation in the adjoint representation, obtained from ∇_μ in Eq.(11) by replacing $\bar{\rho}$ with ad :

$$D_\mu = \partial_\mu + \mathrm{ad}(g\mathcal{A}_\mu). \qquad (3.2.1-21)$$

We now consider the local gauge transformation and its infinitesimal form. A local gauge transformation is specified by a G-valued function u on the spacetime M,

$$M \ni x \mapsto u(x) \in G, \qquad (3.2.1-22)$$

and acts on the matter field $\varphi^\alpha(x)$ as

$$\varphi^\alpha(x) \;\rightarrow\; \varphi'^\alpha(x) = \rho(u(x)^{-1})^\alpha{}_\beta\varphi^\beta(x). \qquad (3.2.1-23)$$

The transformation property of $\mathcal{A}_\mu$ under this local gauge transformation is determined by the covariance property of covariant derivative ∇_μ or $\tilde{\nabla}$ under Eq.(23):

$$\nabla'_\mu\varphi'^\alpha(x) = \rho(u(x)^{-1})^\alpha{}_\beta\nabla_\mu\varphi^\beta(x), \qquad (3.2.1-24)$$

$$(\tilde{\nabla}'\varphi')^\alpha(x) = \rho(u(x)^{-1})^\alpha{}_\beta(\tilde{\nabla}\varphi)^\beta(x), \qquad (3.2.1-25)$$

where

$$\nabla'_\mu = \partial_\mu + \bar{\rho}(g\mathcal{A}'_\mu), \qquad (3.2.1-26)$$

$$\tilde{\nabla}' = d + \bar{\rho}(\omega'). \qquad (3.2.1-27)$$

To ensure Eq.(24) or Eq.(25), we should have[2]

$$\begin{aligned} g\mathcal{A}'_\mu(x) &= u(x)^{-1}g\mathcal{A}_\mu(x)u(x) - \partial_\mu[u(x)^{-1}]u(x) \\ &= \mathrm{Ad}(u(x)^{-1})g\mathcal{A}_\mu(x) + u(x)^{-1}\partial_\mu u(x), \end{aligned} \qquad (3.2.1-28)$$

---

[2]   Note the relation
$$\partial_\mu\rho(u(x)^{-1}) = -\rho(u(x)^{-1})\bar{\rho}(\partial_\mu u(x) \cdot u(x)^{-1})$$
with $(\partial_\mu u)u^{-1} \in LieG$.

or

$$\omega'_x = \mathrm{Ad}(u(x)^{-1})\omega_x + u(x)^{-1}du(x), \qquad (3.2.1-29)$$

where Ad denotes the adjoint representation of G given by $\mathrm{Ad}(u)a \equiv uau^{-1}$. Corresponding to this transformation, Eq.(28) or Eq.(29), the field strength or the curvature $\mathcal{F}_{\mu\nu} = X_a F_{\mu\nu}{}^a$ transforms covariantly as

$$\mathcal{F}'_{\mu\nu}(x) = \mathrm{Ad}(u(x)^{-1})\mathcal{F}_{\mu\nu}(x), \qquad (3.2.1-30)$$

$$\Omega'_x = \mathrm{Ad}(u(x)^{-1})\Omega_x. \qquad (3.2.1-31)$$

Putting $u(x) = \exp t\Lambda^a(x)X_a$ and taking the derivative with respect to a parameter t, we obtain the "infinitesimal" version of the local gauge transformation as follows:

$$\begin{aligned}
\delta\varphi^\alpha(x) &= \frac{d}{dt}\rho\Big(\exp(-t\Lambda^a(x)X_a)\Big)^\alpha{}_\beta\varphi^\beta(x)|_{t=0} \\
&= -\Lambda^a(x)\bar\rho(X_a)^\alpha{}_\beta\varphi^\beta(x) = -(\bar\rho(\Lambda(x))\varphi(x))^\alpha \\
&= i\Lambda^a(x)(\tau_a)^\alpha{}_\beta\varphi^\beta(x), \qquad (3.2.1-32) \\
\delta\mathcal{A}_\mu(x) &= \frac{d}{dt}\Big(\mathrm{Ad}(\exp(-t\Lambda^a(x)X_a))\mathcal{A}_\mu(x) + \frac{1}{g}t\partial_\mu\Lambda^a(x)X_a\Big)|_{t=0} \\
&= -[\Lambda^a(x)X_a, \mathcal{A}_\mu(x)] + \frac{1}{g}\partial_\mu\Lambda^a(x)X_a \\
&= \frac{1}{g}\Big(\partial_\mu\Lambda^a(x) + gf^a{}_{bc}A_\mu{}^b(x)\Lambda^c(x)\Big)X_a \\
&= \frac{1}{g}\nabla_\mu\Lambda^a(x)X_a, \qquad (3.2.1-33) \\
\delta\mathcal{F}_{\mu\nu}(x) &= [\mathcal{F}_{\mu\nu}(x),\ \Lambda(x)] = -(\mathrm{ad}(\Lambda(x)))(\mathcal{F}_{\mu\nu}(x)) \\
&= f^a{}_{bc}F_{\mu\nu}{}^b(x)\Lambda^c(x)X_a. \qquad (3.2.1-34)
\end{aligned}$$

In view of Eq.(34), the local-gauge-invariant Yang-Mills Lagrangian density can be given by

$$\mathcal{L}_{\mathrm{YM}} = -\frac{1}{4}K_{ab}F_{\mu\nu}{}^a F^{\mu\nu b}, \qquad (3.2.1-35)$$

with the aid of the **Killing form** K_{ab} on $LieG$ defined by[3]

$$K_{ab} = -\mathrm{Tr}(\mathrm{ad}(X_a)\mathrm{ad}(X_b)) = -f^c{}_{ad}f^d{}_{bc}. \qquad (3.2.1-36)$$

---

[3]   For the physical reason [see Eq.(37)], the sign convention adopted here for the Killing form K_{ab} is the opposite to the one adopted usually in the mathematical literature.

To be precise, however, Eq.(35) defines a correct Lagrangian density only for the case with G being a *semi-simple* Lie group: Since the condition of semi-simplicity is equivalent to that for the Killing form K_{ab} to be non-degenerate,[4] Eq.(35) does not reproduce the kinetic terms for abelian components which appear in the case of non-semi-simple G, corresponding to the zero eigenvalue of K_{ab}. Since the abelian gauge fields have been fully discussed in Chapter 2, we can disregard them here simply assuming the semi-simplicity of G.

Now, rewriting the kinetic quadratic part,

$$-\frac{1}{4}K_{ab}(\partial_\mu A_\nu{}^a - \partial_\nu A_\mu{}^a)(\partial^\mu A^{\nu b} - \partial^\nu A^{\mu b}),$$

of the first term in Eq.(35) as

$$+\frac{1}{2}K_{ab}\partial_\mu A_.{}^a\partial^\mu A_.{}^b - \frac{1}{2}K_{ab}\partial_\mu A_0{}^a\partial^\mu A_0{}^b$$
$$+\frac{1}{2}K_{ab}\partial_\mu A_\nu{}^a\partial^\nu A^{\mu b}, \qquad (3.2.1-37)$$

we see that the signature of the Killing form K_{ab} is directly related with the signs of state-space *metric* of (transverse) gauge bosons $A_\mu{}^a$. If K_{ab} has both positive and negative eigenvalues, it would be very difficult to eliminate the contributions from negative norm states. Therefore, unless there is some mechanism which makes all the components of $A_\mu{}^a$ unphysical simultaneously, the Killing form Eq.(36) must be of definite sign (more precisely, positive definite) as a bilinear form; this fact implies that our gauge group G should be *compact* (see, e.g. [Ser 65]). Then, the Killing form Eq.(36) can be diagonalized and normalized into the form

$$K_{ab} = \delta_{ab}. \qquad (3.2.1-38)$$

With respect to this diagonalizing basis, we need not distinguish the upper and lower indices in the structure constants $f^a{}_{bc}$, and it can be shown that $K_{ad}f^d{}_{bc} = f_{abc}$ is totally antisymmetric in (a, b, c). Thus, the positions of group indices a, b, c, etc., do not have any significance as long as we are concerned with the Yang-Mills theory with G being compact. As a matter of course, however, such a restriction on the gauge group G as its compactness arises from the specific form Eq.(35) of $\mathcal{L}_{YM}$ and from the assumption on the existence of physical modes in $A_\mu{}^a$. Later

---

[4]   See some textbook on Lie algebras, e.g., [Ser 65]

(in Sec.5.7), we will discuss the gauge theory of local Lorentz transformations whose gauge group $SL(2, \mathbf{C})$ is *not* compact. In that case, gauge field (denoted as $\omega_\mu{}^a{}_b$ in Sec.5.7) does not represent the independent degrees of freedom and its Lagrangian density, Eq.(5.7.3-4) is *not* of the form Eq.(35). The independent degrees of freedom encountered there are represented by the vierbein field $h_\mu{}^a$, in terms of which $\omega_\mu{}^a{}_b$ is expressed by Eq.(5.7.1-6).

Finally, the total Lagrangian density of the system consisting of matter fields φ^α interacting with the gauge field $A_\mu{}^a$ in a local-gauge-invariant manner is given by

$$\mathcal{L}_0 = \mathcal{L}_{\mathrm{YM}} + \mathcal{L}_{\mathrm{M}}(\varphi, \nabla_\mu\varphi). \tag{3.2.1 − 39}$$

Here the local-gauge-invariant matter Lagrangian density $\mathcal{L}_{\mathrm{M}}(\varphi, \nabla_\mu\varphi)$ is obtained from a matter Lagrangian density $\mathcal{L}_{\mathrm{M}}(\varphi, \partial_\mu\varphi)$, invariant under the global action of the group G, by replacing $\partial_\mu\varphi$ in it with the covariant derivative $\nabla_\mu\varphi$.

3.2.2 Local gauge invariance and constraints −Second Noether theorem−

The field equations derived from Eq.(3.2.1-39) with Eq.(3.2.1-35) are

$$D^\mu F_{\mu\nu}{}^a = -g j_\nu{}^a, \tag{3.2.2 − 1}$$

$$(\partial_\mu + ig A_\mu{}^a \tau^a)(\frac{\partial}{\partial(\nabla_\mu\varphi)}\mathcal{L}_{\mathrm{M}}(\varphi, \nabla_\nu\varphi)) = \frac{\partial}{\partial\varphi}\mathcal{L}_{\mathrm{M}}(\varphi, \nabla_\nu\varphi). \tag{3.2.2 − 2}$$

Here, the matter current $j_\nu{}^a$ is defined by

$$\begin{aligned}
g j_\nu{}^a &\equiv -ig\tau^a\varphi\frac{\partial\mathcal{L}_{\mathrm{M}}}{\partial(\partial^\nu\varphi)} \\
&= \frac{\partial\mathcal{L}_{\mathrm{M}}}{\partial A^{\nu a}}
\end{aligned} \tag{3.2.2 − 3}$$

and satisfies the "covariant conservation law" of current,

$$D_\mu j^{\mu a} = \partial_\mu j^{\mu a} + g f^{abc} A_\mu{}^b j^{\mu c} = 0. \tag{3.2.2 − 4}$$

This is a direct consequence of the invariance of *matter Lagrangian density* $\mathcal{L}_{\mathrm{M}}$ under the *global* gauge transformation given by

$$\delta\varphi = -i\epsilon^a\tau^a\varphi, \tag{3.2.2 − 5}$$

$$\delta A_\mu{}^a = f^{abc}\epsilon^b A_\mu{}^c, \tag{3.2.2 − 6}$$

in combination with the field equations, Eq.(2), of matter fields φ and with the definition, Eq.(3), of the matter current $j_\nu{}^a$:

$$
\begin{aligned}
0 &= \delta\mathcal{L}_M(\varphi, \partial_\mu\varphi - igA_\mu{}^a\tau^a\varphi) \\
&= \delta\varphi\frac{\partial\mathcal{L}_M}{\partial\varphi} + \delta\partial_\mu\varphi\frac{\partial\mathcal{L}_M}{\partial\partial_\mu\varphi} + \delta A_\mu{}^a\frac{\partial\mathcal{L}_M}{\partial A_\mu{}^a} \\
&= \partial_\mu\Big(\delta\varphi\frac{\partial\mathcal{L}_M}{\partial\partial_\mu\varphi}\Big) + gf^{abc}\epsilon^a A_\mu{}^b j^{\mu c} = \epsilon^a D_\mu j^{\mu a}.
\end{aligned}
\qquad (3.2.2-7)
$$

[A more straightforward way may be to derive Eq.(4) from Eq.(1) with the aid of the Ricci identity, Eq.(3.2.1-19), with $\bar{\rho} = \mathrm{ad}$:

$$
D_\mu D_\nu \mathcal{F}^{\mu\nu} = \frac{1}{2}[D_\mu, D_\nu]\mathcal{F}^{\mu\nu} = \frac{g}{2}\mathrm{ad}(\mathcal{F}_{\mu\nu})\mathcal{F}^{\mu\nu} = 0. \qquad (3.2.2-8)
$$

In contrast to the above method, however, this derivation cannot be extended to the quantized case, where Eq.(1) should be modified by the gauge fixing and FP ghost terms.] Since the lhs. of Eq.(1) and $F_{\mu\nu}{}^a$ in Eq.(3.2.1-15) are nonlinear in $A_\mu{}^a$,

$$
D_\mu F^{\mu\nu a} = \partial_\mu F^{\mu\nu a} + gf^{abc}A_\mu{}^b F^{\mu\nu c}, \qquad (3.2.2-9)
$$

the Yang-Mills field $A_\mu{}^a$ involves *self-coupling*.

Now, the local gauge invariance of the Lagrangian density, Eq.(3.2.1-39), causes the same trouble as the one encountered in Chapter 2 in the case of abelian gauge theories: The time derivative, $\dot{A}_0{}^a$, of $A_0{}^a$ appears neither in $F_{\mu\nu}{}^a$ nor in the Lagrangian density, Eq.(3.2.1-39), and hence, the field equation, Eq.(1), does not completely determine the time development of the Yang-Mills field $A_\mu{}^a$. With the aid of the second Noether theorem, this situation can be understood as a direct consequence of the local gauge invariance in a more general way independent of such a specific form of the Lagrangian density as Eq.(3.2.1-35). This theorem relates any kind of "local gauge invariance" to the constraints and to the characteristic form of field equation. For simplicity, we consider here the following infinitesimal "local gauge transformation" which generalizes Eqs.(3.2.1-32) and (3.2.1-33) but which does not involve the spacetime transformation :[5]

$$
\delta\varphi^A = \sum_{\alpha=1}^{r}\Big(G^A_\alpha\Lambda^\alpha(x) + T^{A\mu}_\alpha\partial_\mu\Lambda^\alpha(x)\Big). \qquad (3.2.2-10)
$$

---

[5]   The case of local spacetime transformation is discussed in Sec.5.2.4.

Second Noether Theorem : The Lagrangian density $\mathcal{L} = \mathcal{L}(\varphi^A, \partial_\mu \varphi^A)$ with $\varphi^A(A = 1, \cdots, N)$ being generic fields is invariant under the infinitesimal transformation Eq.(10) with arbitrary spacetime-dependent functions $\Lambda^\alpha(x)$ $(\alpha = 1, \ldots, r)$ (of C^2-class) ; i.e.,

$$\delta\mathcal{L} = 0, \tag{3.2.2 - 11}$$

if and only if the following relations hold *identically* (i.e. *irrespective of whether or not* φ^A's are solutions of the field equations):

$$\partial_\mu \left(T^{A\mu}{}_\alpha \frac{\delta\mathcal{L}}{\delta\varphi^A} \right) = G^A{}_\alpha \frac{\delta\mathcal{L}}{\delta\varphi^A}, \tag{3.2.2 - 12}$$

$$\partial_\nu K^{\nu\mu}{}_\alpha + J^\mu{}_\alpha = 0, \tag{3.2.2 - 13}$$

$$K^{\mu\nu}{}_\alpha = -K^{\nu\mu}{}_\alpha. \tag{3.2.2 - 14}$$

Here, $J^\mu{}_\alpha$ and $K^{\nu\mu}{}_\alpha$ are defined by

$$J^\mu{}_\alpha \equiv G^A{}_\alpha \frac{\partial\mathcal{L}}{\partial(\partial_\mu\varphi^A)} + T^{A\mu}{}_\alpha \frac{\delta\mathcal{L}}{\delta\varphi^A}, \tag{3.2.2 - 15}$$

$$K^{\nu\mu}{}_\alpha \equiv T^{A\mu}{}_\alpha \frac{\partial\mathcal{L}}{\partial(\partial_\nu\varphi^A)}, \tag{3.2.2 - 16}$$

and $\delta/\delta\varphi^A$ denotes the Euler derivative.

Proof: This can be proved simply by some reshuffling of terms as follows:

$$\begin{aligned}
\delta\mathcal{L} = &\delta\mathcal{L}(\varphi^A, \partial_\mu\varphi^A) = \delta\varphi^A \frac{\partial\mathcal{L}}{\partial\varphi^A} + \delta(\partial_\mu\varphi^A)\frac{\partial\mathcal{L}}{\partial(\partial_\mu\varphi^A)} \\
= &\left(G^A{}_\alpha \frac{\partial\mathcal{L}}{\partial\varphi^A} + \partial_\mu G^A{}_\alpha \frac{\partial\mathcal{L}}{\partial(\partial_\mu\varphi^A)} \right) \Lambda^\alpha \\
&+ \left(G^A{}_\alpha \frac{\partial\mathcal{L}}{\partial(\partial_\mu\varphi^A)} + T^{A\mu}{}_\alpha \frac{\partial\mathcal{L}}{\partial\varphi^A} + \partial_\nu T^{A\mu}{}_\alpha \frac{\partial\mathcal{L}}{\partial(\partial_\nu\varphi^A)} \right) \partial_\mu\Lambda^\alpha \\
&+ \left(T^{A\nu}{}_\alpha \frac{\partial\mathcal{L}}{\partial(\partial_\mu\varphi^A)} \right) \partial_\mu\partial_\nu\Lambda^\alpha \\
= &\left\{ [G^A{}_\alpha \frac{\delta\mathcal{L}}{\delta\varphi^A} - \partial_\mu(T^{A\mu}{}_\alpha \frac{\delta\mathcal{L}}{\delta\varphi^A})] + \partial_\mu J^\mu{}_\alpha \right\} \Lambda^\alpha \\
&+ (J^\mu{}_\alpha + \partial_\nu K^{\nu\mu}{}_\alpha)\partial_\mu\Lambda^\alpha + \frac{1}{2}(K^{\mu\nu}{}_\alpha + K^{\nu\mu}{}_\alpha)\partial_\mu\partial_\nu\Lambda^\alpha. \quad (3.2.2 - 17)
\end{aligned}$$

We have used here the definitions, Eqs.(15) and (16), of $J^\mu{}_\alpha$ and $K^{\mu\nu}{}_\alpha$ as well as the relation

$$\delta\partial_\mu\varphi^A = \partial_\mu\delta\varphi^A \tag{3.2.2 - 18}$$

following from the assumption that the (infinitesimal) transformation Eq.(10) does not involve the spacetime transformation.[6] Owing to Eq.(17), it is easy to see that the invariance Eq.(11) of $\mathcal{L}$ is equivalent to the relations Eqs.(12)-(14) in view of the arbitrariness of the functions $\Lambda^\alpha(x)$. This is because the lhs. of Eq.(17) vanishes if and only if each coefficient of Λ^α, $\partial_\mu \Lambda^\alpha$ and $\partial_\mu \partial_\nu \Lambda^\alpha$ vanishes. (q.e.d.)

Now, the relation in Eq.(12) means that the N quantities $(\delta/\delta\varphi^A)\mathcal{L}$ ($A = 1, \cdots, N$) are interrelated by r equations ($\alpha = 1, \ldots, r$), and hence, that the number of the independent quantities among $(\delta/\delta\varphi^A)\mathcal{L}$ is $N - r$ in general. This means that the field equations

$$\frac{\delta\mathcal{L}}{\delta\varphi^A} = \frac{\partial\mathcal{L}}{\partial\varphi^A} - \partial_\mu\left(\frac{\partial\mathcal{L}}{\partial(\partial_\mu\varphi^A)}\right) = 0 \qquad (3.2.2 - 19)$$

for N unknown quantities $\varphi^A(A = 1, \cdots, N)$ are not independent of each other, but that only $N - r$ equations are independent among them owing to the *constraints* Eq.(12).

For φ^A satisfying the field equations Eq.(19), the current J^μ_α defined by Eq.(15) become identical to the Noether current associated with the global transformation $\delta\varphi^A$ with $\Lambda^\alpha(x) = \Lambda^\alpha = $ const. in Eq.(10). Their current conservation law

$$\partial_\mu J^\mu_\alpha = 0 \qquad (3.2.2 - 20)$$

automatically follows from Eqs.(13) and (14).

As the antisymmetry, Eq.(14), of $K^{\mu\nu}_\alpha$ suggests, Eq.(13) is closely related to the *Maxwell equation*, Eq.(2.1.1-2). In fact, $K^{\mu\nu}_\alpha$ is nothing but the field strength $F^{\mu\nu}$ [see Eq.(2.1.1-4)] in the familiar case of the abelian gauge theory. It should be noted, however, that Eqs.(13) and (14) are the general consequences directly following from the local gauge invariance of the Lagrangian density $\mathcal{L}$ itself, whatever specific form $\mathcal{L}$ has. In the present case of the Yang-Mills theory, $K^{\mu\nu}_\alpha$ also agrees with the field strength $F^{\mu\nu a}$ given by Eq.(3.2.1-15). Then, the discrepancy between Eqs.(1) and (13), that is, D_μ vs. ∂_μ and "covariantly conserved" $gj^{\nu a}$ [see Eqs.(3) and (4)] vs. "genuinely conserved" J^ν_α [see Eq.(20)], can be resolved by the relation

$$J^{\nu a} = gj^{\nu a} + gf^{abc}A_\mu^{\ b}F^{\mu\nu c}. \qquad (3.2.2 - 21)$$

---

[6]   By considering the Lie derivative (δ_* in Sec.5.2.4), Eq.(18) can be maintained even for the spacetime transformation, while Eq (11) should then be modified by a total divergence term [see Sec 5.2.4].

The second term on the rhs. of Eq.(21) represents the nonlinear self-coupling effect of Yang-Mills fields due to the non-abelian nature. While the field equation, Eq.(1),

$$D_\mu F^{\mu\nu a} = -g j^{\nu a} \qquad (3.2.2-22)$$

is a gauge covariant equation involving gauge covariant quantities only, its equivalent reformulation

$$\partial_\mu F^{\mu\nu a} = -J^{\nu a} = -g j^{\nu a} - g f^{abc} A_\mu{}^b F^{\mu\nu c} \qquad (3.2.2-23)$$

expresses the universal essence common to any kind of local gauge invariant theories.

3.3 GAUGE FIXING AND FADDEEV-POPOV GHOSTS

Owing to the presence of the constaints Eq.(3.2.2-12) directly following from the local gauge invariance, the field equations, Eq.(3.2.2-1) or Eq.(3.2.2-13), cannot determine the time development of the gauge field uniquely. The notion of quantized fields necessarily involves field operators and/or their Green's functions with some specified spacetime dependence determined by field equations (up to divergence problem). To quantize the gauge field, therefore, it is necessary to break the local gauge invariance by some gauge fixing procedure. This is of course the same situation as encountered in the case of abelian gauge theory in Sec.2.2 and Sec.2.3.
As mentioned in Sec.3.1, however, what is new here is that the Lagrangian density, Eq.(3.2.1-39), supplemented by the gauge fixing term with the B-fields B^a cannot define a meaningful quantum theory of Yang-Mills fields with a *unitary* physical S-matrix; we must introduce Faddeev-Popov ghost pairs. Here we try to understand the origin of these ghost fields in a natural way from the viewpoint of operator formalism.

3.3.1 Deformation from abelian to non-abelian

Now, we start from a simple-minded approach to extend Eq.(2.3.1-1) in the abelian gauge theory to the non-abelian one : The Lagrangian density is taken as

$$\mathcal{L}' = \mathcal{L}_0 + \mathcal{L}_{\mathrm{GF}}, \qquad (3.3.1-1)$$

where the gauge fixing term $\mathcal{L}_{\mathrm{GF}}$ added to the original one $\mathcal{L}_0$, Eq.(3.2.1-39), is chosen just in the same way as in the abelian case :

$$\mathcal{L}_{\mathrm{GF}} = B^a \partial_\mu A^{\mu a} + \frac{\alpha}{2} B^a B^a. \qquad (3.3.1-2)$$

For Landau gauge ($\alpha = 0$), this gauge-fixing term can be naturally interpreted in the context of Lagrange multiplier method: B-fields B^a are the Lagrange multipliers implementing the constraints of Lorentz condition

$$\partial_\mu A^{\mu a} = 0. \qquad (3.3.1-3)$$

In the case of *abelian* gauge theory, this type of gauge-fixing constraint or the more general one, Eq.(2.3.1-4),

$$\partial^\mu A_\mu + \alpha B = 0, \qquad (3.3.1-4)$$

remains invariant under the "residual" local gauge transformation

$$\delta A_\mu = \partial_\mu \Lambda, \qquad\qquad (3.3.1-5)$$

$$\delta B = 0, \qquad\qquad (3.3.1-6)$$

with Λ satisfying the d'Alembert equation

$$\square \Lambda = 0. \qquad\qquad (3.3.1-7)$$

In *non-abelian* case, however, the decomposition of B-field into positive and negative frequency parts is not meaningful owing to the nonlinear field equation of B^a,

$$(D_\mu \partial^\mu B)^a = 0, \qquad\qquad (3.3.1-8)^{\cdot}$$

which follows from the field equation for $A_\mu{}^a$,

$$D^\mu F_{\mu\nu}{}^a = -g j_\nu{}^a + \partial_\nu B^a, \qquad\qquad (3.3.1-9)$$

together with Eq.(3.2.2-4). Therefore, the Gupta subsidiary condition, Eq.(2.3.3-11), closely related to the Lorentz condition, cannot be set up consistently in Yang-Mills theory. What can be extended to the general cases is not a particular form of this subsidiary condition itself, but the invariance of the gauge-fixed theory under such a "residual" local gauge transformation as above, which inherits the essence of local gauge invariance to quantum level in the form of Ward-Takahashi identities. This extension can be achieved in the following way with a "Noether procedure" deforming the invariance in abelian case to that in non-abelian one, through which the FP ghosts and BRS transformation will be derived. While the origin of FP ghosts has been expounded in the context of path integral [Fad 67, Fad 69], as is easily found in the standard textbooks on gauge theory, it cannot be denied, however, that they have been objects brought into the operator formalism in a rather *ad hoc* way. Our approach here will show their raison d'être in a way intrinsic in operator formalism as much as possible.

To this end, we first note the important role of B-field in abelian gauge theory as the "generator" of operator gauge transformation as shown by Eqs.(2.3.3-9), (2.3.3-22) and (2.3.3-23). To incorporate such symmetry as this consistently with the non-abelian structure of Yang-Mills theory, we note the "*almost conserved charge*" given by

$$G_\Lambda \equiv \int d\mathbf{x} \left(B^a (D_0 \Lambda)^a - \dot{B}^a \Lambda^a \right) \qquad\qquad (3.3.1-10)$$

as a simple adaptation of Q_Λ [Eq.(2.3.3-21)] to the non-abelian case. By the "**Noether procedure**" we mean deforming G_Λ and the Lagrangian density $\mathcal{L}'$ [Eq.(1)] into a conserved charge and a Lagrangian density having a symmetry generated by the former, respectively. If G_Λ were conserved, the equal-time commutation relations derived from the Lagrangian density $\mathcal{L}'$ and Eq.(4) would qualify it as the generator of infinitesimal local gauge transformation:

$$[iG_\Lambda,\, A_\mu{}^a] = D_\mu\Lambda^a, \qquad (3.3.1-11)$$

$$[iG_\Lambda,\, \varphi^\alpha] = ig\Lambda^a(\tau^a)^\alpha{}_\beta\varphi^\beta. \qquad (3.3.1-12)$$

However, G_Λ is actually *not* conserved:

$$\partial^\mu\Big(B^a(D_\mu\Lambda)^a - \partial_\mu B^a \cdot \Lambda^a\Big) = B^a\partial^\mu(D_\mu\Lambda)^a - (D^\mu\partial_\mu B)^a\Lambda^a. \qquad (3.3.1-13)$$

By using the field equation of B^a, Eq.(8), we obtain

$$\partial^\mu(D_\mu\Lambda)^a = \Box\Lambda^a + gf^{abc}\partial^\mu(A_\mu{}^b\Lambda^c) = 0 \qquad (3.3.1-14)$$

as a condition for the rhs. of Eq.(13) to vanish. Since $A_\mu{}^a$ in Eq.(14) is a *field operator*, however, it is impossible to find a *c-number* function Λ^a satisfying Eq.(14). Furthermore, the transformation involving spacetime-dependent c-number function is not compatible with the invariance of the theory under the spacetime translations. So, we promote the parameter-function $\Lambda^a(x)$ of infinitesimal local gauge transformation to a field operator characterized by Eq.(14). To guarantee Eq.(14), we introduce a Lagrange multiplier η, which modifies the Lagrangian density $\mathcal{L}'$, Eq.(1), into

$$\begin{aligned}\mathcal{L}'' &= \mathcal{L}_0 + \mathcal{L}_{\mathrm{GF}} + \eta^a\partial^\mu(D_\mu\Lambda)^a \\ &\sim \mathcal{L}_0 + \mathcal{L}_{\mathrm{GF}} - \partial^\mu\eta^a \cdot (D_\mu\Lambda)^a.\end{aligned} \qquad (3.3.1-15)$$

Since the added term in Eq.(15) contains $A_\mu{}^b$, the field equation for $A_\mu{}^a$, Eq.(9), is changed into

$$(D^\mu F_{\mu\nu})^a = -gj_\nu{}^a + \partial_\nu B^a - gf^{abc}\partial_\nu\eta^b \cdot \Lambda^c, \qquad (3.3.1-16)$$

and this affects also the field equation for B^a, Eq.(8), as

$$(D^\mu\partial_\mu B)^a = gf^{abc}\partial_\mu\eta^b \cdot (D^\mu\Lambda)^c. \qquad (3.3.1-17)$$

Here we have used the field equation for η obtained from Eq.(15) :

$$(D^\mu \partial_\mu \eta)^a = 0. \qquad (3.3.1 - 18)$$

As a consequence of Eqs.(14) and (17), Eq.(13) is now modified into

$$\partial^\mu \left(B^a (D_\mu \Lambda)^a - \partial_\mu B^a \cdot \Lambda^a \right) = -g f^{abc} \partial_\mu \eta^b \cdot (D^\mu \Lambda)^c \Lambda^a$$
$$= -g f^{abc} \partial_\mu \eta^a \cdot (D^\mu \Lambda)^b \Lambda^c. \quad (3.3.1 - 19)$$

Now, if we are allowed to make here the assumption that Λ^a's and η^b's are *fermionic* fields, a miraculous cancellation takes place in Eq.(19):

$$\text{rhs. of Eq.(19)} = -\partial^\mu \left(\frac{1}{2} g f^{abc} \partial_\mu \eta^a \cdot \Lambda^b \Lambda^c \right). \qquad (3.3.1 - 20)$$

Here use has been made of Eq.(18) and the Jacobi identity involving the anticommutative Λ,

$$\left((A_\mu \times \Lambda) \times \Lambda \right)^a = \frac{1}{2} \left(A_\mu \times (\Lambda \times \Lambda) \right)^a, \qquad (3.3.1 - 21)$$

where

$$(A \times B)^a \equiv f^{abc} A^b B^c. \qquad (3.3.1 - 22)$$

In this case, we obtain, from Eqs.(19) and (20), a current conservation law

$$\partial^\mu \left(B^a (D_\mu \Lambda)^a - \partial_\mu B^a \cdot \Lambda^a + \frac{1}{2} g f^{abc} \partial_\mu \eta^a \cdot \Lambda^b \Lambda^c \right) = 0. \qquad (3.3.1 - 23)$$

Therefore, the charge defined by

$$Q_B \equiv \int d\mathbf{x} \left(B^a (D_0 \Lambda)^a - \dot{B}^a \Lambda^a + \frac{1}{2} g f^{abc} \dot{\eta}^a \Lambda^b \Lambda^c \right). \qquad (3.3.1 - 24)$$

is *conserved* through the time development governed by the Lagrangian density

$$\mathcal{L} \equiv \mathcal{L}_0 + \mathcal{L}_{\text{GF}} + \mathcal{L}_{\text{FP}}, \qquad (3.3.1 - 25)$$

with

$$\mathcal{L}_{\text{FP}} \equiv -\partial^\mu \eta^a \cdot (D_\mu \Lambda)^a. \qquad (3.3.1 - 26)$$

The new term $\mathcal{L}_{FP}$, Eq.(26), is nothing but the Faddeev-Popov term obtained by the path integral method with the standard covariant gauge fixing condition. To conform to the standard convention, therefore, we change the notation for Λ^a into C^a :

$$\Lambda^a(x) \to C^a(x). \qquad (3.3.1-27)$$

In view of its origin seen in the above, Λ^a should represent a *real* quantity, and hence, we take the FP ghost C^a as a *hermitian* field:

$$C^a(x)^\dagger = C^a(x). \qquad (3.3.1-28)$$

Since the field η^a is put in a pair with the *fermionic* field $\Lambda^a = C^a$ in $\mathcal{L}_{FP}$, it should also be fermionic. In view of the anticommutativity of these two objects, the requirement for $\mathcal{L}_{FP}$ in Eq.(26) to be *hermitian* implies that η^a should be *antihermitian*, and hence, we denote it as

$$\eta^a(x) \equiv i\overline{C}^a(x) \qquad (3.3.1-29)$$

in terms of the *hermitian* FP anti-ghost $\overline{C}^a$:

$$\overline{C}^a(x)^\dagger = \overline{C}^a(x). \qquad (3.3.1-30)$$

Thus, we rewrite Eqs.(26) and (24) as

$$\mathcal{L}_{FP} = -i\partial^\mu \overline{C}^a \cdot (D_\mu C)^a = \mathcal{L}_{FP}^{\dagger}, \qquad (3.3.1-31)$$

$$Q_B = \int d\mathbf{x}\,(B^a(D_0 C)^a - \dot{B}^a C^a + i\frac{g}{2}f^{abc}\dot{\overline{C}}^a C^b C^c)$$
$$= Q_B^{\dagger}. \qquad (3.3.1-32)$$

The transformation generated by Q_B is just the **BRS transformation** [Kug 78a, Kug 78c], as will be checked later in Sec.3.4.2 on the basis of the canonical quantization of Eq.(25) with Eq.(31):

$$\delta A_\mu^{\ a} \equiv [iQ_B,\ A_\mu^{\ a}] = (D_\mu C)^a, \qquad (3.3.1-33)$$

$$\delta\varphi^\alpha \equiv [iQ_B,\ \varphi^\alpha] = igC^a(\tau^a)^\alpha_{\ \beta}\varphi^\beta, \qquad (3.3.1-34)$$

$$\delta B^a \equiv [iQ_B,\ B^a] = 0, \qquad (3.3.1-35)$$

$$\delta C^a \equiv \{iQ_B,\ C^a\} = -\frac{g}{2}(C \times C)^a = -\frac{g}{2}f^{abc}C^b C^c, \qquad (3.3.1-36)$$

$$\delta\overline{C}^a \equiv \{iQ_B,\ \overline{C}^a\} = iB^a. \qquad (3.3.1-37)$$

Owing to the supercharge nature of Q_B, the anticommutators appear in Eqs.(36) and (37). Since the fermionic character of $\Lambda^a = C^a$, the assumption crucial for Eq.(20), may still look abrupt, we clarify in the next subsection its intrinsic meaning from geometrical point of view to justify this assumption.

3.3.2 FP ghost as Maurer-Cartan form

Although the local gauge invairance of the theory has been lost through the gauge fixing procedure, it is necessary, or at least desirable, to retain its essence in an infinitesimal form around the gauge-fixed theory. In the usual case, however, the invariance of theory under an infinitesimal transformation can be extended to the invariance under the finite one by "repeating" the former infinitely many times. Therefore, it seems almost impossible to imagine such a symmetry transformation that its infinitesimal form leaves the theory invariant but that its finite version does not. However, this can actually be made possible by prohibiting the "repetition" of the infinitesimal transformation in the following sense.

Heuristically, an infinitesimal local gauge transformation is obtained from a finite one with $u(x) = \exp\left(\Lambda^a(x)X_a\right)$ by expanding it in Λ and by discarding $O(\Lambda^2)$ terms. If we are allowed to take

$$\left(\Lambda^a(x)\right)^2 = 0 \qquad \text{(no summation in } a\text{)} \qquad (3.3.2-1)$$

not an *approximate* relation but as an *exact* one, then the infinitesimal form becomes "exact" by itself being identical with its finite one. The symmetry obtained in this way mimics the infinitesimal form of the original local gauge invariance, whereas it no longer reproduces the finite form which has already been broken by the gauge fixing.

One of the ways to implement Eq.(1) in a mathematically consistent form is to regard Λ^a's as the (differential) forms which constitute a Grassmann algebra equipped with the exterior product (see Appendix $A.1$). We reproduce here the essence of the story in [Bon 83] in its simplified version restricting the spacetime to the flat Minkowski space. In the differential-geometerical interpretation of FP ghost and BRS transformation formulated in [Bon 83], we are concerned with the infinite-dimensional group $\mathcal{G}$ of local gauge transformations together with its Lie algebra $Lie\mathcal{G}$ consisting of infinitesimal local gauge transformations. In the present situation of gauge theory on the flat Minkowski space $M = \mathbf{R}^4$, $\mathcal{G}$ and $Lie\mathcal{G}$ can be identified with the set of (C^∞-) functions on M taking values in the gauge group G and its Lie algebra $LieG$, respectively :[1]

$$\mathcal{G} = C^\infty(M, G) = \Gamma(M \times G), \qquad (3.3.2-2)$$

$$Lie\mathcal{G} = C^\infty(M, LieG) = \Gamma(M \times LieG). \qquad (3.3.2-3)$$

On the Lie algebra $Lie\mathcal{G}$, we can define the **Maurer-Cartan form** θ just as the identity map on it :

$$\theta : Lie\mathcal{G} \longrightarrow Lie\mathcal{G}$$
$$\Lambda \longmapsto \Lambda \qquad (3.3.2-4)$$

This definition may look rather trivial when θ is viewed only within $Lie\mathcal{G}$ identified with the tangent space $T_e(\mathcal{G})$ of $\mathcal{G}$ at its origin e. However, it becomes a non-trivial quantity when it is extended from the origin $e \in \mathcal{G}$ to the whole $\mathcal{G}$ as a $Lie\mathcal{G}$-valued left-invariant 1-form.

First, we note that θ is a functional whose arguments are functions $\Lambda \in Lie\mathcal{G} = C^\infty(M, LieG)$:[2]

$$\theta((\Lambda)) = \Lambda \qquad \text{for } \Lambda \in Lie\mathcal{G}. \qquad (3.3.2-5)$$

Secondly, θ takes its value again in the space $Lie\mathcal{G} = C^\infty(M, LieG)$ of functions:

$$\Big(\theta((\Lambda))\Big)(x) = \Lambda(x). \qquad (3.3.2-6)$$

To take account of this x-dependence, we define θ_x for each $x\epsilon M$ by

$$\theta_x((\Lambda)) \equiv \Big(\theta((\Lambda))\Big)(x) = \Lambda(x). \qquad (3.3.2-7)$$

In view of the Lie algebra structure $LieG$ of the image of the functions $\Lambda \in Lie\mathcal{G} = C^\infty(M, LieG)$, we can decompose θ_x also in reference to a base $\{X_a\}_{1 \le a \le n}$ of $LieG$:

$$\theta_x((\Lambda)) = \sum_a X_a \theta_x{}^a((\Lambda)) = \sum_a X_a \Lambda^a(x), \qquad (3.3.2-8)$$

$$\theta_x{}^a((\Lambda)) = \Lambda^a(x). \qquad (3.3.2-9)$$

---

[1]   For a fiber bundle B with a projection $\pi : B \to M$ (in the present case B is taken as a trivial bundle $M \times G$), $\Gamma(B)$ denotes the space of cross sections σ ($\pi \circ \sigma = id_M$).

[2]   In order to avoid the confusion with the Lie bracket $[\,\cdot\,,\,\cdot\,]$, we use here the symbol $((\cdot))$ for clarifying the functional nature of θ.

Quite similarly to the case of ordinary Lie-algebra cohomology of the finite dimensional Lie algebra $LieG$, we can now define a coboundary operation $\boldsymbol{\delta}$ on $Lie\mathcal{G} = C^\infty(M, LieG)$ as follows. Let $\bar\rho$ be a representation of $Lie\mathcal{G}$ in a vector space V and φ be a V-valued r-form on $Lie\mathcal{G}$. Then, the action of $\boldsymbol{\delta}$ on φ is defined by

$$\boldsymbol{\delta}\varphi((\Lambda_0, \Lambda_1, \cdots, \Lambda_r)) = \sum_i (-)^i \bar\rho(\Lambda_i)\Big(\varphi((\Lambda_0, \cdots, \hat\Lambda_i, \cdots, \Lambda_r))\Big)$$
$$+ \sum_{i<j} (-)^{i+j} \varphi(([\Lambda_i,\ \Lambda_j], \Lambda_0, \cdots, \hat\Lambda_i, \cdots, \hat\Lambda_j, \cdots, \Lambda_r)),$$

$$(3.3.2 - 10)$$

where $\Lambda_0, \cdots, \Lambda_r \in Lie\mathcal{G}$ and the symbol $\hat{\ }$ means that the term is omitted. As usual, $\boldsymbol{\delta}$ is shown to be an **anti-derivation**

$$\boldsymbol{\delta}(\varphi_1 \wedge \varphi_2) = \boldsymbol{\delta}(\varphi_1) \wedge \varphi_2 + (-1)^{r_1} \varphi_1 \wedge \boldsymbol{\delta}(\varphi_2) \quad \text{for } r_i\text{-forms } \varphi_i \ (i = 1, 2) \quad (3.3.2 - 11)$$

satisfying the **nilpotency** property :

$$\boldsymbol{\delta}^2 = 0. \qquad (3.3.2 - 12)$$

The assumption of the left-invariance of θ over $\mathcal{G}$ becomes crucial in determining the representation $\bar\rho$ associated with $\varphi = \theta$: In order to determine the representation $\bar\rho$ in the target space $V = Lie\mathcal{G}$ in the case of Maurer-Cartan form θ given by Eq.(4), we need to know its extension from $Lie\mathcal{G}$ identified with the tangent space $T_e(\mathcal{G})$ of $\mathcal{G}$ at the origin $e \in \mathcal{G}$ to arbitrary $u \in \mathcal{G}$ (in some neighborhood of e). If this extension is done by adopting the 'left invariance' condition, $\bar\rho$ becomes trivial representation : $\bar\rho = 0$. Then, we get

$$\boldsymbol{\delta}\theta((\Lambda_0, \Lambda_1)) = -\theta(([\Lambda_0,\ \Lambda_1])) = -[\Lambda_0,\ \Lambda_1]$$
$$= -\frac{1}{2}[\theta,\ \theta]((\Lambda_0,\ \Lambda_1)), \qquad (3.3.2 - 13)$$

which is nothing but the Maurer-Cartan equation characterizing θ as the Maurer-Cartan form on $\mathcal{G}$:

$$\boldsymbol{\delta}\theta = -\frac{1}{2}[\theta,\ \theta], \qquad (3.3.2 - 14)$$

or equivalently,

$$\boldsymbol{\delta}\theta_x^{\ a} = -\frac{1}{2} f^a_{\ bc} \theta_x^{\ b} \theta_x^{\ c}. \qquad (3.3.2 - 15)$$

By rewriting $\theta_x{}^a$ as $gC^a(x)$,

$$\theta_x{}^a \Rightarrow gC^a(x), \qquad (3.3.2-16)$$

it can easily be seen that this equation reproduces the BRS transformation law, Eq.(3.3.1-36), of the FP ghost :

$$\boldsymbol{\delta}C^a = -\frac{g}{2}f^{abc}C^bC^c = -\frac{g}{2}(C \times C)^a. \qquad (3.3.2-17)$$

Considering the Yang-Mills field $A_\mu{}^a$ as a function on $\mathcal{G}$ given by its local gauge transformation,

$$\mathcal{G} \ni u \mapsto u^{-1}\mathcal{A}u + \frac{1}{g}u^{-1}du \equiv \hat{\mathcal{A}}(u), \qquad (3.3.2-18)$$

we obtain

$$\begin{aligned}
\boldsymbol{\delta}\mathcal{A}((\Lambda)) &= \frac{d}{dt}\hat{\mathcal{A}}(e^{t\Lambda})|_{t=0} = \frac{1}{g}(d\Lambda + g[\mathcal{A}, \ \Lambda]) \\
&= (d \circ C + g[\mathcal{A}, \ C])((\Lambda)),
\end{aligned} \qquad (3.3.2-19)$$

which reproduces also the BRS transformation property Eq.(3.3.1-33) of $A_\mu{}^a$ if $d \circ C$ is identified with $\partial_\mu C \, dx^\mu$.

It is noteworthy that many important relations are condensed in the very simple equation, Eq.(12), expressing the nilpotency of (BRS-)coboundary operation $\boldsymbol{\delta}$. On the assumption of the BRS transformation property of A_μ, Eq.(3.3.1-33),

$$\boldsymbol{\delta}A_\mu{}^a = (D_\mu C)^a, \qquad (3.3.2-20)$$

the logical relations can be summarized in the following way :

i) First assuming the validity of Jacobi identity, Eq.(3.2.1-5), i.e.,

$$f^d{}_{ab}f^e{}_{dc} + f^d{}_{bc}f^e{}_{da} + f^d{}_{ca}f^e{}_{db} = 0, \qquad (3.3.2-21)$$

we obtain the equality

$$\boldsymbol{\delta}^2 A_\mu{}^a = (D_\mu(\boldsymbol{\delta}C + \frac{1}{2}gC \times C))^a, \qquad (3.3.2-22)$$

which implies the equivalence of the nilpotency condition, $\boldsymbol{\delta}^2 A_\mu{}^a = 0$, for $\boldsymbol{\delta}$ on $A_\mu{}^a$ with the Maurer-Cartan equation, Eq.(17), for the FP ghost C^a (on the assumption of the absence of constant term in $\boldsymbol{\delta}C^a$).

ii) On the other hand, if Eq.(17) is assumed, then the equation

$$\boldsymbol{\delta}^2 C^a = \frac{g^2}{6}(f^e_{\ bc}f^a_{\ ed} + f^e_{\ cd}f^a_{\ eb} + f^e_{\ db}f^a_{\ ec})C^b C^c C^d \qquad (3.3.2-23)$$

asserts that the validity of the Jacobi identity is equivalent to the nilpotency condition, $\boldsymbol{\delta}^2 C^a = 0$, of $\boldsymbol{\delta}$ on the FP ghost C^a [Nak 78d]. Namely, we obtain the following relationship :

$$\boldsymbol{\delta} A_\mu^{\ a} = (D_\mu C)^a$$

$$\Downarrow$$

$$\text{nilpotency}: \boldsymbol{\delta}^2 A_\mu^{\ a} = 0 \iff \text{Maurer-Cartan}: \boldsymbol{\delta} C^a = -\frac{1}{2}(C \times C)^a$$

$$\text{i)} \Uparrow \qquad \text{ii)} \Downarrow$$

$$\text{Jacobi identity} \iff \text{nilpotency}: \boldsymbol{\delta}^2 C^a = 0$$

$$(3.3.2-24)$$

iii) The BRS transformation and its nilpotency for B and $\overline{C}$ are trivial.[3]

iv) As for the matter field φ and its BRS transformation property, Eq.(3.3.1-34), i.e.,

$$\boldsymbol{\delta}\varphi^\alpha = igC^a(\tau^a)^\alpha_{\ \beta}\varphi^\beta, \qquad (3.3.2-25)$$

the equality [Nak 78d]

$$\boldsymbol{\delta}^2 \varphi = \frac{1}{2}g^2([\tau^a, \ \tau^b] - if^c_{\ ab}\tau^c)C^a C^b \varphi \qquad (3.3.2-26)$$

concludes the equivalence of the nilpotency condition, $\boldsymbol{\delta}^2 \varphi = 0$, on φ with the equality for τ^a,

$$[\tau_a, \ \tau_b] = if^c_{\ ab}\tau_c, \qquad (3.3.2-27)$$

expressing that the matrices $(-i)\tau^a$ give a *representation* of the Lie algebra $LieG$. In this context, it should be recalled [see Eq.(3.2.1-8)] that the Jacobi identity is nothing but the condition for the matrices $\text{ad}(X_a) = (\text{ad}(X_a)^c_{\ b}) = (f^c_{\ ab})$ $\quad (a = 1, \cdots, n)$ to constitute the adjoint *representation* of $LieG$:

$$[\text{ad}(X_a), \ \text{ad}(X_b)] = f^c_{\ ab}\text{ad}(X_c). \qquad (3.3.2-28)$$

---

[3]   Later in Sec.3.5.2, we will see that, if the theory contains a *bare mass term* of Yang-Mills field, BRS transformation should be deformed in this sector so as to retain the invariance and that this deformation violates the nilpotency [Oji 82].

On the assumption of Eq.(20), therefore, we can conclude that the three properties, nilpotency Eq.(12) of $\boldsymbol{\delta}$, Maurer-Cartan equation Eq.(17), and the preservation of Lie algebra structure (especially Jacobi identity), are in such a mutual relation that *any one of the three properties follows from the other two.*

In this way, the FP ghost C^a and BRS transformation can be accommodated in the theory not as artifacts but as natural notions related to the gauge degrees of freedom: They are to be interpreted, respectively, as the Maurer-Cartan form $\theta^a = gC^a$ and as the coboundary operation $\boldsymbol{\delta}$ acting on the forms on the Lie algebra $Lie\mathcal{G}$ of infinitesimal local gauge transformations. Therefore, the anticommuting property of C^a is seen to derive from the fact that it is a differential form constituting a Grassmann algebra of left-invariant forms on $Lie\mathcal{G}$. At the same time, it is also important to note that as the Maurer-Cartan form the FP ghost C^a represents *all* infinitesimal local gauge transformations in $\mathcal{G}$ in a *generic* way,

$$C^a(x)(\!(\Lambda)\!) = \frac{1}{g}\theta_x{}^a(\!(\Lambda)\!) = \frac{1}{g}\Lambda^a(x), \qquad (3.3.2-29)$$

but that it should *not* be identified with any *particular* $\Lambda \in Lie\mathcal{G}$. Thus, the BRS invariance of the gauge-fixed action defined by the Lagrangian density, Eq.(3.3.1-25), supplemented by Eqs.(3.3.1-2) and (3.3.1-26), can properly be regarded as reflecting the invariance of the theory under the *infinitesimal* local gauge transformations.

3.4 QUANTIZATION OF YANG-MILLS THEORY

We apply here the method of canonical quantization to the gauge-fixed Lorentz-invariant Lagrangian density of Yang-Mills theory.[1] Then the algebraic structure of conserved charges related to the FP ghosts and the BRS transformation can be explicitly constructed. The essential roles of this algebraic structure are found in its relation with the subsidiary condition specifying physical states and with the symmetry properties of Green's functions in the form of Slavnov-Taylor or Ward-Takahashi identities.

3.4.1 Canonical quantization of Yang-Mills theory

The Lagrangian density for quantum non-abelian gauge theory obtained in the previous discussion is summarized as follows :

$$\mathcal{L} = \mathcal{L}_0(A_\mu, \ \varphi) + \mathcal{L}_{\mathrm{GF}} + \mathcal{L}_{\mathrm{FP}}, \qquad (3.4.1-1)$$

$$\mathcal{L}_0(A_\mu, \ \varphi) = -\frac{1}{4}F_{\mu\nu}{}^a F^{\mu\nu a} + \mathcal{L}_{\mathrm{M}}(\varphi, \nabla_\mu\varphi), \qquad (3.4.1-2)$$

$$\mathcal{L}_{\mathrm{GF}} = -\partial^\mu B^a \cdot A_\mu{}^a + \frac{\alpha}{2}B^a B^a, \qquad (3.4.1-3)$$

$$\mathcal{L}_{\mathrm{FP}} = -i\partial^\mu \overline{C}^a \cdot (D_\mu C)^a, \qquad (3.4.1-4)$$

where

$$F_{\mu\nu}{}^a \equiv \partial_\mu A_\nu{}^a - \partial_\nu A_\mu{}^a + g(A_\mu \times A_\nu)^a, \qquad (3.4.1-5)$$

$$\nabla_\mu\varphi \equiv (\partial_\mu - iA_\mu{}^a g\tau^a)\varphi. \qquad (3.4.1-6)$$

Since Eq.(2) is rewritten in non-Laudau gauges ($\alpha \neq 0$) as

$$\mathcal{L}_{\mathrm{GF}} = -\frac{1}{2\alpha}(\partial^\mu A_\mu)^2 + \frac{\alpha}{2}\left(B + \frac{1}{\alpha}\partial^\mu A_\mu\right)^2 - \partial^\mu(B^a A_\mu{}^a), \qquad (3.4.1-7)$$

the present gauge fixing is equivalent to a frequently used form of covariant gauge fixing term, at the level of field equations and of Feynman diagrams. The introduction of the B-fields B^a will, however, be quite important in the following discussion of operator formalism, not only for the sake of treating the Landau gauge $\alpha = 0$

---

[1]   The presentation of the quantum Yang-Mills theory is based on [Kug 79d].

on the same footing as non-Landau ones, but also from conceptual viewpoint : It allows us to determine the form of BRS transformation independently of specific forms of Lagrangian densities, and hence, the *nilpotency* of the BRS transformation is ensured without the use of field equations, which is impossible in the formulation without B-fields.[2]

In this connection, it is very important to note that the gauge fixing and FP ghost terms, $\mathcal{L}_{\text{GF}}$ and $\mathcal{L}_{\text{FP}}$, can be added up into such a remarkable form [Oji 80a] as

$$\mathcal{L}_{\text{GF}} + \mathcal{L}_{\text{FP}} = i\boldsymbol{\delta}(\partial^\mu \overline{C}^a \cdot A_\mu{}^a - \frac{\alpha}{2}\overline{C}^a B^a). \qquad (3.4.1-8)$$

This shows that, in spite of the drastic changes of the theory due to the introduction of FP ghosts, the difference, $\mathcal{L} - \mathcal{L}_0 = \mathcal{L}_{\text{GF}} + \mathcal{L}_{\text{FP}}$, between the original Lagrangian density $\mathcal{L}_0$ and the total quantum one $\mathcal{L}$, is just a **BRS-coboundary** term of such a form as $\boldsymbol{\delta}(*)$. Thus, the BRS invariance of $\mathcal{L}$ is a natural consequence of local gauge invariance of the original Lagrangian density $\mathcal{L}_0$ as well as of the nilpotency of BRS anti-derivation $\boldsymbol{\delta}$. This observation evidently leads to an extension of the procedure for attaining the BRS invariant theory with arbitrary choice of gauge fixing conditions F : Namely, it suffices to replace $\partial^\mu A_\mu{}^a$ by $F[A]^a$ and then to take

$$\mathcal{L}_{\text{GF}} + \mathcal{L}_{\text{FP}} = i\boldsymbol{\delta}\left(\overline{C}^a(F[A]^a - \frac{\alpha}{2}B^a)\right). \qquad (3.4.1-9)$$

To be explicit, however, we restrict the choice of gauge fixing condition to the above one, Eq.(3).

As explained earlier, the factor i is introduced in front of the FP ghost term Eq.(4) in order for both FP ghosts, C^a and $\overline{C}^a$, to be hermitian fields [Kug 78a]:

$$C^{a\dagger} = C^a, \quad \overline{C}^{a\dagger} = \overline{C}^a. \qquad (3.4.1-10)$$

Hermiticity properties of other fields are taken as usual. It should be stressed that, only when this hermiticity assignment, Eq.(10), to the FP ghosts is adopted, the Lagrangian density $\mathcal{L}$ becomes hermitian, and hence, the total S-matrix becomes (pseudo-)unitary :

$$\mathcal{L}^\dagger = \mathcal{L}, \quad S^\dagger S = SS^\dagger = 1. \qquad (3.4.1-11)$$

---

[2]   In the context of effective action, the necessity of introducing an auxiliary field for the manifest nilpotency was first noticed in [Bec 76b].

Now, the Euler-Lagrange equations for $A_\mu{}^a$, B^a, C^a and $\overline{C}^a$ are given by

$$(D^\mu F_{\mu\nu})^a = \partial_\nu B^a - g j_\nu{}^a - ig(\partial_\nu \overline{C} \times C)^a, \qquad (3.4.1-12)$$

$$\partial^\mu A_\mu{}^a + \alpha B^a = 0, \qquad (3.4.1-13)$$

$$\partial^\mu (D_\mu C)^a = 0, \qquad (3.4.1-14)$$

$$(D^\mu \partial_\mu \overline{C})^a = 0, \qquad (3.4.1-15)$$

where the matter current $j_\mu{}^a$ is defined by

$$j_\mu{}^a \equiv -i(\tau^a \varphi)^\alpha \frac{\partial \mathcal{L}}{\partial(\partial^\mu \varphi^\alpha)}. \qquad (3.4.1-16)$$

The field equation directly governing the time development of B-field can be obtained by applying the covariant differentiation D^ν to Eq.(12) and by using Eqs.(3.2.2-4) and (15),

$$(D^\mu \partial_\mu B)^a = ig(\partial_\mu \overline{C} \times D^\mu C)^a, \qquad (3.4.1-17)$$

which agrees also with the BRS transform of Eq.(15), showing the internal consistency of the theory.

In order to give canonical (anti)commutation relations, we need the canonical conjugate momenta, which we define by

$$\pi_B^a \equiv \partial \mathcal{L}/\partial \dot{B}^a = -A_0{}^a, \qquad (3.4.1-18)$$

$$\pi_{A_k}^a \equiv \partial \mathcal{L}/\partial \dot{A}_k{}^a = F_{0k}{}^a, \quad (k = 1, 2, 3) \qquad (3.4.1-19)$$

$$\pi_C^a \equiv (\partial/\partial \dot{C}^a)\mathcal{L} = +i\dot{\overline{C}}^a, \qquad (3.4.1-20)$$

$$\pi_{\overline{C}}^a \equiv (\partial/\partial \dot{\overline{C}}^a)\mathcal{L} = -i(D_0 C)^a = -i(\dot{C} + gA_0 \times C)^a, \qquad (3.4.1-21)$$

$$\pi_\varphi^\alpha \equiv (\partial/\partial \dot{\varphi}^\alpha)\mathcal{L}, \qquad (3.4.1-22)$$

For these momentum variables $\pi_\Phi^I (= B^a, A_k{}^a, C^a, \overline{C}^a, \varphi^\alpha)$, we take the following canonical (anti)commutation relations as usual :

$$[\pi_\Phi^I(x), \ \Phi_J(y)]_{\mp,0} = -i\delta^I{}_J \delta(\mathbf{x} - \mathbf{y}),$$
$$[\pi_\Phi^I(x), \ \pi_\Phi^J(y)]_{\mp,0} = [\Phi_I(x), \ \Phi_J(y)]_{\mp,0} = 0, \qquad (3.4.1-23)$$

where the anticommutators $(+)$ are taken only if both are fermion fields. Although the FP ghosts C^a and $\overline{C}^a$ should be taken as fermionic fields as explained in Sec.3.2, we adopt here the convention, just for simplicity, of taking *commutators* $(-)$ in Eq.(23) between any one of FP ghost fields and any one of fermion matter fields in φ^α.[3]

Finally, we add the following three remarks :

First, we note the importance of the hermiticity property, Eq.(10), of the FP ghosts for guaranteeing the whole consistency of the present formulation. For instance, if we had adopted such a *wrong* hermiticity assignment as[4]

$$C^{a\dagger} = i\overline{C}^a, \quad \overline{C}^{a\dagger} = iC^a, \tag{3.4.1 - 24}$$

then not only the hermiticity of the Lagrangian density Eq.(11) would be violated,

$$\mathcal{L}^\dagger - \mathcal{L} = ig\partial^\mu A_\mu{}^a(\overline{C} \times C)^a - ig\partial^\mu(A_\mu{}^a(\overline{C} \times C)^a) \neq 0, \tag{3.4.1 - 25}$$

but also fatal inconsistency would arise as will be seen later.

Second, in Eq.(18), we take the time-component of gauge field $A_0{}^a$ not as canonical coordinates but as the momentum variables conjugate to the B^a's. If we treat the $A_0{}^a$'s also as independent canonical variables, Eq.(18) and $\varphi_1^a \equiv \pi_B^a + A_0{}^a = 0$, together with $\varphi_2^a \equiv \pi_0{}^a \equiv \partial\mathcal{L}/\partial\dot{A}_0{}^a = 0$, become *second class constraints* in Dirac's treatment of constrained systems (see Chapter 1 Addendum 1-A). Then, canonical (anti)commutation relations should be determined according to the definition of Dirac brackets given by

$$\begin{aligned}
(\alpha, \beta)_{\mathrm{D}} &\equiv (\alpha, \beta)_{\mathrm{P}} - (\alpha, \varphi_i^a)_{\mathrm{P}}(C^{-1})_{ij}(\varphi_j^a, \beta)_{\mathrm{P}}, \\
\delta^{ab}C_{ij} &\equiv (\varphi_i^a, \varphi_j^b)_{\mathrm{P}}, \ (i, j = 1, 2)
\end{aligned} \tag{3.4.1 - 26}$$

instead of the usual Poisson brackets $(\alpha, \beta)_{\mathrm{P}}$. This Dirac procedure, however, gives just the same canonical (anti)commutation relations as those obtained by the

[3]   In the presence of FP ghost number conservation explained later, either commutator or anticommutator can be adopted without altering physics owing to the Klein transformation related to FP ghost charge.

[4]   Historically, the wrong assignment Eq.(24) (see, for instance, [Cur 76b], etc.) prevailed because the necessity of FP ghosts were found originally in the Feynman-diagrammatic approach [Fey 63, DeW 67], in which the relation between C^a and $i\overline{C}^a$ is quite similar to that between a complex scalar field ϕ and its conjugate $\phi^\dagger$. Note also that the Feynman path-integral approach can say nothing about hermiticity property in the indefinite-metric theory.

above simplified treatment.[5] Further, even if our starting gauge fixing Lagrangian density $\mathcal{L}_{GF}$, Eq.(3), is changed by the following replacement :

$$-\partial^\mu B^a \cdot A_\mu{}^a \to \omega B^a \cdot \partial^\mu A_\mu{}^a + (\omega - 1)\partial^\mu B^a \cdot A_\mu{}^a \qquad (3.4.1-27)$$

with an arbitrary real ω, this Dirac's method reproduces the same results. In particular, in the $\omega = 1$ case, we may take the B-fields as non-canonical as done in Sec.2.3.2.

Finally, we note here that, as noted in Explanatory Notes, the left-differentiation convention have been adopted in Eqs.(20)-(22) with respect to the anticommuting number such as the FP ghosts or fermion matter fields in φ^α; that is, the differential operator $(\partial/\partial\xi)$ is supposed to be an anti-derivation

$$(\partial/\partial\xi)AB = [(\partial/\partial\xi)A]B + (-1)^{p_A}A[(\partial/\partial\xi)B], \qquad (3.4.1-28)$$

where A and B are any monomials in the commuting and anticommuting numbers and p_A is the number of factors anticommuting with ξ contained in A. [Note that we are taking the convention that the fermion matter fields in φ^α, if any, *commute* with the FP ghosts.] Corresponding to this convention of left-differentiation, the Hamiltonian density $\mathcal{H}$ should be constructed as

$$\mathcal{H} = \dot{\Phi}_I \pi_\Phi^I - \mathcal{L} \qquad (3.4.1-29)$$

but not as $\mathcal{H} = \pi_\Phi^I \dot{\Phi}_I - \mathcal{L}$. This is because the variation of Hamiltonian density $\delta\mathcal{H} = \delta\dot{\Phi}_I \pi_\Phi^I + \dot{\Phi}_I \delta\pi_\Phi^I - \delta\mathcal{L}$ has to be independent of the velocity variation $\delta\dot{\Phi}_I$, which can be materialized by $\delta\mathcal{L} = \delta\Phi_I(\partial/\partial\Phi_I)\mathcal{L} + \delta\dot{\Phi}_I(\partial/\partial\dot{\Phi}_I)\mathcal{L}$ due to the left-differentiation Eq.(28).

3.4.2 BRS charge, FP ghost charge and BRS algebra

Now the canonical (anti)commutation relations, Eqs.(3.4.1-23) combined with Eqs.(3.4.1-18)-(3.4.1-22), set up in the previous subsection allow us to ensure that the BRS charge Q_B, Eq.(3.3.1-32), actually generates the BRS transformation.

---

[5]   An analogous situation indeed occurs in the canonical anticommutation relation of Dirac field ψ : In the usual treatment, the variable $i\psi^\dagger$ is not taken as a coordinate but as a momentum variable π_ψ conjugate to ψ.

Since it is a *global* symmetry transformation in contrast to the original *local* gauge transformation, we can utilize here the general consequence of the *first Noether theorem* in Sec.1.2.4 : Namely, without direct computation, the (anti)commutation relations, Eqs.(3.3.1-33)-(3.3.1-37), giving the BRS transformation on the field operators can be guaranteed to hold. All what one should do for this purpose is just to check that the Noether current deriving from the BRS invariance of the total Lagrangian density $\mathcal{L}$, Eqs.(3.4.1-1)-(3.4.1-4), reproduces the BRS charge Q_B defined by Eq.(3.3.1-32). The former is given by

$$
\begin{aligned}
J_B{}^\mu &= \sum_{\Phi_I = A_\nu, \varphi, B, C, \bar{C}} \boldsymbol{\delta}\Phi_I \frac{\partial \mathcal{L}}{\partial(\partial_\mu \Phi_I)} \\
&= (D_\nu C)^a \frac{\partial \mathcal{L}}{\partial(\partial_\mu A_\nu{}^a)} + i(C^a g T^a \varphi)^\alpha \frac{\partial \mathcal{L}}{\partial(\partial_\mu \varphi^\alpha)} \\
&\quad - \frac{1}{2} g (C \times C)^a \frac{\partial \mathcal{L}}{\partial(\partial_\mu C^a)} + iB^a \frac{\partial \mathcal{L}}{\partial(\partial_\mu \overline{C}^a)}.
\end{aligned} \qquad (3.4.2-1)
$$

With the aid of the field equation, Eq.(3.4.1-12), of $A_\mu{}^a$, it can be rewritten as

$$
J_{B\mu} = B^a (D_\mu C)^a - \partial_\mu B^a \cdot C^a + \frac{i}{2} g \partial_\mu \overline{C}^a \cdot (C \times C)^a - \partial^\nu (F_{\mu\nu}{}^a C^a). \qquad (3.4.2-2)
$$

Then it is evident that this Noether current $J_{B\mu}$ agrees with the conserved current derived in Sec.3.3.1 [see Eq.(3.3.1-23) combined with Eqs.(3.3.1-27) and (3.3.1-29)], aside from the total divergence term on rhs. of Eq.(2) which is trivially conserved owing to the antisymmetry of $F_{\mu\nu}{}^a$ in μ and ν. Unless the product of $F_{\mu\nu}{}^a$ and FP ghost C^a generates a *massless bound-state spectrum*, this divergence term does not contribute at the spatial infinity to the conserved charge obtained as a volume integral of J_{B0}. Then it reproduces just the BRS charge Q_B given by Eq.(3.3.1-32):

$$
Q_B = \int d\boldsymbol{x} [B^a (D_0 C)^a - \dot{B}^a C^a + \frac{i}{2} g \dot{\overline{C}}^a (C \times C)^a]. \qquad (3.4.2-3)
$$

Thus, Eqs.(3.3.1-33)-(3.3.1-37), i.e.,

$$
[iQ_B,\ \Phi_I(x)]_{\mp} = \boldsymbol{\delta}\Phi_I(x) \qquad \text{for } \Phi_I = A_\mu{}^a, \varphi, B^a, C^a, \text{ and } \overline{C}^a, \qquad (3.4.2-4)
$$

have been established.

Next, we have another important conserved charge, FP ghost charge Q_C, in our system defined by Eq.(3.4.1-1) : It is easily seen that the FP ghost number is conserved. Unlike the usual case of fermion number conservation, however, the conservation of FP ghost number is *not* due to the invariance under a *phase* transformation

$$C^a \to e^{i\theta} C^a,$$
$$\overline{C}^a \to e^{-i\theta} \overline{C}^a.$$

In fact, such a phase transformation is incompatible with our fundamental hermiticity assignment, Eq.(3.4.1-10), to C^a and $\overline{C}^a$. Instead, an invariance exists under the *scale* transformation,

$$C^a \to e^{\theta} C^a,$$
$$\overline{C}^a \to e^{-\theta} \overline{C}^a. \qquad (3.4.2-5)$$

with a real parameter θ, which is fully consistent with the hermiticity of C^a and $\overline{C}^a$. The corresponding conserved current and charge are given, respectively, by

$$J_{C\mu} = i(\overline{C}^a (D_\mu C)^a - \partial_\mu \overline{C}^a \cdot C^a); \qquad \partial^\mu J_{C\mu} = 0, \qquad (3.4.2-6)$$

$$Q_C = i \int d\mathbf{x} (\overline{C}^a (D_0 C)^a - \dot{\overline{C}}^a C^a) = Q_C^\dagger. \qquad (3.4.2-7)$$

This charge Q_C, called **FP ghost charge**, indeed generates the above transformation, Eq.(5), on the FP ghost fields and leaves all other fields invariant :

$$[\, iQ_C, \ C^a(x)] = C^a(x),$$
$$[\, iQ_C, \ \overline{C}^a(x)] = -\overline{C}^a(x). \qquad (3.4.2-8)$$

The **FP ghost number** N_{FP} is then identified with the eigenvalue of the operator Q_C *multiplied by* i : $N_{\text{FP}} = $ eigenvalue of iQ_C. Although it may sound strange that the *hermitian* operator Q_C has pure imaginary eigenvalues, $-i$ times integers, this is fully consistent in the presence of *indefinite metric* : Whenever pure imaginary eigenvalues $-iN_{\text{FP}}$ appear, they are in complex conjugate pairs with $+iN_{\text{FP}}$. As will be seen in Chapter 4, this fact is very important not only for guaranteeing the hermiticity of Q_C, but also for materializing the norm-cancellation mechanism to "confine" unphysical ghosts with negative norms.

Now, from the definitions of Q_B and Q_C, it follows that they constitute quite a simple algebraic structure characterized by the following relations:

$$\{\, Q_B, \ Q_B\} = 2(Q_B)^2 = 0, \qquad (3.4.2-9)$$

$$[\, iQ_C, \ Q_B] = Q_B, \qquad (3.4.2-10)$$

$$[\, Q_C, \ Q_C] = 0. \qquad (3.4.2-11)$$

This superalgebra consisting of BRS charge Q_B and FP ghost charge Q_C is called **BRS algebra** in what follows. Since Eq.(11) is trivial, it suffices to verify Eqs.(9) and (10). To do this, we note the following interesting relation between the local currents, $J_{B\mu}$ and $J_{C\mu}$, of BRS charge Q_B and FP ghost charge Q_C,

$$J_{B\mu} = -\boldsymbol{\delta} J_{C\mu} - \partial^\nu (F_{\mu\nu}{}^a C^a), \qquad (3.4.2-12)$$

which can be obtained by comparing Eqs.(2) and (6) in the light of BRS transformation properties of $\overline{C}^a, C^a$ and $D_\mu C^a$:

$$\boldsymbol{\delta}\overline{C}^a = iB^a, \qquad (3.4.2-13)$$

$$\boldsymbol{\delta}(D_\mu C)^a = \boldsymbol{\delta}^2 A_\mu{}^a = 0, \qquad (3.4.2-14)$$

$$\boldsymbol{\delta}C^a = -\frac{g}{2}(C \times C)^a. \qquad (3.4.2-15)$$

Eq.(10), expressing the fact that the BRS charge Q_B carries the FP ghost number $N_{\text{FP}} = 1$, follows from Eq.(12) as its volume-integral form:

$$\begin{aligned}
Q_B &= \int d\mathbf{x} J_{B0} = -\int d\mathbf{x}\Big(\boldsymbol{\delta} J_{C0} + \partial^i (F_{0i}{}^a C^a)\Big) \\
&= -\boldsymbol{\delta} Q_C = -[\, iQ_B,\ Q_C].
\end{aligned} \qquad (3.4.2-16)$$

The remarkable *nilpotency* property, Eq.(9), of the BRS charge Q_B, can be derived from the same property, Eq.(3.3.2-12), of the BRS transformation $\boldsymbol{\delta}$ by using Eq.(16):

$$2i(Q_B)^2 = \{iQ_B,\ Q_B\} = \boldsymbol{\delta} Q_B = \boldsymbol{\delta}(-\boldsymbol{\delta} Q_C) = -\boldsymbol{\delta}^2 Q_C = 0. \qquad (3.4.2-17)$$

Although the algebraic structure characterized by Eqs.(9)-(11) looks very simple, it will be seen in Chapter 4 to play a vitally important role in combination with the above characteristic feature of the spectrum of Q_C in ensuring the physical S matrix unitarity through the "confinement" of unphysical particles with negative norms.

Finally, some remarks are in order. The hermiticity assignment, Eq.(3.4.1-10), to FP ghosts plays an important role also here in assuring the consistency of the formulation. First, the charges Q_B and Q_C are hermitian only when this assignment is adopted :

$$Q_B^\dagger = Q_B, \qquad Q_C^\dagger = Q_C. \qquad (3.4.2-18)$$

If such a *wrong* hermiticity assignment as $C^{a\dagger} = \overline{C}^a$ (to be precise, $C^{a\dagger} = i\overline{C}^a$ as in Eq.(3.4.1-24)) is adopted, $Q_B{}^\dagger$ and $Q_C{}^\dagger$ become quite different quantities having no simple algebraic relations with the original Q_B and Q_C, and, moreover, are not assured to be conserved by equations of motion. This contradicts the conservation of Q_B following from $(d/dt)\langle\alpha|Q_B{}^\dagger|\beta\rangle = \langle\beta|(d/dt)Q_B|\alpha\rangle^* = 0$, which should hold as long as Q_B is well-defined. Further, the above wrong assignment is also incompatible with the BRS transformation Eq.(4) itself : In fact, since the transformed field $C^{a\prime} = C^a + \lambda\delta C^a$ should have the same hermiticity property as the original one C^a, the relation $(\lambda\delta C^a)^\dagger = \lambda i\delta\overline{C}^a$ is required to hold in the conventional case, while

$$(\delta C)^{a\dagger} = -\frac{g}{2}(C \times C)^{a\dagger} \neq -B^a = i\delta\overline{C}^a. \qquad (3.4.2-19)$$

As for the assignment $C^{a\dagger} = C^a$, $\overline{C}^{a\dagger} = \overline{C}^a$, the BRS transformation is fully consistent with it, as long as λ is assumed to be "anticommuting number", as usual, obeying the rule

$$(\lambda\mathcal{R})^\dagger = \mathcal{R}^\dagger\lambda^\dagger \qquad (3.4.2-20)$$

for arbitrary operator $\mathcal{R}$, and be "purely imaginary":

$$\lambda^\dagger = -\lambda. \qquad (3.4.2-21)$$

Incidentally, the usage of such "anticommuting numbers" λ belonging to a Grassmann algebra unifies the treatment of "super Lie bracket" structure $[\,\cdot\,,\,\cdot\,]_\mp$ appearing in the BRS transformation Eq.(4) into Lie bracket as

$$[i\lambda Q_B, \Phi_I(x)] = \lambda\delta\Phi_I(x), \qquad (3.4.2-22)$$

which may be just a matter of convenience, however.

3.4.3 Subsidiary condition, quantum Maxwell equation and symmetries

As explained in Chapter 2, the total state vector space $\mathcal{V}$ in a covariant formulation of gauge theory necessarily contains negative norm states: i.e., $\mathcal{V}$ has an *indefinite metric*. In order to ensure a physically meaningful interpretation of the theory, we should pick up consistently a *physical subspace* $\mathcal{V}_{\text{phys}} = \{|\text{phys}\rangle\}$ so that it satisfies the requirement of time invariance and norm-positivity. In abelian gauge theories, the physical subspace $\mathcal{V}_{\text{phys}}$ is specified by a concise subsidiary condition, Eq.(2.3.3-11), i.e.,

$$B^{(+)}(x)|\text{phys}\rangle = 0. \qquad (3.4.3 - 1)$$

In spite of all the complications in non-abelian case due to the nonlinear equation of B-field, Eq.(3.4.1-17), the introduction of FP ghosts as well as the BRS transformation brings about a great simplification in an astonishing way: The physical subspace $\mathcal{V}_{\text{phys}} = \{|\text{phys}\rangle\}$ in this case turns out to be specified by a more elegant **subsidiary condition** proposed by Kugo and Ojima [Kug 78a, Kug 78c][6]:

$$Q_B|\text{phys}\rangle = 0; \qquad \mathcal{V}_{\text{phys}} \equiv \{|\Phi\rangle; \ Q_B|\Phi\rangle = 0\}. \qquad (3.4.3 - 2)$$

This condition essentially expresses the gauge invariance of physical states belonging to $\mathcal{V}_{\text{phys}}$. It is analogous to the above subsidiary condition, Eq.(1): In the abelian case, the field $B(x)$ [or more precisely, the conserved charge $\int d\mathbf{x}(\Lambda(x)\dot{B}(x) - \dot{\Lambda}(x)B(x))$ with $\Box\Lambda = 0$] represents a generator of (a class of residual) local gauge transformations, while the BRS charge Q_B is shown in Sec. 3.3.2 to be the generator of a quantum version of infinitesimal (non-abelian) local gauge transformation. In fact, we can now show that the condition, Eq.(2), really reduces in the abelian case to the above familiar condition, Eq.(1), owing to the special feature of abelian gauge theories [Kug 78a] : Since the structure constant vanishes in this case, not only B-field but also the FP ghost fields C and $\overline{C}$ satisfy *free* field equation as is seen from Eqs.(3.4.1-14) and (3.4.1-15) :

$$\Box B = \Box C = \Box \overline{C} = 0. \qquad (3.4.3 - 3)$$

For the same reason, the BRS charge given by Eq.(3.4.2-3) takes such a simple form as

$$Q_B = \int d\mathbf{x} : (B\dot{C} - \dot{B}C) := i\sum_k (C_k{}^\dagger B_k - B_k{}^\dagger C_k), \qquad (3.4.3 - 4)$$

---

[6]   Digression : Regarding the priority of the subsidiary condition, Eq.(2), one of the authors (I.O.) in collaboration with T.Kugo once made a misjudgement in [Kug 79d] by attributing it to the paper by Curci and Ferrari [Cur 76c]. What they actually did in their paper was *not* to propose a subsidiary condition characterizing the physical subspace $\mathcal{V}_{\text{phys}}$ in the Heisenberg picture, but they just observed that the *transverse modes* of Yang-Mills field satisfies this equality without clarifying in what picture they were working. Further, because of the lack of B-fields, it was not possible in their paper to guarantee the nilpotency of BRS transformation and BRS charge.

where $B_k{}^\dagger$ (B_k) and $C_k{}^\dagger$ (C_k) denote creation (annihilation) operators of B and C fields, respectively, corresponding to some complete set of wave packets $\{g_k\}$ satisfying d'Alembert equation (see Appendix $A.5$) and $:\cdots:$ means the Wick normal ordering. Although FP ghosts are indispensable for treating the BRS transformation, they are totally decoupled from other fields and are irrelevant degrees of freedom in the *standard* formulation of abelian gauge theories (at least in its covariant linear gauges). Namely, the total state vector space $\mathcal{V}$ in the abelian case is decomposed in a *time-invariant* manner into the tensor product :

$$\mathcal{V} = \mathcal{V}' \otimes \mathcal{V}_{\text{FP}}. \qquad (3.4.3-5)$$

Here, $\mathcal{V}_{\text{FP}}$ is the Fock space of FP ghosts and antighosts, generated by the creation and annihilation operators of C and $\overline{C}$ fields from the FP ghost-Fock vacuum $|0\rangle_{\text{FP}}$. $\mathcal{V}'$ is the state vector space in the standard formulation without FP ghosts, as is discussed in Chapter 2. It can be identified in the larger space $\mathcal{V}$, Eq.(5), with its time-invariant subspace $\mathcal{V}' \otimes |0\rangle_{\text{FP}}$ containing *no* FP ghosts, which is just characterized by the additional subsidiary conditions

$$C^{(+)}(x)|\text{phys}\rangle = \overline{C}^{(+)}(x)|\text{phys}\rangle = 0. \qquad (3.4.3-6)$$

Thus, putting $|\text{phys}\rangle \equiv |f\rangle \otimes |0\rangle_{\text{FP}}$, we see that the subsidiary condition, Eq.(2), with Q_B given by Eq.(4) reduces in this subspace, $\mathcal{V}' \otimes |0\rangle_{\text{FP}}$, to

$$Q_B(|f\rangle \otimes |0\rangle_{\text{FP}}) = i \sum_k B_k|f\rangle \otimes |C_k\rangle_{\text{FP}} = 0. \qquad (3.4.3-7)$$

Here we have denoted $|C_k\rangle \equiv C_k{}^\dagger|0\rangle$, etc. By the linear independence of $|C_k\rangle$, we obtain

$$B_k|f\rangle = 0 \qquad \text{for all } k, \qquad (3.4.3-8)$$

which is nothing but the familiar condition, $B^{(+)}(x)|f\rangle = 0$ (for all x), in the abelian case.[7] In this way, the present subsidiary condition, Eq.(2), gives a natural extension of Gupta condition in the abelian cases to any type of gauge theories in a much simpler form than its prototype. The condition, Eq.(2), provides an important basis in the present formalism on which we develop all the discussions hereafter.

---

[7]   In [Nis 87], the Gupta condition, Eq.(8), is derived from the subsidiary conditions Eqs.(2) and (6) as their *consistency condition*.

We should note here that the BRS charge Q_B in the above has been implicitly assumed *not* to be broken spontaneously. This assumption is equivalent to one of the following statements:

(i) $Q_B |0\rangle = 0,$ $(3.4.3 - 9)$

(ii)   The vacuum is physical : $|0\rangle \in \mathcal{V}_{\text{phys}}.$ $(3.4.3 - 10)$

The first simple equation, Eq.(9), will be seen in the next subsection to play a crucial role in the derivation of Ward-Takahashi identities. Violation of.this equality causes the breakdown of Ward-Takahashi identities, which is closely related to the appearance of (non-abelian) gauge anomaly [Bau 84, Bau 85] endangering the consistent renormalization of a gauge theory. In this sense, the assumption of Eq.(9) is almost an inevitable requirement for the consistent formulation of a gauge theory.

We shall see in Chapter 4 how the physical S-matrix unitarity is assured generally by the subsidiary condition, Eq.(2), together with some explicit examples. This simple condition turns out to be really sufficient for ensuring the physical S-matrix unitarity. However, if one prefers specifying the physical subspace as small as possible, one may add one more subsidiary condition [Kug 78a, Kug 78c]:

$$Q_C|\text{phys}\rangle = 0,   \qquad (3.4.3 - 11)$$

where Q_C is the conserved FP ghost charge given by Eq.(3.4.2-7). This additional condition works only in reducing the physical subspace to the vanishing-FP-ghost-number sector, and it does *not* change the genuine physical contents of the theory at all. Therefore, the condition Eq.(11), will not be imposed in this book unless it is explicitly mentioned.

Although the gauge-fixing term is to break the local gauge invariance, our choice of it, Eq.(3.4.1-3), does not break the symmetry of the theory under the *global* gauge transformation. Therefore, the latter is inherited from $\mathcal{L}_0(A_\mu, \varphi)$ to the total Lagrangian density $\mathcal{L}$ given by Eq.(3.4.1-1). Then, the Noether current of the global gauge transformation,

$$J_\mu{}^a = (A^\nu \times F_{\nu\mu})^a + j_\mu{}^a + (A_\mu \times B)^a - i(\overline{C} \times D_\mu C)^a + i(\partial_\mu \overline{C} \times C)^a, \;(3.4.3 - 12)$$

is conserved ($\partial^\mu J_\mu{}^a = 0$) corresponding to the invariance of $\mathcal{L}$ under the global gauge transformation given by

$$[i\epsilon^a Q^a, \; \Phi^b(x)] = -f^{abc}\epsilon^a \Phi^c(x) \qquad \text{for } \Phi^b = A_\mu{}^b, B^b, C^b \text{ and } \overline{C}^b,$$
$$[i\epsilon^a Q^a, \; \varphi(x)] = -i\epsilon^a \tau^a \varphi(x).   \qquad (3.4.3 - 13)$$

The global ('color' or 'flavor') charge Q^a is given by $Q^a = \int dx\, J_0{}^a$ and the matter current $j_\mu{}^a$ is defined by Eq.(3.4.1-16). Note that this global invariance would not be manifest (if any in physical sector) under *asymmetric* gauge-fixing choices such as R_ξ-gauges [Fujk 72]. As was first noted in [Oji 78], the field equation of $A_\mu{}^a$, Eq.(3.4.1-12), can now be rewritten into the following remarkable form by the use of the BRS charge Q_B and Eqs.(3.3.1-33), (3.3.1-37) and (12) :

$$\partial^\nu F_{\nu\mu}{}^a + g J_\mu{}^a = \{Q_B,\ D_\mu \overline{C}^a\}. \qquad (3.4.3-14)$$

This is just the **quantum Maxwell equation** in the non-abelian case, corresponding to Eq.(2.3.1-3) in abelian case. It can also be interpreted as the quantum version of Eq.(3.2.2-13), one of the important results of second Noether theorem, whose modification is due to the gauge fixing and FP ghost terms in Eq.(3.3.1-25) added to the original local-gauge-invariant Lagrangian density $\mathcal{L}_0$. The form of this modification on rhs. of Eq.(14), expressed as an anticommutator of the BRS charge Q_B, clearly shows that the classical Maxwell-type equation

$$\langle f_1 | (\partial^\nu F_{\nu\mu}{}^a + g J_\mu{}^a) | f_2 \rangle = 0 \qquad (3.4.3-15)$$

holds for any physical states $|f_1\rangle, |f_2\rangle \in \mathcal{V}_{\text{phys}}$ specified by Eq.(2). Eq.(14) will play an important role in the discussion of observables in Sec.4.3 and of the relation between color confinement and spontaneous symmetry breakdown in Sec.4.4. It should be kept in mind that the $J_\mu{}^a$'s in QCD are the currents of global color transformation which is supposed to be unbroken.

Aside from other possible continuous symmetries such as chiral symmetry, flavor symmetry, etc., the present system described by $\mathcal{L}$, Eq. (3.4.1-1), has also the basic discrete symmetries, P (space inversion, or parity transformation), C (charge conjugation) and T (time reversal) if the original Lagrangian $\mathcal{L}_0(A_\mu, \varphi)$ has. Since it is not intended in this book to discuss discrete symmetries, we only note here the **PCT invariance**: If the matter Lagrangian density, $\mathcal{L}_M(\varphi, \partial_\mu \varphi)$, defines a PCT invariant theory with a suitable PCT transformation law for the matter fields φ^α, then $\mathcal{L}$, Eq.(3.4.1-1), is invariant under the following *anti-linear* PCT transformation [Kug 79d]:

$$\text{PCT}: \quad \begin{aligned} A_\mu^{a,\text{PCT}}(x) &= -A_\mu{}^a(-x), \\ B^{a,\text{PCT}}(x) &=\ \ B^a(-x), \\ C^{a,\text{PCT}}(x) &= -C^a(-x), \\ \overline{C}^{a,\text{PCT}}(x) &=\ \ \overline{C}^a(-x). \end{aligned} \qquad (3.4.3-16)$$

[The extraordinary minus sign in front of FP ghost field C^a should be noticed.] Then, we can safely suppose the existence of the anti-unitary PCT operator Θ satisfying

$$\Theta|0\rangle = |0\rangle, \qquad (3.4.3-17)$$

$$\Theta^2 = 1, \qquad (3.4.3-18)$$

$$\Theta\Phi_I(x)\Theta = \Phi_I{}^{\text{PCT}}(x), \qquad (3.4.3-19)$$

where $\Phi_I{}^{\text{PCT}}(x)$ stands generically for the PCT-transformed fields in Eq.(16) supplemented by the PCT-transformed matter fields $\varphi^{\alpha,\text{PCT}}(x)$. Under the PCT transformation, the BRS charge Q_B, the FP ghost charge Q_C and the generators Q^a of the global gauge transformation behave as follows [Kug 79d]:

$$\Theta Q_B\Theta = Q_B, \qquad (3.4.3-20)$$

$$\Theta Q_C\Theta = -Q_C, \qquad (3.4.3-21)$$

$$\Theta Q^a\Theta = -Q^a. \qquad (3.4.3-22)$$

Note that the BRS transformation Eq.(3.4.2-4) is consistent with the PCT transformation, Eqs.(16) and (20).

3.4.4 BRS invariance and Ward-Takahashi identities

As is well known, **Ward-Takahashi(WT) identities** are, in general, the identities among Green's functions following from the existence of a symmetry. Namely, such an equality as

$$\langle 0|\delta\mathcal{R}|0\rangle = \langle 0|[iQ,\ \mathcal{R}]_{\mp}|0\rangle = 0 \qquad (3.4.4-1)$$

holds for any physical quantity $\mathcal{R}$, when the vacuum is invariant,[8]

$$Q|0\rangle = 0, \qquad (3.4.4-2)$$

under an infinitesimal symmetry transformation δ generated by a Noether charge Q:

$$\delta\mathcal{R} = [iQ,\ \mathcal{R}]_{\mp}. \qquad (3.4.4-3)$$

---

[8]   For a *spontaneously* broken symmetry, the vacuum is *not* invariant, but a modified version of Eq.(1) is valid with a nonvanishing rhs. due to the massless NG boson.

Choosing as $\mathcal{R}$ a product of field operators $\Phi_1(x_1)\Phi_2(x_2)\cdots\Phi_n(x_n)$, we can easily obtain an equation among Green's functions when the explicit form of the infinitesimal symmetry transformation $\delta(\Phi_I)$ is specified for each field Φ_I.

Although the local gauge invariance at the classical level has been broken in the quantum theory through the gauge fixing procedure, WT identities hold inheriting its essence, without which the renormalizability cannot be guaranteed. In the non-abelian case, WT identities of this sort [Tay 71, Sla 72] are called **Slavnov-Taylor identities**. As already mentioned in Sec. 3.1, the discovery of BRS symmetry [Bec 76a] has simplified very much the derivation of these identities which were derived previously by 't Hooft and Veltman [tHo 72] in a very complicated diagrammatic way. As shown in Sec.3.3.2, the BRS symmetry is a *global* gauge symmetry summarizing the essence of local gauge invariance, and hence, they naturally follow, in the same way as Eq.(1), from the invariance of the vacuum under the BRS transformation :

$$Q_B|0\rangle = 0 \qquad\qquad (3.4.4-4)$$

$$\Downarrow$$

$$\langle 0|\boldsymbol{\delta}\mathcal{R}|0\rangle = \langle 0|[iQ_B,\ \mathcal{R}]_{\mp}|0\rangle = 0. \qquad (3.4.4-5)$$

From the conceptual point of view, what is more important is that the complicated Slavnov-Taylor identities are now condensed in such simple operator equation as Eq.(5) (in combination with the BRS transformation law, Eq.(3.4.2-4), for the individual fields). This is just one of the most important virtues of *operator formalism*, which will be fully understood in the discussion developed in the next chapter. Furthermore, the subsidiary condition, Eq.(3.4.3-2), allows *any physical states* to take the place of the vacuum $|0\rangle$ in the above discussion, Eq.(4) $\Rightarrow$ Eq.(5), and hence, we obtain

$$\langle f|\boldsymbol{\delta}\mathcal{R}|g\rangle = 0 \qquad \text{for } \forall|f\rangle, |g\rangle \in \mathcal{V}_{\text{phys}}. \qquad (3.4.4-6)$$

However, the physical significance of Eq.(6) has not been so much clarified as yet.

Although all the essence of Slavnov-Taylor identities is contained in Eq.(5), their explicit forms will be necessary in the practical calculations for the analysis of concrete models. For this purpose, it is convenient to formulate these identities as the equations for generating functional Γ (effective action) of one-particle-irreducible (or proper) vertices as follows. To define Γ, we need a source functional $\mathcal{S}$ defined by

$$\mathcal{S}[J,K] \equiv \int d^4x \left(J^{\mu a} A^a_{\ \mu} + J_\varphi^{\ \alpha} \varphi^\alpha + J_C^{\ a} C^a + J_{\overline{C}}^{\ a} \overline{C}^a + J_B^{\ a} B^a \right.$$
$$\left. + K^{\mu a} (D_\mu C)^a + i K_\varphi^{\ \alpha} (g C^a \tau^a)^\alpha_{\ \beta} \varphi^\beta - \frac{1}{2} K_C^{\ a} g(C \times C)^a \right), \quad (3.4.4-7)$$

where $J_\mu, J_B, K_C, J_\varphi^{\ \alpha}$ for bosonic φ^α as well as $K_\varphi^{\ \beta}$ for fermionic φ^β are c-number sources and $J_C, J_{\overline{C}}, K_\mu, K_\varphi^{\ \alpha}$ for bosonic φ^α and $J_\varphi^{\ \beta}$ for fermionic φ^β are "anticommuting c-numbers", respectively. [The use of "anticommuting c-number sources" is just for simplifying the notation.] To control the *nonlinearity* of BRS transformation for the fields $A_\mu^{\ a}, \varphi^\alpha$, and C^a, we have introduced the source terms K_I for the "BRS-derivatives", $\boldsymbol{\delta}\Phi_I$, in addition to the usual source terms J_I for Φ_I. [Here the Φ_I stands generically for the fields $A_\mu^{\ a}, \varphi^\alpha, B^a, C^a$ and $\overline{C}^a$.] Because of the nilpotency of $\boldsymbol{\delta}$ expressed as

$$\boldsymbol{\delta}(D_\mu C)^a = \boldsymbol{\delta}^2 A^a_{\ \mu} = 0, \qquad (3.4.4-8)$$

$$\boldsymbol{\delta} i g C^a (\tau^a \varphi)^\alpha = \boldsymbol{\delta}^2 \varphi^\alpha = 0, \qquad (3.4.4-9)$$

$$\boldsymbol{\delta}(-\frac{g}{2})(C \times C)^a = \boldsymbol{\delta}^2 C^a = 0, \qquad (3.4.4-10)$$

$$\boldsymbol{\delta} B^a = -i\boldsymbol{\delta}^2 \overline{C}^a = 0, \qquad (3.4.4-11)$$

the higher BRS-derivative terms are unnecessary. Then, applying Eq.(5) to the case with $\mathcal{R} = \text{Texp} i\mathcal{S}[J,K]$, we obtain[9]

$$0 = \langle 0 | \left[i Q_B, \ \text{Texp} i\mathcal{S}[J,K] \right] | 0 \rangle$$
$$= i \int d^4x \, \langle 0 | \text{T}(J^{\mu a} (D_\mu C)^a + i J_\varphi^{\ \alpha} g C^a (\tau^a \varphi)^\alpha $$
$$+ \frac{g}{2} J_C^{\ a} (C \times C)^a - i J_{\overline{C}}^{\ a} B^a) \text{exp} i\mathcal{S}[J,K] | 0 \rangle. \qquad (3.4.4-12)$$

Taking the derivatives of this equation with respect to sources J_I at $\dot{J}_I \equiv 0$, one can easily obtain the Slavnov-Taylor identities for Green's functions of field operators. [Here the left-differentiation convention, Eq.(3.4.1-28), should be adopted

---

[9] According to Eq.(6), the vacuum $|0\rangle$ here can also be replaced by any physical states characterized by the subsidiary condition Eq.(3.4.3-2). In that case, the interpretation of the effective action Γ as the generating functional of one-particle-irreducible vertices in the following should be modified, of course.

for "anticommuting c-number sources".] To formulate these identities in a concise form, we introduce the generating functional W of connected Green's functions and effective action (or the generating functional of one-particle irreducible vertices) Γ connected with each other through **Legendre transformation** (see, for instance, [Aber 73], [Jac 74], etc.):

$$\exp i W[J, K] \equiv \langle 0|\mathrm{T}\exp i \mathcal{S}[J, K]|0\rangle, \qquad (3.4.4 - 13)$$

$$\Gamma[\Phi, K] \equiv W[J, K] - J_I \Phi_I, \qquad (3.4.4 - 14)$$

$$\Phi_I \equiv (\delta/\delta J_I) W[J, K]. \qquad (3.4.4 - 15)$$

[Here, the integration $\int d^4x$ over the whole space-time is omitted.] By the abuse of notation, the (anti)commuting c-number arguments Φ_I of Γ are denoted by the same symbols as the corresponding Heisenberg field operators, which will not, however, cause any confusion.[10] Now, we rewrite Eq.(12) utilizing the relations dual to Eq.(15) in Legendre transformation,

$$(\delta/\delta\Phi_I)\Gamma = \begin{cases} -J_I & \text{for bosonic fields,} \\ +J_I & \text{for fermioninc fields,} \end{cases} \qquad (3.4.4 - 16)$$

to give a concise expression for Slavnov-Taylor identities in terms of effective action Γ [Bec 76a] as follows :

$$\frac{\delta\Gamma}{\delta A_\mu{}^a}\frac{\delta\Gamma}{\delta K^{\mu a}} + \frac{\delta\Gamma}{\delta\varphi^\alpha}\frac{\delta\Gamma}{\delta K_\varphi{}^\alpha} + \frac{\delta\Gamma}{\delta C^a}\frac{\delta\Gamma}{\delta K_C{}^a} + i\frac{\delta\Gamma}{\delta\overline{C}^a}B^a = 0. \qquad (3.4.4 - 17)$$

From the field equations, Eqs.(3.4.1-13) and (3.4.1-14), as well as canonical commutation relations, Eqs.(3.4.1-23), one can derive the following equations:

$$\frac{\delta\Gamma}{\delta B^a} = \partial^\mu A_\mu{}^a + \alpha B^a, \qquad (3.4.4 - 18)$$

$$\partial_\mu\frac{\delta\Gamma}{\delta K^{\mu a}} + i\frac{\delta\Gamma}{\delta\overline{C}^a} = 0. \qquad (3.4.4 - 19)$$

For instance, the derivation of Eq.(18) is given as follows: In the identity following from Eq.(3.4.1-13),

$$\langle 0|\mathrm{T}(\partial^\mu A_\mu{}^a(x) + \alpha B^a(x))\exp i J_I \Phi_I|0\rangle = 0, \qquad (3.4.4 - 20)$$

---

[10]   To perform the Legendre transformation, Eq.(14), one should solve Eq.(15) in favor of J_I as functionals of Φ_I's. In general, the solution to this equation may not be unique as in the cases with spontaneous symmetry breaking, but we simply assume that, even in such cases, some suitable branch has been chosen to ensure the uniqueness of solution.

we can take out the derivative ∂^μ from the inside of T-product, using the canonical commutation relation between $A_0{}^a$ and B^a. Then, we obtain the equality

$$\partial^\mu\langle 0|\mathrm{T}A_\mu{}^a(x)\mathrm{exp}iJ_I\Phi_I|0\rangle + \langle 0|\mathrm{T}\alpha B^a(x)\mathrm{exp}iJ_I\Phi_I|0\rangle = -J_B(x), \qquad (3.4.4-21)$$

the Legendre transform of which is just Eq.(18). Eq.(19) can be obtained almost the same way. Supplemented with the equalities, Eqs.(18) and (19), the Slavnov-Taylor identity, Eq.(17), for Γ provides us with a useful tool for the analysis of the asymptotic fields in Chapter 4. They will be referred to as the Γ-**WT identities**.

3.5 SUPERALGEBRA STRUCTURES INVOLVING BRS CHARGE

It has been found that the BRS symmetry is accompanied by another symmetry, called anti-BRS symmetry, satisfying also nilpotency in the relation of mirror image. The Lie superalgebra structure of BRS algebra can be extended so as to include this new symmetry and also some other symmetries related to these symmetries. The effect of introducing the bare mass term of Yang-Mills field is also investigated.

3.5.1 Anti-BRS symmetry

In the BRS transformation, Eqs.(3.3.1-33)-(3.3.1-37), FP ghost and FP antighost play quite asymmetric roles, and, as noted in Sec.3.4.1, they cannot be related by the hermitian conjugation. As the Maurer-Cartan form on the group $\mathcal{G}$ of local gauge transformations, the FP ghost is directly related to the infinitesimal local gauge transformation as is seen in Eq.(3.3.2-29), whereas the origin of the FP antighost should be traced back to its role as a *Lagrange multiplier* η in Eq.(3.3.1-15) for keeping the gauge fixing condition unchanged under the BRS transformation.

However, the difference of their field equations is easily seen to consist only in the term proportional to $\partial^{\mu} A_{\mu}{}^{a}$,

$$\partial^{\mu} D_{\mu} - D_{\mu} \partial^{\mu} = g(\partial^{\mu} A_{\mu}) \times, \qquad (3.5.1-1)$$

which *vanishes* in the Landau gauge. Therefore, FP ghosts C^{a} and FP antighosts $\overline{C}^{a}$ can be interchanged in this special case. In fact, the Lagrangian density $\mathcal{L}$ given by Eqs.(3.4.1-1)-(3.4.1-4) with $\alpha = 0$ is invariant under the following **FP conjugation** $\mathcal{C}_{\text{FP}}$ [1] [Kug unp.a, Oji 80a] defined by

$$\mathcal{C}_{\text{FP}} : \begin{cases} A_{\mu}{}^{a} & \rightarrow \quad \mathcal{C}_{\text{FP}} A_{\mu}{}^{a} = A_{\mu}{}^{a}, \\ \varphi^{\alpha} & \rightarrow \quad \mathcal{C}_{\text{FP}} \varphi^{\alpha} = \varphi^{\alpha}, \\ B^{a} & \rightarrow \quad \mathcal{C}_{\text{FP}} B^{a} = B^{a} - ig(\overline{C} \times C)^{a}, \\ C^{a} & \rightarrow \quad \mathcal{C}_{\text{FP}} C^{a} = \overline{C}^{a}, \\ \overline{C}^{a} & \rightarrow \quad \mathcal{C}_{\text{FP}} \overline{C}^{a} = -C^{a}. \end{cases} \qquad (3.5.1-2)$$

---

[1] $\mathcal{C}_{\text{FP}}$ is an automorphism of the algebra of field operators satisfying the property: $\mathcal{C}_{\text{FP}}{}^{2} \Phi = e^{\pm \pi Q_{C}} \Phi e^{\pm \pi Q_{C}}$ for any fields Φ.

The second term on rhs. of the transformation law for B^a is necessary for cancelling the change of FP ghost Lagrangian density $\mathcal{L}_{FP}$ under the interchange between FP ghosts and antighosts. Although this change is proportional to $\partial^\mu A_\mu{}^a$, it cannot, of course, be taken as vanishing *in the Lagrangian density* contrary to the case of *the field equations*.

Then, combining the BRS transformation $\boldsymbol{\delta}$ with this FP conjugation $\mathcal{C}_{FP}$, we define **anti-BRS transformation** [Cur 76a, Oji 80a] by

$$\overline{\boldsymbol{\delta}} \equiv \mathcal{C}_{FP}\boldsymbol{\delta}\mathcal{C}_{FP}^{-1}, \qquad (3.5.1-3)$$

which induces the following transformations on the primary fields:

$$\overline{\boldsymbol{\delta}}A_\mu{}^a = (D_\mu\overline{C})^a, \qquad (3.5.1-4)$$

$$\overline{\boldsymbol{\delta}}\varphi^\alpha = ig\overline{C}^a(\tau^a\varphi)^\alpha, \qquad (3.5.1-5)$$

$$\overline{\boldsymbol{\delta}}B^a = -g(\overline{C} \times B)^a, \qquad (3.5.1-6)$$

$$\overline{\boldsymbol{\delta}}C^a = -g(\overline{C} \times C)^a - iB^a, \qquad (3.5.1-7)$$

$$\overline{\boldsymbol{\delta}}\overline{C}^a = -\frac{g}{2}(\overline{C} \times \overline{C})^a. \qquad (3.5.1-8)$$

In the Landau gauge ($\alpha = 0$), the invariance under this anti-BRS transformation immediately follows from the invariances of the theory under the BRS transformation $\boldsymbol{\delta}$ and the FP conjugation $\mathcal{C}_{FP}$. Although the Landau gauge condition has been indispensable up to this point for ensuring the invariance under the FP conjugation $\mathcal{C}_{FP}$, this restriction can now be lifted as far as the invariance under anti-BRS transformation itself is concerned: Namely, the anti-BRS invariance turns out to survive in non-Landau gauges ($\alpha \neq 0$) irrespective of the breakdown of FP conjugation invariance:

$$\overline{\boldsymbol{\delta}}\mathcal{L} = 0. \qquad (3.5.1-9)$$

This can be checked by a direct calculation, but, from the theoretical point of view, it will be more useful to base it on the following observations. First, we note that the anti-BRS $\overline{\boldsymbol{\delta}}$ is *nilpotent* just in the same way as the BRS $\boldsymbol{\delta}$:

$$\overline{\boldsymbol{\delta}}^2 = 0. \qquad (3.5.1-10)$$

It can also be checked from Eqs.(4)-(8) as well as the definition of BRS transformation, Eqs.(3.3.1-33)-(3.3.1-37), that the following remarkable anticommutativity holds between $\boldsymbol{\delta}$ and $\overline{\boldsymbol{\delta}}$:

$$\boldsymbol{\delta}\overline{\boldsymbol{\delta}} + \overline{\boldsymbol{\delta}}\boldsymbol{\delta} = 0. \qquad (3.5.1-11)$$

Using these relations as well as the following ones,

$$\partial^\mu \overline{C}^a \cdot A_\mu{}^a = (D^\mu \overline{C})^a A_\mu{}^a - g(A^\mu \times \overline{C})^a A_\mu{}^a = \frac{1}{2}\overline{\boldsymbol{\delta}}(A^\mu A_\mu), \quad (3.5.1-12)$$

$$B^a B^a = B^a(i\overline{\boldsymbol{\delta}}C + ig\overline{C} \times C)^a = i\overline{\boldsymbol{\delta}}(B^a C^a), \qquad (3.5.1-13)$$

we can rewrite $\mathcal{L}_{\mathrm{GF}} + \mathcal{L}_{\mathrm{FP}}$ given by Eq.(3.4.1-8) further in such forms as

$$\begin{aligned}
\mathcal{L}_{\mathrm{GF}} + \mathcal{L}_{\mathrm{FP}} &= i\boldsymbol{\delta}(\partial^\mu \overline{C}^a \cdot A_\mu{}^a - \frac{\alpha}{2}\overline{C}^a B^a) \\
&\cdot= \frac{i}{2}\boldsymbol{\delta}\overline{\boldsymbol{\delta}}(A^\mu A_\mu) + i\frac{\alpha}{2}\overline{\boldsymbol{\delta}}(B^a C^a) \\
&= -i\overline{\boldsymbol{\delta}}(\partial^\mu C^a \cdot A_\mu{}^a - \frac{\alpha}{2}C^a B^a). \qquad (3.5.1-14)
\end{aligned}$$

Hence the anti-BRS invariance of the Lagrangian density $\mathcal{L}$ trivially follows from Eq.(10).

In view of the remarkable parallelism between $\boldsymbol{\delta}$ and $\overline{\boldsymbol{\delta}}$ found in Eq.(14), it may be instructive to give some interpretations to the form of anti-BRS transformation, Eqs.(4)-(8), in its relation to that of the original BRS transformation, Eqs.(3.3.1-33)-(3.3.1-37) :

i) As for the transformation law for $A_\mu{}^a, \varphi^\alpha$ and $\overline{C}^a$, given respectively by Eqs.(4),(5) and (8), it is obvious that they correspond, just through the interchange between C^a and $\overline{C}^a$, to those for $A_\mu{}^a, \varphi^\alpha$ and C^a, in Eqs.(3.3.1-33),(3.3.1-34) and (3.3.1-36).

ii) The forms of Eqs.(6) and (7) may be more naturally understood with the aid of "anti-B field" $\overline{B}^a$, so to speak, defined by

$$\overline{B}^a \equiv -B^a + ig(\overline{C} \times C)^a, \qquad (3.5.1-15)$$

or, more symmetrically in such a form of *constraint* as

$$B^a + \overline{B}^a - ig(\overline{C} \times C)^a = 0. \qquad (3.5.1-16)$$

Then, Eq.(7) is rewritten as

$$\overline{\boldsymbol{\delta}}C^a = i\overline{B}^a, \qquad (3.5.1-17)$$

and it can be shown that Eq.(6) is equivalent to

$$\overline{\boldsymbol{\delta}}\,\overline{B}^a = 0, \qquad (3.5.1-18)$$

from which the parallelism with BRS transformation laws for C^a and B^a becomes clear. With this notation, FP conjugation of B^a in Eq.(2) can be seen as the interchange between B^a and $\overline{B}^a$:

$$\mathcal{C}_{\mathrm{FP}} B^a = -\overline{B}^a,$$
$$\mathcal{C}_{\mathrm{FP}} \overline{B}^a = -B^a. \qquad (3.5.1-19)$$

iii) In the relation with (infinitesimal) global gauge transformation given by

$$\delta A_\mu{}^a = (\epsilon \times A_\mu)^a, \qquad (3.5.1-20)$$

$$\delta\varphi^\alpha = -i\epsilon^a(\tau^a \varphi)^\alpha, \qquad (3.5.1-21)$$

$$\delta B^a = (\epsilon \times B)^a, \qquad (3.5.1-22)$$

$$\delta C^a = (\epsilon \times C)^a, \qquad (3.5.1-23)$$

$$\delta\overline{C}^a = (\epsilon \times \overline{C})^a, \qquad (3.5.1-24)$$

the form of anti-BRS transformation, Eqs.(4)-(8), has a stronger similarity to it than that of BRS transformation does : Except for the term $-iB^a$ in Eq.(7) and the factor $\frac{1}{2}$ in Eq.(8), the anti-BRS transformation can be reproduced from the above Eqs.(20)-(24) by replacing ϵ^a with $-g\overline{C}^a$. This should be compared with the behavior of B^a and $\overline{C}^a$ under the BRS transformation which are quite different from Eqs.(22) and (24). However, this similarity brings more nonlinearity in the anti-BRS transformation than the BRS one, which complicates the analysis of WT identities following from the anti-BRS invariance.

iv) The parallelism between BRS and anti-BRS transformations is also seen in the quantum Maxwell equation, Eq.(3.4.3-15), as follows:

$$\partial^\nu F_{\nu\mu}{}^a + gJ_\mu{}^a = \{Q_B, (D_\mu\overline{C})^a\} = -i\boldsymbol{\delta}\boldsymbol{\bar{\delta}} A_\mu{}^a. \qquad (3.5.1-25)$$

On account of the anti-BRS symmetry, we have its Noether current:

$$J_{\overline{B}}^\mu = (D_\nu\overline{C})^a \frac{\partial\mathcal{L}}{\partial(\partial_\mu A_\nu{}^a)} + ig(\overline{C}^a\tau^a\varphi)^\alpha \frac{\partial\mathcal{L}}{\partial(\partial_\mu\varphi^\alpha)} - \frac{1}{2}g(\overline{C}\times\overline{C})^a \frac{\partial\mathcal{L}}{\partial(\partial_\mu\overline{C}^a)}$$
$$+ i\overline{B}^a \frac{\partial\mathcal{L}}{\partial(\partial_\mu C^a)} - g(\overline{C}\times B)^a \frac{\partial\mathcal{L}}{\partial(\partial_\mu B^a)}. \qquad (3.5.1-26)$$

Quite similarly to the case of $J_{B\mu}$, Eq.(3.4.2-12), this current can also be rewritten with the aid of the field equation, Eq.(3.4.1-12), as

$$J_{\overline{B}\mu} = B^a(D_\mu\overline{C})^a - \partial_\mu B^a \cdot \overline{C}^a + \frac{i}{2}g(\overline{C}\times\overline{C})^a(D_\mu C)^a - \partial^\nu(F_{\mu\nu}{}^a\overline{C}^a)$$
$$= \bar{\boldsymbol{\delta}} J_{C\mu} - \partial^\nu(F_{\mu\nu}{}^a\overline{C}^a). \qquad (3.5.1-27)$$

Here $J_{C\mu}$ is the current of FP ghost charge given by Eq.(3.4.2-6), the conservation law of which concludes again that of $J_{\overline{B}\mu}$:

$$\partial^\mu J_{\overline{B}\mu} = 0. \qquad (3.5.1-28)$$

In the same way as the BRS charge Q_B, the anti-BRS charge $\overline{Q}_B$ is given by

$$\overline{Q}_B \equiv \int d\mathbf{x} J_{\overline{B}0} = \int d\mathbf{x}(B^a(D_0\overline{C})^a - \dot{B}^a\overline{C}^a + \frac{i}{2}g(\overline{C}\times\overline{C})^a(D_0C)^a)$$
$$= \overline{\boldsymbol{\delta}}Q_C = [i\overline{Q}_B,\ Q_C\,] = -[iQ_C,\ \overline{Q}_B\,], \qquad (3.5.1-29)$$

which generates the anti-BRS transformation :

$$[i\overline{Q}_B,\ \Phi\,]_\mp = \overline{\boldsymbol{\delta}}\Phi. \qquad (3.5.1-30)$$

It satisfies the nilpotency

$$\overline{Q}_B^2 = \frac{1}{2i}\overline{\boldsymbol{\delta}}\overline{Q}_B = \frac{1}{2i}\overline{\boldsymbol{\delta}}^2 Q_C = 0, \qquad (3.5.1-31)$$

and is shown to anticommute with Q_B,

$$\{Q_B,\ \overline{Q}_B\,\} = -i\overline{\boldsymbol{\delta}}\overline{Q}_B = -i\int d\mathbf{x}\,\boldsymbol{\delta}\left(B^a(D_0\overline{C})^a - \dot{B}^a\overline{C}^a + \frac{i}{2}g(\overline{C}\times\overline{C})^a(D_0C)^a\right)$$
$$= 0, \qquad (3.5.1-32)$$

from which the anticommutativity, Eq.(11), between $\boldsymbol{\delta}$ and $\overline{\boldsymbol{\delta}}$ is reproduced :

$$(\boldsymbol{\delta}\overline{\boldsymbol{\delta}} + \overline{\boldsymbol{\delta}}\boldsymbol{\delta})\Phi = -[\{Q_B,\ \overline{Q}_B\,\},\ \Phi\,] = 0. \qquad (3.5.1-33)$$

It may be useful to summarize the relations involving the three charges, $Q_B, \overline{Q}_B$ and Q_C, as follows:

$$\boldsymbol{\delta}^2 Q_C = -2iQ_B^2 = 0$$

$$\boldsymbol{\delta}\ \nearrow$$

$$\boldsymbol{\delta}Q_C = -[iQ_C,\ Q_B] = -Q_B$$

$$\boldsymbol{\delta}\ \nearrow \qquad\qquad\qquad \overline{\boldsymbol{\delta}}\ \searrow$$

$$Q_C \qquad\qquad\qquad \boldsymbol{\delta}Q_B = \boldsymbol{\delta}\overline{Q}_B = i\{Q_B,\ \overline{Q}_B\,\} = 0 \quad (3.5.1-34)$$

$$\overline{\boldsymbol{\delta}}\ \searrow \qquad\qquad\qquad \boldsymbol{\delta}\ \nearrow$$

$$\overline{\boldsymbol{\delta}}Q_C = -[iQ_C,\ \overline{Q}_B] = \overline{Q}_B$$

$$\overline{\boldsymbol{\delta}}\ \searrow$$

$$\overline{\boldsymbol{\delta}}^2 Q_C = 2i\overline{Q}_B^2 = 0$$

We call this Lie superalgebra structure **double BRS algebra**.[2]

According to the general discussion in Sec.3.4.4, we can derive a new set of WT identities from the assumption of anti-BRS invariance of the vacuum:

$$\overline{Q}_B|0\rangle = 0. \qquad (3.5.1-35)$$

These identities can be formulated in the similar way to Γ-WT identities in Sec.3.4.4 by introducing the source terms for the first "anti-BRS derivatives", $\overline{\delta}\Phi_I$, of the primary fields Φ_I. Since they are not used in this book, however, we omit the detailed discussion and just present some of simple examples:

$$\langle 0|\mathrm{T}((\overline{C} \times C)^a(x)B^b(y))|0\rangle = \langle 0|\mathrm{T}(C^a(x)(\overline{C} \times B)^b(y))|0\rangle, \quad (3.5.1-36)$$

$$\langle 0|\mathrm{T}(C^a(x)(D_\mu\overline{C})^b(y))|0\rangle$$
$$= -\delta^{ab}\partial_\mu D_F(x-y) - \langle 0|\mathrm{T}(g(\overline{C} \times C)^a(x)A_\mu^{\ b}(y))|0\rangle. \qquad (3.5.1-37)$$

Finally, we add a comment on the relevance of anti-BRS symmetry to the problem of extracting physical contents of the theory. In view of the parallelism between BRS and anti-BRS symmetry visualized in Eq.(34), the superalgebra structure (called **anti-BRS algbra**) involving anti-BRS charge $\overline{Q}_B$ and FP ghost charge Q_C is the same, up to a minor change of sign, as the BRS algebra, Eqs.(3.4.2-9)-(3.4.2-11), of Q_B and Q_C. As will be seen in Chapter 4, a cancellation mechanism (called **quartet mechanism**) to "confine" negative-norm states works (almost) exclusively on the basis of the BRS algebra structure, which can be just repeated for the anti-BRS algebra without changing the physical contents of the theory, at least at the level of asymptotic states and fields. Because of the anticommutativity, Eq.(32), between Q_B and $\overline{Q}_B$, the subsidiary condition

$$\overline{Q}_B|\mathrm{phys}\rangle = 0, \qquad (3.5.1-38)$$

can be imposed in addition, or even as an alternative, to the original one, $Q_B|\mathrm{phys}\rangle = 0$, without causing any inconsistency nor changes of the physical

---

[2] This is usually called the "**extended BRS algebra**", about which many interesting discussions have been done, see for instance, [Bon 81b, Qui 81, Hoy 83], and also other references cited there. However, it is *not* the maximally extended structure containing the charges, Q_B, $\overline{Q}_B$ and Q_C; we will see in the next subsection, Sec.3.5.2, that this structure can be embedded in a more extended Lie superalgebra structure which we call *extended double BRS algebra*.

consequences. In this sense, anti-BRS symmetry may look as redundant, but, from the theoretical point of view, it will be useful in combination with the BRS symmetry, in clarifying the detailed structure of the theory : For instance, we will see in Chapter 4 that it helps to negate the possibility of BRS singlet pairs which endangers the quartet mechanism of BRS algebra for "confining" negative norms. In view of the anticommutativity, Eq.(11), between $\boldsymbol{\delta}$ and $\overline{\boldsymbol{\delta}}$, the analogy to the two anticommuting exterior differentials, d' and d'', in the complex geometry (or, in a more general context of double complexes in cohomology theory) may also be interesting, as is discussed, for instance, in [Thie 79, Qui 81, Hoy 83]. In the next subsection, we place the BRS and anti-BRS symmetries in a more general context of extending the BRS algebra as well as its deformation by mass parameter.

3.5.2 Extension of BRS algebra and its deformation by mass term

Just in parallel to Eq.(3.5.1-13), i.e.,

$$B^a B^a = -i\boldsymbol{\delta}(B^a \overline{C}^a) = i\overline{\boldsymbol{\delta}}(B^a C^a), \qquad (3.5.2-1)$$

expressing $B^a B^a$ as a BRS- and anti-BRS coboundary term, one can easily verify its FP-conjugate version for $\overline{B}^a \overline{B}^a$ as follows:

$$\overline{B}^a \overline{B}^a = -i\overline{\boldsymbol{\delta}}(\overline{B}^a C^a) = i\boldsymbol{\delta}(\overline{B}^a \overline{C}^a). \qquad (3.5.2-2)$$

This observation allows us to add to the Lagrangian density $\mathcal{L}$ given by Eq.(3.4.1-1) the term proportional to $\overline{B}^a \overline{B}^a = (B - ig\overline{C} \times C)^2$ which contains a *quartic* term in FP ghosts and antighosts. In [Bau 82], it has been shown by the Feynman-diagrammatical analysis that the most general form of Lorentz invariant renormalizable Lagrangian densities, $\tilde{\mathcal{L}}$, of Yang-Mills fields can be uniquely determined by the requirement of BRS and anti-BRS invariance, allowing an additional arbitrary parameter β:

$$\begin{aligned}
\tilde{\mathcal{L}} &= \mathcal{L} + \frac{\beta}{2}\overline{B}^a \overline{B}^a \\
&= \mathcal{L}_0 + i\boldsymbol{\delta}(\partial^\mu \overline{C}^a \cdot A_\mu{}^a - \frac{\alpha}{2}B^a \overline{C}^a + \frac{\beta}{2}\overline{B}^a \overline{C}^a) \\
&= \mathcal{L}_0 - i\overline{\boldsymbol{\delta}}(\partial^\mu C^a \cdot A_\mu{}^a - \frac{\alpha}{2}B^a C^a + \frac{\beta}{2}\overline{B}^a C^a). \qquad (3.5.2-3)
\end{aligned}$$

It is true that the case with $\beta \neq 0$ is not so convenient in perturbative calculations owing to the nonlinear contributions of FP ghosts. From the theoretical point of view, however, it is useful for clarifying the position of the standard formulation in more general contexts. For instance, while in the previous subsection the FP conjugation $\mathcal{C}_{\mathrm{FP}}$ was allowed only in Landau gauge ($\alpha = 0$), the theory given by the Lagrangian density Eq.(3) is now invariant under it as long as the two parameters α and β are equal. Furthermore, to understand the vital importance of nilpotency of the (anti-)BRS symmetry in confining the unphysical particles, it will be instructive to know what would happen if the nilpotency is violated. We clarify the unique characters of BRS algebra in the following two directions: i) At the cost of special gauge choice, both the BRS and anti-BRS algebras in Landau gauge will be shown to be just small parts of a larger (super)algebra [Nak 80f], called *extended double BRS algebra*. ii) We will see how the (anti-)BRS symmetry is deformed by the introduction of *bare mass term* of the field $A_\mu{}^a$ which damages the nilpotency [Oji 82]. This will clarify the special character of the BRS algebra structure in the *massless* case which guarantees the ghost-confining mechanism to work, and also will serve the above mentioned purpose to contrast the situations with and without the nilpotency in Chapter 4.

By rearranging the result of [Cur 76a, Cur 76b] in terms of B^a and $\overline{B}^a$, the general form of renormalizable Lagrangian density of a vector field can be given by

$$
\begin{aligned}
\mathcal{L}_m =& \mathcal{L}_0 - \partial^\mu B^a \cdot A_\mu{}^a - i\partial^\mu \overline{C}^a \cdot D_\mu C^a \\
& + \frac{\alpha}{2} B^2 + \frac{\beta}{2} \overline{B}^2 \\
& + m^2 \left(\frac{1}{2} A_\mu{}^a A^{\mu a} + i(\alpha + \beta) \overline{C}^a C^a \right).
\end{aligned}
\tag{3.5.2 - 4}
$$

Corresponding to the addition of the term proportional to m^2, the BRS and anti-BRS transformations of B-field are to be modified as

$$
\boldsymbol{\delta} B^a = m^2 C^a,
\tag{3.5.2 - 5}
$$

$$
\overline{\boldsymbol{\delta}} B^a = m^2 \overline{C}^a - g(\overline{C} \times B)^a,
\tag{3.5.2 - 6}
$$

through which $\mathcal{L}_m$ becomes invariant under the modified transformations, *if* the condition

$$
m^2(\alpha - \beta) = 0
\tag{3.5.2 - 7}
$$

is satisfied. Of course, $\tilde{\mathcal{L}}$ in Eq.(3) is reproduced if $m^2 = 0$; In the case of $m^2 \neq 0$ the two gauge parameters, α and β, are required to be identical. Since the case of $\alpha \neq \beta$ with $m^2 = 0$ does not bring anything new, we here choose the case of $\alpha = \beta$. Contrary to the case with $m^2 = 0$ observed in Sec.3.5.1, both the invariances in the latter case are attained by non-trivial cancellations among different terms, where the term $\overline{B}^2$ in Eq.(4) is indispensable except for the Landau gauge case, $\alpha = \beta = 0$, in which the Lagrangian density reduces to

$$\mathcal{L}_m = \mathcal{L}_0 - \partial^\mu B^a \cdot A_\mu{}^a - i\partial^\mu \overline{C}^a \cdot D_\mu C^a + m^2 A_\mu{}^a A^{\mu a}/2. \qquad (3.5.2-8)$$

Although the terms proportional to m^2 in Eqs.(5) and (6) are crucial for preserving the invariances, they *violate* the nilpotency of the modified BRS and anti-BRS transformations :

$$\boldsymbol{\delta}^2 \overline{C}^a = i\boldsymbol{\delta} B^a = im^2 C^a \neq 0, \qquad (3.5.2-9)$$

$$\overline{\boldsymbol{\delta}}^2 C^a = -im^2 \overline{C}^a \neq 0, \qquad (3.5.2-10)$$

which leads to the breakdown of physical S-matrix unitarity as will be seen in Chapter 4. In spite of these modifications, the expressions for the corresponding BRS and anti-BRS currents are just the same as those in the case with $m^2 = 0$, Eqs.(3.4.2-2) and (3.5.1-27), and are connected by the same relations with the FP ghost current whose expression is also unchanged. Irrespective of the validity of nilpotency, therefore, we continue to use the same symbols for these currents as well as the corresponding integrated charges. Then, the obstruction terms for the nilpotency of Q_B and $\overline{Q}_B$ as well as for their mutual anticommutativity due to $m^2 \neq 0$ can be explicitly given by

$$\begin{aligned} \{Q_B, Q_B\} &= 2Q_B^2 = -i\boldsymbol{\delta} Q_B \\ &= -m^2 Q_{CC} \equiv -im^2 \int dx(C^a(D_0 C)^a - \dot{C}^a C^a) \\ &= -m^2 Q_{CC}{}^\dagger, \qquad (3.5.2-11) \end{aligned}$$

$$\begin{aligned} \{\overline{Q}_B, \overline{Q}_B\} &= 2\overline{Q}_B^2 = -i\overline{\boldsymbol{\delta}} \overline{Q}_B \\ &= -m^2 Q_{\overline{C}\overline{C}} \equiv -im^2 \int dx(\overline{C}^a(D_0 \overline{C})^a - \dot{\overline{C}}^a \overline{C}^a) \\ &= -m^2 Q_{\overline{C}\overline{C}}{}^\dagger, \qquad (3.5.2-12) \end{aligned}$$

$$\{Q_B, \overline{Q}_B\} = -i\boldsymbol{\delta}\overline{Q}_B = -i\overline{\boldsymbol{\delta}} Q_B = -m^2 Q_C. \qquad (3.5.2-13)$$

Since the charges on lhs. are all conserved and hermitian, the charges, $Q_{CC}, Q_{\bar{C}\bar{C}}$, appearing on rhs. should also be conserved and hermitian. Together with FP ghost charge Q_C, these charges constitute the Lie algebra of 3-dimensional Lorentz group[3] $SO(2,1)$ ($\cong SL(2,\mathbf{R}) \cong Sp(2,\mathbf{R})$),

$$[iQ_C/2, \; Q_{CC}/2\,] = Q_{CC}/2, \qquad\qquad (3.5.2-14)$$

$$[iQ_C/2, \; Q_{\bar{C}\bar{C}}/2\,] = -Q_{\bar{C}\bar{C}}/2, \qquad\qquad (3.5.2-15)$$

$$[iQ_{CC}/2, \; Q_{\bar{C}\bar{C}}/2\,] = -Q_C, \qquad\qquad (3{:}5.2-16)$$

under which Q_B and $\overline{Q}_B$ behave as a 2-component spinor:

$$[iQ_C/2, \; Q_B\,] = \frac{1}{2}Q_B, \qquad\qquad (3.5.2-17)$$

$$[iQ_C/2, \; \overline{Q}_B\,] = -\frac{1}{2}\overline{Q}_B; \qquad\qquad (3.5.2-18)$$

$$[iQ_{CC}/2, \; Q_B\,] = 0, \quad [iQ_{CC}/2, \; \overline{Q}_B\,] = -Q_B; \qquad (3.5.2-19)$$

$$[iQ_{\bar{C}\bar{C}}/2, \; Q_B\,] = \overline{Q}_B, \quad [iQ_{\bar{C}\bar{C}}/2, \; \overline{Q}_B\,] = 0. \qquad (3.5.2-20)$$

From Eqs.(11)-(20), five charges, $Q_B, \overline{Q}_B, Q_C/2, Q_{CC}/2$ and $Q_{\bar{C}\bar{C}}/2$, are seen to constitute a superalgebra extension of the Lie algebra of 3-dimensional Lorentz group. From this point of view, the combined structure of BRS and anti-BRS algebras for $m^2 = 0$ given by Eq.(3.5.1-34) is to be interpreted as its degenerate case obtained through a Lie superalgebra version of "*group contraction*" procedure of taking the limit of $m^2 \to 0$.

In addition to the n conserved charges Q^a associated with the global gauge group G ($n =$ dimension of G), we have $3n$ charges given by

$$P(\chi^a) \equiv \int d\mathbf{x} A_0{}^a, \qquad\qquad (3.5.2-21)$$

$$P(C^a) \equiv \int d\mathbf{x} D_0 C^{\,a}, \qquad\qquad (3.5.2-22)$$

$$P(\overline{C}^a) \equiv \int d\mathbf{x} D_0 \overline{C}^{\,a}, \qquad\qquad (3.5.2-{\cdot}23)$$

---

[3] In [Nis 84, Nis 85a], the algebraic structure determined by the commutation relations Eqs.(14)-(16) is identified with the Lie algebra associated with $SU(2)$ consisting of angular momenta in 3 dimensions. However, this is valid only as a *complex* Lie algebra. Because of the hermiticity of the charges Q_C, Q_{CC} and of $Q_{\bar{C}\bar{C}}$, they should be related to a *real* Lie algebra, in which context (anti-hermitian) operators, $X_1 = iQ_C/2$, $X_2 = i(Q_{CC} + Q_{\bar{C}\bar{C}})/4$ and $X_3 = i(Q_{CC} - Q_{\bar{C}\bar{C}})/4$ can be chosen as a basis of the Lie algebra of non-compact group $SL(2,\mathbf{R})$ satisfying $[X_1, X_2] = X_3, [X_2, X_3] = X_1, [X_3, X_1] = -X_2$

which are conserved in the case of Landau gauge ($\alpha = 0$), as can be seen from the field equations for B^a, C^a and $\overline{C}^a$

$$\partial^\mu A_\mu{}^a + \alpha(B^a - \overline{B}^a) = 0, \qquad (3.5.2-24)$$

$$\partial^\mu D_\mu C^a + 2\alpha m^2 C^a - g\alpha(\overline{B} \times C)^a = 0, \qquad (3.5.2-25)$$

$$D_\mu \partial^\mu \overline{C}^a + 2\alpha m^2 \overline{C}^a + g\alpha(\overline{B} \times \overline{C})^a = 0, \qquad (3.5.2-26)$$

and from Eq.(3.5.1-1). Thus, the Landau gauge choice gives us $5 + 4n$ conserved charges, $Q_C; Q_{CC}, Q_{\bar{C}\bar{C}}; Q_B, \overline{Q}_B; P(B^a), P(\chi^a), P(C^a)$, and $P(\overline{C}^a)$, where $P(B^a)$ is defined by

$$P(B^a) \equiv gQ^a + m^2 P(\chi^a). \qquad (3.5.2-27)$$

We call the algebraic structure consisting of these charges **extended double BRS algebra**, which is clarified now.

First, as for the transformation properties of field operators under these generators, what is new is only those involving $Q_{CC}, Q_{\bar{C}\bar{C}}, P(\chi^a), P(C^a)$ and $P(\overline{C}^a)$. We list up the non-vanishing (anti)commutators only:

$$[iQ_{CC}/2, B^a] = -ig(C \times C)^a/2, \qquad (3.5.2-28)$$

$$[iQ_{CC}/2, \overline{C}^a] = -C^a; \qquad (3.5.2-29)$$

$$[iQ_{\bar{C}\bar{C}}/2, B^a] = ig(\overline{C} \times \overline{C})^a/2, \qquad (3.5.2-30)$$

$$[iQ_{\bar{C}\bar{C}}/2, C^a] = \overline{C}^a; \qquad (3.5.2-31)$$

$$[iP(\chi^a), B^b] = -\delta^{ab}; \qquad (3.5.2-32)$$

$$\{iP(C^a), \overline{C}^b\} = i\delta^{ab}; \qquad (3.5.2-33)$$

$$[iP(\overline{C}^a), B^b] = -gf^{abc}\overline{C}^c, \qquad (3.5.2-34)$$

$$\{iP(\overline{C}^a), C^b\} = -i\delta^{ab}. \qquad (3.5.2-35)$$

Eqs.(32), (33) and (35) tell us that the generators $P(\chi^a), P(C^a)$ and $P(\overline{C}^a)$ suffer from spontaneous symmetry breaking.

Next, the Lie superalgebra structure of the extended double BRS algebra is specified by the (anti)commutation relations, Eqs.(11)-(20), and those involving $P(\chi^a), P(C^a)$ and $P(\overline{C}^a)$, the non-vanishing combinations of which are given by

$$[iQ_C, \ P(C^a)] = P(C^a), \qquad (3.5.2-36)$$

$$[iQ_C, \ P(\overline{C}^a)] = -P(\overline{C}^a); \qquad (3.5.2-37)$$

$$[iQ_{CC}, \ P(\overline{C}^a)] = -2P(C^a), \tag{3.5.2 − 38}$$

$$[iQ_{\bar{C}\bar{C}}, \ P(C^a)] = 2P(\overline{C}^a); \tag{3.5.2 − 39}$$

$$[iQ_B, \ P(\chi^a)] = P(C^a), \tag{3.5.2 − 40}$$

$$[iQ_B, \ P(B^a)] = m^2 P(C^a), \tag{3.5.2 − 41}$$

$$\{Q_B, \ P(\overline{C}^a)\} = P(B^a); \tag{3.5.2 − 42}$$

$$[i\overline{Q}_B, \ P(\chi^a)] = P(\overline{C}^a), \tag{3.5.2 − 43}$$

$$[i\overline{Q}_B, \ P(B^a)] = m^2 P(\overline{C}^a) \tag{3.5.2 − 44}$$

$$\{\overline{Q}_B, \ P(C^a)\} = -P(B^a); \tag{3.5.2 − 45}$$

$$[iP(B^a), \ P(X^b)] = -gf^{abc}P(X^c)$$
$$\text{for } X^a = B^a, \chi^a, C^a \text{ and } \overline{C}^a; \tag{3.5.2 − 46}$$

$$\{P(C^a), \ P(\overline{C}^b)\} = gf^{abc}P(\chi^c). \tag{3.5.2 − 47}$$

Finally, we note that the extended double BRS algebra has originally been introduced from the analogy of the construction of the symmetry generators in quantum gravity (see Sec.5.5.2) in the *massless* case in [Nak 80f], where the conservation of Q_{CC} and $Q_{\bar{C}\bar{C}}$ cannot be guaranteed by the anticommutation relations Eqs.(11) and (12). Actually, this follows directly from Eqs.(25) and (26) with $\alpha = 0$ irrespective of whether $m = 0$ or not: Namely, the equalities

$$\partial^\mu (D_\mu X)^a = D_\mu (\partial^\mu X)^a = \partial^\mu (D_\mu Y)^a = D_\mu (\partial^\mu Y)^a = 0 \tag{3.5.2 − 48}$$

implies the conservation of another current:

$$\partial^\mu [X^a (D_\mu Y)^a - \partial_\mu X^a \cdot Y^a] = 0. \tag{3.5.2 − 49}$$

Another independent route of finding the existence of Q_{CC} and $Q_{\bar{C}\bar{C}}$ is an explicitly broken inhomogeneous $OSp(4, 2)$ symmetry, which combines Poincaré algebra and the double BRS algebra, although one must forbid mixing both algebras. As discussed in Eqs.(14)-(20), Q_{CC} and $Q_{\bar{C}\bar{C}}$ are necessary for having an inhomogeneous $Sp(2, \mathbf{R})$ [Del 82a].

CHAPTER FOUR

PHYSICAL CONTENTS OF
QUANTIZED YANG-MILLS THEORY

On the basis of BRS symmetry, a norm-cancellation mechanism called "quartet mechanism" is found to confine unphysical particles with negative norms. The unitarity of physical S-matrix is ensured by it on a general ground. By the requirement of consistency with probabilistic interpretation, the notion of an observable is defined, which is shown to be invariant under the BRS transformation as well as the unbroken gauge groups. The quantum Maxwell equation is shown to play important roles in the general structural analysis of gauge theory, such as the confinement problem in QCD.

4.1 PROOF OF THE POSITIVITY IN PHYSICAL SUBSPACE

In this section, we analyze the structure of state vector space $\mathcal{V}$ from the viewpoint of BRS algebra given by Eqs.(3.4.2-9)-(3.4.2-11): In terms of the representations of this algebra, any state and particle are classified into physical and unphysical categories, identified with BRS-singlet representations with vanishing FP-ghost number and with FP-conjugate pairs of BRS-doublets, respectively. It can be proved on quite a general ground that the unphysical particles violating norm-positivity are "confined" in the sense that they appear in the physical subspace $\mathcal{V}_{\text{phys}}$ only in *zero-norm combinations*. Thus, given an explicit model, we can, in principle, answer definitely what kind of particles should appear in the physical world. If these physical particles are shown to satisfy norm-positivity, then the standard probabilistic interpretation of the theory is guaranteed in a consistent manner (at least in the scattering-theoretical context).

4.1.1 General aspects

In order to formulate a physically meaningful theory of *scattering processes* in gauge theory, one should be able to define a **physical S-matrix** between physical states with *positive norm* and should verify that it satisfies the *genuine unitarity* without spoiling the standard probabilistic interpretation of quantum theory. As is shown by Theorem $A.2$-2 in Appendix $A.2$, this requirement can be summarized into the following three conditions.

(i) Hermiticity of Hamiltonian:

$$H^\dagger = H. \tag{4.1.1 - 1}$$

(ii) Time invariance of physical subspace: Physical subspace $\mathcal{V}_{\mathrm{phys}}$ specified by the subsidiary condition should be invariant under the time development, namely,

$$H\mathcal{V}_{\mathrm{phys}} \subseteq \mathcal{V}_{\mathrm{phys}} \tag{4.1.1 - 2}$$

should hold for the Hamiltonian $H (= P_0$: generator of time translation).

(iii) Positive semi-definiteness: The indefinite inner product $\langle \cdot | \cdot \rangle$ in the total state vector space $\mathcal{V}$ should be *positive semi-definite* when restricted to the physical subspace $\mathcal{V}_{\mathrm{phys}}$,

$$|\Psi\rangle \in \mathcal{V}_{\mathrm{phys}} \;\Rightarrow\; \langle \Psi | \Psi \rangle \geq 0. \tag{4.1.1 - 3}$$

If asymptotic completeness is postulated in the total state vector space $\mathcal{V}$, the first requirement (i) can be reformulated as

(i$'$) (Pseudo-)unitarity of total S-matrix: The total S-matrix S should be unitary with respect to the indefinite inner product $\langle \cdot | \cdot \rangle$ of the total space $\mathcal{V}$

$$S^\dagger S = SS^\dagger = 1. \tag{4.1.1 - 4}$$

In this case, the condition (ii) can be rewritten as

(ii$'$) $$S\mathcal{V}_{\mathrm{phys}} = \mathcal{V}_{\mathrm{phys}}, \tag{4.1.1 - 5}$$

or equivalently

(ii$''$) $$\mathcal{V}_{\mathrm{phys}}{}^{\mathrm{in}} = \mathcal{V}_{\mathrm{phys}}{}^{\mathrm{out}}, \tag{4.1.1 - 6}$$

where $\mathcal{V}_{\mathrm{phys}}{}^{\mathrm{in,out}}$ are the physical subspaces of the Fock space of in- and out-states.

Now, the validity of the first criterion (i′), the unitarity of total S-matrix S, $S^\dagger S = S S^\dagger = 1$, holds in $\mathcal{V}$ with respect to its indefinite inner product $\langle \cdot | \cdot \rangle$, because of the hermiticity assignment, Eq.(3.4.1-10), to the FP ghosts which guarantees the hermiticity, Eq.(3.4.1-11), of Lagrangian density $\mathcal{L}$. Next, the second criterion, (ii′), i.e., the time invariance of physical subspace $\mathcal{V}_{\text{phys}}$, is satisfied automatically because $\mathcal{V}_{\text{phys}}$ is defined by the subsidiary condition, Eq.(3.4.3-2), i.e.

$$Q_B |\text{phys}\rangle = 0, \qquad (4.1.1-7)$$

in terms of the *conserved* (and *scalar*) charge Q_B, and hence, is manifestly invariant under the time evolution (or more precisely, is Poincaré invariant).

What remains to prove now is only the third criterion (iii), i.e., the *positive semi-definiteness* of the inner product in the physical subspace $\mathcal{V}_{\text{phys}}$, which is quite a nontrivial problem. For its complete solution, detailed analysis is necessary of the inner product structure of the total state vector space $\mathcal{V}$ and of the physical subspace $\mathcal{V}_{\text{phys}}$ in each concrete model. However, it is also possible to extract the essential features of inner product structure to a considerable extent just from the information about irreducible representations of the BRS algebra defined by Eqs.(3.4.2-9)-(3.4.2-11). On the basis of such analysis, we will find in Sec.4.1.4 a very general norm-cancellation mechanism, called **quartet mechanism**. By this mechanism, the unphysical particles having non-positive norms are made undetectable completely in the physical world described by the physical Hilbert space, $H_{\text{phys}} \equiv$ (completion of) $\mathcal{V}_{\text{phys}}/\mathcal{V}_0$, which is defined as the quotient space of $\mathcal{V}_{\text{phys}}$ with respect to its zero-norm subspace $\mathcal{V}_0$.

4.1.2 Q_B-cohomology and representations of BRS algebra

First, we summarize some basic points instrumental for the following discussion :

i) Nilpotency of BRS charge Q_B :

$$Q_B^2 = 0. \qquad (4.1.2-1)$$

ii) Q_B changes FP ghost number N_{FP} by 1 :

$$[iQ_C,\ Q_B] = Q_B. \qquad (4.1.2-2)$$

iii) Because of the fact that the non-vanishing FP-ghost charge is carried only by the fields C^a and $\overline{C}^a$ with the eigenvalues $\mp i$ of Q_C [Eq.(3.4.2-8)], it can be assumed that the spectrum of FP-ghost charge Q_C is given by $i\mathbf{Z}$. Its eigenstates $|\Psi_N\rangle$ with eigenvalues $-iN$ (N: FP-ghost number) satisfy such strange orthogonality relations as

$$\langle\Psi_N|\Psi_M\rangle \propto \delta_{N,-M}, \qquad (4.1.2-3)$$

owing to the hermiticity of Q_C [Eq.(3.4.2-7)]. We can also assume that the total state vector space $\mathcal{V}$ is decomposed into a *direct sum* of sectors $\mathcal{V}_N$ with definite FP-ghost numbers N, but, because of Eq.(3), this is *not* an orthogonal decomposition. To get an orthogonal decomposition, it is necessary to combine the sector $\mathcal{V}_N$ with another sector $\mathcal{V}_{-N}$ having FP-ghost number $-N$ of *opposite sign*.

    iv) Non-degeneracy of inner product $\langle\cdot|\cdot\rangle$ in $\mathcal{V}$:

$$\langle\Psi|\Phi\rangle = 0 \text{ for } \forall|\Psi\rangle \in \mathcal{V} \Longrightarrow |\Phi\rangle = 0. \qquad (4.1.2-4)$$

To understand that the property iv) can be postulated without loss of generality, we note that a state $|\chi\rangle$ orthogonal to *all* states belonging to $\mathcal{V}$ gives *no* contribution to the inner product $\langle\cdot|\cdot\rangle$ and that, because of the hermiticity of the Hamiltonian H, the totality of such vectors $|\chi\rangle$ constitute a *time-invariant subspace* $\mathcal{V}^\perp$,

$$|\chi\rangle \in \mathcal{V}^\perp \Leftrightarrow \langle\Psi|\chi\rangle = 0 \text{ for } \forall|\Psi\rangle \in \mathcal{V}$$
$$\Rightarrow \langle\Psi|(H|\chi\rangle) = ((\langle\Psi|H)|\chi\rangle) = 0 \text{ for } \forall|\Psi\rangle \in \mathcal{V} \qquad (4.1.2-5)$$
$$\Leftrightarrow H|\chi\rangle \in \mathcal{V}^\perp.$$

Therefore, such a subspace $\mathcal{V}^\perp$ can be totally neglected by shifting everything to the quotient space $\mathcal{V}/\mathcal{V}^\perp$. In combination with iii), this property iv) implies the following important consequence:

    iv') For any eigenstate $|\Psi_N\rangle$ of iQ_C with *non-vanishing* eigenvalue $N_{FP} = N \neq 0$, there exists always some FP-conjugate state, $|\Psi_{-N}\rangle$, with the *opposite* FP-ghost number $-N$ and non-orthogonal to it :

$$\langle\Psi_{-N}|\Psi_N\rangle = 1 \neq 0. \qquad (4.1.2-6)$$

It is also important to note that

    v) the existence of a non-vanishing eigenvalue of Q_C is inseparably connected with the *indefiniteness* of inner product as a consequence of the properties iii) and iv'):

$$|\psi_\lambda\rangle \equiv |\Psi_N\rangle + \lambda|\Psi_{-N}\rangle \Rightarrow \langle\psi_\lambda|\psi_\lambda\rangle = 2\text{Re}\lambda \gtreqless 0. \qquad (4.1.2-7)$$

BRS-singlets and Q_B-cohomology

Because of the nilpotency of the BRS charge Q_B, any state can be classified into its irreducible representations with dimensions up to two, namely, *singlet* or *doublet* representations. We call a singlet representation of Q_B a **BRS-singlet** and a doublet one a **BRS-doublet**. If a state $|\pi\rangle$ is *not* annihilated by Q_B,

$$Q_B|\pi\rangle \equiv |\delta\rangle \neq 0, \qquad (4.1.2-8)$$

then the two vectors $|\pi\rangle$ and $|\delta\rangle$ constitute a BRS-doublet. For convenience' sake, we call such a $|\pi\rangle$ with non-vanishing image of Q_B a **parent state**, and its image $Q_B|\pi\rangle \neq 0$ a **daughter state**. When $|\Phi\rangle$ is annihilated by Q_B satisfying the subsidiary condition, Eq.(3.4.3-2), i.e.,

$$Q_B|\Phi\rangle = 0, \qquad (4.1.2-9)$$

it may or may not be accompanied by a parent state. In the former case, let its parent be $|\pi\rangle$,

$$|\Phi\rangle = Q_B|\pi\rangle. \qquad (4.1.2-10)$$

Then the state $|\Phi\rangle$ also belongs to a BRS-doublet as a daughter state, while, in the latter case, it constitutes a BRS-singlet.

Owing to the nilpotency itself, the definitions of a BRS-singlet and a parent state in a BRS-doublet necessarily involve the *ambiguity* of adding arbitrary daughter states to them. However, unambiguous description of the behavior of BRS-singlets can be attained by switching over to the space of equivalence classes where two BRS-singlets identified if their difference is given by a daughter state. Since this is just in agreement with the notion of *cohomology group*, it will be convenient to introduce some elementary terminology in the theory of cohomology: The nilpotent BRS charge Q_B can be interpreted as a *coboundary operation* acting on the state vectors in $\mathcal{V}$ viewed as *cochains* graded by the FP-ghost number N_{FP} [except for such a difference due to iii) that N_{FP} runs over all the integers including *negative* ones, in contrast to the usual case of cochains with non-negative degrees]. Then, a physical state $|\Phi\rangle$ annihilated by Q_B can legitimately be called a $\boldsymbol{Q_B}$-**cocycle**, and a daughter state given as the image of Q_B is a $\boldsymbol{Q_B}$-**coboundary**. The physical subspace $\mathcal{V}_{phys}$ defined by the subsidiary condition Eq.(9) is just the space $Z(Q_B, \mathcal{V})$ of Q_B-cocycles:

$$Z(Q_B, \mathcal{V}) \equiv \mathrm{Ker} Q_B = \mathcal{V}_{phys}. \qquad (4.1.2-11)$$

It readily follows from the nilpotency of Q_B that any Q_B-coboundary $\in \mathrm{Im}Q_B = Q_B \mathcal{V} \equiv B(Q_B, \mathcal{V})$ is a Q_B-cocycle :

$$B(Q_B, \mathcal{V}) \subseteq Z(Q_B, \mathcal{V}). \qquad (4.1.2 - 12)$$

The obstruction to the *converse* of this statement is just described by the **Q_B-cohomology group**,[1]

$$H(Q_B, \mathcal{V}) \equiv Z(Q_B, \mathcal{V})/B(Q_B, \mathcal{V}), \qquad (4.1.2 - 13)$$

defined as the quotient vector space of $Z(Q_B, \mathcal{V})$ with respect to its subspace $B(Q_B, \mathcal{V})$ of Q_B-coboundaries. Namely, $H(Q_B, \mathcal{V})$ is obtained as the space of equivalence classes in $Z(Q_B, \mathcal{V})$, where all the Q_B-coboundaries are regarded to be vanishing. Since a BRS-singlet is a Q_B-cocycle *without* being a Q_B-coboundary, the totality of BRS-singlets are described in this space $H(Q_B, \mathcal{V})$ without the above-mentioned "ambiguities" due to the Q_B-coboundary terms.

Next, we note that the hermiticity of Q_B asserts the *orthogonality* of any Q_B-coboundary to all the Q_B-cocycles, or physical states,

$$
\begin{aligned}
&|\delta\rangle = Q_B|\pi\rangle \in B(Q_B, \mathcal{V}) \\
&\Longrightarrow \langle\Phi|\delta\rangle = \langle\Phi|(Q_B|\pi\rangle) = ((\langle\Phi|Q_B)|\pi\rangle = 0 \qquad (4.1.2 - 14) \\
&\text{for } \forall|\Phi\rangle \in Z(Q_B, \mathcal{V}) = \mathrm{Ker}Q_B = \mathcal{V}_{\mathrm{phys}},
\end{aligned}
$$

according to which the space $B(Q_B, \mathcal{V})$ of all Q_B-coboundaries is contained in the *isotropic subspace* $Z(Q_B, \mathcal{V}) \cap Z(Q_B, \mathcal{V})^{\perp}$ of $Z(Q_B, \mathcal{V})$,

$$B(Q_B, \mathcal{V}) \subseteq Z(Q_B, \mathcal{V}) \cap Z(Q_B, \mathcal{V})^{\perp} \equiv \mathcal{V}_0. \qquad (4.1.2 - 15)$$

Thus, Q_B-coboundaries give *no* contribution to the inner product in $Z(Q_B, \mathcal{V})$, and hence, the inner product structure of $Z(Q_B, \mathcal{V})$ can be transferred consistently to the quotient space $H(Q_B, \mathcal{V}) = Z(Q_B, \mathcal{V})/B(Q_B, \mathcal{V})$,

$$\langle\hat{\Phi}|\hat{\Psi}\rangle \equiv \langle\Phi|\Psi\rangle \quad \text{for } |\hat{\Phi}\rangle, |\hat{\Psi}\rangle \in H(Q_B, \mathcal{V}), \qquad (4.1.2 - 16)$$

---

[1]   We reserve such names as *BRS-cocycle, BRS-coboundary, and BRS-cohomology,* etc for the cohomology of BRS anti-derivation $\boldsymbol{\delta}$ defined *on the algebra of field operators,* using the names Q_B-cocycle and Q_B-coboundary, etc here for the cohomology of BRS charge Q_B on $\mathcal{V}$.

where $|\hat{\Phi}\rangle \equiv |\Phi\rangle + B(Q_B, \mathcal{V}) \in H(Q_B, \mathcal{V})$, etc. By this definition, Q_B-cohomology $H(Q_B, \mathcal{V})$ can be identified with the space of BRS-singlets inclusive of the inner product structure.

Representations of BRS algebra and inner product structure

From the identity of the physical subspace $\mathcal{V}_{\mathrm{phys}}$ with $Z(Q_B, \mathcal{V})$ [Eq.(11)] and from the orthogonality of $B(Q_B, \mathcal{V})$ to $Z(Q_B, \mathcal{V})$ [Eq.(15)], it clearly follows that all the physical contents of the theory should be contained in this space $H(Q_B, \mathcal{V})$ of BRS-singlets. In view of iii), iv') and v), however, its inner product may still be indefinite or can be degenerate. We need to analyze more closely the structure of $H(Q_B, \mathcal{V})$ including its inner product structure, discriminating between physical and unphysical objects. For this purpose, it is important to investigate the representations of BRS algebra structure characterized by Eqs.(3.4.2-9)-(3.4.2-11), and we combine the above discussion on the Q_B-cohomology with the properties iii) and iv') relating the spectrum of FP-ghost charge Q_C with the inner product structure of the state vector space $\mathcal{V}$. According to iii), irreducible representations of BRS algebra can be given by the members of BRS-singlets or -doublets with definite FP-ghost numbers N_{FP}, namely, eigenstates of FP-ghost charge Q_C. In connection with the inner product structure, however, the peculiar (non-)orthogonality relations, Eq.(3), among the eigenstates of Q_C require us not to separate those states which constitute FP-conjugate pairs having FP-ghost numbers $\pm N_{\mathrm{FP}}$ with opposite signs. Then, BRS singlets should be divided into two classes according to the FP-ghost number N_{FP},

(I) BRS-singlet with $N_{\mathrm{FP}} = 0$

and

(II) BRS-singlet with $N_{\mathrm{FP}} \neq 0$.

The latter one is divided further into two subclasses :

(IIa) **unpaired singlet**= BRS-singlet with $N_{\mathrm{FP}} \neq 0$ whose FP-conjugate state is an unphysical parent state in a BRS-doublet,

(IIb) **singlet pair**= FP-conjugate pair of two BRS-singlets.

In the case of a BRS-doublet, the property iv') requires the existence of *another* BRS-doublet in an FP-conjugate relation to the original one, which leads to the definition of a quartet:

(III) **quartet**=FP-conjugate pair of BRS-doublets.

It can be seen in the following that this classification exhausts all the possible representations of BRS algebra in indefinite inner product spaces. Although our main interest lies in the space $H(Q_B, \mathcal{V})$ of BRS-singlets, the analysis of the roles of these three classes and of their inner product structures is indispensable for guaranteeing the consistency of the theory because of the possible interplay between different classes as is seen below. Now we begin with the case (III).

Quartet = FP-conjugate pair of BRS-doublets

We pick up a BRS-doublet consisting of two states, $|\pi_N\rangle$ and $Q_B|\pi_N\rangle \equiv |\delta_{N+1}\rangle$, carrying FP-ghost numbers N and $N + 1$, respectively. Because of the nilpotency i), the daughter state $|\delta_{N+1}\rangle$ has a vanishing norm,

$$\langle\delta_{N+1}|\delta_{N+1}\rangle = \langle\pi_N|Q_B^2|\pi_N\rangle = 0, \qquad (4.1.2 - 17)$$

and hence, the non-degeneracy iv') requires the existence of its FP-conjugate $|\pi_{-N-1}\rangle$ *not* orthogonal to it:

$$\langle\pi_{-N-1}|\delta_{N+1}\rangle = \langle\pi_{-N-1}|(Q_B|\pi_N\rangle) = ((\langle\pi_{-N-1}|Q_B)|\pi_N\rangle \neq 0. \qquad (4.1.2 - 18)$$

As a convention, we normalize the value of this inner product to unity. From Eq.(18), it follows that the states $|\pi_{-N-1}\rangle$ and $Q_B|\pi_{-N-1}\rangle \equiv |\delta_{-N}\rangle (\neq 0)$ constitute another BRS-doublet. In this way, whenever a BRS-doublet appears, it is necessarily accompanied by another BRS-doublet FP-conjugate to it. We call this FP-conjugate pair of two BRS-doublets a **quartet**.

BRS-singlet with $N_{\mathrm{FP}} \neq 0$

Next, we consider the case (II). If $H(Q_B, \mathcal{V})$ contains a BRS-singlet state $|\sigma_N\rangle$ in the class (II) with FP ghost number $N_{\mathrm{FP}} = N \neq 0$, then the property iv') [see Eq.(6)] tells us that there exists in $\mathcal{V}$ its FP-conjugate state $|\Psi_{-N}\rangle$ satisfying Eq.(6). If there is a *BRS-singlet* state among such $|\Psi_{-N}\rangle$'s, the same argument as v) [Eq.(7)] concludes the existence of *negative norms* in $H(Q_B, \mathcal{V})$. This is the case of (IIb), a **singlet pair**, FP-conjugate pair of BRS-singlets. If there is *no* BRS-singlet among its FP-conjugates, $|\sigma_N\rangle$ belongs to the class (IIa) of an **unpaired singlet**. In this case, any FP-conjugate state, $|\Psi_{-N}\rangle$, should belong to a certain BRS-doublet as a parent state; the possibility of being a daughter state, $|\Psi_{-N}\rangle = Q_B|\pi_{-N-1}\rangle$, is excluded because the relation

$$0 \neq \langle\Psi_{-N}|\sigma_N\rangle = ((\langle\pi_{-N-1}|Q_B)|\sigma_N\rangle = \langle\pi_{-N-1}|(Q_B|\sigma_N\rangle) \qquad (4.1.2 - 19)$$

contradicts the premise of $|\sigma_N\rangle$ being a BRS-singlet. Then, $|\hat{\sigma}_N\rangle \in H(Q_B, \mathcal{V})$ cannot find its FP-conjugate in $H(Q_B, \mathcal{V})$, and hence, the inner product in $H(Q_B, \mathcal{V})$ is *degenerate*. Although purely algebraic arguments do not exclude this possibility, an unpaired singlet in $H(Q_B, \mathcal{V})$ causes *no* trouble in itself (except for such context as *anomalies*) because of its *orthogonality to all* the BRS-singlet states. Therefore, we can neglect it in the present context.

Singlet pairs

In view of the orthogonality, Eq.(3), those states belonging to the class (II) seem to be expelled easily by imposing an additional subsidiary condition $Q_C|\text{phys}\rangle = 0$ to pick up the states of class (I). However, this is not the case because, once an *FP-conjugate pair* of BRS-singlets happen to exist, they can sneak into the space of class (I), damaging its positivity. To see this explicitly, let $\sigma_k^\dagger(\sigma_k)$ and $\bar{\sigma}_k^\dagger(\bar{\sigma}_k)$ denote *fermionic* creation (annihilation) operators of FP-conjugate pair of BRS-singlets with FP ghost numbers $\pm N \neq 0$, which generate a Fock space from the vacuum state $|0\rangle$ being assumed to be invariant under Q_B and Q_C, i.e., $Q_B|0\rangle = Q_C|0\rangle = 0$. Namely, they are supposed to satisfy

$$\{Q_B,\ \sigma_k\} = \{Q_B,\ \bar{\sigma}_k\} = 0; \qquad (4.1.2-20)$$

$$[iQ_C,\ \sigma_k] = N\sigma_k, \qquad (4.1.2-21)$$

$$[iQ_C,\ \bar{\sigma}_k] = -N\bar{\sigma}_k, \qquad (4.1.2-22)$$

and, in accordance with Eq.(3),

$$\{\bar{\sigma}_k,\ \sigma_l{}^\dagger\} = \{\sigma_k,\ \bar{\sigma}_l{}^\dagger\} = \delta_{kl}, \qquad (4.1.2-23)$$

$$\{\sigma_k,\ \sigma_l{}^\dagger\} = \{\bar{\sigma}_k,\ \bar{\sigma}_l{}^\dagger\} = 0. \qquad (4.1.2-24)$$

Then, a state defined by $|v\rangle \equiv \sigma_k{}^\dagger\bar{\sigma}_k{}^\dagger|0\rangle$ explicitly shows that such subsidiary conditions as $Q_B|\text{phys}\rangle = Q_C|\text{phys}\rangle = 0$ cannot prevent the invasion of σ- and $\bar{\sigma}$-particles into the space of BRS-singlets with *vanishing* FP-ghost number picked by these conditions, and that this causes the breakdown of norm-positivity :

$$Q_B|v\rangle = Q_C|v\rangle = 0, \qquad (4.1.2-25)$$

$$\langle v|v\rangle = -1 < 0. \qquad (4.1.2-26)$$

Thus, the existence of singlet pairs makes it impossible to formulate a consistent theory.

While this kind of possibility cannot either be denied in the present discussion of the BRS algebra, we will *never* encounter its concrete examples in the actual models of gauge theories. This is because there exist some evidences supporting their absence.

A) Absence of singlet pairs at the Heisenberg-operator level [Nak 79f, Kug 80]: This is a conclusion derived from a result in Sec.4.3.2, on the **BRS-cohomology** of anti-derivation $\boldsymbol{\delta}$ acting on the algebra of *Heisenberg. fields*. On the basis of the simple relation, Eq.(3.3.1-37), between $\overline{C}^a$ and B^a in the BRS transformation, it is shown that the non-trivial cohomology can be supplied only by the field operators carrying *non-negative* FP-ghost numbers (e.g., $f^{abc}C^aC^bC^c$ with $N_{\mathrm{FP}} = 3$ and $\epsilon^{a_1a_2\cdots a_n}C^{a_1}C^{a_2}\cdots C^{a_n}$ with $N_{\mathrm{FP}} = n = \dim G$) and that those with *negative* ones involving the field $\overline{C}^a$ are *trivial* [Sul 77, Nak 79f, Kug 80]:

$$\boldsymbol{\delta}(\Phi) = [iQ_B,\ \Phi]_{\mp} = 0 \text{ and } [iQ_C,\ \Phi] = N\Phi \text{ with } N < 0$$
$$\Longrightarrow \exists\Psi \text{ such that } \Phi = \boldsymbol{\delta}(\Psi).$$

Owing to this result, the existence of singlet pairs is negated among the states in $\mathcal{V}$ generated by the *Heisenberg fields* from the vacuum which constitute a *dense* subspace in $\mathcal{V}$ owing the cyclicity assumption [see Appendix A.2].

B) Absence of singlet pairs in the subspace annihilated by both Q_B and $\overline{Q}_B$ [Bon 81b]: This statement is based upon an interesting expression [Bon 81b] for the local Noether current $J_{C\mu}$ of Q_C,

$$J_{C\mu} = -i\boldsymbol{\delta}(\overline{C}^aA_\mu{}^a) - i\overline{\boldsymbol{\delta}}(C^aA_\mu{}^a)$$
$$= \{Q_B,\ \overline{C}^aA_\mu{}^a\} + \{\overline{Q}_B,\ C^aA_\mu{}^a\}. \qquad (4.1.2-27)$$

If it is allowed to derive from Eq.(27) such a global relation for the *volume-integrated* charge as

$$Q_C = \{Q_B,\ \overline{\mathcal{R}}\} + \{\overline{Q}_B,\ \mathcal{R}\}, \qquad (4.1.2-28)$$

the non-orthogonality $\langle \Psi_{-N}|\Psi_N\rangle \neq 0$ with $N \neq 0$ is easily seen to imply $Q_B|\Psi_N\rangle \neq 0$ or $\overline{Q}_B|\Psi_N\rangle \neq 0$, from which the absence of singlet pairs can be concluded. Actually, this claim cannot be justified in its naive *global* form, because both the currents, $\overline{C}^aA_\mu{}^a$ and $C^aA_\mu{}^a$, are non-conserved, and hence, $\overline{\mathcal{R}}$ and $\mathcal{R}$ are *not* well defined global charges. However, we can verify its *localized* version restricting $|\Psi_N\rangle$ or $|\Psi_{-N}\rangle$ to those states which are generated from the vacuum by Heisenberg fields smeared with test functions with *compact supports*.[3] Let $|\Psi_N\rangle$ be such a localized state,

$$|\Psi_N\rangle = \Psi_N(h)|0\rangle, \qquad (4.1.2-29)$$

where $\Psi_N(h) = \int d^4x \, \Psi_N(x) h(x)$ is a Heisenberg field operator smeared with a test function $h(x)$ whose support is compact. Using the notation in Eq.(A.3-5) of Appendix A.3, we obtain from Eq.(27)

$$
\begin{aligned}
0 \neq {}& -iN\langle\Psi_{-N}|\Psi_N\rangle = \langle\Psi_{-N}|Q_C\Psi_N(h)|0\rangle \\
={}& \langle\Psi_{-N}| \, [Q_C, \, \Psi_N(h)] \, |0\rangle \\
={}& \langle\Psi_{-N}| \, [(Q_C)_R, \, \Psi_N(h)] \, |0\rangle \\
={}& \langle\Psi_{-N}| \, [\{Q_B, (\overline{\mathcal{R}})_R\} + \{\overline{Q}_B, (\mathcal{R})_R\}, \, \Psi_N(h)] \, |0\rangle \\
={}& \langle\Psi_{-N}| \, \Big([Q_B, \, [(\overline{\mathcal{R}})_R, \, \Psi_N(h)]_\mp]_\pm + [(\overline{\mathcal{R}})_R, \, [Q_B, \, \Psi_N(h)]_\mp]_\pm \Big)|0\rangle \\
& + \langle\Psi_{-N}| \, \Big([\overline{Q}_B, \, [(\mathcal{R})_R, \, \Psi_N(h)]_\mp]_\pm + [(\mathcal{R})_R, \, [\overline{Q}_B, \, \Psi_N(h)]_\mp]_\pm \Big)|0\rangle \\
={}& (\langle\Psi_{-N}|Q_B) \, [(\overline{\mathcal{R}})_R, \, \Psi_N(h)]_\mp \, |0\rangle + \langle\Psi_{-N}| \, [(\overline{\mathcal{R}})_R, \, [Q_B, \, \Psi_N(h)]_\mp]_\pm \, |0\rangle \\
& + (\langle\Psi_{-N}|\overline{Q}_B) \, [(\mathcal{R})_R, \, \Psi_N(h)]_\mp|0\rangle + \langle\Psi_{-N}| \, [(\mathcal{R})_R, \, [\overline{Q}_B, \, \Psi_N(h)]_\mp]_\pm \, |0\rangle,
\end{aligned}
$$
$$(4.1.2-30)$$

which asserts that, if $[Q_B, \, \Psi_N(h)]_\mp = [\overline{Q}_B, \, \Psi_N(h)]_\mp = 0$, then $Q_B|\Psi_{-N}\rangle \neq 0$ or $\overline{Q}_B|\Psi_{-N}\rangle \neq 0$.

Owing to the difficulties of the infinite-dimensional topology in an indefinite inner product space, we cannot extend the properties of the dense subspace generated by local Heisenberg fields, directly to the total state space $\mathcal{V}$ containing asymptotic states appearing in the asymptotic limit of *infinite* past or future. However, the above results show at least that singlet pairs are quite unnatural objects difficult to be materialized. Since their existence totally invalidates a physically meaningful interpretation of a theory, it is reasonable for us to omit the case (IIb) of singlet pairs. Then, the relevant possibilities are exhausted by the two cases, (I) BRS-singlets with $N_{\mathrm{FP}} = 0$ and (III) quartets.

---

[3]   In this version, B) becomes a rather weaker statement than A).

4.1.3 Multi-particle structure and asymptotic fields

The example of a singlet pair mentioned in Sec.4.1.2, shows that the "conspiracy" between *many particles* contained in a state may endanger the control of physicality by the subsidiary condition(s) of global type without inspecting the *particle contents*. In view of this observation, it is also necessary for ensuring the consistency to clarify how the quartet particles behave in the multi-particle states belonging to $H(Q_B, \mathcal{V})$. To this end, we need the postulate of *asymptotic completeness* (see Sec.1.4.3) which claims that any state in the total space $\mathcal{V}$ can be described in terms of asymptotic fields.[4]

On the assumption that BRS charge Q_B and FP ghost charge Q_C constituting the BRS algebra are not broken spontaneously, that is, $Q_B|0\rangle = Q_C|0\rangle = 0$, we apply the classification of the representations of BRS algebra in Sec.4.1.2 to the *one-particle asymptotic states* and translate it into the properties of creation and annihilation operators of asymptotic fields. Then, we can analyze the structure of the total state space $\mathcal{V}$ in more detail as the Fock space of the asymptotic fields. If the presence of the particles belonging to the class (IIb) of singlet pairs can be negated safely, what remains to be investigated is the cases of (I) BRS-singlets with $N_{FP} = 0$ and of (III) quartets. It is clear that the case (IIa) of unpaired singlets can be neglected, because any physical state in $\mathcal{V}_{phys}$ containing the particles of this class are orthogonal to the whole $\mathcal{V}_{phys}$.

(a) *BRS-singlets with vanishing FP-ghost number=genuine physical particles.*

Let $\varphi_k{}^\dagger$ and φ_k denote generically the creation and annihilation operators of BRS-singlets particles with $N_{FP} = 0$, where k stands collectively for all the relevant indices or quantum numbers to discriminate different modes, such as energy-momentum (or indices of wave packets), spin and internal quantum numbers, etc. By this definition, we have

$$Q_B\varphi_k{}^\dagger|0\rangle = 0 = [Q_B, \ \varphi_k{}^\dagger]|0\rangle, \qquad (4.1.3-1)$$

$$Q_C\varphi_k{}^\dagger|0\rangle = 0 = [Q_C, \ \varphi_k{}^\dagger]|0\rangle, \qquad (4.1.3-2)$$

---

[4]   Although the asymptotic fields are *free*, they should not be confused with the free fields in the *interaction picture*. In contrast to the perturbation theory based upon the latter, the assumption of asymptotic completeness does not exclude the appearance of *bound states* which requires the non-perturbative treatment

from which (anti)commutativity of $\varphi_k{}^\dagger$ with Q_B and Q_C follows:

$$[Q_B,\ \varphi_k{}^\dagger] = [Q_B,\ \varphi_k] = 0, \qquad\qquad (4.1.3-3)$$

$$[Q_C,\ \varphi_k{}^\dagger] = [Q_C,\ \varphi_k] = 0, \qquad\qquad (4.1.3-4)$$

where we have used the hermiticity of Q_B and Q_C.[5] Thus, φ_k is a BRS- or gauge-invariant operator with vanishing FP ghost number. The particles described by $\varphi_k{}^\dagger$ and φ_k can freely appear in the physical states belonging to $\mathcal{V}_{\mathrm{phys}}$. Owing to the algebraic nature of our general arguments, however, we cannot guarantee the positivity of their norms, because their metric structure depends on the details of individual models.[6] Since any consistent theory cannot be developed without it, we assume here the norm-positivity of the states $\varphi_k{}^\dagger|0\rangle$,

$$\langle 0|\varphi_k\varphi_l{}^\dagger|0\rangle = \langle 0|[\varphi_k,\varphi_l{}^\dagger]_\mp|0\rangle$$
$$= +\delta_{kl}, \qquad\qquad (4.1.3-5)$$

where l runs over the same set of indices as k does. Owing to the Greenberg-Robinson theorem in Appendix $A.4$ asserting that the (anti)commutators of asymptotic fields (with their energy-momentum spectrum inside the lightcone) are c-$numbers$, Eq.(5) is equivalent to

$$[\varphi_k,\varphi_l{}^\dagger]_\mp = +\delta_{kl}. \qquad\qquad (4.1.3-6)$$

On this assumption, we call a BRS-singlet particle with $N_{\mathrm{FP}} = 0$ a **genuine physical particle**.

---

[5]   Eq.(1) by itself admits such a possibility that $[Q_B,\varphi_k{}^\dagger]$ is a linear combination of some annihilation operators. This can, however, be negated by the equality $\langle 0|[Q_B,\varphi_k{}^\dagger] = \langle 0|\varphi_k{}^\dagger Q_B = 0$ following from the hermiticity of Q_B. The same remark applies to the case of Q_C.

[6]   It should be noted that *no proof* of the positivity has been given in 4-dimensional models, except for the trivial case of free fields and their small perturbations. It is known in the discussion of Bethe-Salpeter equation [Nak 69] that the norm-positivity can be violated at certain critical couplings, even if the small coupling guarantees the positivity.

(b) *Quartets=unphysical particles.*

According to the definition in Sec.4.1.2, let an FP-conjugate pair of BRS-doublets, $|\pi_{k,N}\rangle$, $|\delta_{k,N+1}\rangle \equiv Q_B|\pi_{k,N}\rangle$ and $|\pi_{k,-N-1}\rangle$, $|\delta_{k,-N}\rangle \equiv Q_B|\pi_{k,-N-1}\rangle$, be one-particle states constituting a *quartet* characterized by the relation

$$\langle \pi_{k,-N-1}|\delta_{k,N+1}\rangle = \langle \pi_{k,-N-1}|(Q_B|\pi_{k,N}\rangle)$$
$$= ((\langle \pi_{k,-N-1}|Q_B)|\pi_{k,N}\rangle) = \langle \delta_{k,-N}|\pi_{k,N}\rangle = 1, \tag{4.1.3-7}$$

where the meaning of k is the same as above. Similarly to the case (a) of genuine physical particles, we can translate Eq.(7) into the (anti)commutation relations of creation and annihilation operators corresponding to these one-particle states. For the sake of later convenience, we introduce the following notation :

$$|\pi_{k,N}\rangle \equiv \chi_k^{\dagger}|0\rangle, \qquad\qquad |\delta_{k,N+1}\rangle = Q_B|\pi_{k,N}\rangle \equiv -i\gamma_k^{\dagger}|0\rangle;$$
$$|\delta_{k,-N}\rangle = Q_B|\pi_{k,-N-1}\rangle \equiv -\beta_k^{\dagger}|0\rangle, \quad |\pi_{k,-N-1}\rangle \equiv -\overline{\gamma}_k^{\dagger}|0\rangle. \tag{4.1.3-8}$$

(The annihilation operators are defined by the hermitian conjugates of these creation operators as usual.) Since the FP ghost numbers of parent states are given by N and $-N-1$, we can assume without loss of generality that N is an even number (if necessary, by interchanging the roles of N and $-N-1$). Then, repeating the same arguments as the above case (a), we can derive from the definition, Eq.(8), the following BRS transformation properties for χ and γ,

$$[iQ_B, \chi_k] = \gamma_k, \tag{4.1.3-9}$$

$$\{iQ_B, \overline{\gamma}_k\} = i\beta_k, \tag{4.1.3-10}$$

together with those for β and $\overline{\gamma}$ which follow from the nilpotency of Q_B,

$$[Q_B, \beta_k] = \{Q_B, \gamma_k\} = 0. \tag{4.1.3-11}$$

We can utilize the orthogonality relation Eq.(4.1.2-3) (expressing the FP-ghost number conservation), the FP-conjugate relation Eq.(7) (which is a kind of "WT identity") and the Greenberg-Robinson theorem (Appendix A.4), to determine the (anti)commutation relations for the quartet particles $(\tau_{kA})_{A=1,\cdots,4} \equiv (\chi_k, \beta_k, \gamma_k, \overline{\gamma}_k)$. They can be summarized in a matrix form as follows:

$$([\tau_{k,A}, \tau_{l,B}]_{\mp})_{A,B=1,\cdots,4} = (\langle 0|\tau_{k,A}\tau_{l,B}^{\dagger}|0\rangle)_{A,B=1,\cdots,4}$$

$$
\equiv (\eta_{kA,lB}) = \begin{array}{c} \\ \chi_k \\ \beta_k \\ \gamma_k \\ \overline{\gamma}_k \end{array} \begin{array}{cccc} \chi_l{}^\dagger & \beta_l{}^\dagger & \gamma_l{}^\dagger & \overline{\gamma}_l{}^\dagger \\ \left(\begin{array}{cccc} \omega_{kl} & -\delta_{kl} & 0 & 0 \\ -\delta_{kl} & 0 & 0 & 0 \\ 0 & 0 & 0 & i\delta_{kl} \\ 0 & 0 & -i\delta_{kl} & 0 \end{array} \right) \end{array}. \qquad (4.1.3-12)
$$

All the vanishing matrix elements are due to the mismatching of FP-ghost numbers, except for the case of $\langle 0|\beta_k\beta_l{}^\dagger|0\rangle = 0$ which is due to the nilpotency of $Q_B{}^2 = 0$. In the case of $N = 0$, the matrix element $\omega_{kl} = \langle 0|\chi_k\chi_l{}^\dagger|0\rangle$ cannot be fixed by our algebraic arguments, but fortunately we never need its specific value in what follows.

In order to get a better insight into the notion of quartets, we present here its typical example of **elementary quartets** which always appear in the theory, containing as their members each "elementary" FP ghost pair, C^a and $\overline{C}^a$ [Kug unp.b, Kug 79d]. Their existence can surely be concluded from the *exact* forms of the following Green's functions given by

$$
\langle 0|TB^a(x)B^b(y)|0\rangle = 0, \qquad (4.1.3-13)
$$

$$
\langle 0|TA_\mu{}^a(x)B^b(y)|0\rangle = i\langle 0|T(D_\mu C)^a(x)\overline{C}^b(y)|0\rangle
$$
$$
= -i\delta^{ab}\partial_\mu D_F(x-y) = -\delta^{ab} \int \frac{d^4p}{(2\pi)^4} e^{-ip\cdot(x-y)} \frac{p_\mu}{p^2+i0}, \qquad (4.1.3-14)
$$

They can be derived from the WT identities, and the field equation of C^a as well as the canonical anticommutation relation for $\overline{C}^a$ as follows. First, from the BRS-invariance of the vacuum, $Q_B|0\rangle = 0$, and the BRS transformation Eqs.(3.3.1-33), (3.3.1-35), and (3.3.1-37), we obtain the following WT identities :

$$
0 = \langle 0|\{Q_B, T\overline{C}^a(x)B^b(y)\}|0\rangle = \langle 0|TB^a(x)B^b(y)|0\rangle, \qquad (4.1.3-15)
$$

$$
0 = \langle 0|\{Q_B, T(A_\mu{}^a(x)\overline{C}^b(y))\}|0\rangle
$$
$$
= \langle 0|TA_\mu{}^a(x)B^b(y)|0\rangle - i\langle 0|T(D_\mu C)^a(x)\overline{C}^b(y)|0\rangle. \qquad (4.1.3-16)
$$

Then, the field equation of C^a,

$$
\partial^\mu(D_\mu C)^a = 0, \qquad (4.1.3-17)
$$

and the canonical anticommutation relation,

$$i\{(D_0 C)^a(x),\ \overline{C}^b(y)\}_0 = -\{\pi_C^a(x),\ \overline{C}^b(y)\}_0 = i\delta^{ab}\delta(\boldsymbol{x} - \boldsymbol{y}), \qquad (4.1.3-18)$$

yield such a simple differential equation as

$$\begin{aligned}
\partial_x{}^\mu \langle 0|\mathrm{T}(D_\mu C)^a(x)\overline{C}^b(y)|0\rangle &= \langle 0|\mathrm{T}\partial^\mu(D_\mu C)^a(x)\overline{C}^b(y)|0\rangle \\
&\quad + \delta(x_0 - y_0)\langle 0|\{(D_0 C)^a(x),\ \overline{C}^b(y)\}|0\rangle \\
&= \delta^{ab}\delta^4(x - y). \qquad (4.1.3-19)
\end{aligned}$$

Therefore, we obtain

$$\begin{aligned}
\mathrm{F.T.}(\langle 0|\mathrm{T}(D_\mu C)^a \overline{C}^b|0\rangle)(p) &\equiv \int d^4x\, e^{ip\cdot(x-y)} \langle 0|\mathrm{T}(D_\mu C)^a(x)\overline{C}^b(y)|0\rangle \\
&= i\delta^{ab} p_\mu/(p^2 + i0), \qquad (4.1.3-20)
\end{aligned}$$

$$\mathrm{F.T.}(\langle 0|\mathrm{T}A_\mu{}^a B^b|0\rangle)(p) = -\delta^{ab} p_\mu/(p^2 + i0). \qquad (4.1.3-21)$$

The pole terms in Eqs.(20) and (21) imply the existence of massless asymptotic fields χ^a, β^a, γ and $\overline{\gamma}^a$ showing up in the asymptotic limit of the Heisenberg fields $A_\mu{}^a, B^a, C^a$ and $\overline{C}^a$ in the sense of LSZ (see Sec.1.4.4):

$$A^{\mathrm{as}}{}_\mu{}^{,a}(x) = \partial_\mu \chi^a(x) + \cdots, \qquad (4.1.3-22)$$

$$B^{\mathrm{as}a}(x) = \beta^a(x) + \cdots, \qquad (4.1.3-23)$$

$$(D_\mu C)^{\mathrm{as}a}(x) = \partial_\mu \gamma^a(x) + \cdots, \qquad (4.1.3-24)$$

$$\overline{C}^{\mathrm{as}a}(x) = \overline{\gamma}^a(x) + \cdots, \qquad (4.1.3-25)$$

where $\Phi^{\mathrm{as}}(x)$ denotes the asymptotic field of a Heisenberg field $\Phi(x)$ defined by the weak operator limit of infinite-past or future corresponding to 'as'= 'in' or 'out', and the dots $(\cdots)$ stand for other possible asymptotic fields irrelevant to the poles at $p^2 = 0$ appearing in Eqs.(20) and (21).[7] According to Theorem $A.4\text{-}4$ in Appendix $A.4$, any unbroken symmetry transformation induced on the asymptotic fields is *linear* and, also is related by Eq.($A.4\text{-}14$) to that on the Heisenberg fields. Thus, the BRS transformation properties for the asymptotic fields $\chi^a, \beta^a, \gamma^a$ and $\overline{\gamma}^a$ can be inferred from those for $A_\mu{}^a, B^a, C^a$ and $\overline{C}^a$ as follows:

$$\begin{aligned}
[iQ_B, \chi^a(x)] &= \gamma^a(x), & \{iQ_B, \gamma^a(x)\} &= 0, \\
\{iQ_B, \overline{\gamma}^a(x)\} &= i\beta^a(x), & [iQ_B, \beta^a(x)] &= 0.
\end{aligned} \qquad (4.1.3-26)$$

---

[7]   The asymptotic field $\chi^a(x)$ may not be covariant, in general.

When rewritten in terms of the creation and annihilation operators, these equations are just identical with the BRS-transformation properties Eqs.(9)-(11) of the quartets discussed in the above. Their mutual (anti)commutation relations are proved also to be identical with Eq.(12) by essentially the same reasoning as above: Eqs.(20) and (21) resulting from the WT identities, Eqs.(15) and (16), correspond to Eq.(7), and, the application of Greenberg-Robinson theorem again yields the following 4-dimensional commutation relations :

$$[\chi^a(x),\ \beta^b(y)] = i\{\gamma^a(x),\ \overline{\gamma}^b(y)\} = -i\delta^{ab}D(x-y). \qquad (4.1.3-27)$$

Similarly to the case in Eq.(12), all the other (anti)commutators for $\chi^a, \beta^a, \gamma^a$ and $\overline{\gamma}^a$ except for $[\chi^a(x),\ \chi^b(y)]$ are shown to vanish, for instance,

$$[\beta^a(x),\ \beta^b(y)] = \{\gamma^a(x),\ \gamma^b(y)\} = \{\overline{\gamma}^a(x),\ \overline{\gamma}^b(y)\} = 0. \qquad (4.1.3-28)$$

Rewriting the (anti)commutation relations, Eqs.(27) and (28), in terms of the creation and annihilation operators, one can easily see that they reproduce Eq.(12). In this way, we have just checked on a general ground the existence of *elementary quartet*. Its members, $\beta^a(x)$ and $\overline{\gamma}^a(x)$, are given by the asymptotic fields of the Heisenberg fields, $B^a(x)$ and $\overline{C}^a(x)$, and γ^a is the asymptotic field contained in $(D_\mu C)^a(x)$. In the examples in Sec.4.2, $\chi^a(x)$ will be identified with the longitudinal component of $A_\mu{}^a(x)$ in the Yang-Mills theory with the unbroken global gauge symmetry, and, in the case of Higgs-Kibble model with spontaneously broken symmetry of $SU(2)$, Goldstone mode arising from the Higgs scalar field plays the role of χ^a. The behavior of these elementary quartets will play an important role in Sec.4.4 in determining whether a generator of the global gauge symmetry suffers from spontaneous breakdown or not.

4.1.4 Quartet mechanism
–Poincaré lemma for unphysical particle sectors–

In order to guarantee the consistency of the theory, we should verify that any *unphysical particles* belonging to the *quartet* representations of BRS algebra do not appear in the physical world through scattering processes, in such a way that the assumption on the norm-positivity of genuine physical particles directly leads to the norm-positivity of the physical state space. To this end, we consider the case that there exist a variety of quartets characterized by the BRS transformation property, Eqs.(4.1.3-9)-(4.1.3-10), and by the inner product structure given by Eq.(4.1.3-12), and that all the other particles belong to the category of genuine physical particles with *positive norm*. Since, as explained before, the existence of *singlet pairs* makes it totally impossible to formulate any consistent theory and since the *unpaired singlets* decouple from all the physical quantities because of their orthogonality to any physical states, BRS-singlets with $N_{\mathrm{FP}} \neq 0$ can be neglected here. Then, inspite of the existence of negative norms in the total (asymptotic) Fock space $\mathcal{V}$ due to the unphysical particles, the physical subspace $\mathcal{V}_{\mathrm{phys}}$ specified by the subsidiary condition, $Q_B|\mathrm{phys}\rangle = 0$, can be shown to be *positive semi-definite*: Any members of quartets appear here in $\mathcal{V}_{\mathrm{phys}}$ only in *zero-norm combinations*. Thus, any quartet members can never be detected with finite probability in $\mathcal{V}_{\mathrm{phys}}$: Quartets are always confined! To prove it is the subject of this subsection.

First, we note that any one-particle asymptotic states of genuine physical particles can be assumed to be *orthogonal* to those of unphysical particles. This is always possible by taking advantage of the "ambiguity" of adding BRS-coboundary terms in the definition of BRS-singletness as follows. Let $|\varphi_i\rangle$ be in the former category, and $|\chi_k\rangle$, $|\gamma_k\rangle = -iQ_B|\chi_k\rangle$, $|\overline{\gamma}_k\rangle$ and $|\beta_k\rangle = Q_B|\overline{\gamma}_k\rangle$ constitute quartets with inner product structure described in Eq.(4.1.3-12). The orthogonality between $|\varphi_i\rangle$ and all the daughter states, $|\gamma_k\rangle = -iQ_B|\chi_k\rangle$ and $|\beta_k\rangle = Q_B|\overline{\gamma}_k\rangle$, follows directly from Eq.(4.1.2-14) or Eq.(4.1.2-15). Since the FP-ghost number $-(N+1)$ of $|\overline{\gamma}_k\rangle$ is *odd* integer by our convention [see below Eq.(4.1.3-8)], this state is also orthogonal to $|\varphi_i\rangle$. Therefore, the only possible case of non-orthogonality, $\langle \chi_k|\varphi_i\rangle \neq 0$, occurs between $|\varphi_i\rangle$ and $|\chi_k\rangle$ if the FP-ghost number N of the latter vanishes. In this case, we define a state $|\varphi_i{'}\rangle$ by[8]

$$|\varphi_i{'}\rangle \equiv |\varphi_i\rangle + \sum_l |\beta_l\rangle\langle\chi_l|\varphi_i\rangle = |\varphi_i\rangle + Q_B(\sum_l \langle\chi_l|\varphi_i\rangle|\overline{\gamma}_l\rangle). \qquad (4.1.4-1)$$

Since this new state differs from the original one, $|\varphi_i\rangle$, only by a Q_B-coboundary term, no changes arise in its relations with Q_B and with the inner product structure, other than its orthogonality with $|\chi_k\rangle$:

$$\langle\chi_k|\varphi_i{'}\rangle = \langle\chi_k|\varphi_i\rangle + \sum_l \langle\chi_k|\beta_l\rangle\langle\chi_l|\varphi_i\rangle$$

---

[8]   To avoid the difficulty of the infinite sum, we need to assume that only finite terms in $\langle\chi_l|\varphi_i\rangle$ are non-vanishing.

$$= \langle \chi_k | \varphi_i \rangle + \sum_l (-\delta_{kl}) \langle \chi_l | \varphi_i \rangle = 0. \qquad (4.1.4-2)$$

By replacing $|\varphi_i\rangle$ with $|\varphi_i'\rangle$, therefore, one-particle asymptotic states of genuine physical particles are now *orthogonal* to those of unphysical particles belonging to quartets. In terms of creation and annihilation operators, this orthogonality can be translated into the *commutativity* between genuine physical particles and unphysical particles. In what follows, we use the notation φ_i for genuine physical particles instead of φ_i' with the understanding that the above orthogonalization procedure has already been performed.

For the analysis of the Q_B-cohomological structure of the Fock space $\mathcal{V}$ of asymptotic fields, it is very important to decompose it into the orthogonal direct sum according to the number of unphysical particles: The sector of states containing n unphysical particles in sum aside from arbitrary number of genuine physical particles φ_i, is called n-**unphysical-particle sector**. With the aid of the inverse of the metric matrix, η, in Eq.(4.1.3-12), we can now define [Kug 78a, Kug 78c] the projection operator $P^{(n)}$ onto the n-unphysical-particle sector by a recursion equation,

$$\begin{aligned}
P^{(n)} &= \frac{1}{n} \sum_{k,l} \sum_{A,B=1}^{4} (\eta^{-1})_{kA,lB} \; \tau_{k,A}{}^\dagger P^{(n-1)} \tau_{l,B} \\
&= \frac{1}{n} \sum_k \{ -\beta_k{}^\dagger P^{(n-1)} \chi_k - \chi_k{}^\dagger P^{(n-1)} \beta_k - \sum_l \omega_{k,l} \beta_k{}^\dagger P^{(n-1)} \beta_l \\
&\qquad + i \gamma_k{}^\dagger P^{(n-1)} \bar{\gamma}_k - i \bar{\gamma}_k{}^\dagger P^{(n-1)} \gamma_k \}, \qquad (4.1.4-3)
\end{aligned}$$

with $n = 1, 2, \cdots$. Here the index k (or l) is understood to stand for the one to discriminate different kind of quartets as well as other relevant indices, because we are considering the case in which there exist arbitrary variety of quartets aside from many kinds of genuine physical particles. As the initial data with $n = 0$, $P^{(0)}$ is the projection operator onto the *zero-unphysical-particle sector*, which is just the subspace spanned solely by the genuine physical particles, φ_i. We denote this subspace by $\mathcal{H}_{\text{phys}}$:

$$\mathcal{H}_{\text{phys}} \equiv P^{(0)} \mathcal{V} = \text{Fock space of genuine physical particles } \varphi_i. \qquad (4.1.4-4)$$

By the assumption, $\mathcal{H}_{\text{phys}}$ has positive definite inner product, and the projection operator $P^{(0)}$ can be given by

$$P^{(0)} = \sum_{m=0}^{\infty} \sum_{i_1, \cdots, i_m} \frac{1}{m!} \varphi_{i_1}{}^\dagger \varphi_{i_2}{}^\dagger \cdots \varphi_{i_m}{}^\dagger |0\rangle \langle 0| \varphi_{i_m} \cdots \varphi_{i_2} \varphi_{i_1}, \qquad (4.1.4-5)$$

because of the diagonal metric structure, $\langle\varphi_i|\varphi_j\rangle = [\varphi_i, \varphi_j{}^\dagger]_\mp = \delta_{ij}$, [Eq.(4.1.3-6)].
Note that the $P^{(n)}$'s are projection operators *orthogonal* to one another, satisfying
the *completeness relation* (on the assumption of asymptotic completeness):

$$(P^{(n)})^2 = P^{(n)} = P^{(n)\dagger}, \tag{4.1.4 - 6}$$

$$P^{(n)} P^{(m)} = P^{(m)} P^{(n)} = \delta_{mn} P^{(n)}, \tag{4.1.4 - 7}$$

$$\sum_{n=0}^{\infty} P^{(n)} = 1. \tag{4.1.4 - 8}$$

Using the BRS transformation properties, Eqs.(4.1.3-9)-(4.1.3-11), of quartets, we can verify the important BRS-invariance properties of $P^{(n)}$ [Kug 78a, Kug 78c],

$$[Q_B, P^{(n)}] = 0 \quad \text{for } n = 0, 1, 2, \cdots. \tag{4.1.4 - 9}$$

This can be proved by induction. First, the BRS-invariance of $P^{(0)}$, $[Q_B, P^{(0)}] = 0$ trivially follows from the definition of genuine physical particles φ_i. Indeed, it implies the commutativity, $[Q_B, \varphi_i] = 0$, of φ_i with Q_B, and the expression Eq.(5) for $P^{(0)}$ leads to $Q_B P^{(0)} = P^{(0)} Q_B = 0$ by $Q_B|0\rangle = 0$. To proceed from $n - 1$ to n, we calculate the commutator $[Q_B, P^{(n)}]$ using the BRS transformation property, Eqs.(4.1.3.-9)-(4.1.3-11), of quartets, the recursion equation Eq.(3) for $P^{(n)}$ and the induction hypothesis at $n - 1$,

$$[Q_B, P^{(n)}] = \frac{1}{n} \sum_k \{\beta_k{}^\dagger P^{(n-1)}(-i\gamma_k) + (-i\gamma_k{}^\dagger)P^{(n-1)}\beta_k$$
$$+ i\gamma_k{}^\dagger P^{(n-1)}\beta_k + i\beta_k{}^\dagger P^{(n-1)}\gamma_k\} = 0. \tag{4.1.4 - 10}$$

This finishes the proof of Eq.(9). By virtue of this equation Eq.(9) together with the BRS transformation property, Eqs.(4.1.3.-9)-(4.1.3-11), we can rewrite $P^{(n)}$ into the following remarkable form :

$$P^{(n)} = \{iQ_B, R^{(n)}\} = \boldsymbol{\delta}(R^{(n)}) \quad \text{for } n \geq 1;$$
$$R^{(n)} \equiv i\frac{1}{n}\sum_k (\overline{\gamma}_k{}^\dagger P^{(n-1)}\chi_k + \chi_k{}^\dagger P^{(n-1)}\overline{\gamma}_k + \sum_l \omega_{kl}\beta_k{}^\dagger P^{(n-1)}\overline{\gamma}_l).$$
$$\tag{4.1.4 - 11}$$

Combining this with the completeness relation Eq.(8), we obtain the most important relation giving a resolution of identity,

$$1 = \sum_{n=0}^{\infty} P^{(n)} = P^{(0)} + \{iQ_B, \sum_{n=1}^{\infty} R^{(n)}\} = P^{(0)} + \boldsymbol{\delta}(R), \tag{4.1.4 - 12}$$

which states equivalently that the projection $P^{(0)\perp}$ onto the orthogonal complement of the subspace $\mathcal{H}_{\text{phys}} = P^{(0)}\mathcal{V}$ of genuine physical particles is a Q_B-coboundary:

$$P^{(0)\perp} \equiv 1 - P^{(0)} = \boldsymbol{\delta}(R). \qquad (4.1.4-13)$$

From this relation, a *Poincaré lemma* can be easily derived for all n-unphysical-particle-sectors with $n \neq 0$. Namely, the Q_B-cocycle condition on a state $|f\rangle \in P^{(n)}\mathcal{V}$ with $n \neq 0$ implies that it is just a Q_B-coboundary:[9]

$$
\begin{aligned}
& Q_B|f\rangle = 0 \quad \text{for } |f\rangle \in P^{(n)}\mathcal{V} \\
& \Longleftrightarrow |f\rangle \in Z(Q_B, \mathcal{V}) \cap P^{(n)}\mathcal{V} \subset (P^{(0)}\mathcal{V})^\perp \\
& \Longrightarrow |f\rangle = P^{(0)\perp}|f\rangle = \boldsymbol{\delta}(R)|f\rangle = iQ_B R|f\rangle \in B(Q_B, \mathcal{V}) \cap P^{(n)}\mathcal{V}.
\end{aligned}
$$

$$(4.1.4-14)$$

From Eq.(14), we can derive the following consequences:

(i) Owing to the isotropic property, Eq.(4.1.2-15), of Q_B-coboundaries $\in B(Q_B, \mathcal{V})$, we obtain

$$\langle f|P^{(n)}|g\rangle = 0 \qquad \text{for } n \geq 1 \qquad (4.1.4-15)$$

for any physical states $\forall |f\rangle, \forall |g\rangle \in \mathcal{V}_{\text{phys}}$ satisfying $Q_B|f\rangle = Q_B|g\rangle = 0$. Therefore, the inner products in $\mathcal{V}_{\text{phys}}$ are totally determined by the components of state vectors projected out onto its subspace $\mathcal{H}_{\text{phys}}$ containing no unphysical particles and all other components in $\mathcal{V}_{\text{phys}}{}^\perp$ give no contributions:

$$\langle f|g\rangle = \langle f|P^{(0)}|g\rangle = \langle P^{(0)}f|\; P^{(0)}g\rangle. \qquad (4.1.4-16)$$

Thus, the physical subspace $\mathcal{V}_{\text{phys}}$ has positive semi-definite inner product, as long as the norm-positivity of genuine physical particles is assured. This completes the proof of the third physicality criterion (iii) in Sec.4.1.1, and hence, the physical S-matrix is proved to be unitary. We note here also the following isomorphism:

$$\mathcal{V}_{\text{phys}}/\mathcal{V}_0 \cong \mathcal{H}_{\text{phys}} = P^{(0)}\mathcal{V}, \qquad (4.1.4-17)$$

Further, from the completeness relation Eq.(8), the Poincaré lemma Eq.(14), the nilpotency $Q_B{}^2 = 0$, and from $P^{(0)}\mathcal{V}_0 = 0$ due to the positive definiteness of

---

[9]   This important property has first been remarked by K. Fujikawa in a footnote of his paper [Fujk 78] giving comments on [Kug 78a, Kug 78c]

genuine physical particles in $\mathcal{H}_{\text{phys}}$, we obtain

$$\mathcal{V}_0 = \sum_{n=0}^{\infty} P^{(n)} \mathcal{V}_0 = \sum_{n=1}^{\infty} P^{(n)} \mathcal{V}_0 \subseteq \sum_{n=1}^{\infty} P^{(n)} \mathcal{V}_{\text{phys}} \subseteq Q_B \mathcal{V} \subseteq \mathcal{V}_0, \qquad (4.1.4-18)$$

and hence, the following equalities:

$$\mathcal{V}_0 = \sum_{n=1}^{\infty} P^{(n)} \mathcal{V}_{\text{phys}} = Q_B \mathcal{V}. \qquad (4.1.4-19)$$

(ii) It may be instructive to see explicitly what type of combinations of unphysical particles appear in the physical subspace $\mathcal{V}_{\text{phys}}$. We present a complete list of them up to 2-unphysical-particle sector in Table I. For this purpose, we can use Eq.(15) which asserts that any state in $P^{(n)} \mathcal{V}_{\text{phys}}$ ($n \geq 1$) can be obtained as a Q_B-coboundary, $Q_B |\Psi^{(n)}\rangle$, with some n-unphysical-particle state $|\Psi^{(n)}\rangle$. All the states in Table I are constructed by this method: E.g., $Q_B \bar{\gamma}_k{}^\dagger \bar{\gamma}_l{}^\dagger |\alpha\rangle = (\beta_k{}^\dagger \bar{\gamma}_l{}^\dagger - \bar{\gamma}_k{}^\dagger \beta_l{}^\dagger)|\alpha\rangle$.

(iii) In a scattering process starting from an initial state $|i\rangle \in \mathcal{V}_{\text{phys}}(= \mathcal{V}_{\text{phys}}{}^{\text{in}})$, any final states $|f\rangle$ [10] appearing after the scattering should remain in the physical subspace $\mathcal{V}_{\text{phys}}(= \mathcal{V}_{\text{phys}}{}^{\text{out}})$ owing to the conservation of Q_B. On the contrary, the subspace $\mathcal{H}_{\text{phys}} = P^{(0)} \mathcal{V}_{\text{phys}}$ is *not* invariant under the time evolution:

$$\mathcal{H}_{\text{phys}}{}^{\text{in}} = P^{(0)\,\text{in}} \mathcal{V}_{\text{phys}} \neq P^{(0)\,\text{out}} \mathcal{V}_{\text{phys}} = \mathcal{H}_{\text{phys}}{}^{\text{out}}. \qquad (4.1.4-20)$$

Namely, even when we start from an initial state $|i\rangle \in \mathcal{H}_{\text{phys}}{}^{\text{in}}$ containing *no* unphysical particles, the final state $|f\rangle$ generally has non-vanishing components

Table I. List of unphysical-particle states belonging to $\mathcal{V}_{\text{phys}}$.

($|\alpha\rangle \in \mathcal{H}_{\text{phys}}$ is an arbitrary state containing physical particles only.)

	$N_{\text{FP}} = -1$	$N_{\text{FP}} = 0$	$N_{\text{FP}} = 1$	$N_{\text{FP}} = 2$
$P^{(1)}$	–	$\beta_k{}^\dagger |\alpha\rangle$	$\gamma_k{}^\dagger |\alpha\rangle$	–
$P^{(2)}$	$(\beta_k{}^\dagger \bar{\gamma}_l{}^\dagger - \bar{\gamma}_k{}^\dagger \beta_l{}^\dagger)|\alpha\rangle$	$\beta_k{}^\dagger \beta_l{}^\dagger |\alpha\rangle$ $(\beta_k{}^\dagger \chi_l{}^\dagger + i\bar{\gamma}_k{}^\dagger \gamma_l{}^\dagger)|\alpha\rangle$	$\beta_k{}^\dagger \gamma_l{}^\dagger |\alpha\rangle$ $(\chi_k{}^\dagger \gamma_l{}^\dagger + \gamma_k{}^\dagger \chi_l{}^\dagger)|\alpha\rangle$	$\gamma_k{}^\dagger \gamma_l{}^\dagger |\alpha\rangle$

[10]   For definiteness, the initial and final states, $|i\rangle$ and $|f\rangle$, should be described respectively in reference to the *in-* and *out-state bases* : $|i\rangle = |i \text{ in }\rangle$ and $|f\rangle = |f \text{ out }\rangle$. Since any state vector $|\alpha\rangle$ in the Heisenberg picture does *not* change in itself through the time evolution, the state change through the scattering process is treated as a change of base without the actual change of states: $|\alpha\rangle = |f \text{ out }\rangle = |i \text{ in }\rangle = \sum_{n=0}^{\infty} P^{(n)}{}_{\text{out}} |i \text{ in }\rangle$.

$P^{(n)\ \text{out}}|f\rangle$ of n-unphysical-particle states which are really produced by such processes as FP-ghost pair creations. The important point is that such unphysical particles are produced *only in zero-norm combinations* as assured by Eq.(15). If the wrong hermiticity assignment Eq.(3.4.1-24) were adopted, this would not be the case. We here present an example showing its incorrectness. Since the elementary quartets are shown to always exist, the state $(\beta_k{}^\dagger \chi_l{}^\dagger + i\bar{\gamma}_k{}^\dagger \gamma_l{}^\dagger)|\alpha\rangle \equiv |2\rangle$ in Table I with χ, β, γ and $\bar{\gamma}$ belonging to some elementary quartet will be produced through the interaction, but with zero-norm $\langle 2|2\rangle = 0$ in our hermiticity assignment. We note the following correspondence between the FP ghosts with our hermiticity assignment Eq.(3.4.1-10) and the wrong one Eq.(3.4.1-24) (the latter of which are denoted by lowercases for distinction),

$$C^{\text{as}}(x) = \sum_k (C_k g_k(x) + C_k{}^\dagger g_k{}^*(x)); \quad \overline{C}^{\text{as}}(x) = \sum_k (\overline{C}_k g_k(x) + \overline{C}_k{}^\dagger g_k{}^*(x)),$$

$$\updownarrow \qquad\qquad\qquad\qquad \updownarrow$$

$$c^{\text{as}}(x) = \sum_k (c_k g_k(x) + \bar{c}_k{}^\dagger g_k{}^*(x)); \quad -i\bar{c}^{\text{as}}(x) = -i\sum_k (\bar{c}_k g_k(x) + c_k{}^\dagger g_k{}^*(x)),$$

$$(4.1.4 - 21)$$

where $\{g_k(x)\}$ is a complete system of massless wave packets [see Appendix A.5] and the indices for internal symmetry are omitted. The contrast between the two different hermiticity assignments, Eqs.(3.4.1-10) and (3.4.1-24), can be exhibited more clearly in terms of creation and annihilation operators:

$$C_k \leftrightarrow c_k, \quad C_k{}^\dagger \leftrightarrow \bar{c}_k{}^\dagger, \quad \overline{C}_k \leftrightarrow -i\bar{c}_k, \quad \overline{C}_k{}^\dagger \leftrightarrow -ic_k{}^\dagger; \qquad (4.1.4 - 22)$$

$$\{C_k, \overline{C}_l{}^\dagger\} = -\{\overline{C}_k, C_l{}^\dagger\} = i\delta_{kl} \leftrightarrow \{c_k, c_l{}^\dagger\} = -\{\bar{c}_k, \bar{c}_l{}^\dagger\} = -\delta_{kl}.$$

$$(4.1.4 - 23)$$

In view of these relations, the state corresponding in the latter assignment to the above zero-norm physical state $|2\rangle$ is seen to be given by

$$|\tilde{2}\rangle = (\beta_k{}^\dagger \chi_l{}^\dagger + c_k{}^\dagger \bar{c}_l{}^\dagger |\alpha\rangle, \qquad (4.1.4 - 24)$$

which has *negative norm* (for $k \neq l$)!:

$$\langle \tilde{2}|\tilde{2}\rangle = \langle\alpha|\bar{c}_l c_k c_k{}^\dagger \bar{c}_l{}^\dagger|\alpha\rangle = -\langle\alpha|\alpha\rangle < 0. \qquad (4.1.4 - 25)$$

From the viewpoint of the Poincaré lemma expressing $|\tilde{2}\rangle$ as a Q_B-coboundary,

$$|\tilde{2}\rangle = -iQ_B c_k{}^\dagger \chi_l{}^\dagger|\alpha\rangle, \qquad (4.1.4 - 26)$$

Eq.(25) explicitly shows a result of the non-hermiticity of the BRS charge due to the wrong hermiticity assignment: $Q_B{}^\dagger \neq Q_B$ and $Q_B{}^\dagger Q_B \neq Q_B{}^2 = 0$.

(iv) For any Green's functions of gauge-(BRS-)invariant operators, any intermediate states containing unphysical particles do not contribute to the their matrix elements between two physical states. For instance, let $\Phi_i(x)$ $(i = 1, 2)$ be BRS-invariant field operators,

$$[Q_B, \ \Phi_i(x)] = 0. \tag{4.1.4 - 27}$$

Then we have

$$Q_B \Phi_i(x)|\text{phys}\rangle = 0 \implies \Phi_i(x)|\text{phys}\rangle \in \mathcal{V}_{\text{phys}}, \tag{4.1.4 - 28}$$

and hence, the contribution of n-unphysical-particle intermediate states with $n \neq 0$ to the matrix elements of $\Phi_i(x)$ sandwiched by two physical state $|f_i\rangle$ $(i = 1, 2)$ vanishes by Eq.(15):

$$\langle f_1|\Phi_1(x)P^{(n)}\Phi_2(y)|f_2\rangle = 0 \quad \text{for } n \neq 0. \tag{4.1.4 - 29}$$

Thus, we have proved that the unphysical particles, i.e., any quartet members, cannot be detected in the physical subspace $\mathcal{V}_{\text{phys}}$ because they can appear there only in zero-norm combinations. We call this norm-cancellation mechanism **quartet mechanism** [Kug 79d].

(v) To understand the crucial role of the nilpotency of Q_B in quartet mechanism, we present here an explicit example of the unitarity violation caused by the *breakdown of nilpotency* in massive Yang-Mills theory [Oji 82] formulated in Sec.3.5.2. Although some techniques explained in the next section are required for the analysis of asymptotic fields, we state just the result without going into details: The asymptotic fields $A^{\text{as}}{}_\mu{}^a, B^{\text{as}\ a}, C^{\text{as}\ a}$ and $\overline{C}^{\text{as}\ a}$ of the corresponding Heisenberg fields can be shown to satisfy the following field equations and 4-dimensional (anti)commutation relations,[11]

$$(\Box + m_r{}^2)A^{\text{as}}{}_\mu{}^a = (1 - \alpha_r)\partial_\mu B^{\text{as}\ a}, \tag{4.1.4 - 30}$$

$$(\Box + \alpha_r m_r{}^2) \begin{bmatrix} B^{\text{as}\ a} \\ C^{\text{as}\ a} \\ \overline{C}^{\text{as}\ a} \end{bmatrix} = 0; \tag{4.1.4 - 31}$$

---

[11]   The normalizations of $A^{\text{as}}{}_\mu$ and B^{as} are different from those in Sec.2.4.2 by a factor $\sqrt{k}$.

$$[A^{as}{}_\mu{}^a(x),\ B^{as\ b}(y)] = -i\delta^{ab}\partial_\mu\Delta(x-y;\alpha_r m_r{}^2), \qquad (4.1.4-32)$$

$$[B^{as\ a}(x),\ B^{as\ b}(y)] = -im_r{}^2\delta^{ab}\Delta(x-y;\alpha_r m_r{}^2), \qquad (4.1.4-33)$$

$$\{C^{as\ a}(x),\ \overline{C}^{as\ b}(y)\} = -\delta^{ab}\Delta(x-y;\alpha_r m_r{}^2), \qquad (4.1.4-34)$$

where m_r and α_r are the renormalized quantities. We define the Proca field $U^{as}{}_\mu{}^a$ satisfying

$$(\Box + m_r{}^2)U^{as}{}_\mu{}^a = \partial^\mu U^{as}{}_\mu{}^a = 0 \qquad (4.1.4-35)$$

by

$$U^{as}{}_\mu{}^a \equiv A^{as}{}_\mu{}^a - \partial_\mu B^{as\ a}/m_r{}^2, \qquad (4.1.4-36)$$

and introduce the creation and annihilation operators $U_\alpha{}^\dagger, U_\beta$ and $B_k{}^\dagger$, etc. of the asymptotic fileds with respect to suitable complete sets of wave packets. Then, the matrix η representing the inner product corresponding to Eq.(4.1.3-12) is given by

$$(\eta_{kA,lB}) = \begin{array}{c} \\ U_\alpha \\ B_k \\ C_k \\ \overline{C}_k \end{array}\begin{array}{cccc} U_\beta{}^\dagger & B_l{}^\dagger & C_l{}^\dagger & \overline{C}_l{}^\dagger \\ \left(\begin{array}{cccc} \delta_{\alpha\beta} & 0 & 0 & 0 \\ 0 & -m_r{}^2\delta_{k,l} & 0 & 0 \\ 0 & 0 & 0 & i\delta_{kl} \\ 0 & 0 & -i\delta_{kl} & 0 \end{array}\right). \end{array} \qquad (4.1.4-37)$$

The BRS and anti-BRS transformations for these asymptotic fields are given respectively by

$$[iQ_B{}^r,\ U_\alpha] = 0, \qquad\qquad [iQ_B{}^r,\ B_k] = m_r{}^2 C_k,$$
$$\{iQ_B{}^r,\ C_k\} = 0, \qquad\qquad \{iQ_B{}^r,\ \overline{C}_k\} = iB_k;$$
$$(4.1.4-38)$$

$$[i\overline{Q}_B{}^r,\ U_\alpha] = 0, \qquad\qquad [i\overline{Q}_B{}^r,\ B_k] = m_r{}^2\overline{C}_k,$$
$$\{i\overline{Q}_B{}^r,\ C_k\} = -iB_k, \qquad \{i\overline{Q}_B{}^r,\ \overline{C}_k\} = 0,$$
$$(4.1.4-39)$$

where $Q_B{}^r$ and $\overline{Q}_B{}^r$ denote the renormalized BRS and anti-BRS charges and where the indices of internal degrees of freedom are omitted. Using Eqs.(38) and (39), we can easily check that the state $|\Psi\rangle$ defined by

$$|\Psi\rangle \equiv (B_k{}^\dagger B_l{}^\dagger - im_r{}^2 C_k{}^\dagger\overline{C}_l{}^\dagger + im_r{}^2\overline{C}_k{}^\dagger C_l{}^\dagger)|0\rangle \qquad (4.1.4-40)$$

is annihilated by both $Q_B{}^r$ and $\overline{Q}_B{}^r$:

$$Q_B{}^r|\Psi\rangle = \overline{Q}_B{}^r|\Psi\rangle = 0. \qquad (4.1.4-41)$$

Because of Eq.(37), however, its norm turns out to be negative,

$$\langle\Psi|\Psi\rangle = -m_r{}^4(1+\delta_{kl}) < 0, \qquad (4.1.4-42)$$

which clearly shows the breakdown of positive semi-definiteness of the subspace defined by such subsidiary conditions as Eq.(41). By the same argument as (iii), such a combination of unphysical particles as in $|\Psi\rangle$ given by Eq.(40) are created in final states of scattering processes starting from an initial state without any unphysical particles, as long as the gauge group is non-abelian (and semisimple). Only in the case of abelian gauge group, special situation arises: FP ghosts are totally *decoupled* owing to the vanishing structure constant, which allows us to impose the additional subsidiary conditions

$$C^{(+)}(x)|\text{phys}\rangle = \overline{C}^{(+)}(x)|\text{phys}\rangle = 0, \qquad (4.1.4-43)$$

just in the same manner as in the $m_r = 0$ case. Since these conditions prohibit the appearance of all the unphysical particles B, C and $\overline{C}$ with indefinite metric, massive vector meson theory can be formulated consistently only in the case with an abelian gauge group.

4.1.5 Comments on subsidiary conditions in cases with non-simple gauge groups

Now, we discuss the problem of arbitrariness in the choice of the subsidiary conditions in cases with *non-simple* gauge group G. In such a case, G can be decomposed into two factor groups,

$$G = G_1 \times G_2, \qquad (4.1.5-1)$$

where G_1 and G_2 may possibly be still non-simple. Consider such a gauge fixing term that it can be decomposed into two parts, each of which is related exclusively to one of the two gauge groups, G_1 and G_2, being independent of the other. [Of course, this includes our previous choice, Eq.(3.4.1-3), of gauge fixing term invariant under the global gauge group G.] Then, in accordance with the decomposition

Eq.(1), the BRS charge Q_B and FP ghost charge Q_C for the total group G are decomposed into the sum of those for the groups G_1 and G_2 as

$$Q_B = Q_B{}^{(1)} + Q_B{}^{(2)}, \qquad (4.1.5-2)$$

$$Q_C = Q_C{}^{(1)} + Q_C{}^{(2)}, \qquad (4.1.5-3)$$

and each of these charges is separately conserved. As has been seen in the previous subsection, Sec.4.1.4, the single subsidiary condition given by Eq.(4.1.1.-7), $Q_B|\text{phys}\rangle = 0$, is sufficient to pick up the physical subspace $\mathcal{V}_{\text{phys}}$ so as for the isomorphism, Eq.(4.1.4-17), $\mathcal{V}_{\text{phys}}/\mathcal{V}_0 \cong \mathcal{H}_{\text{phys}}(= \text{zero-unphysical-particle sector})$, to hold. However, we have here two different BRS algebras consisting, respectively, of $Q_B{}^{(1)}$ and $Q_C{}^{(1)}$ and of $Q_B{}^{(2)}$ and $Q_C{}^{(2)}$, whose existence implies that there are several possibilities in choosing subsidiary conditions; for example, one can define physical subspaces $\mathcal{V}_{\text{phys}}{}^{B_1,B_2}$ and $\mathcal{V}_{\text{phys}}{}^{B_1,C_1,B_2,C_2}$ by

$$\mathcal{V}_{\text{phys}}{}^{B_1,B_2} \ni |\text{phys}\rangle \;\Leftrightarrow\; Q_B{}^{(1)}|\text{phys}\rangle = Q_B{}^{(2)}|\text{phys}\rangle = 0; (4.1.5-4)$$

$$\mathcal{V}_{\text{phys}}{}^{B_1,C_1,B_2,C_2} \ni |\text{phys}\rangle \;\Leftrightarrow\; Q_B{}^{(1)}|\text{phys}\rangle = Q_C{}^{(1)}|\text{phys}\rangle$$
$$= Q_B{}^{(2)}|\text{phys}\rangle = Q_C{}^{(2)}|\text{phys}\rangle = 0, (4.1.5-5)$$

and so on, where the superscripts are attached to $\mathcal{V}_{\text{phys}}$'s for distinguishing the different subsidiary conditions to be imposed. It is not difficult to see the following inclusion relations among these possibilities:

$$\mathcal{V}_{\text{phys}} \equiv \mathcal{V}_{\text{phys}}{}^{B} \supseteq \left\{ \begin{array}{c} \mathcal{V}_{\text{phys}}{}^{B,C} \\ \mathcal{V}_{\text{phys}}{}^{B_1,B_2} \supseteq \left\{ \begin{array}{c} \mathcal{V}_{\text{phys}}{}^{B_1,B_2,C} \\ \mathcal{V}_{\text{phys}}{}^{B_1,B_2,C_1} \\ \mathcal{V}_{\text{phys}}{}^{B_1,B_2,C_2} \end{array} \right\} \end{array} \right\} \supseteq \mathcal{V}_{\text{phys}}{}^{B_1,B_2,C_1,C_2} \supseteq \mathcal{H}_{\text{phys}}.$$

$$(4.1.5-6)$$

Note here that

$$\mathcal{V}_{\text{phys}}{}^{B,C_i} = \mathcal{V}_{\text{phys}}{}^{B_1,B_2,C_i} \quad \text{for } i = 1,2, \qquad (4.1.5-7)$$

because from the commutation relations

$$[iQ_C{}^{(i)}, Q_B] = Q_B{}^{(i)}, \quad \text{for } i = 1,2, \qquad (4.1.5-8)$$

it follows that the condition $Q_B|\text{phys}\rangle = Q_C{}^{(1)}|\text{phys}\rangle = 0$, for instance, necessarily implies $Q_B{}^{(1)}|\text{phys}\rangle = [iQ_C{}^{(1)}, Q_B]|\text{phys}\rangle = 0$ and $Q_B{}^{(2)}|\text{phys}\rangle = (Q_B -$

$Q_B{}^{(1)})|\text{phys}\rangle = 0$. As an illustration, we cite some simple states belonging to the above different choices of physical subspaces

$$\mathcal{V}_{\text{phys}}{}^B \ni (\chi^{(1)}{}_k{}^\dagger \gamma^{(2)}{}_l{}^\dagger + \gamma^{(1)}{}_k{}^\dagger \chi^{(2)}{}_l{}^\dagger)|\alpha\rangle,$$

$$\mathcal{V}_{\text{phys}}{}^{B,C} \ni (\beta^{(1)}{}_k{}^\dagger \gamma^{(2)}{}_l{}^\dagger + i\overline{\gamma}^{(1)}{}_k{}^\dagger \gamma^{(2)}{}_l{}^\dagger)|\alpha\rangle,$$

$$\mathcal{V}_{\text{phys}}{}^{B_1,B_2} \ni (\gamma^{(1)}{}_k{}^\dagger \gamma^{(2)}{}_l{}^\dagger |\alpha\rangle,$$

$$\mathcal{V}_{\text{phys}}{}^{B_1,B_2,C} \ni \gamma^{(1)}{}_k{}^\dagger (\beta^{(2)}{}_l{}^\dagger \overline{\gamma}^{(2)}{}_m{}^\dagger - \overline{\gamma}^{(2)}{}_l{}^\dagger \beta^{(2)}{}_m{}^\dagger)|\alpha\rangle,$$

$$\mathcal{V}_{\text{phys}}{}^{B_1,B_2,C_1} \ni \beta^{(1)}{}_k{}^\dagger \gamma^{(2)}{}_l{}^\dagger |\alpha\rangle,$$

$$\mathcal{V}_{\text{phys}}{}^{B_1,B_2,C_2} \ni \gamma^{(1)}{}_k{}^\dagger \beta^{(2)}{}_l{}^\dagger |\alpha\rangle,$$

$$\mathcal{V}_{\text{phys}}{}^{B_1,B_2,C_1,C_2} \ni \beta^{(1)}{}_k{}^\dagger \beta^{(2)}{}_l{}^\dagger |\alpha\rangle.$$

It is important to stress that the differences among these choices are only relevant to the size of zero-norm subspace $\mathcal{V}_0$ in each $\mathcal{V}_{\text{phys}}$. Indeed, it is evident from the arguments in the preceding section that the isomorphism

$$\mathcal{V}_{\text{phys}}{}^X / \mathcal{V}_0{}^X \cong \mathcal{H}_{\text{phys}} \tag{4.1.5 − 9}$$

holds for any physical subspaces in Eq.(6), denoted by $\mathcal{V}_{\text{phys}}{}^X$ generically. We prefer, however, our original choice, $\mathcal{V}_{\text{phys}}(=\mathcal{V}_{\text{phys}}{}^B)$ specified by a single subsidiary condition, Eq.(4.1.1-7), $Q_B|\text{phys}\rangle = 0$. Aside from the fact that it is the simplest choice and makes the theoretical analysis easy, it allows us to take a wide variety of gauge fixing conditions: The gauge fixing terms with respect to the gauge group G_1 need not be invariant under G_2 gauge transformations and vice versa. In fact, in the cases when the degrees of freedom related to G_1 and G_2 are mixed with each other by the gauge fixing terms, the individual BRS charges $Q_B{}^{(1)}$ and $Q_B{}^{(2)}$ can no longer be defined separately. For instance, this is the case in the 't Hooft-Feynman gauge in the electroweak theory [tHo 71]. Therefore, we always take only one subsidiary condition, Eq.(4.1.1-7), hereafter.

4.2 SCATTERING THEORETICAL ANALYSIS OF SOME MODELS

In the previous section, quartet mechanism of "confining" unphysical particles has been presented in a general but abstract way. In this section, we exhibit its essence in a more explicit form through some model examples [Kug 79a, Kug 79b]. The models treated here are:

I. Higgs-Kibble model with spontaneous symmetry breaking,

II. pure Yang-Mills theory without spontaneous symmetry breaking.

Assuming the asymptotic completeness with respect to the "elementary" fields, we analyze the properties of asymptotic fields of the "elementary" particles in detail by the following method: First, we determine the general forms of the 2-point functions of Heisenberg fields by the requirements of Lorentz covariance and BRS invariance with the aid of Γ-WT identities. Since the Greenberg-Robinson theorem assures that the (anti)commutators between asymptotic fields are c-numbers, the discrete pole parts contained in these 2-point functions are seen to determine uniquely the (anti)commutation relations of asymptotic fields. Because of the *irreducibility* due to the asymptotic completeness, the obtained (anti)commutation relations determine the field equations obeyed by these asymptotic fields. According to the general result in Appendix A.4 ensuring that an unbroken conserved charge generates a *linear* transformation on asymptotic fields, we can also determine their behavior under the BRS transformation. Putting together all these results, we choose a complete set of the mutually independent modes, physical ones and unphysical ones. Then, the properties required in the previous section are explicitly shown to be satisfied; namely, all the physical modes have positive norm and can be made orthogonal to all the unphysical modes which fall into quartets satisfying the BRS transformation property, Eqs.(4.1.3-9)-(4.1.3-11), and the inner product structure, Eq.(4.1.3-12).

4.2.1 Higgs-Kibble model

Now, we analyze, from the viewpoint of BRS invariance, the structure of **Higgs-Kibble model** [tHo 71] as a typical and the simplest Yang-Mills theory with spontaneous breakdown of the gauge symmetry [Kug 79a]. The Lagrangian density $\mathcal{L}$ governing the dynamics is obtained from the general form, Eqs.(3.4.1-1)-(3.4.1-

6), by taking $SU(2)$ as the gauge group and by specifying the contents of matter fields and their Lagrangian density $\mathcal{L}_M$. Let $\Psi(x)$ denote a complex scalar field belonging to a spinor representation of the internal $SU(2)$ group. The matter Lagrangian density $\mathcal{L}_M$ of $\Psi(x)$ minimally coupled to $SU(2)$ Yang-Mills field $A_\mu{}^a$ is given by

$$\mathcal{L}_M = (\partial^\mu \Psi - ig\frac{\sigma^a}{2}A^{\mu a}\Psi)^\dagger(\partial_\mu \Psi - ig\frac{\sigma^b}{2}A_\mu{}^b\Psi) - V(\Psi^\dagger\Psi), \qquad (4.2.1-1)$$

where σ^a's denote the isospin matrices giving the spinor representation of $SU(2)$. The "potential term" $V(\Psi^\dagger\Psi)$ describing the $SU(2)$-invariant mass and quartic coupling terms of the field $\Psi(x)$ is supposed to be so chosen that its vacuum expectation value is given by

$$\langle 0|\Psi(x)|0\rangle = \frac{1}{\sqrt{2}}\begin{pmatrix} v \\ 0 \end{pmatrix}. \qquad (4.2.1-2)$$

Then, it is convenient to reparametrize the field Ψ as

$$\Psi(x) = \frac{1}{\sqrt{2}}\Big[(v + \varphi(x)) + i\chi^a(x)\sigma^a\Big]\begin{pmatrix} 1 \\ 0 \end{pmatrix}, \qquad (4.2.1-3)$$

in which χ^a $(a = 1,2,3)$ will be shown in the following to represent the (unphysical) massless NG bosons associated to the spontaneous breakdown of the global $SU(2)$ due to the vacuum condensation given by Eq.(2) and $\varphi(x)$ the (real) massive Higgs scalar. The BRS transformations for these matter fields φ and χ^a are given by

$$\boldsymbol{\delta}\varphi = [iQ_B,\ \varphi] = -\frac{g}{2}\chi^a C^a, \qquad (4.2.1-4)$$

$$\boldsymbol{\delta}\chi^a = [iQ_B,\ \chi^a] = \frac{g}{2}[(v + \varphi)C^a + (\chi \times C)^a]. \qquad (4.2.1-5)$$

We note here such peculiarity of this model that it retains still another global $SU(2)$ symmetry remaining *unbroken* after the spontaneous breakdown of the global one associated with the local $SU(2)$ gauge symmetry. With respect to this $SU(2)$ symmetry, φ and χ^a are isosinglet and isotriplet, respectively.

To determine the structure of asymptotic fields in this model, we utilize the Γ-WT identities given by Eqs.(3.4.4-17)-(3.4.4-19). The first one, Eq.(3.4.4-17), is adapted to this case as

$$\frac{\delta\Gamma}{\delta A_\mu{}^a}\frac{\delta\Gamma}{\delta K^{\mu a}} + \frac{\delta\Gamma}{\delta\varphi}\frac{\delta\Gamma}{\delta K_\varphi} + \frac{\delta\Gamma}{\delta\chi^a}\frac{\delta\Gamma}{\delta K_{\chi^a}} + \frac{\delta\Gamma}{\delta C^a}\frac{\delta\Gamma}{\delta K_{C^a}} + i\frac{\delta\Gamma}{\delta\bar{C}^a}B^a = 0, \qquad (4.2.1-6)$$

and the second and third ones, Eqs.(3.4.4-18) and (3.4.4-19), remain unchanged:

$$\frac{\delta\Gamma}{\delta B^a} = \partial^\mu A_\mu{}^a + \alpha B^a, \tag{4.2.1-7}$$

$$\partial_\mu \frac{\delta\Gamma}{\delta K^{\mu a}} + i\frac{\delta\Gamma}{\delta\overline{C}^a} = 0, \tag{4.2.1-8}$$

where K_φ and $K_\chi{}^a$ are respectively the source functions coupled to the BRS transforms, $\boldsymbol{\delta}\varphi$ and $\boldsymbol{\delta}\chi^a$, of φ and χ^a given by Eqs.(4) and (5). These Γ-WT identities give us useful information on the structure of 2-point functions. Taking the derivatives of Eq.(7) with respect to $A_\mu{}^b$, B^b and χ^b, we obtain[1]

$$\Gamma^{(2)}{}_{A_\mu B}{}^{ab}(p) \equiv \int d^4x \, e^{ip(x-y)} \frac{\delta^2\Gamma}{\delta A^{\mu a}(x)\delta B^b(y)}\bigg|_0 = i\delta^{ab}p_\mu, \tag{4.2.1-9}$$

$$\Gamma^{(2)}{}_{BB}{}^{ab}(p) = \alpha\delta^{ab}, \tag{4.2.1-10}$$

$$\Gamma^{(2)}{}_{B\chi}{}^{ab}(p) = 0. \tag{4.2.1-11}$$

In view of Lorentz covariance, the remaining global $SU(2)$ symmetry, the FP ghost number conservation and Eqs.(9)-(11), we can give the general form of one-particle-irreducible 2-vertices or the inverse propagators:

$$\Gamma^{(2)}{}_{\Phi_i\Phi_j}{}^{ab}(p) = \delta^{ab} \times$$

	A_ν	B	χ	C	$\overline{C}$
A_μ	$F_1(p^2)(\eta_{\mu\nu} - p_\mu p_\nu/p^2)$ $+F_2(p^2)p_\mu p_\nu/p^2$	ip_μ	$iF_3(p^2)p_\mu$	0	0
B	$-ip_\nu$	α	0	0	0
χ	$-iF_3(p^2)p_\nu$	0	$F_4(p^2)p^2$	0	0
C	0	0	0	0	$iF_5(p^2)p^2$
$\overline{C}$	0	0	0	$-iF_5(p^2)p^2$	0

$$\tag{4.2.1-12}$$

where $1/p^2$ should properly be understood as $1/(p^2 + i0)$. We have omitted here the parts containing φ which is seen to be decoupled from other fields because of its isosinglet property. Further we define

$$\text{F.T.}\frac{\delta^2\Gamma}{\delta K^{\mu a}\delta C^b}\bigg|_0 = -i\delta^{ab}p_\mu F_6(p^2), \tag{4.2.1-13}$$

$$\text{F.T.}\frac{\delta^2\Gamma}{\delta K_\chi{}^a\delta C^b}\bigg|_0 = \delta^{ab}F_7(p^2). \tag{4.2.1-14}$$

---

[1] $|_0$ means the evaluation of derivatives with all the arguments set equal to zero.

Differentiating Eq.(8) with respect to $C^b(y)$, we obtain an identity

$$\partial_\mu^x \left(\frac{\delta^2\Gamma}{\delta K^{\mu a}(x)\delta C^b(y)} \right)\bigg|_0 = i \frac{\delta^2\Gamma}{\delta\overline{C}^a(x)\delta C^b(y)}\bigg|_0, \qquad (4.2.1-15)$$

which implies

$$F_6(p^2) = F_5(p^2). \qquad (4.2.1-16)$$

Similarly, the derivative of Eq.(6) with respect to C^b yields an equality at $C = \overline{C} = B = \varphi = 0$:

$$\left(\frac{\delta\Gamma}{\delta A_\mu{}^a} \frac{\delta^2\Gamma}{\delta K^{\mu a}\delta C^b} + \frac{\delta\Gamma}{\delta\chi^a} \frac{\delta^2\Gamma}{\delta K_\chi{}^a\delta C^b} \right)\bigg|_{B=\varphi=C=\overline{C}=0} = 0, \qquad (4.2.1-17)$$

Substituting the definitions of $\delta^2\Gamma/\delta A_\mu{}^a\delta A_\nu{}^b$, $\delta^2\Gamma/\delta A_\mu{}^a\delta\chi^b$, etc. given by Eqs.(12), (13) and (14) into the derivatives of Eq.(17) with respect to $A_\mu{}^a$ and χ^a, we can derive the following equations,

$$F_2(p^2)F_6(p^2) = F_3(p^2)F_7(p^2), \qquad (4.2.1-18)$$

$$F_3(p^2)F_6(p^2) = F_4(p^2)F_7(p^2), \qquad (4.2.1-19)$$

which further imply

$$F_2(p^2)F_4(p^2) = F_3(p^2)^2. \qquad (4.2.1-20)$$

Using Eqs.(16) and (20), we obtain the propagators given as the inverse matrix of Eq.(12):

$$i^{-1}\text{F.T.}\langle 0|\text{T}\Phi_i{}^a\Phi_j{}^b|0\rangle = (\Gamma^{(2)})^{(-1)}{}_{ij}{}^{ab} = \delta^{ab} \times$$

	A_ν	B	χ	C	$\overline{C}$
A_μ	$(\eta_{\mu\nu} - p_\mu p_\nu/p^2)F_1^{-1}$ $-\alpha p_\mu p_\nu/(p^2)^2$	ip_μ/p^2	$i\alpha p_\mu F_3/(p^2)^2 F_4$	0	0
B	$-ip_\nu/p^2$	0	$-F_3/p^2 F_4$	0	0
χ	$-i\alpha p_\nu F_3/(p^2)^2 F_4$	$-F_3/p^2 F_4$	$(p^2 - \alpha F_2)/(p^2)^2 F_4$	0	0
C	0	0	0	0	$i/p^2 F_5$
$\overline{C}$	0	0	0	$-i/p^2 F_5$	0

$$(4.2.1-21)$$

From Eq.(21), we can derive the following vacuum expectation values of (anti)commutators:

$$\langle 0|[A_\mu^{\ a}(x),\ A_\nu^{\ b}(y)]|0\rangle = \delta^{ab}\Big[-iZ_3(\eta_{\mu\nu} + m^{-2}\partial_\mu\partial_\nu)\Delta(x-y;m^2)$$
$$+ iL\partial_\mu\partial_\nu D(x-y) - i\alpha\partial_\mu\partial_\nu E(x-y)$$
$$- i\int_{+0}^{\infty} ds\sigma(s)(\eta_{\mu\nu} + s^{-1}\partial_\mu\partial_\nu)\Delta(x-y;s)\Big], \qquad (4.2.1-22)$$

$$\langle 0|[A_\mu^{\ a}(x),\ B^b(y)]|0\rangle = -i\delta^{ab}\partial_\mu D(x-y), \qquad (4.2.1-23)$$

$$\langle 0|[B^a(x),\ B^b(y)]|0\rangle = 0, \qquad (4.2.1-24)$$

$$\langle 0|[A_\mu^{\ a}(x),\ \chi^b(y)]|0\rangle = \delta^{ab}\alpha\partial_\mu\Big[-iM_1 D(x-y) + iM_2 E(x-y)$$
$$- i\int_{+0}^{\infty} ds\sigma_{A\chi}(s)\Delta(x-y;s)\Big], \qquad (4.2.1-25)$$

$$\langle 0|[B^a(x),\ \chi^b(y)]|0\rangle = \delta^{ab}\Big[-iM_2 D(x-y)$$
$$- i\int_{+0}^{\infty} ds\sigma_{B\chi}(s)\Delta(x-y;s)\Big], \qquad (4.2.1-26)$$

$$\langle 0|[\chi^a(x),\ \chi^b(y)]|0\rangle = \delta^{ab}\Big[(Z_\chi - \alpha M_3)iD(x-y) + i\alpha M_2^{\ 2}E(x-y)$$
$$+ i\int_{+0}^{\infty} ds\sigma_\chi(s)\Delta(x-y;s)\Big], \qquad (4.2.1-27)$$

$$\langle 0|\{C^a(x),\ \overline{C}^b(y)\}|0\rangle = \delta^{ab}\Big[-\tilde{Z}_3 D(x-y)$$
$$- \int_{+0}^{\infty} ds\tilde{\sigma}(s)\Delta(x-y;s)\Big], \qquad (4.2.1-28)$$

where $E(x)$ denotes the dipole-ghost invariant D function defined by

$$E(x) \equiv -\frac{\partial}{\partial m^2}\Delta(x;m^2)\Big|_{m^2=0}, \quad \Box E(x) = D(x). \qquad (4.2.1-29)$$

(See Sec.2.3.4 for some of its properties.) In addition, we have

$$\langle 0|[\varphi(x),\ \varphi(y)]|0\rangle = iZ_\varphi\Delta(x-y;m_\varphi^{\ 2}) + i\int_{+0}^{\infty} ds\sigma_\varphi(s)\Delta(x-y;s), \quad (4.2.1-30)$$

and all the vacuum expectation values of (anti)commutators other than Eqs.(22)-(28) and (30) vanish owing to the global $SU(2)$ symmetry and/or the FP ghost number conservation. In Eq.(22), the mass m^2 is defined as the pole of $F_1(s)^{-1}$, i.e., $F_1(m^2) = 0$, and other quantities appearing in the above are defined as follows:[2]

$$Z_3^{\ -1} \equiv -\frac{d}{ds}F_1(s)\Big|_{s=m^2}, \quad \tilde{Z}_3^{\ -1} \equiv F_5(0), \quad Z_\chi^{\ -1} \equiv F_4(0),$$

---

[2]   We note the relations $Z_3 = 1 - \int_{+0}^{\infty} ds\sigma(s)$ and $L = Z_3/m^2 + \int_{+0}^{\infty} ds\sigma(s)/s$.

$$L^{-1} \equiv F_1(0), \quad M_1 \equiv -\frac{d}{ds}\mathrm{Re}\left(\frac{F_3(s)}{F_4(s)}\right)\Big|_{s=0},$$

$$M_2 \equiv \frac{F_3(0)}{F_4(0)}, \quad M_3 \equiv -\frac{d}{ds}\mathrm{Re}\left(\frac{F_2(s)}{F_4(s)}\right)\Big|_{s=0},$$

$$\sigma(s) \equiv \frac{1}{\pi}\mathrm{Im}(\frac{1}{F_1(s)}) - Z_3\delta(s - m^2),$$

$$\sigma_{B\chi}(s) \equiv -s\sigma_{A\chi}(s) \equiv -\frac{1}{\pi s}\mathrm{Im}\left(\frac{F_3(s)}{F_4(s)}\right),$$

$$\sigma_\chi(s) \equiv -\frac{1}{\pi s}\mathrm{Im}(\frac{1}{F_4(s)}) + \alpha\frac{1}{\pi s^2}\mathrm{Im}\left(\frac{F_2(s)}{F_4(s)}\right),$$

$$\tilde{\sigma}(s) \equiv -\frac{1}{\pi s}\mathrm{Im}(\frac{1}{F_5(s)}). \tag{4.2.1 - 31}$$

Assuming the LSZ asymptotic conditions, we define asymptotic fields by the weak limit of infinite past or infinite future $x_0 \to \pm\infty$:[3],[4]

$$A_\mu(x) \xrightarrow[x^0 \to \pm\infty]{} Z_3^{1/2}A^{\mathrm{as}}{}_\mu(x), \quad B(x) \to Z_B^{1/2}B^{\mathrm{as}}(x),$$

$$\chi(x) \to Z_\chi^{1/2}\chi^{\mathrm{as}}(x), \qquad \varphi(x) \to Z_\varphi^{1/2}\varphi^{\mathrm{as}}(x),$$

$$C(x) \to \tilde{Z}_3^{1/2}C^{\mathrm{as}}(x), \qquad \overline{C}(x) \to \tilde{Z}_3^{1/2}\overline{C}^{\mathrm{as}}(x), \tag{4.2.1 - 32}$$

where the renormalization constant Z_B is taken as $Z_B \equiv L^{-1} = F_1(0)$ just for convenience. Since these asymptotic fields are, of course, supposed to have their supports in the forward lightcone in the momentum space, their (anti)commutation relations should be c-numbers according to the Greenberg-Robinson theorem [Appendix A.4]. Then, the discrete spectrum parts of Eqs.(22)-(28) and (30) determine the (anti)commutation relations of asymptotic fields as follows:

$$[A^{\mathrm{as}}{}_\mu(x), A^{\mathrm{as}}{}_\nu(y)] = -i(\eta_{\mu\nu} + m^{-2}\partial_\mu\partial_\nu)\Delta(x - y; m^2)$$

$$+iK\partial_\mu\partial_\nu D(x - y) - i\alpha_r\partial_\mu\partial_\nu E(x - y), \tag{4.2.1 - 33}$$

$$[A^{\mathrm{as}}{}_\mu(x), B^{\mathrm{as}}(y)] = -i\sqrt{K}\partial_\mu D(x - y), \tag{4.2.1 - 34}$$

$$[B^{\mathrm{as}}(x), B^{\mathrm{as}}(y)] = 0, \tag{4.2.1 - 35}$$

$$[A^{\mathrm{as}}{}_\mu(x), \chi^{\mathrm{as}}(y)] = -i\alpha_r N\partial_\mu D(x - y)$$

---

[3]   We will often omit the $SU(2)$-indices a of the fields $A_\mu{}^a, B^a, \chi^a, C^a$ and $\overline{C}^a$.

[4]   Here all the Z-factors are assumed to be positive; this is always true in the perturbation theory.

$$+ i\alpha_r \sqrt{K^{-1}} \partial_\mu E(x - y), \qquad (4.2.1 - 36)$$

$$[B^{as}(x), \chi^{as}(y)] = - iD(x - y), \qquad (4.2.1 - 37)$$

$$[\chi^{as}(x), \chi^{as}(y)] = i(1 - 2\alpha_r N \sqrt{K^{-1}})D(x - y)$$
$$+ i\alpha_r K^{-1} E(x - y), \qquad (4.2.1 - 38)$$

$$\{C^{as}(x), \overline{C}^{as}(y)\} = - D(x - y), \qquad (4.2.1 - 39)$$

$$[\varphi^{as}(x), \varphi^{as}(y)] = i\Delta(x - y; m_\varphi{}^2), \qquad (4.2.1 - 40)$$

and all the (anti)commutators of other combinations vanish. Here we have defined the constants

$$K \equiv L/Z_3 = (Z_3 Z_B)^{-1}, \quad \alpha_r \equiv \alpha/Z_3, \qquad (4.2.1 - 41)$$

$$N \equiv (Z_3/Z_\chi)^{1/2} M_1 = (\sqrt{K} Z_3/Z_\chi) M_3/2, \qquad (4.2.1 - 42)$$

with the aid of the WT relation Eq.(20) at $p^2 = 0$, i.e. $F_2(0)F_4(0) = F_3{}^2(0)$, as well as the equalities

$$F_1(0) = F_2(0), \qquad (4.2.1 - 43)$$

$$\frac{d}{ds}\left(\mathrm{Re}\frac{F_2(s)}{F_4(s)}\right)\Bigg|_{s=0} = 2\frac{F_3(0)}{F_4(0)}\frac{d}{ds}\left(\mathrm{Re}\frac{F_3(s)}{F_4(s)}\right)\Bigg|_{s=0}. \qquad (4.2.1 - 44)$$

Eq.(43) is valid as long as the inverse propagator $\Gamma^{(2)}{}_{A_\mu A_\nu}(p)$ is not singular at $p^2 = 0$. Eq.(44) is due to the equality $F_2(p^2)/F_4(p^2) = (F_3(p^2)/F_4(p^2))^2$ following from Eq.(20) on the reasonable assumption $\mathrm{Im}(F_3(s)/F_4(s))|_{s=0} = 0$. These two assumptions are satisfied at any order of the perturbation theory. The last equality in Eq.(42) is a consequence of Eq.(44).

Now, our assumption of asymptotic completeness means that the asymptotic fields $A^{as}{}_\mu, B^{as}, C^{as}, \overline{C}^{as}, \chi^{as}$ and φ^{as} are *irreducible* or "complete" without any other asymptotic fields corresponding to bound states. Therefore, we can deduce from the (anti)commutation relations Eqs.(33)-(40) the following field equations of the asymptotic fields on the basis of their irreducibility :

$$\square B^{as} = \square C^{as} = \square \overline{C}^{as} = (\square + m_\varphi{}^2)\varphi = 0, \qquad (4.2.1 - 45)$$

$$\square \chi^{as} = -\alpha_r K^{-1} B^{as}. \qquad (4.2.1 - 46)$$

Eq.(46) shows that χ^{as} is a *dipole ghost* field except for the Landau gauge case ($\alpha_r = 0$). Separating the unphysical modes, we can pick up from $A^{as}{}_\mu$ the massive Proca field denoted by $U^{as}{}_\mu$,

$$U^{as}{}_\mu \equiv A^{as}{}_\mu - (\sqrt{K} - \alpha_r N)\partial_\mu B^{as} - \sqrt{K}\partial_\mu \chi^{as}, \qquad (4.2.1 - 47)$$

which is seen to satisfy the following field equations and commutation relations:[5]

$$(\Box + m^2)U^{\mathrm{as}}{}_\mu = 0, \quad \partial^\mu U^{\mathrm{as}}{}_\mu = 0, \qquad\qquad (4.2.1-48)$$

$$[U^{\mathrm{as}}{}_\mu(x),\ U^{\mathrm{as}}{}_\nu(y)] = -i(\eta_{\mu\nu} + m^{-2}\partial_\mu\partial_\nu)\Delta(x-y;m^2), \quad (4.2.1-49)$$

and the other commutators of $U^{\mathrm{as}}{}_\mu$ all vanish. Thus $A^{\mathrm{as}}{}_\mu$ satisfies

$$(\Box + m^2)A^{\mathrm{as}}{}_\mu = [(\sqrt{K} - \alpha_r N)m^2 - \alpha_r \sqrt{K^{-1}}]\partial_\mu B^{\mathrm{as}}$$
$$+ \sqrt{K}m^2\partial_\mu\chi^{\mathrm{as}}, \qquad\qquad (4.2.1-50)$$

$$\partial^\mu A^{\mathrm{as}}{}_\mu + \alpha_r\sqrt{K^{-1}}B^{\mathrm{as}} = 0. \qquad\qquad (4.2.1-51)$$

From all the information obtained above, we can now construct the Fock space of asymptotic fields, which is identified with the total state vector space $\mathcal{V}$ on the assumption of asymptotic completeness. The creation (annihilation) operators for the Proca field $U^{\mathrm{as}}{}_\mu$ and the real Higgs scalar φ^{as} are defined as usual by using suitable complete sets of wave packets, and are denoted by $U_\alpha{}^\dagger (U_\alpha)$ and $\varphi_\sigma{}^\dagger(\varphi_\sigma)$, respectively. Since the NG boson χ^{as} is a dipole-ghost field as is seen in Eq.(46), we introduce some manipulation to treat it. Let us define a field operator[6] $\tilde{\chi}^{\mathrm{as}}$ by

$$\tilde{\chi}^{\mathrm{as}} \equiv \chi^{\mathrm{as}} + \alpha_r K^{-1}\mathcal{E}^{(1/2)}B^{\mathrm{as}}, \qquad\qquad (4.2.1-52)$$

where the operator $\mathcal{E}^{(1/2)}$ is defined by Eq.$(A.5\text{-}3)$ or $(A.5\text{-}4)$ in Appendix $A.5$. By Eq.$(A.5\text{-}4)$, $\Box\mathcal{E}^{(1/2)}f = f$ valid for $f(x)$ satisfying $\Box f = 0$, and by Eq.(46), the $\tilde{\chi}^{\mathrm{as}}$ is a simple-pole field:

$$\Box\tilde{\chi}^{\mathrm{as}} = 0. \qquad\qquad (4.2.1-53)$$

Since all the four fields $B^{\mathrm{as}}, \tilde{\chi}^{\mathrm{as}}, C^{\mathrm{as}}$ and $\overline{C}^{\mathrm{as}}$ are massless simple-pole fields, we can define their creation and annihilation operators in the usual manner by using a common wave packet system $\{g_k(x)\}$ [see Appendix $A.5$ for the properties of $\{g_k(x)\}$]; especially, the annihilation operators for $\tilde{\chi}^{\mathrm{as}}$ field are defined by

$$\chi^{\mathrm{as}}{}_k \equiv (g_k, \tilde{\chi}^{\mathrm{as}}) \equiv i\int d\mathbf{x}\left(g_k^*(x)\partial_0\tilde{\chi}^{\mathrm{as}}(x) - \partial_0 g_k^*(x)\cdot\tilde{\chi}^{\mathrm{as}}(x)\right). \qquad (4.2.1-54)$$

---

[5]   In deriving Eq.(49), we need the relation $L = Z_\chi/M_2{}^2$, which is guaranteed by the WT relation Eq.(20) and the equality Eq.(43). Note also that Eq.(42) is indispensable for the consistency of Eq.(49) with Eqs.(47), (36) and (38).

[6]   This $\tilde{\chi}^{\mathrm{as}}$ is non-covariant and non-local (with respect to the asymptotic fields).

Then, one can easily show with the aid of Eqs.(A.5-30) and (52) that

$$\chi^{as}{}_k = (g_k, \chi^{as}) - \alpha_r K^{-1}(h_k, B^{as}), \qquad (4.2.1-55)$$

$$\chi^{as}(x) = \sum_k [\chi^{as}{}_k g_k(x) - \alpha_r K^{-1} B^{as}{}_k h_k(x) + \text{h.c.}], \qquad (4.2.1-56)$$

where the dipole wave packet system $\{h_k\}$ is defined by Eq.(A.5-28) in Appendix
A.5 by $h_k = \mathcal{E}^{(1/2)} g_k$.

The 4-dimensional (anti)commutation relations, Eqs.(38)-(40) and (49) in
the above lead to the following (anti)commutation relations between the creation
and annihilation operators:

$$(\eta_{k,l}) = \begin{array}{c} \\ U_\alpha \\ \varphi_\rho \\ \chi_k \\ B_k \\ C_k \\ \overline{C}_k \end{array} \begin{array}{cccccc} U_\beta{}^\dagger & \varphi_\sigma{}^\dagger & \chi_l{}^\dagger & B_l{}^\dagger & C_l{}^\dagger & \overline{C}_l{}^\dagger \\ \left(\begin{array}{cccccc} \delta_{\alpha\beta} & 0 & 0 & 0 & 0 & 0 \\ 0 & \delta_{\rho\sigma} & 0 & 0 & 0 & 0 \\ 0 & 0 & \omega\delta_{kl} & -\delta_{kl} & 0 & 0 \\ 0 & 0 & -\delta_{kl} & 0 & 0 & 0 \\ 0 & 0 & 0 & 0 & 0 & i\delta_{kl} \\ 0 & 0 & 0 & 0 & -i\delta_{kl} & 0 \end{array} \right) \end{array}. \qquad (4.2.1-57)$$

In deriving Eq.(57), we have utilized Eq.(A.5-30) in Appendix A.5. These
(anti)commutation relations completely determine the inner product structure of
the total state vector space $\mathcal{V}$. We note here that the modes U_α and φ_ρ have positive norm and the inner product structure of the other four $(\chi^{as}{}_k, B^{as}{}_k, C^{as}{}_k, \overline{C}^{as}{}_k)$
is just the same as in the case of quartet $(\chi_k, \beta_k, \gamma_k, \overline{\gamma}_k)$ given by Eq.(4.1.3-12).
Indeed, the following analysis of BRS transformation property of these asymptotic
fields assures that the massive vector boson U_α and the Higgs scalar φ_ρ are genuine
physical particles belonging to BRS-singlets and $(\chi^{as}{}_k, B^{as}{}_k, C^{as}{}_k, \overline{C}^{as}{}_k)$ constitute
quartet representations.

To verify this statement, we apply to the BRS transformation the general
results in Appendix A.4 on the behavior of asymptotic fields under the symmetry
transformation: If a conserved charge Q generating a symmetry transformation
on Heisenberg fields Φ_l, denoted by

$$[iQ, \ \Phi_l(x)]_\mp = \delta\Phi_l(x), \qquad (4.2.1-58)$$

is not broken spontaneously, then it induces a *linear* transformation on the asymptotic fields. The transformation property of asymptotic fields φ^{as}, corresponding to Φ_i,

$$[iQ,\ \varphi^{\mathrm{as}}{}_i(x)]_{\mp} = (\delta\Phi)^{\mathrm{as}}{}_i(x), \qquad (4.2.1-59)$$

can be determined by picking up the discrete mass spectrum contained in such Green's functions as $\langle 0|\mathrm{T}(\delta\Phi)_i(x)\Phi_j(y)|0\rangle$:

$$\text{Discrete spectrum in } \langle 0|\mathrm{T}(\delta\Phi)_i(x)\Phi_j(y)|0\rangle$$
$$= \langle 0|\mathrm{T}(\delta\Phi)^{\mathrm{as}}{}_i(x)\varphi^{\mathrm{as}}{}_j(y)|0\rangle, \qquad (4.2.1-60)$$

In the present case of BRS transformation, the transformation property of Heisenberg fields is given by Eqs.(3.3.1-33), (3.3.1-35)-(3.3.1-37), (4) and (5). Thus, we can easily obtain

$$\boldsymbol{\delta}A_{\mu}(x) = [iQ_B, A_{\mu}(x)] = D_{\mu}C(x)$$
$$\longrightarrow (\boldsymbol{\delta}A_{\mu})^{\mathrm{as}}(x) = F_6(0)\tilde{Z}_3{}^{1/2}\partial_{\mu}C^{\mathrm{as}}(x), \quad (4.2.1-61)$$
$$\boldsymbol{\delta}\varphi(x) = -\frac{g}{2}\chi(x)\cdot C(x)$$
$$\longrightarrow (\boldsymbol{\delta}\varphi)^{\mathrm{as}}(x) = 0, \qquad (4.2.1-62)$$
$$\boldsymbol{\delta}\chi(x) = \frac{g}{2}[(v+\varphi)C + \chi \times C](x)$$
$$\longrightarrow (\boldsymbol{\delta}\chi)^{\mathrm{as}}(x) = F_7(0)\tilde{Z}_3{}^{1/2}C^{\mathrm{as}}(x), \qquad (4.2.1-63)$$
$$\boldsymbol{\delta}B(x) = 0 \longrightarrow (\boldsymbol{\delta}B)^{\mathrm{as}}(x) = 0, \qquad (4.2.1-64)$$
$$\boldsymbol{\delta}C(x) = -\frac{g}{2}(C \times C)(x)$$
$$\longrightarrow (\boldsymbol{\delta}C)^{\mathrm{as}}(x) = 0, \qquad (4.2.1-65)$$
$$\boldsymbol{\delta}\overline{C}(x) = iB(x) \longrightarrow (\boldsymbol{\delta}\overline{C})^{\mathrm{as}}(x) = iZ_B{}^{1/2}B^{\mathrm{as}}(x). \quad (4.2.1-66)$$

Eqs.(61) and (63) are due to the fact that there are no (bound-state) one-particle poles contributing to $\boldsymbol{\delta}A_{\mu}$ and $\boldsymbol{\delta}\chi$ other than that of "elementary" FP ghost C^{as} because of our assumption of asymptotic completeness. The coefficients $F_6(0)$ and $F_7(0)$ arise from

$$\langle 0|\mathrm{T}\boldsymbol{\delta}A_{\mu}\cdot\overline{C}|0\rangle = \frac{\delta^2\Gamma}{\delta K^{\mu}\delta C}\langle 0|\mathrm{T}C\overline{C}|0\rangle, \qquad (4.2.1-67)$$

$$\langle 0|\mathrm{T}\boldsymbol{\delta}\chi\cdot\overline{C}|0\rangle = \frac{\delta^2\Gamma}{\delta K_{\chi}\delta C}\langle 0|\mathrm{T}C\overline{C}|0\rangle, \qquad (4.2.1-68)$$

in combination with the definitions Eqs.(13) and (14). Eqs.(62) and (65) follow from the absence of bound state poles in the composite channels $\chi \cdot C$ and $C \times C$. Thus, from Eqs.(59), (61)-(66) and the definitions of asymptotic fields in Eq.(32), we obtain

$$[Q_B{}^r, \; U^{as}{}_\mu(x)] = [Q_B{}^r, \; \varphi^{as}(x)] = 0, \qquad (4.2.1-69)$$

$$[Q_B{}^r, \; \chi^{as}(x)] = -iC^{as}(x), \qquad (4.2.1-70)$$

$$\{Q_B{}^r, \; \overline{C}^{as}(x)\} = B^{as}(x), \qquad (4.2.1-71)$$

$$[Q_B{}^r, \; B^{as}(x)] = \{Q_B{}^r, \; C^{as}(x)\} = 0, \qquad (4.2.1-72)$$

where $Q_B{}^r$ is the renormalized BRS charge defined by

$$Q_B{}^r \equiv (Z_3/Z_B)^{1/2} Q_B. \qquad (4.2.1-73)$$

To derive Eqs.(69)-(72), we have used the definition, Eq.(36), of $U^{as}{}_\mu$, and the relations

$$\frac{\tilde{Z}_3 F_7(0)}{(Z_B Z_\chi)^{1/2}} = \frac{\tilde{Z}_3 F_6(0)}{(Z_B Z_3 K)^{1/2}} = 1, \qquad (4.2.1-74)$$

following from the WT relations, Eqs.(16),(18) and (19) together with the definitions $Z_B = F_1(0), Z_\chi{}^{-1} = F_4(0), \tilde{Z}_3{}^{-1} = F_5(0)$ in Eq.(31) and $K = (Z_B Z_3)^{-1}$ in Eq.(41) and the relation $F_1(0) = F_2(0)$ stated in Eq.(43).

As is expected, Eq.(69) indicates that the Proca field $U^{as}{}_\mu$ and the real Higgs scalar field φ^{as} are genuine physical particles in BRS-singlets having positive norm by Eq.(57). The BRS transformation property Eqs.(70)-(72) is nothing but that of quartet, Eqs.(4.1.2-8)-(4.1.2-10), and hence the NG boson χ^{as}, the B-field B^{as}, and the FP ghosts C^{as} and $\overline{C}^{as}$ (for each of omitted group indices $a = 1, 2, 3$) are found to belong to quartet representation. Thus we have finished the proof of the unitarity of physical S-matrix defined on $H_{phys} \equiv \mathcal{V}_{phys}/\mathcal{V}_0$ which is isomorphic to the Hilbert space $\mathcal{H}_{phys}$ spanned solely by the physical particles $U^{as}{}_\mu$ and φ^{as}. Just similarly to the situation [Nak 73a] encountered in abelian Higgs model (see Sec.2.4.3), the Higgs mechanism in the present non-abelian case also is understood without any inconsistency with the Goldstone theorem : The NG bosons surely exist but belong to some undetectable unphysical quartet, while the gauge bosons acquire non-vanishing mass and are physical.

We add two comments here:

(i) By the help of the present assumption of asymptotic completeness, the BRS charge Q_B is expressed in terms of asymptotic fields as

$$Q_B{}^r = \int d\mathbf{x} : B^{\mathrm{as}}(x) \cdot \partial_0 C^{\mathrm{as}}(x) - \partial_0 B^{\mathrm{as}}(x) \cdot C^{\mathrm{as}}(x) :$$
$$= i \sum_k (C^{\mathrm{as}}{}_k{}^\dagger \cdot B^{\mathrm{as}}{}_k - B^{\mathrm{as}}{}_k{}^\dagger \cdot C^{\mathrm{as}}{}_k). \qquad (4.2.1-75)$$

This form in fact reproduces the asymptotic transformations Eqs.(69)-(73) and the original transformations on Heisenberg fields as is proved generally in Appendix A.4. It is interesting to note that this expression Eq.(75) for $Q_B{}^r$ has just the same form as that of abelian case Eq.(3.4.3-4). This again shows that our present formulation provides a natural extension of the manifestly covariant canonical formalism of abelian gauge fields.

(ii) By comparing Eq.(47) with Eq.(4.1.3-22), the field χ^a belonging to the "elementary" quartets found in the general discussion in Sec.4.1.3 is given in the present model explicitly by

$$\chi^a \text{ in Eq.(4.1.3-22)} = Z_3^{1/2}[\sqrt{K}\chi^{\mathrm{as}a} + (\sqrt{K} - \alpha_r N)B^{\mathrm{as}a}]$$
$$\propto (\chi^{\mathrm{as}a} + \boldsymbol{\delta}(*)). \qquad (4.2.1-76)$$

Namely, the unphysical NG bosons $\chi^{\mathrm{as}a}$ can be identified with χ^a in "elementary quartets" except for the shift by a BRS-coboundary term and rescaling due to renormalization.

4.2.2 Pure Yang-Mills theory without spontaneous symmetry breaking

Next, we analyze the pure Yang-Mills theory (with a compact simple Lie group G as its gauge group) assuming that no spontaneous symmetry breaking takes place [Kug 79b]. Of course, the massless Yang-Mills theory is believed to suffer from serious *infrared divergences* which may have profound meaning in connection with the confinement mechanism of quarks. We, however, neglect the infrared problem here for simplicity and make a formal analysis of asymptotic fields.

The Lagrangian density is given by Eqs.(3.4.1-1)-(3.4.1-6) with $\mathcal{L}_M$ being omitted. The propagators are given by the same form as Eq.(4.2.1-21) in the preceding model, where the group index a should be understood to run over $a = 1, 2, \cdots, n(= \dim(G))$, and the NG boson field χ^a is to be discarded. The invariant function corresponding to $F_1(p^2)^{-1}$ in Eq.(4.2.1-21) now has a massless pole, and hence we rewrite it as

$$F_1(p^2) \equiv p^2 \tilde{F}_1(p^2). \qquad (4.2.2-1)$$

Through the same procedure as in the previous subsection, we find that the asymptotic fields given by

$$A_\mu(x) \rightarrow Z_3^{1/2} A^{as}{}_\mu(x), \quad B(x) \rightarrow Z_3^{-1/2} B^{as}(x),$$
$$C(x) \rightarrow \tilde{Z}_3^{1/2} C^{as}(x), \qquad \overline{C}(x) \rightarrow \tilde{Z}_3^{1/2} \overline{C}^{as}(x), \qquad (4.2.2-2)$$

satisfy the following (anti)commutation relations:

$$[A^{as}{}_\mu(x), A^{as}{}_\nu(y)] = -i(\eta_{\mu\nu} - K\partial_\mu\partial_\nu)D(x-y)$$
$$+ i(1-\alpha_r)\partial_\mu\partial_\nu E(x-y), \qquad (4.2.2-3)$$
$$[A^{as}{}_\mu(x), B^{as}(y)] = -i\partial_\mu D(x-y), \qquad (4.2.2-4)$$
$$[B^{as}(x), B^{as}(y)] = 0, \qquad (4.2.2-5)$$
$$\{C^{as}(x), \overline{C}^{as}(y)\} = -D(x-y). \qquad (4.2.2-6)$$

Here we have introduced the following constants

$$Z_3^{-1} \equiv \tilde{F}_1(0), \quad K \equiv -Z_3^{-1}\frac{d}{ds}\text{Re}\left(\frac{1}{\tilde{F}_1(s)}\right)\bigg|_{s=0}, \qquad (4.2.2-7)$$

and other quantities in the same way as before:

$$\tilde{Z}_3^{-1} \equiv F_5(0), \quad \alpha_r \equiv \alpha/Z_3. \qquad (4.2.2-8)$$

Z_B is taken in the present case as $Z_B \equiv Z_3^{-1}$. Eq.(2) together with the assumption of asymptotic completeness lead to the field equations of asymptotic fields:

$$\Box B^{as} = \Box C^{as} = \Box \overline{C}^{as} = 0, \qquad (4.2.2-9)$$
$$\Box A^{as}{}_\mu = (1-\alpha_r)\partial_\mu B^{as}, \qquad (4.2.2-10)$$
$$\partial^\mu A^{as}{}_\mu + \alpha_r B^{as} = 0. \qquad (4.2.2-11)$$

We can now construct the asymptotic Fock space. Noting that the vector field $A^{as}{}_\mu$ is in general a *dipole* field except for the case of Feynman gauge ($\alpha_r = 1$), we extract a (non-covariant and non-local) simple-pole field $\dot{\tilde{A}}^{\mathrm{as}}{}_\mu$ from $A^{\mathrm{as}}{}_\mu$ in a way similar to Eq.(4.2.1-52):

$$\tilde{A}^{\mathrm{as}}{}_\mu(x) \equiv A^{\mathrm{as}}{}_\mu(x) - (1 - \alpha_r)\partial_\mu \mathcal{E}^{(1/2)} B^{\mathrm{as}}(x), \tag{4.2.2 - 12}$$

Then, by using $\Box \mathcal{E}^{1/2} B^{\mathrm{as}} = B^{\mathrm{as}}$ [see Eq.(A.5-4) in Appendix A.5], Eqs.(10) and (11) can be simplified into

$$\Box \tilde{A}^{\mathrm{as}}{}_\mu = 0, \tag{4.2.2 - 13}$$

$$\partial^\mu \tilde{A}^{\mathrm{as}}{}_\mu + B^{\mathrm{as}} = 0. \tag{4.2.2 - 14}$$

Then, we can introduce the creation and annihilation operators of scalar fields $B^{\mathrm{as}}, C^{\mathrm{as}}, \overline{C}^{\mathrm{as}}$ in reference to a common system $\{g_k\}$ of massless wave packets, and those of massless vector field $A^{\mathrm{as}}{}_\mu$ using a wave packet system $\{f_{k,\sigma}{}^\mu\}$ closely related to the former one. [These two wave packet systems, $\{g_k\}$ and $\{f_{k,\sigma}{}^\mu\}$, are constructed in Appendix A.5, and their mutual relationships simplify the formulae below.]: The annihilation operator $A^{\mathrm{as}}{}_k$ is defined by

$$A^{\mathrm{as}}{}_{k,\sigma} \equiv (f_{k,\sigma}{}^\mu, \tilde{A}^{\mathrm{as}}{}_\mu) = i \int d\mathbf{x} \left(f_{k,\sigma}{}^{\mu*}(x)\partial_0 \tilde{A}^{\mathrm{as}}{}_\mu(x) - \partial_0 f_{k,\sigma}{}^{\mu*}(x) \cdot \tilde{A}^{\mathrm{as}}{}_\mu(x) \right), \tag{4.2.2 - 15}$$

$$\tilde{A}^{\mathrm{as}}{}_\mu(x) = \sum_{k,\sigma}(A^{\mathrm{as}}{}_{k,\sigma} f_k{}^\sigma{}_\mu(x) + \mathrm{h.c.}), \tag{4.2.2 - 16}$$

and, $B^{\mathrm{as}}{}_k$, $C^{\mathrm{as}}{}_k$ and $\overline{C}^{\mathrm{as}}{}_k$ by

$$B^{\mathrm{as}}{}_k \equiv (g_k, B^{\mathrm{as}}); \quad B^{\mathrm{as}}(x) = \sum_k (B^{\mathrm{as}}{}_k g_k(x) + \mathrm{h.c.}), \tag{4.2.2 - 17}$$

and so on. These modes $A^{\mathrm{as}}{}_{k,\sigma}$ ($\sigma = 1, 2, L, S$) and $B^{\mathrm{as}}{}_k$ are not all mutually independent, as is evident from Eq.(14) implying the relation[7]

$$B^{\mathrm{as}}{}_k = A^{\mathrm{as}}{}_k{}^S \equiv \sum_\sigma A^{\mathrm{as}}{}_{k,\sigma}(= A^{\mathrm{as}}{}_{k,L}), \tag{4.2.2 - 18}$$

where $\tilde{\eta}$ is the metric of polarizations defined by

$$\tilde{\eta} \equiv \begin{array}{c} \\ 1 \\ 2 \\ L \\ S \end{array} \begin{array}{cccc} 1 & 2 & L & S \\ \left(\begin{array}{cccc} -1 & 0 & 0 & 0 \\ 0 & -1 & 0 & 0 \\ 0 & 0 & 0 & 1 \\ 0 & 0 & 1 & 0 \end{array} \right) \end{array}, \tag{4.2.2 - 19}$$

---

[7]   It is important to note the positions of the indices $\sigma = 1, 2, L$ and S; e.g., $A^{\mathrm{as}}{}_{k,L} = A^{\mathrm{as}}{}_k{}^S \neq A^{\mathrm{as}}{}_{k,S} = A^{\mathrm{as}}{}_k{}^L$.

and where Eqs.(A.5-26) and (A.5-27) in Appendix A.5 have been used. Thus, the scalar polarization modes $A^{\mathrm{as}}{}_k{}^S$ are nothing but the modes represented by $B^{\mathrm{as}}{}_k$. The asymptotic fields $A^{\mathrm{as}}{}_\mu$ and B^{as} are fully described by the transverse modes $A^{\mathrm{as}}{}_{k,\sigma=1,2}(=-A^{\mathrm{as}}{}_k{}^{\sigma=1,2})$, the longitudinal modes $A^{\mathrm{as}}{}_k{}^L(=A^{\mathrm{as}}{}_{k,S})$ and $B^{\mathrm{as}}{}_k$. Indeed, owing to Eqs.(12), (16) and (18), $A^{\mathrm{as}}{}_\mu(x)$ can be expressed by

$$A^{\mathrm{as}}{}_\mu(x) = \{ \sum_{k;\sigma=1,2} A^{\mathrm{as}}{}_{k,\sigma} f_k{}^\sigma{}_\mu(x) + \sum_k A^{\mathrm{as}}{}_k{}^L \partial_\mu g_k(x)$$
$$+ \sum_k B^{\mathrm{as}}{}_k [f_{k,S,\mu}(x) + (1-\alpha_r)\partial_\mu h_k(x)] \} + \mathrm{h.c.}, (4.2.2-20)$$

where the dipole wave packets $h_k(x)$ are defined by Eq.(A.5-28): $\mathcal{E}^{(1/2)} g_k(x) \equiv h_k(x)$. Noting that the commutator Eq.(3) is rewritten by virtue of the identity Eq.(A.5-8) into

$$[\tilde{A}^{\mathrm{as}}{}_\mu(x), \tilde{A}^{\mathrm{as}}{}_\nu(y)] = -i(\eta_{\mu\nu} - K\partial_\mu\partial_\nu)D(x-y), \qquad (4.2.2-21)$$

we obtain from Eqs.(21) and (4)-(6) the commutation relations of the creation and annihilation operators characterized by the following metric matrix η:

$$(\eta_{kl}) = \begin{array}{c} A_{k,\sigma=1,2} \\ A_k{}^L \\ B_k \\ C_k \\ \overline{C}_k \end{array} \begin{pmatrix} \delta_{kl}\delta_{\sigma\tau} & 0 & 0 & 0 & 0 \\ 0 & -K\delta_{kl} & -\delta_{kl} & 0 & 0 \\ 0 & -\delta_{kl} & 0 & 0 & 0 \\ 0 & 0 & 0 & 0 & i\delta_{kl} \\ 0 & 0 & 0 & -i\delta_{kl} & 0 \end{pmatrix}, \qquad (4.2.2-22)$$

with column headers $A_{l,\tau=1,2}{}^\dagger \quad A_l{}^L{}^\dagger \quad B_l{}^\dagger \quad C_l{}^\dagger \quad \overline{C}_l{}^\dagger$

where Eqs.(A.5-23) and (A.5-24) have been used.

In contrast to the usual LSZ reduction formula [Leh 55] for the transverse modes,

$$\langle A_{k,\sigma=1,2}, \; \alpha \text{ out}|\mathrm{T}(\cdots)|\beta \text{ in}\rangle = \langle \alpha \text{ out}|\mathrm{T}(\cdots)A^{\mathrm{in}}{}_{k,\sigma=1,2}|\beta \text{ in}\rangle$$
$$+ i \int d^4x Z_3^{-1/2} f_{k,\sigma}{}^\mu{}^*(x)\Box^x \langle \alpha \text{ out}|\mathrm{T}(A_\mu(x)\cdots)|\beta \text{ in}\rangle, \quad (4.2.2-23)$$

the corresponding formula for the longitudinal modes is given [Nak 74b, Kug 79b] by

$$\langle A_k{}^L \; \alpha \text{ out}|\mathrm{T}(\cdots)|\beta \text{ in}\rangle = \langle \alpha \text{ out}|\mathrm{T}(\cdots)A^{\mathrm{in}}{}_k{}^L|\beta \text{ in}\rangle$$

$$+ i \int d^4x Z_3^{-1/2} f_k{}^{L,\mu*}(x) \square^x \langle \alpha \text{ out}|T(A_\mu(x) \cdots)|\beta \text{ in} \rangle$$

$$- i(1 - \alpha_r) \int d^4x Z_3^{+1/2} [f_k{}^{L,\mu*}(x) \partial_\mu^x - h_k{}^*(x) \square^x]$$

$$\times \langle \alpha \text{ out}|T(B(x) \cdots)|\beta \text{ in} \rangle. \qquad (4.2.2-24)$$

The BRS transformations of asymptotic fields are determined in the same way as the one in the preceding subsection. In terms of the renormalized BRS charge defined by $Q_B{}^r = \tilde{Z}_3{}^{1/2} Z_B{}^{-1/2} Q_B = \tilde{Z}_3{}^{1/2} Z_3{}^{1/2} Q_B$, they are given by

$$[Q_B{}^r, \; A^{as}{}_\mu(x)] = -i\partial_\mu C^{as}(x), \qquad (4.2.2-25)$$

$$\{Q_B{}^r, \; \overline{C}^{as}(x)\} = B^{as}(x), \qquad (4.2.2-26)$$

$$[Q_B{}^r, \; B^{as}(x)] = \{Q_B{}^r, \; C^{as}(x)\} = 0, \qquad (4.2.2-27)$$

which lead to

$$[Q_B{}^r, \; A^{as}{}_{k,\sigma=1,2}] = 0; \qquad (4.2.2-28)$$

$$[Q_B{}^r, \; A^{as}{}_k{}^L] = -iC^{as}{}_k, \quad \{Q_B{}^r, \; \overline{C}^{as}{}_k\} = B^{as}{}_k,$$

$$[Q_B{}^r, \; B^{as}{}_k] = \{Q_B{}^r, \; C^{as}{}_k\} = 0. \qquad (4.2.2-29)$$

From the (anti)commutation relations given by Eq.(22) and the above BRS transformation properties, we see that the transverse modes $A^{as}{}_{k,\sigma=1,2}$ are really genuine physical particles belonging to BRS-singlets with positive norm and that the longitudinal modes $A^{as}{}_k{}^L$ together with the scalar modes $B^{as}{}_k$ and FP ghosts $C^{as}{}_k$ and $\overline{C}^{as}{}_k$ constitute quartets. Therefore, the physical S-matrix unitarity has been proved also in the case of pure Yang-Mills theory, provided that the difficult problem of infrared divergence can be neglected.

Finally we add a comment. The above construction of physical transverse modes clearly depends on the choice of Lorentz frame to which we refer, and hence, the Hilbert space $\mathcal{H}_{\text{phys}}$ spanned by the transverse modes is *not* Lorentz invariant. However, our proof of unitarity of the physical S-matrix S_{phys} defined on $\overline{\mathcal{V}_{\text{phys}}/\mathcal{V}_0} = H_{\text{phys}}$ has Lorentz invariant meaning, because the physical subspace $\mathcal{V}_{\text{phys}}$ is specified by a Lorentz invariant condition in terms of the Lorentz-scalar charge Q_B and hence its positive semi-definiteness of metric as well as the spaces $\mathcal{V}_{\text{phys}}, \mathcal{V}_0$ and H_{phys} have Lorentz-invariant meaning, although the proof is given in reference to a specific Lorentz frame. As is easily seen, the Lorentz invariance of the transverse modes belonging to $\mathcal{H}_{\mathrm{phys}}$ is violated only by zero-norm contributions belonging to $\mathcal{V}_0$ which give no contribution to the physical amplitudes occurring in $\overline{\mathcal{V}_{\mathrm{phys}}/\mathcal{V}_0} = H_{\mathrm{phys}}$.

4.3 OBSERVABLES IN YANG-MILLS THEORY

In the previous section, we analyzed the structure of asymptotic states and asymptotic fields to describe the scattering theoretical aspects of gauge theory. The relevant physical quantity treated there is only the physical S-matrix, which has been proved to be a *unitary* operator in the Hilbert space of physical states, $H_{\text{phys}} = \overline{\mathcal{V}_{\text{phys}}/\mathcal{V}_0} (= \text{completion of } \mathcal{V}_{\text{phys}}/\mathcal{V}_0)$: Although such unphysical particles as FP ghosts may be produced in the final states with non-vanishing S-matrix elements from the initial states containing *no* unphysical particles, they can appear in $\mathcal{V}_{\text{phys}}$ only in the *zero-norm combinations* belonging to $\mathcal{V}_0$, as has been proved explicitly in Sec.4.1.4. Because of the orthogonality of any zero-norm physical states in $\mathcal{V}_0$ to all the physical states in $\mathcal{V}_{\text{phys}}$, these unphysical particles make no contribution to the transition amplitudes between two physical states. Therefore, we can describe in H_{phys} consistently all the physical scattering processes in the quantum theory of non-abelian gauge fields, because zero-norm physical states containing unphysical particles can safely be neglected. However, the S-matrix is *not* the only physically measurable quantity. To describe every physical process in this Hilbert space H_{phys} consistently with the standard principles of quantum theory, we need to have many other physical quantities to be measured in the physical world, and need to clarify what physical contents can be extracted from H_{phys} in terms of them. Since the present covariant operator formalism of gauge theories contains many unphysical objects such as B-field and FP ghosts, it is very important to give a definite criterion to discriminate observable quantities from unobservable ones. To this end, we propose the notion of **observables** from a very general viewpoint of guaranteeing the probabilistic interpretation of quantum theory in the measuring processes, and then this general requirement will be found to lead to a concise and natural criterion characterizing observables by BRS invariance.

4.3.1 Observables and BRS invariance as their criterion

We start the analysis of observables in gauge theories with the problem of consistency between the description of scattering processes and that of measuring processes of physical quantities. As remarked above, any zero-norm physical state

$|\chi\rangle \in \mathcal{V}_0$ is orthogonal to all the physical states in $\mathcal{V}_{\text{phys}}$ [Eq.$(A.2\text{-}7)$ in Appendix A.2],

$$\langle \chi | \Phi \rangle = 0 \qquad \text{for } |\Phi\rangle \in \mathcal{V}_{\text{phys}}, \quad |\chi\rangle \in \mathcal{V}_0, \qquad (4.3.1-1)$$

and hence, $|\chi\rangle \in \mathcal{V}_0$ containing unphysical particles gives no contribution to the inner product or *transition amplitudes* between any pair of physical states. This has allowed us to transfer the framework of our discussion about scattering processes to the quotient space $H_{\text{phys}} = \overline{\mathcal{V}_{\text{phys}}/\mathcal{V}_0}$, where all the states $|\chi\rangle$ belonging to $\mathcal{V}_0$ are regarded as undetectable null objects:

$$|\hat{\chi}\rangle = |\chi\rangle + \mathcal{V}_0 = \hat{0} \qquad \text{in } H_{\text{phys}} \qquad (\text{for } |\chi\rangle \in \mathcal{V}_0). \qquad (4.3.1-2)$$

This situation can be paraphrased in a more general manner as follows. Defining the **transition probability** $T(\Phi_1|\Phi_2)$ between two physical states $|\Phi_1\rangle$, $|\Phi_2\rangle \in \mathcal{V}_{\text{phys}}$ by

$\underline{\text{Definition } 4.3.1-1} \quad T(\Phi_1|\Phi_2) \equiv |\langle \Phi_1|\Phi_2\rangle|^2, \qquad\qquad (4.3.1-3)$

we obtain, from Eq.(1), the following relation:

$$T(\Phi_1 + \chi_1|\Phi_2 + \chi_2) = T(\Phi_1|\Phi_2) \qquad \text{for } |\Phi_i\rangle \in \mathcal{V}_{\text{phys}}, \quad |\chi_i\rangle \in \mathcal{V}_0. \qquad (4.3.1-4)$$

Namely, the transition probability $T(\Phi_1|\Phi_2)$ in $\mathcal{V}_{\text{phys}}$ does not depend on the choice of the representative vectors $|\Phi_i\rangle \in \mathcal{V}_{\text{phys}}$ in the equivalence class $|\hat{\Phi}_i\rangle = |\Phi_i\rangle + \mathcal{V}_0$ with respect to $\mathcal{V}_0$, and it is a function $\hat{T}$ depending on pairs of equivalence classes $|\hat{\Phi}_i\rangle \in H_{\text{phys}} = \overline{\mathcal{V}_{\text{phys}}/\mathcal{V}_0}$:

$$\hat{T} : H_{\text{phys}} \times H_{\text{phys}} \longrightarrow \mathbf{R}_+$$
$$\hat{T}(\hat{\Phi}_1|\hat{\Phi}_2) \equiv T(\Phi_1|\Phi_2) = |\langle \Phi_1|\Phi_2\rangle|^2 = |\langle \hat{\Phi}_1|\hat{\Phi}_2\rangle|^2. \qquad (4.3.1-5)$$

This guarantees that such problems as the scattering processes described by the transition probabilities can be formulated completely in the positive definite Hilbert space H_{phys}.

Besides the transition probability, however, there are many physical quantities to be measured, for instance, the energy-momentum operator P_μ, and so on. If we want to describe, in the Hilbert space H_{phys}, every physical process consistently according to the standard principles of quantum theory, any state

$|\chi\rangle \in \mathcal{V}_0$ which corresponds to the null vector in H_{phys} should make no physical effects in the measurement of physical quantities. Now, since the starting point of the present formulation is not the quantum theory given in H_{phys} but the field theory formulated in the state vector space $\mathcal{V}$ with an indefinite inner product, we should first write down, in terms of the operators $\mathcal{R}$ in $\mathcal{V}$, what condition can guarantee the consistency of the theory realized in H_{phys}. If a zero-norm physical state $|\chi\rangle \in \mathcal{V}_0$ were transformed by a physical quantity $\mathcal{R}$ into a state $|\chi'\rangle = \mathcal{R}|\chi\rangle$ such that

$$\langle\Phi|\chi'\rangle = \langle\Phi|\mathcal{R}|\chi\rangle \neq 0, \quad \text{for some } |\Phi\rangle \in \mathcal{V}_{\text{phys}}, \qquad (4.3.1-6)$$

then the measurement of $\mathcal{R}$ could not be described consistently in H_{phys}, because the state $|\chi\rangle \in \mathcal{V}_0$ regarded as the null vector in H_{phys} would make a non-vanishing contribution in Eq.(6). So, we require a physical quantity $\mathcal{R}$ to satisfy the following equality:

$$\langle\Phi|\mathcal{R}|\chi\rangle = \langle\chi|\mathcal{R}|\Phi\rangle = 0, \quad \text{for } \forall|\Phi\rangle \in \mathcal{V}_{\text{phys}}, \ \forall|\chi\rangle \in \mathcal{V}_0. \qquad (4.3.1-7)$$

As was shown in [Str 74], the condition Eq.(7) agrees with the one which guarantees the usual connection between the transition probability and the expectation values of observables in the quantum theory. In the usual quantum theory formulated in the Hilbert space with a positive definite inner product, we know that the relation

$$T(\Psi|\Phi) = |\langle\Psi|\Phi\rangle|^2 = \langle\Phi|(|\Psi\rangle\langle\Psi|)|\Phi\rangle = E(P_\Psi|\Phi) \qquad (4.3.1-8)$$

holds, where the expectation value of the observable $\mathcal{R}$ in the state $|\Phi\rangle$ is denoted as

$$E(\mathcal{R}|\Phi) = \langle\Phi|\mathcal{R}|\Phi\rangle \qquad (4.3.1-9)$$

and P_Ψ denotes the projection operator onto the state $|\Psi\rangle$

$$P_\Psi = |\Psi\rangle\langle\Psi|. \qquad (4.3.1-10)$$

Namely, the transition probability $T(\Psi|\Phi)$ is nothing but the expectation value $E(P_\Psi|\Phi)$ of a special type of observable P_Ψ corresponding to the yes-no question about the state $|\Psi\rangle$. Conversely, since every observable $\mathcal{R}$ which is a self-adjoint operator admits the spectral decomposition[1]

$$\mathcal{R} = \sum_n a_n P_{\varphi_n} = \sum_n a_n |\varphi_n\rangle\langle\varphi_n| \qquad (4.3.1-11)$$

with $\mathcal{R}|\varphi_n\rangle = a_n|\varphi_n\rangle$, the expectation value $E(\mathcal{R}|\Phi)$ can be reconstructed from the knowledge of the transition probabilities:

$$E(\mathcal{R}|\Phi) = \sum_n a_n \langle\Phi|\varphi_n\rangle\langle\varphi_n|\Phi\rangle = \sum_n a_n T(\varphi_n|\Phi). \qquad (4.3.1-12)$$

Thus, in order to maintain the relations, Eqs.(8) and (12), with Eq.(9) also in our H_{phys}, Eq.(4) should imply the equality

$$E(\mathcal{R}|\Phi + \chi) = E(\mathcal{R}|\Phi) \qquad \text{for } |\Phi\rangle \in \mathcal{V}_{\text{phys}}, \ |\chi\rangle \in \mathcal{V}_0, \qquad (4.3.1-13)$$

which is really equivalent to Eq.(7). Thus, the conditions Eqs.(7) and (13) are the equivalent expressions of a necessary condition for the consistent measurement of a physical quantity $\mathcal{R}$.

It can easily be checked that the energy-momentum operator P_μ satisfies Eq.(7) as follows. First, since the BRS charge Q_B given by Eq.(3.3.1-32) is a translationally invariant Lorentz scalar, we obtain

$$[Q_B, P_\mu] = 0, \qquad (4.3.1-14)$$

as a consequence of which the state $P_\mu|\Phi\rangle$ with $|\Phi\rangle \in \mathcal{V}_{\text{phys}}$ belongs to $\mathcal{V}_{\text{phys}}$:

$$Q_B P_\mu|\Phi\rangle = [Q_B, P_\mu]|\Phi\rangle + P_\mu Q_B|\Phi\rangle = 0 \Longrightarrow P_\mu|\Phi\rangle \in \mathcal{V}_{\text{phys}}. \qquad (4.3.1-15)$$

Then, we obtain, from Eqs.(1) and (15),

$$\langle\chi|P_\mu|\Phi\rangle = \langle\Phi|P_\mu|\chi\rangle = 0 \qquad \text{for } |\Phi\rangle \in \mathcal{V}_{\text{phys}}, \ |\chi\rangle \in \mathcal{V}_0. \qquad (4.3.1-16)$$

---

[1]   Precisely speaking, Eq.(11) should be written in general as

$$\mathcal{R} = \int \lambda dP(\lambda), \qquad dP(\lambda) : \text{spectral measure}$$

in order to treat $\mathcal{R}$ with continuous spectrum as well as discrete one.

In what follows, we call any operator $\mathcal{R}$ (hermitian or not) satisfying Eq.(7) **observable**.

<u>Definition 4.3.1 − 2</u>   An operator $\mathcal{R}$ is called an *observable* if it satisfies

$$\langle\chi|\mathcal{R}|\Phi\rangle = \langle\Phi|\mathcal{R}|\chi\rangle = 0 \qquad\qquad (4.3.1-17)$$

or equivalently

$$\langle\Phi+\chi|\mathcal{R}|\Phi+\chi\rangle = \langle\Phi|\mathcal{R}|\Phi\rangle \equiv E(\mathcal{R}|\Phi) \qquad\qquad (4.3.1-18)$$

for any $|\Phi\rangle \in \mathcal{V}_{\text{phys}}, |\chi\rangle \in \mathcal{V}_0$.

Now, as easily understood from the above argument of the "observability" of P_μ, the concept of observable is closely related to gauge invariance, since the BRS charge Q_B in Eq.(14) is essentially a generator of "local" gauge transformation in quantum theory. In [Str 74], the following four notions of the gauge invariance are introduced for the operators in QED:

(i) *gauge independence* : $\langle\Phi_1+\chi_1|\mathcal{R}|\Phi_2+\chi_2\rangle = \langle\Phi_1|\mathcal{R}|\Phi_2\rangle$
   for $|\Phi_i\rangle \in \mathcal{V}_{\text{phys}}, |\chi_i\rangle \in \mathcal{V}_0,$ $\qquad\qquad (4.3.1-19)$

(ii) *weak gauge invariance* : $\mathcal{R}\mathcal{V}_0 \subseteq \mathcal{V}_0, \mathcal{R}^\dagger\mathcal{V}_0 \subseteq \mathcal{V}_0,$ $\qquad (4.3.1-20)$

(iii) *gauge invariance* : $\mathcal{R}\mathcal{V}_{\text{phys}} \subseteq \mathcal{V}_{\text{phys}}, \mathcal{R}^\dagger\mathcal{V}_{\text{phys}} \subseteq \mathcal{V}_{\text{phys}},$ $\qquad (4.3.1-21)$

(iv) *strict gauge invariance* : gauge invariance (iii) augmented
   by the condition $[\mathcal{R}, \partial^\nu F_{\nu\mu} + ej_\mu] = 0.$ $\qquad (4.3.1-22)$

One can easily see that the statement are stronger in the increasing order of numbering, and that the weakest one (i) is identical with our definition, Eq.(7), of observables. Note that the condition (iii) allows us to define an operator $\hat{\mathcal{R}}$ in $H_{\text{phys}} = \overline{\mathcal{V}_{\text{phys}}/\mathcal{V}_0}$ by the equation

$$\hat{\mathcal{R}}|\hat{\Phi}\rangle = (\mathcal{R}|\Phi\rangle)^\wedge, \qquad |\Phi\rangle \in \mathcal{V}_{\text{phys}}, \ |\hat{\Phi}\rangle = |\Phi\rangle + \mathcal{V}_0 \in \mathcal{V}_{\text{phys}}/\mathcal{V}_0. \qquad (4.3.1-23)$$

In the covariant canonical operator formalism of QED where a physical state $|\text{phys}\rangle \in \mathcal{V}_{\text{phys}}$ is specified by the Gupta condition

$$B^{(+)}(x)|\text{phys}\rangle = 0, \qquad\qquad (4.3.1-24)$$

the condition (iv) is equivalent to the following one:

$$(iv') \qquad [\mathcal{R}, \ B(x)] = 0. \qquad\qquad (4.3.1 - 25)$$

This is due to the quantum Maxwell equation

$$\partial^\nu F_{\nu\mu} = -ej_\mu + \partial_\mu B. \qquad\qquad (4.3.1 - 26)$$

Furthermore, the above four conditions in the abelian cases are distinct from one another, namely, each latter one is truly stronger than the former one. For example, the energy-momentum P_μ satisfies (iii) but not (iv):

$$B^{(+)}(x)P_\mu \mathcal{V}_{\text{phys}} = [B^{(+)}(x), \ P_\mu]\mathcal{V}_{\text{phys}} = i\partial_\mu B^{(+)}(x)\mathcal{V}_{\text{phys}} = 0,(4.3.1 - 27)$$

$$[B(x), \ P_\mu] = i\partial_\mu B(x) \neq 0. \qquad\qquad (4.3.1 - 28)$$

Contrary to the case of QED, the weakest condition (i) in the present situation with the BRS charge is really equivalent to a stronger one (iii), which is rewritten equivalently as

$$[Q_B, \ \mathcal{R}]_{\mp}\mathcal{V}_{\text{phys}} = [Q_B, \ \mathcal{R}^\dagger]_{\mp}\mathcal{V}_{\text{phys}} = 0. \qquad\qquad (4.3.1 - 29)$$

Proposition 4.3.1 − 3 [Oji 78]   In the gauge theory with the BRS charge Q_B, the condition (i) for an operator $\mathcal{R}$ implies the condition (iii).

Proof: If an operator $\mathcal{R}$ satisfies (i), or its equivalent Eq.(7), we obtain for any vector $|f\rangle \in \mathcal{V}$ and for any $|\Phi\rangle \in \mathcal{V}_{\text{phys}}$,

$$\langle f|(Q_B\mathcal{R}|\Phi\rangle) = ((\langle f|Q_B)\mathcal{R}|\Phi\rangle = 0, \qquad\qquad (4.3.1 - 30)$$

because the state $|\chi\rangle \equiv Q_B|f\rangle$ belongs to $\mathcal{V}_0$. Since the inner product of $\mathcal{V}$ is assumed to be non-degenerate [see Sec.4.1.2], Eq.(30) concludes

$$Q_B\mathcal{R}\mathcal{V}_{\text{phys}} = 0, \qquad\qquad (4.3.1 - 31)$$

which is nothing but the condition $\mathcal{R}\mathcal{V}_{\text{phys}} \subseteq \mathcal{V}_{\text{phys}}$. The condition $\mathcal{R}^\dagger\mathcal{V}_{\text{phys}} \subseteq \mathcal{V}_{\text{phys}}$ follows in the same way, and hence, we arrive at the condition (iii).   (q.e.d.)

Thus, in the present operator formalism of gauge theory, the three notions of gauge invariance (i)-(iii) are all equivalent. This criterion can be further sharpened for *local* observables in the following way, as is proved after Lemma 4.3.1-5.

Proposition 4.3.1 − 4 [Oji 78]   If $\mathcal{R}$ is a smeared local observable, namely, an operator[2] $\mathcal{R} \in \mathcal{F}(\mathcal{O})$ satisfying one (and hence, all) of the conditions (i)-(iii) [i.e., Eqs.(19), (20), (21), (29)], then it satisfies the equality

$$[iQ_B, \ \mathcal{R}]_{\mp} = \boldsymbol{\delta}\mathcal{R} = 0, \qquad (4.3.1 - 32)$$

which implies, conversely, (i)-(iii). Namely, a smeared local operator $\mathcal{R} \in \mathcal{F}(\mathcal{O})$ is an observable if and only if it satisfies Eq.(32).

By comparing Eq.(32) with Eq.(25) in view of the corresponding subsidiary conditions, Eqs.(3.4.3-2) and (2.3.3-11), the BRS invariance of $\mathcal{R}$, Eq.(32), should be interpreted as the one for strict gauge invariance. Namely, the condition for a local operator to be an *observable* is just identical with the condition of BRS invariance as *strict gauge invariance* in the present formalism. The proof of the above proposition can be easily made by using the following lemma.

Lemma 4.3.1 − 5 [Oji 79]   If a smeared *local* operator $\mathcal{R} \in \mathcal{F}(\mathcal{O})$ satisfies the condition

$$Q_B \mathcal{R}|0\rangle = 0, \qquad (4.3.1 - 33)$$

then it is a BRS invariant operator satisfying Eq.(32), namely, $\mathcal{R}$ is a smeared local observable. If we denote the set of smeared local observables belonging to $\mathcal{F}(\mathcal{O})$ by $\mathcal{A}(\mathcal{O})$, then the following equality holds:

$$\mathcal{F}(\mathcal{O})|0\rangle \cap \mathcal{V}_{\text{phys}} = \mathcal{A}(\mathcal{O})|0\rangle. \qquad (4.3.1 - 34)$$

Proof: First, we note that, if necessary, $\mathcal{R}$ given as a polynomial in smeared field operators can be decomposed into the components with *even* FP ghost number N_{FP} and the one with *odd* N_{FP}, and hence that, because of the linear independence of the sectors with different N_{FP}, Eq.(33) holds for both of these components. Therefore, we can take $\mathcal{R}$ as either of these components with even or odd N_{FP}, and hence, the BRS derivative of $\mathcal{R}$ which can be given by $\boldsymbol{\delta}\mathcal{R} = [iQ_B, \ \mathcal{R}]_{\mp}$ is a *local* operator satisfying the local (anti)commutativity. Since we have assumed that the vacuum $|0\rangle$ is a physical state, namely, $Q_B|0\rangle = 0$, we obtain from Eq.(33)

$$\boldsymbol{\delta}\mathcal{R}|0\rangle = [iQ_B, \mathcal{R}]_{\mp}|0\rangle = iQ_B \mathcal{R}|0\rangle = 0. \qquad (4.3.1 - 35)$$

---

[2] $\mathcal{F}(\mathcal{O})$ is the polynomial algebra generated by field operators smeared with test functions with compact supports in a finite spacetime region $\mathcal{O}$. [See Eq.(A.2-21) in Appendix A.2]

Then, $\delta\mathcal{R}$ is a smeared local operator $\in \mathcal{F}(\mathcal{O})$ *annihilating the vacuum*, which vanishes by itself because of the separating property of the vacuum [see Corollary A.2-6 in Appendix A.2]:

$$\delta\mathcal{R} = 0. \qquad (4.3.1-36)$$

(q.e.d.)

Proof of Proposition 4.3.1-4: Let $\mathcal{R}$ be a smeared local observable satisfying Eq.(29), then it satisfies Eq.(33) because of Eq.(3.4.3-2). By Lemma 4.3.1-5, we obtain Eq.(32).   (q.e.d.)

In the above, it is worth while remarking that the very modest requirement (i) [Eq.(17) or (13)] for the natural relations, Eqs.(8) and (12), between the transition probabilities and the expectation values in the ordinary quantum theory to be preserved leads us to the condition (iii) of gauge invariance Eq.(29) for a physical quantity $\mathcal{R}$. Especially, this requirement for a smeared local observable $\mathcal{R}$ is reduced to the salient algebraic condition Eq.(32) of BRS invariance, which can be examined directly with the aid of the canonical (anti)commutation relations without recourse to the dynamical information of Green's functions, etc. This shows the pertinence of the present formulation, especially, in the choice of the state vector space $H_{\mathrm{phys}} = \overline{\mathcal{V}_{\mathrm{phys}}/\mathcal{V}_0}$ in which every physical process should be described. In this context, it may be instructive to note another evidence for the consistency of the choice of the subsidiary condition, Eq.(3.4.3-2). Just reversing the direction of the above arguments, let us select the observable $\mathcal{R}$ by Eq.(36) according to the principle of BRS invariance, and require the observables to be represented in a Hilbert space H. Then, Q_B is an observable,

$$\delta Q_B = 2i Q_B{}^2 = 0, \qquad (4.3.1-37)$$

whose representation $\hat{Q}_B$ in H with a positive definite inner product is nothing but 0,

$$\hat{Q}_B = 0, \qquad (4.3.1-38)$$

because of the nilpotency, Eq.(3.4.2-17), and hermiticity, Eq.(3.4.2-18), of Q_B:

$$\hat{Q}_B{}^\dagger \hat{Q}_B = (Q_B{}^\dagger Q_B)^\wedge = \hat{Q}_B^2 = 0. \qquad (4.3.1-39)$$

Thus, the subsidiary condition, $Q_B|\mathrm{phys}\rangle = 0$, can be said to follow almost directly from the principle of gauge invariance expressed by Eq.(32) for the observable.

Finally, we remark that the condition Eq.(36) picking up smeared local observbles $\mathcal{R}$ is just the *cocycle condition* with respect to the BRS-coboundary operation $\boldsymbol{\delta}$ satisfying the nilpotency, and also that any operator of the form $\boldsymbol{\delta}(M) = [iQ_B, M]_{\mp}$ vanishes in $\mathcal{V}_{\text{phys}}$ and $H_{\text{phys}} = \overline{\mathcal{V}_{\text{phys}}/\mathcal{V}_0}$ owing to the subsidiary condition. This shows the relevance to the (smeared) local observables of the cohomological aspects of $\boldsymbol{\delta}$ quite similarly to the case of Q_B-cohomology in the discussion of physical states in Sec.4.2. In the next subsection, we discuss the structure of (smeared local) observables in more detail adopting such terminology as **BRS-cocycles**, **BRS-coboundaries**, and **BRS-cohomology**, respectively, for cocycles, coboundaries and cohomology with respect to $\boldsymbol{\delta}$ acting on the Heisenberg fields.

4.3.2 Quantum Maxwell equation and structure of local observables
–Local observables as group invariants–

In the covariant canonical operator formalism of QED, the field strength $F_{\mu\nu}$ and the electromagnetic current j_μ satisfy the condition (iv) of strict gauge invariance which is equivalent to Eq.(4.3.1-25):

$$[F_{\mu\nu}(x),\ B(y)] = -i(\partial_\mu\partial_\nu - \partial_\nu\partial_\mu)D(x-y) = 0, \qquad (4.3.2-1)$$

$$[j_\mu(x),\ B(y)] = -e^{-1}[\partial^\nu F_{\nu\mu}(x) - \partial_\mu B(x),\ B(y)] = 0, \qquad (4.3.2-2)$$

and hence, they are (strictly gauge invariant) observables. For these observables, we obtain the quantum Maxwell equation in H_{phys}:

$$\partial^\nu \hat{F}_{\nu\mu} = -e\hat{j}_\mu, \qquad (4.3.2-3)$$

from the field equation

$$\partial^\nu F_{\nu\mu} = -ej_\mu + \partial_\mu B, \qquad (4.3.2-4)$$

or

$$\langle\Phi_1|(\partial^\nu F_{\nu\mu} + ej_\mu)|\Phi_2\rangle = 0 \qquad \text{for } |\Phi_i\rangle \in \mathcal{V}_{\text{phys}}\ (i = 1, 2). \qquad (4.3.2-5)$$

On the contrary, the field equation, Eq.(3.4.1-12), of the Yang-Mills field,

$$(D^\mu F_{\mu\nu})^a = \partial_\nu B^a - gj_\nu{}^a - ig(\partial_\nu\overline{C} \times C)^a, \qquad (4.3.2-6)$$

contains an unphysical term, $-ig(\partial_\mu \overline{C} \times C)^a$, which does not vanish even in the matrix elements between physical states. Hence, the classical field equation, Eq.(3.2.2-22), can never be reproduced in the non-abelian case. As remarked in Sec.3.4.3, more natural form connecting classical and quantum cases is given by the **quantum Maxwell equation**, Eq.(3.4.3-14), i.e.,

$$\partial^\nu F_{\nu\mu}{}^a + g J_\mu{}^a = \{Q_B, (D_\mu \overline{C})^a\}, \qquad (4.3.2-7)$$

owing to which the classical Maxwell-type equation, Eq.(3.4.3-15), holds as the matrix elements between physical states in a manner quite similar to Eq.(5):

$$\langle \Phi_1 | (\partial^\nu F_{\nu\mu}{}^a + g J_\mu{}^a) | \Phi_2 \rangle = 0 \qquad \text{for } |\Phi_i\rangle \in \mathcal{V}_{\text{phys}}. \qquad (4.3.2-8)$$

However, Eq.(8) differs from Eq.(5) in an important point: Although the matter current $j_\mu{}^a$ defined by Eq.(3.4.1-16) is *covariantly conserved* but is *not* genuinely conserved,

$$(D^\mu j_\mu)^a = 0 \implies \partial^\mu j_\mu{}^a = -g(A^\mu \times j_\mu)^a \neq 0, \qquad (4.3.2-9)$$

and hence, it has been replaced in Eq.(7) by the genuinely conserved Noether current $J_\mu{}^a$ of the global gauge symmetry. As is evident from its definition given by Eq.(3.4.3-12), i.e.,

$$
\begin{aligned}
J_\mu{}^a &= j_\mu{}^a + (A^\nu \times F_{\nu\mu})^a + (A_\mu \times B)^a - i(\overline{C} \times D_\mu C)^a + i(\partial_\mu \overline{C} \times C)^a \\
&= [j_\mu{}^a + (A^\nu \times F_{\nu\mu})^a] - \{Q_B, (A_\mu \times \overline{C})^a\} + i(\partial_\mu \overline{C} \times C)^a, \quad (4.3.2-10)
\end{aligned}
$$

this $J_\mu{}^a$ contains the same unphysical term, $(\partial_\mu \overline{C} \times C)^a$, as the above. Consequently, *neither* the Noether current $J_\mu{}^a$ *nor* the field strength $F_{\mu\nu}{}^a$ are observables in the non-abelian case in contrast to the abelian case,

$$[Q_B, F_{\mu\nu}{}^a] = ig(C \times F_{\mu\nu})^a \neq 0, \qquad (4.3.2-11)$$

$$[Q_B, J_\mu{}^a] = -i\partial^\nu(C \times F_{\mu\nu})^a \neq 0. \qquad (4.3.2-12)$$

Therefore, Eq.(8) for the matrix elements cannot yield such an equation as Eq.(3) for the *operators in* H_{phys} in the non-abelian cases. Since the $J_\mu{}^a$ itself cannot be observed in the physical world, we need not worry about the appearance of the unphysical term $(\partial_\mu \overline{C} \times C)^a$ in it. We note here that this consequence is not an accidental situation, but that it can be viewed in a more general context. Namely, the following theorem asserting the *group invariance* of smeared local observables can be proved in the present formulation:

Theorem 4.3.2 − 1 [Str 76, Oji 78, Str 78]   Every smeared local observable $\mathcal{R}$ commutes in H_{phys} with the global charge Q^a of the *unbroken* global gauge symmetry[3]:

$$[\hat{Q}^a,\ \hat{\mathcal{R}}] = 0 \qquad \text{in } H_{\mathrm{phys}}. \qquad (4.3.2 - 13)$$

In order to prove this theorem, we should first examine the global charge operator Q^a of the global gauge symmetry defined (formally) by

$$Q^a \equiv \int d\mathbf{x}\, J_0{}^a. \qquad (4.3.2 - 14)$$

Lemma 4.3.2 − 2   If a conserved current J_μ of the form

$$J_\mu = \partial^\nu K_{\nu\mu} \qquad (4.3.2 - 15)$$

with $K_{\nu\mu}$ being a local (or anti-local) antisymmetric tensor field yields a charge Q suffering from *no* spontaneous symmetry breaking,

$$Q = \int d\mathbf{x}\, J_0 = \int d\mathbf{x}\, \partial^i K_{i0}; \quad Q|0\rangle = 0, \qquad (4.3.2 - 16)$$

then the charge Q should vanish, $Q = 0$.

Proof: Let φ be a smeared local operator belonging to $\mathcal{F}(\mathcal{O})$: $\varphi \in \mathcal{F}(\mathcal{O})$. Taking a sufficiently large $R > 0$, we obtain

$$\begin{aligned} Q\varphi|0\rangle &= [Q,\ \varphi]_\mp|0\rangle \pm \varphi Q|0\rangle \\ &= [Q_R,\ \varphi]_\mp|0\rangle \end{aligned} \qquad (4.3.2 - 17)$$

because of the assumption that Q is an unbroken charge annihilating the vacuum, $Q|0\rangle = 0$. Since Q_R is, roughly speaking,[4] the volume integral of J_0 within the

---

[3]   The whole group symmetry may be broken spontaneously, in which case Eq.(13) holds for the charges of the unbroken subgroup of the remaining symmetry.

[4]   As for the precise definition of Q_R, see Eqs.(A 3-5), (A.3-7) and (A.3-8) in Appendix A.3.

region $|\mathbf{x}| < R$, the (anti)commutator $[Q_R, \varphi]_{\mp}$ vanishes for a sufficiently large $R > 0$ owing to the local (anti)commutativity of K_{i0} and φ:

$$[Q_R, \varphi]_{\mp} \sim \int d\mathbf{x}\, \partial_x{}^i [K_{i0}(x), \varphi]_{\mp} = \int dS_x{}^i [K_{i0}(x), \varphi]_{\mp} = 0, \qquad (4.3.2-18)$$

or more precisely,

$$[Q_R, \varphi]_{\mp} = \int d^4x\, \alpha_T(x^0) f_R(\mathbf{x}) [\partial_x{}^i K_{i0}(x), \varphi]_{\mp}$$

$$= -\int d^4x\, \alpha_T(x^0) \partial_x{}^i f_R(\mathbf{x}) [K_{i0}(x), \varphi]_{\mp} = 0, \quad (4.3.2-19)$$

because $\mathrm{supp} f_R \subseteq \{\mathbf{x} \in \mathbf{R}^3; R \leq |\mathbf{x}| \leq 2R\}$. Thus, we obtain for $\varphi|0\rangle \in \mathcal{F}(\mathcal{O})|0\rangle$

$$Q(\varphi|0\rangle) = 0, \qquad (4.3.2-20)$$

and hence,

$$\langle\Psi|Q(\varphi|0\rangle) = 0 \quad \text{for } \forall|\Psi\rangle \in \mathcal{V}. \qquad (4.3.2-21)$$

By virtue of the Reeh-Schlieder theorem [Ree 61] (Theorem A.2-5 in Appendix A.2)

$$\mathcal{V} = \overline{\mathcal{F}(\mathcal{O})|0\rangle}^w, \qquad (4.3.2-22)$$

we arrive at the conclusion

$$\langle\Psi|Q|\Phi\rangle = 0 \quad \text{for } \forall|\Psi\rangle, \forall|\Phi\rangle \in \mathcal{V}. \qquad (4.3.2-23)$$

which is nothing but the statement: $Q = 0$. \hfill (q.e.d.)

Applying this Lemma 4.3.2-2 to the conserved current $[Q_B, J_\mu{}^a]$ in Eq.(12) $[K_{\nu\mu} \to -i(C \times F_{\nu\mu})^a]$, we conclude:

Corollary 4.3.2 – 3   The global charge Q^a given by Eq.(14) is a (non-local) *observable*

$$[Q_B, Q^a] = 0, \qquad (4.3.2-24)$$

as long as the global gauge symmetry corresponding to this charge is not broken spontaneously.

Now we prove Theorem 4.3.2-1.

Proof of Theorem 4.3.2-1: Let $\mathcal{R}$ be a smeared local observable belonging to $\mathcal{F}(\mathcal{O})$:

$$\boldsymbol{\delta}\mathcal{R} = 0. \qquad (4.3.2-25)$$

As has been shown in the proof of Lemma 4.3.1-5, $\mathcal{R}$ can be assumed without loss of generality to satisfy either $[Q_B,\ \mathcal{R}] = 0$ or $\{Q_B,\ \mathcal{R}\} = 0$. Then, the quantum Maxwell equation, Eq.(7), tells us the equality

$$
\begin{aligned}
[gJ_0{}^a,\ \mathcal{R}] &= [-\partial^\iota F_{\iota 0}{}^a + \{Q_B,\ (D_0\overline{C})^a\},\ \mathcal{R}] \\
&= -[\partial^\iota F_{\iota 0}{}^a,\ \mathcal{R}] + [Q_B,\ [(D_0\overline{C})^a,\ \mathcal{R}]_\mp]_\pm + [(D_0\overline{C})^a,\ [Q_B,\ \mathcal{R}]_\mp]_\pm \\
&= -[\partial^\iota F_{\iota 0}{}^a,\ \mathcal{R}] + [Q_B,\ [(D_0\overline{C})^a,\ \mathcal{R}]_\mp]_\pm \qquad (4.3.2-26)
\end{aligned}
$$

corresponding to the cases $[Q_B,\ \mathcal{R}]_\mp = 0$. Thus, for any $R > 0$, we obtain

$$
\begin{aligned}
[gQ_R{}^a,\ \mathcal{R}] &= \int d^4x\,\alpha_T(x^0)\partial^\iota f_R(\boldsymbol{x})[F_{\iota 0}{}^a(x),\ \mathcal{R}] \\
&\quad + \left[Q_B,\ \int d^4x\,\alpha_T(x^0)f_R(\boldsymbol{x})[(D_0\overline{C})^a(x),\ \mathcal{R}]_\mp\right]_\pm. (4.3.2-27)
\end{aligned}
$$

By taking $R > 0$ sufficiently large, the first term vanishes by the local commutativity, and hence, we obtain

$$
\begin{aligned}
\langle\Phi|[gQ_R{}^a,\ \mathcal{R}]|\Psi\rangle &= \langle\Phi|\left[Q_B,\ \int d^4x\,\alpha_T(x^0)f_R(\boldsymbol{x})[(D_0\overline{C})^a(x),\ \mathcal{R}]_\mp\right]_\pm|\Psi\rangle \\
&= 0 \qquad \text{for } |\Phi\rangle, |\Psi\rangle \in \mathcal{V}_{\text{phys}}. \qquad (4.3.2-28)
\end{aligned}
$$

Since we have assumed that the symmetry corresponding to Q^a is not broken spontaneously, we obtain from Eq.(28)

$$\langle\Phi|[gQ^a,\ \mathcal{R}]|\Psi\rangle = 0 \quad \text{for } |\Phi\rangle, |\Psi\rangle \in \mathcal{V}_{\text{phys}}. \qquad (4.3.2-29)$$

This concludes the following equality for the *observable* Q^a (see Corollary 4.3.2-3):

$$[g\hat{Q}^a,\ \hat{\mathcal{R}}] = 0 \ \text{ in } H_{\text{phys}}, \qquad (4.3.2-30)$$

which is just what to be verified.   (q.e.d.)

Now, we investigate the structure of the observables in more detail. The *canonical* energy-momentum tensor $T_{\mu\nu}$ is given by[5],[6]

$$
\begin{aligned}
T_{\mu\nu} =& \partial_\nu \Phi_I \frac{\partial \mathcal{L}}{\partial(\partial^\mu \Phi_I)} - \eta_{\mu\nu} \mathcal{L} \\
=& -F_{\mu\lambda}{}^a \partial_\nu A^{\lambda\,a} - A_\mu{}^a \partial_\nu B^a - i\partial_\nu \overline{C}^a \cdot (D_\mu C)^a - i\partial_\mu \overline{C}^a \cdot (\partial_\nu C)^a \\
&+ \partial_\nu \varphi^\alpha \frac{\partial \mathcal{L}_M}{\partial(\partial^\mu \varphi^\alpha)} - \eta_{\mu\nu} \mathcal{L},
\end{aligned}
\tag{4.3.2 - 31}
$$

and satisfies the commutation relation

$$
[iQ_B,\ T_{\mu\nu}] = \partial^\rho(\partial_\nu C^a \cdot F_{\rho\mu}{}^a) \neq 0.
\tag{4.3.2 - 32}
$$

Hence, in view of the criterion Eq.(4.3.1-32) for the local observable, $T_{\mu\nu}$ is *not* an observable. There exists, however, a physically more reasonable definition of the energy-momentum tensor, namely, the symmetric energy-momentum tensor $\Theta_{\mu\nu}$ defined by[7]

$$
\begin{aligned}
\Theta_{\mu\nu} \equiv& \left. h_{\mu a} \frac{\delta(h\tilde{\mathcal{L}})}{\delta h^\nu{}_a} \right|_{h^\lambda{}_b = \delta^\lambda{}_b} = \left. h_{\mu a} \left[\frac{\partial(h\tilde{\mathcal{L}})}{\partial h^\nu{}_a} - \partial_\lambda \left(\frac{\partial(h\tilde{\mathcal{L}})}{\partial(\partial_\lambda h^\nu{}_a)} \right) \right] \right|_{h^\lambda{}_b = \delta^\lambda{}_b} \\
=& \Theta_{\nu\mu},
\end{aligned}
\tag{4.3.2 - 33}
$$

$$
\partial^\mu \Theta_{\mu\nu} = 0,
\tag{4.3.2 - 34}
$$

where $\tilde{\mathcal{L}}$ is the local-Lorentz invariant Lagrangian density obtained from the original one $\mathcal{L}$, Eq.(3.4.1-1), by replacing the flat Minkowski metric $\eta_{\mu\nu}$ and the derivatives ∂_μ by

$$
\eta_{\mu\nu} \ \rightarrow \ h_\mu{}^a \eta_{ab} h_\nu{}^b
\tag{4.3.2 - 35}
$$

----

[5]   In Eq.(31), the summation of Φ_I should include all the fields $A_\mu{}^a, B^a, C^a, \overline{C}^a$ and φ^α. In particular, the contribution from the conjugate $\overline{\psi}$ of Dirac spinor ψ should be summed equally along with ψ, if they appear among the matter fields φ^α.

[6]   If we adopt the gauge fixing term $\mathcal{L}'_{GF} = B^a \partial^\mu A_\mu{}^a + \alpha B^a B^a/2$, Eqs.(31) and (32) read as

$$
T_{\mu\nu}' = (\eta_{\mu\lambda} B^a - F_{\mu\lambda}{}^a)\partial_\nu A^{\lambda\,a} - i\partial_\nu \overline{C}^a \cdot (D_\mu C)^a - i\partial_\mu \overline{C}^a \cdot \partial_\nu C^a + \partial_\nu \varphi^\alpha \frac{\partial \mathcal{L}_M}{\partial(\partial^\mu \varphi^\alpha)} - \eta_{\mu\nu} \mathcal{L}',
$$

and

$$
[iQ_B,\ T_{\mu\nu}'] = \partial^\rho(\partial_\nu C^a \cdot F_{\rho\mu}{}^a) + \partial_\nu(B^a(D_\mu C)^a) - \eta_{\mu\nu}\partial^\rho(B^a(D_\rho C)^a).
$$

[7]   See, for instance, [Wei 72].

and

$$\partial_\mu \;\to\; \nabla_\mu \;(\text{general- and local-Lorentz covariant differentiation}), \qquad (4.3.2-36)$$

respectively. Here $h_\mu{}^a$ is the vierbein component (see Sec.5.7.1) and $\eta_{ab} = \text{diag.}(+1,-1,-1,-1)$. Further, h in Eq.(33) is defined by

$$h \equiv \det(h_\mu{}^a). \qquad (4.3.2-37)$$

This definition of symmetric energy-momentum tensor $\Theta_{\mu\nu}$ agrees with the one obtained by the Belinfante method adding to the canonical $T_{\mu\nu}$ the spin angular-momentum density term $\mathcal{S}_{\rho\mu\nu} \equiv -i(s_{\mu\nu})^I{}_J \Phi^J (\partial/\partial(\partial^\rho \Phi^I))\mathcal{L}$ (see Sec.1.2.5)

$$\Theta_{\mu\nu} = T_{\mu\nu} + \frac{1}{2}\partial^\rho(\mathcal{S}_{\rho\mu\nu} + \mathcal{S}_{\mu\nu\rho} + \mathcal{S}_{\nu\mu\rho}), \qquad (4.3.2-38)$$

and is found to be

$$\Theta_{\mu\nu} = \Theta^{\text{phys}}{}_{\mu\nu} - \{Q_B, \; \partial_\mu \overline{C}^a \cdot A_\nu{}^a + \partial_\nu \overline{C}^a \cdot A_\mu{}^a$$
$$+ \eta_{\mu\nu}(\frac{1}{2}\alpha \overline{C}^a B^a - \partial^\rho \overline{C}^a \cdot A_\rho{}^a)\}, \qquad (4.3.2-39)$$

$$\Theta^{\text{phys}}{}_{\mu\nu} \equiv F_{\mu\rho}{}^a F^\rho{}_\nu{}^a + \frac{1}{4}\eta_{\mu\nu}F_{\rho\sigma}{}^a F^{\rho\sigma a} + \Theta_{M\mu\nu}. \qquad (4.3.2-40)$$

$\Theta_{M\mu\nu}$ is the contribution from the matter Lagrangian density $\mathcal{L}_M$; for instance, in the case of QCD with

$$\mathcal{L}_M = \overline{q}\Big(\frac{i}{2}\gamma^\mu(\overrightarrow{\partial}_\mu - \overleftarrow{\partial}_\mu) + g\frac{\lambda^a}{2}A_\mu{}^a\gamma^\mu - m\Big)q, \qquad (4.3.2-41)$$

it is given by

$$\Theta_{M\mu\nu} = \frac{1}{2}\overline{q}\Big(\frac{i}{2}\gamma_\mu(\overrightarrow{\partial}_\nu - \overleftarrow{\partial}_\nu) + g\frac{\lambda^a}{2}A_\nu{}^a\gamma_\mu\Big)q + (\mu \leftrightarrow \nu). \qquad (4.3.2-42)$$

Contrary to the case of $T_{\mu\nu}$ in Eq.(32), we can show the BRS invariance of $\Theta_{\mu\nu}$ and $\Theta^{\text{phys}}{}_{\mu\nu}$ given by Eqs.(39) and (40),

$$[Q_B, \; \Theta^{\text{phys}}{}_{\mu\nu}] = [Q_B, \; \Theta_{\mu\nu}] = 0. \qquad (4.3.2-43)$$

Thus, $\Theta_{\mu\nu}$ is an observable as it should be, and hence, the energy-momentum operator $P_\mu = \int d\mathbf{x}\, T_{0\mu} = \int d\mathbf{x}\, \Theta_{0\mu}$ as well as the Lorentz generators $M_{\mu\nu} = \int d\mathbf{x}\,(x_\mu \Theta_{0\nu} - x_\nu \Theta_{0\mu})$ are observables:

$$[Q_B, \; P_\mu] = [Q_B, \; M_{\mu\nu}] = 0. \qquad (4.3.2-44)$$

It is important to note that Eq.(44) ensures the Poincaré covariance of the theory in H_{phys}. Owing to the observability, Eq.(43), of $\Theta_{\mu\nu}$, the energy-momentum can be observed also through the *local* measurement:

$$(P_\mu)_R \equiv \int d^4x \, \alpha_T(x^0) f_R(\mathbf{x}) \Theta_{0\mu}(x)$$

$$[Q_B, \, (P_\mu)_R] = 0. \qquad\qquad (4.3.2-45)$$

Here, we remark such a structural feature of $\Theta_{\mu\nu}$ exhibited in Eqs.(39) and (40) that it consists of the following two parts. The first part $\Theta^{\text{phys}}_{\mu\nu}$ which is an observable in itself contains no such unphysical fields as $C^a, \overline{C}^a$ and B^a, and is identical with the energy-momentum tensor derived from the Lagrangian density $\mathcal{L}_0$ without the gauge fixing terms nor FP ghosts. The second one contains unphysical fields $C^a, \overline{C}^a$ and B^a in such a form that it vanishes in the physical subspace $\mathcal{V}_{\text{phys}}$, as is physically reasonable. It should be noted, however, that this unphysical second term, $\Theta_{\mu\nu} - \Theta^{\text{phys}}_{\mu\nu}$, with no contribution in the physical world H_{phys}, plays an essential role in the conservation of $\Theta_{\mu\nu}$ at the level of $\mathcal{V}$. Without this term, $\Theta^{\text{phys}}_{\mu\nu}$ cannot be conserved by itself as an operator in $\mathcal{V}$:

$$\partial^\mu \Theta^{\text{phys}}_{\mu\nu} = \{Q_B, \, \partial^\mu \overline{C}^a \cdot F_{\mu\nu}{}^a\} \neq 0. \qquad\qquad (4.3.2-46)$$

As an operator in H_{phys}, $\hat{\Theta}^{\text{phys}}_{\mu\nu}$ coincides with $\hat{\Theta}_{\mu\nu}$:

$$\hat{\Theta}_{\mu\nu} = \hat{\Theta}^{\text{phys}}_{\mu\nu}, \qquad\qquad (4.3.2-47)$$

and, of course, it is conserved: $\partial^\mu \hat{\Theta}^{\text{phys}}_{\mu\nu} = \partial^\mu \hat{\Theta}_{\mu\nu}$ in the Hilbert space H_{phys}. This situation is similar to the one encountered in the analysis of the S-matrix in Sec.4.2: Such unphysical particles as FP ghost pairs turn out to be produced easily in the final states but it is only in zero-norm combinations, and hence, they do not appear in the physical world H_{phys}. At the very beginning of quantization, we have also observed that the local-gauge-invariant Lagrangian density $\mathcal{L}_0$, Eq.(3.4.1-2), cannot be quantized without adding the gauge fixing and FP ghost terms, $\mathcal{L}_{\text{GF}} + \mathcal{L}_{\text{FP}} = i\boldsymbol{\delta}(\partial^\mu \overline{C}^a \cdot A_\mu{}^a - \alpha \overline{C}^a B^a/2)$, which are combined into a *BRS-coboundary term vanishing* in $\mathcal{V}_{\text{phys}}$. In short, if one simply neglects such unphysical fields as $C^a, \overline{C}^a$ and B^a in the present formulation, the invariance under the time evolution or Lorentz transformations cannot be attained owing to the leakage terms of *zero norm*, which make no effects once we arrive at the physical description in H_{phys}. However, since one cannot reach H_{phys} directly without passing through the underlying V and V_{phys}, such unphysical fields as $C^a, \overline{C}^a$ and B^a are indispensable for formulating the theory in a consistent and Lorentz-covariant manner at every step.

In the case of QCD, some other examples of local observables (hermitian or not) are given by

$$F_{\mu\nu}{}^a F_{\rho\sigma}{}^a,$$

$$\overline{q}\Gamma q,$$

$$\overline{q}\Gamma(i\partial_\mu + g\frac{\lambda^a}{2}A_\mu{}^a)q, \quad (-i\partial_\mu \overline{q}\Gamma + \overline{q}g\frac{\lambda^a}{2}A_\mu{}^a)\Gamma q; \qquad (4.3.2-48)$$

$$B^a = \boldsymbol{\delta}(\overline{C}^a), \quad (D_\mu C)^a = \boldsymbol{\delta}(A_\mu{}^a), \quad -\frac{g}{2}(C \times C)^a = \boldsymbol{\delta}(C^a),$$

$$igC^a\frac{\lambda^a}{2}\Gamma q = \boldsymbol{\delta}(\Gamma q), \quad -ig\overline{q}\Gamma\frac{\lambda^a}{2}C^a = \boldsymbol{\delta}(\overline{q}\Gamma), \qquad (4.3.2-49)$$

where Γ is one of the Dirac matrices (or some product of them belonging to the Clifford algebra generated by Dirac matrices) commuting with λ^a's. The first group, Eq.(48), consists of local gauge invariant color singlets containing none of such unphysical fields as $B^a, C^a, \overline{C}^a$. Although the observables in the second group, Eq.(49), are color non-singlets, they are all *trivial* ones vanishing in H_{phys} as BRS-coboundaries, without any contradiction to Theorem 4.3.2-1.

These examples, Eqs.(39), (46), (48) and (49), suggest that a non-trivial (smeared) local observable $\mathcal{R}$ can be decomposed into a sum of a trivial observable given by a BRS-coboundary term and a local-gauge-invariant operator $F(A_\mu{}^a, \varphi^\alpha)$ consisting of gauge field $A_\mu{}^a$ and of matter fields φ^α excluding any of $B^a, C^a, \overline{C}^a$; if the latter unphysical fields appears in $\mathcal{R}$, they can be confined into the BRS-coboundary term giving no contribution to the expectation values of $\mathcal{R}$ in the physical world.

Because of the *non-(anti)commutativity* of field operators smeared over different times and of their causal dependence due to *field equations*, however, it is not so easy to prove this natural expectation in a satisfactory way. In [Kug 80], this difficulty was circumvented by imposing the assumption that the interacting Heisenberg fields can completely be described by the field operators $\int d\boldsymbol{x}\Phi_I(t=0, \boldsymbol{x})f_I(\boldsymbol{x})$ and their canonical conjugate variables $\int d\boldsymbol{x}\pi_{\Phi_I}(t=0, \boldsymbol{x})f_I(\boldsymbol{x})$, restricted on a spacelike 3-dimensional hyperplane at $t = 0$ and smeared with "test functions" $f_I(\boldsymbol{x})$ on that hyperplane. Although it is difficult to judge the legitimacy of this assumption in the general context of relativistic quantum field theory, it allows us to neglect the complications due to the field equations, and the problem of non-(anti)commutativity is reduced to that only between Φ_I and Π_{Φ_I} arising from the equal-time canonical (anti)commutation relations. On this assumption, the above expectation can be justified, in a somewhat modified form, as the following theorem obtained in [Kug 80], which can be almost reduced to the results obtained in the context of *free graded differential algebras* [Sul 77][8].

Theorem $4.3.2 - 4$ Let $\mathcal{R}$ be a smeared local observable in the above sense, i.e., a BRS-cocycle consisting of the field operators smeared on the 3-dimensional hyperplane at $t = 0$ with a definite FP-ghost number N_{FP}. Then, the following statements hold :

(i) [Poincaré lemma for BRS-cohomology] If $N_{\mathrm{FP}} < 0$, then $\mathcal{R}$ is a *trivial* observable given by a BRS-coboundary, namely, there exists some smeared local operator M such that

$$\mathcal{R} = [iQ_B,\ M]_{\mp} = \boldsymbol{\delta}(M). \qquad (4.3.2 - 50)$$

In a word, the N-th BRS-cohomology with $N = N_{\mathrm{FP}} < 0$ vanishes identically: $H^N(\boldsymbol{\delta}) = 0$.

(ii) If $N_{\mathrm{FP}} = 0$, then $\mathcal{R}$ can be expressed in such a form as

$$\mathcal{R} = F_{\mathrm{inv}}(A_k,\ \varphi, \pi_{A_k}, \pi_\varphi) + \boldsymbol{\delta}(M), \qquad (4.3.2 - 51)$$

where F_{inv} is a *local-gauge-invariant operator* given as a polynomial in $A_k{}^a, \varphi^\alpha$ and their conjugate variables without containing any of $B^a, C^a, \overline{C}^a$.

(iii) If $N_{\mathrm{FP}} > 0$, then an expression similar to (ii) holds for $\mathcal{R}$,

$$\mathcal{R} = P[I_\iota(C), F_{\mathrm{inv},\jmath}(A_k,\ \varphi, \pi_{A_k}, \pi_\varphi)] + \boldsymbol{\delta}(M), \qquad (4.3.2 - 52)$$

but here P is a polynomial in local-gauge-invariant operators $F_{\mathrm{inv},\jmath}$ *and* in **invariant polynomials** $I_\iota(C)$ on $LieG$ consisting of FP ghost C^a (with no spatial derivatives).

Remark: The possible invariant polynomials $I_\iota(C)$ are determined intrinsically by the structure of the Lie algebra $LieG$, and they give the basis of *cohomology*

---

[8]   Roughly speaking, a free graded differential algebra is a Grassmann algebra equipped with an anti-derivation, free of any algebraic relations (besides the graded commutativity).

of Lie algebra, $H^*(LieG)$, *of LieG (at each point* $\boldsymbol{x}$). For instance, if the gauge group G is semi-simple, there is no invariant polynomial with $N_{\mathrm{FP}} = 1, 2$, and hence, possible smeared local observables $\mathcal{R}$ with $N_{\mathrm{FP}} = 1, 2$ are only trivial ones represented by BRS-coboundaries.

Since the proof of the statements in the above Theorem 4.3.2-4 involves some tedious technicalities not much relevant to the main context of this book[9], we note here only some important lemmas useful in the proof, without going into details. First, to resolve the non-(anti)commutativity between the canonical variables and their conjugates, we need the following invariance property under the "quantizing deformation":

<u>Lemma 4.3.2 − 5</u>   When the canonical quantization is performed by imposing the (anti)commutation relations with a parameter ϵ given by

$$[\pi_{\Phi_I}(\boldsymbol{x}), \Phi_J(\boldsymbol{y})]_{\mp} = -i\epsilon\delta_J{}^I\delta(\boldsymbol{x} - \boldsymbol{y}), \qquad (4.3.2 - 53)$$

the statements in Theorem 4.3.2-4 about the smeared local observable $\mathcal{R}$ are independent of the parameter ϵ. This lemma makes it possible to reduce the proof to the (anti)commutative classical situation with $\epsilon = 0$.

Now, our strategy is to "sweep out" step by step the unphysical fields B^a, C^a and $\overline{C}^a$ (and their canonical conjugates) contained in $\mathcal{R}$ into BRS-coboundary terms $\boldsymbol{\delta}(M)$, taking advantage of the BRS-cocycle condition on $\mathcal{R}$. In this process, the next two lemmas play a crucial role:

<u>Lemma 4.3.2 − 6</u> [Nak 79f]   Let b^α's be independent commutative variables and $G^{[\alpha_1\alpha_2\cdots\alpha_n]}$ be polynomials in b^α's totally antisymmetric in the indices α,'s. If the equalities

$$b^{\alpha_1}G^{[\alpha_1\alpha_2\cdots\alpha_n]} = 0, \qquad (4.3.2 - 54)$$

(with summation over the repeated index α_1) hold identically, then $G^{[\alpha_1\cdots\alpha_n]}$ can be written in the form

$$G^{[\alpha_1\cdots\alpha_n]} = b^{\alpha_0}H^{[\alpha_0\alpha_1\cdots\alpha_n]} \qquad (4.3.2 - 55)$$

with some polynomials $H^{[\alpha_0\alpha_1\cdots\alpha_n]}$ totally antisymmetric in $n + 1$ indices $\alpha_0, \alpha_1, \cdots, \alpha_n$.

---

[9]   There are also some points in the proof of [Kug 80], such as 'Proposition 3' there, difficult to reformulate into a mathematically meaningful form.

A similar result holds in the case of anticommutative Grassmannian variables if the total antisymmetry property is replaced by the total symmetry.

If $N_{\mathrm{FP}} < 0$, we expand $\mathcal{R}$ into a power series of $\overline{C}^a$ and, using the BRS transformation law $\boldsymbol{\delta}(\overline{C}^a) = iB^a$, we can utilize the BRS-cocycle condition and the above lemma to terminate the procedure to "sweeping out" B^a and $\overline{C}^a$. Namely, we obtain the following proposition.

<u>Proposition 4.3.2 − 7</u> [Sul 77, Nak 79]   Let b^α's and a^α's be, respectively, commuting (or, anticommuting) and anticommuting (or, commuting) variables, related through the BRS transformation

$$\boldsymbol{\delta} a^\alpha = b^\alpha. \qquad (4.3.2 - 56)$$

If these variables b^α and a^α are all independent of one another as well as of the other variables denoted generically by φ, then any BRS-invariant polynomials $G(a^\alpha, b^\alpha, \varphi)$ in a^α, b^α and φ such that each term in G contains at least one factor of a's or b's, is a BRS-coboundary, i.e., there exists a polynomial $M(a^\alpha, b^\alpha, \varphi)$ such that

$$G = \boldsymbol{\delta}(M). \qquad (4.3.2 - 57)$$

Namely, the variables b^α's and a^α's constrained by Eq.(56) generate a cohomologically trivial (graded-commutative differential) subalgebra, which just corresponds to the **contractible algebra** according to the terminology of [Sul 77].

If $\mathcal{R}$ is not smeared, Eq.(51) is sharpened as follows.

<u>Proposition 4.3.2 − 8</u> [Kug 80]   Any (strictly) local observable $\mathcal{R}$ with positive FP-ghost number $N_{\mathrm{FP}} > 0$ can be written in the form

$$\mathcal{R} = \sum_\imath I_\imath(C) F_{\mathrm{inv},\imath}(A_k, \varphi, \pi_{A_k}, \pi_\varphi) + \boldsymbol{\delta}(M), \qquad (4.3.2 - 58)$$

where $F_{\mathrm{inv},\imath}$ are local-gauge-invariant operators without containing any of $B^a, C^a, \overline{C}^a$, and $I_\imath(C)$'s are invariant polynomials in the FP-ghost fields C^a of degree N_{FP} with no spatial derivatives.

<u>Proposition 4.3.2 − 9</u> [Kug 80]   If $\mathcal{R}$ is a (strictly) local operator consisting solely of the FP ghosts C^a with *no spatial derivatives*, then the BRS-cocycle condition on $\mathcal{R}$ is equivalent to its invariance under the global gauge group G. This implies that any such G-singlet $\mathcal{R}$ is necessarily a *non-trivial* BRS-cocycle.

Now, we note that the case (iii) in Theorem 4.3.2-4 is closely related to the disastrous possibility of the existence of *singlet pair states* in Q_B-cohomology mentioned in Sec.4.1.2. However, it can be shown that the spacetime exterior differential $d\mathcal{R}$ of a BRS-cocycle $[\boldsymbol{\delta}(\mathcal{R}) = 0]$ with FP-ghost number $N_{\text{FP}} > 0$ is a BRS-coboundary: Namely, there exists a form $\mathcal{R}'$ satisfying

$$d\mathcal{R} = \boldsymbol{\delta}(\mathcal{R}'). \qquad (4.3.2-59)$$

Therefore, as long as a state $|\Psi\rangle = \mathcal{R}|0\rangle$ generated from the vacuum by the (strictly) local observable $\mathcal{R}$ belonging to the case (iii) in Theorem 4.3.2-4 does not contain zero energy-momentum spectrum so that the spacetime differential d is invertible, $|\Psi\rangle$ is a Q_B-coboundary state having its parent state "d^{-1}"$\boldsymbol{\delta}(\mathcal{R}')|0\rangle$ (which is, however, highly non-local, in general), and hence, it should belong to a BRS-doublet. Thus, the problem of unpaired singlets is only relevant to the situation involving degenerate vacua with zero energy-momentum spectrum, whose existence is omitted in the present context. This observation suggests also the close connection of unpaired singlets with the problem of **gauge anomaly**: The latter can be formulated as such a local quantity that it is a BRS-*coboundary* in the *global* sense having a *non-local* parent, but that it is *locally* a non-trivial BRS-*cocycle* because of the absence of its *local* parent [Bon 83].

We can see, in the following derivation of Eq.(59) from the BRS-cocycle condition $\boldsymbol{\delta}(\mathcal{R}) = 0$, that the above parallelism is *not* accidental from the viewpoint of **double complex**:

1) The above (iii) in Theorem 4.3.2-4 tells us that, when we neglect the spacetime dependence, the BRS-cohomology with $N = N_{\text{FP}} > 0$ can simply be identified with the Lie-algebra cohomology of $LieG$, through the identification of the FP-ghosts with the Maurer-Cartan forms on $LieG$.

2) When *both* the gauge and spacetime degrees of freedom are to be taken into account, we need to investigate a double complex equipped with two exterior differentials, the BRS $\boldsymbol{\delta}$ and the spacetime differential d, acting on the cochains $\varphi^{r,s}$ with the corresponding gradations, $r, s \geq 0$ (as for the double complex, see Appendix A.1). In this context, the *total* differential Δ is defined, on the *total complex* of the double complex consisting of cochains $(\varphi^{N,0}, \varphi^{N-1,1}, \cdots, \varphi^{0,N})$, by

$$(\Delta\varphi)^{N+1} \equiv (\boldsymbol{\delta}\varphi^{N,0}, \; \boldsymbol{\delta}\varphi^{N-1,1} + (-1)^N d\varphi^{N,0}, \; \boldsymbol{\delta}\varphi^{N-2,2} + (-1)^{N-1} d\varphi^{N-1,1},$$
$$\cdots, \boldsymbol{\delta}\varphi^{0,N} - d\varphi^{1,N-1}, \; d\varphi^{0,N})$$
$$\text{for } \varphi^N = (\varphi^{N,0}, \varphi^{N-1,1}, \cdots, \varphi^{0,N}). \qquad (4.3.2-60)$$

3) In the flat Minkowski spacetime considered here, the spacetime dependence of the FP-ghost in BRS-cocycles $\mathcal{R}$ can be controlled by the help of Poincaré lemma due to its topological triviality.

4) Combining 2) and 3), we obtain an *isomorphism* between the cohomology group $H^N(\Delta)$ for Δ and the Lie-algebra cohomology $H^N(LieG)$ for $N > 0$:

$$H^N(\Delta) \simeq H^N(LieG). \qquad (4.3.2-61)$$

5) Since the *non-trivial spacetime dependence* of the BRS-cocycles $\mathcal{R}$ ($\boldsymbol{\delta}\mathcal{R} = 0$) appearing in (iii) of Theorem 4.3.2-4 implies $d\mathcal{R} \neq 0$, in general, the Δ-cocycle $(\mathcal{R}^{N,0}, \cdots, \mathcal{R}^{0,N})$ in the double complex corresponding to $\mathcal{R} = \mathcal{R}^{N,0}$ cannot satisfy the Δ-cocycle condition,

$$\Delta(\mathcal{R}^{r,s}) = 0, \qquad (4.3.2-62)$$

without having its non-vanishing 'descendent' components $\mathcal{R}^{r,N-r}$ with $r < N$. Namely, in the correspondence between the BRS-cohomology *with* spacetime dependence and the $LieG$-cohomology *without* it through the isomorphism of the latter to the Δ-cohomology, the non-vanishing term $d\mathcal{R} \neq 0$ appearing in $\Delta(\mathcal{R}, 0, \cdots, 0) = (\boldsymbol{\delta}(\mathcal{R}), d\mathcal{R}(\neq 0), 0, \cdots, 0)$ for the spacetime-dependent BRS-cocycle $\mathcal{R} = \mathcal{R}(C^a(x))$ should be compensated properly within the same Δ-cohomology class as $\Delta\Big(\mathcal{R}(\omega^a), 0, \cdots, 0\Big) = \Big(\boldsymbol{\delta}\mathcal{R}(\omega^a), d\mathcal{R}(\omega^a) = 0, 0, \cdots, 0\Big)$: This is realized by adding the 'descendent' components, by virtue of the Poincaré lemma for the spacetime d. Now, the component-wise expression for this equation is nothing but the **descent equations** frequently appearing in the context of gauge anomaly (see, for instance, [Zum 84]):

$$\boldsymbol{\delta}\mathcal{R}^{N,0} = 0,$$
$$\boldsymbol{\delta}\mathcal{R}^{N-1,1} + (-1)^N d\mathcal{R}^{N,0} = 0,$$
$$\boldsymbol{\delta}\mathcal{R}^{N-2,2} + (-1)^{N-1} d\mathcal{R}^{N-1,1} = 0, \qquad (4.3.2-63)$$
$$\cdots$$
$$d\mathcal{R}^{0,N} = 0.$$

Especially, the second equality claims that the spacetime differential, $d\mathcal{R} = d\mathcal{R}^{N,0}$, of the BRS-cocycle $\mathcal{R}$ should a BRS-coboundary, which is just what we have sought.

Finally, we comment that the above double-complex formulation reproduces the formulae (evaluated at the point $(x^\mu, \bar{\theta}) = (x^\mu, 0)$ with the unphysical $\bar{\theta}$ taken to be 0) appearing in the **superfield formulation** of the BRS symmetry [Bon 81a, Hirs 81]:[10] Namely, putting a superfield $A_\mu{}^a dx^\mu + d\bar\theta C^a$ in correspondence with a cochain $(C^a \tau^a, A_\mu{}^a \tau^a dx^\mu)$ in the total complex of the double complex, we can see that the exterior differential $dx^\mu \partial/\partial x^\mu + d\bar\theta \partial/\partial\bar\theta|_{\bar\theta=0}$ in the former context corresponds to the total coboundary operator Δ in the latter. For instance, we obtain a formula

$$\Delta(C^a \tau^a, A_\mu{}^a \tau^a dx^\mu) = (0, 0, \frac{1}{2} F_{\mu\nu}{}^a \tau^a dx^\mu \wedge dx^\nu)$$
$$- \frac{g}{2}[(C^a \tau^a, A_\mu{}^a \tau^a dx^\mu), (C^b \tau^b, A_\nu{}^b \tau^b dx^\nu)], \qquad (4.3.2-64)$$

as a unified expression for the BRS transformation laws, Eqs.(3.3.1-33) and (3.3.1-36), of $A_\mu{}^a$ and C^a, and for the definition, Eq.(3.2.1-15), of the field strength $F_{\mu\nu}{}^a$. By rewriting Eq.(64) as

$$\Delta(C^a \tau^a, A_\mu{}^a \tau^a dx^\mu) + \frac{g}{2}[(C^a \tau^a, A_\mu{}^a \tau^a dx^\mu), (C^b \tau^b, A_\nu{}^b \tau^b dx^\nu)]$$
$$= (0, 0, \frac{1}{2} F_{\mu\nu}{}^a \tau^a dx^\mu \wedge dx^\nu), \qquad (4.3.2-65)$$

one can easily see that the first two 0's in rhs. just represent the "soul-flatness" condition characterizing the BRS transformation in the context of superfield formulation.

4.3.3 Characterization of localized physical states as group invariants —Absence of localized colored physical states—

A remarkable consequence of Theorem 4.3.2-1 given in Sec.4.3.2 is the following theorem stating the absence of *localized* colored (or charged) physical states in QCD (QED).

<u>Theorem 4.3.3 − 1</u> [Fer 77, Oji 79]   Let $|\Phi\rangle$ be a localized physical state, namely, a physical state $|\Phi\rangle \in \mathcal{V}_{\text{phys}}$ expressible by a suitable smeared local operator $\varphi \in \mathcal{F}(\mathcal{O})$ with finite space-time region $\mathcal{O}$ in such a form as

$$|\Phi\rangle = \varphi|0\rangle. \qquad (4.3.3-1)$$

---

[10]   In most of the literature on the superfield formulation of BRS symmetry, the anti-BRS symmetry is also treated on the same footing as the former However, we restrict our discussion to the BRS case only just for simplicity of the presentation

Then, $|\hat{\Phi}\rangle \in H_{\text{phys}}$ is invariant,

$$\hat{Q}^a|\hat{\Phi}\rangle = 0 \qquad\qquad (4.3.3-2)$$

under any *unbroken* global gauge transformation with Q^a as its generator.

Proof: Since $|\Phi\rangle = \varphi|0\rangle$ is a physical state: $Q_B|\Phi\rangle = Q_B\varphi|0\rangle = 0$, Lemma 4.3.1-5 tells us that φ is a smeared local observable. Then, by virtue of Theorem 4.3.2-1, φ satisfies, for the unbroken charge Q^a,

$$[\hat{Q}^a, \hat{\varphi}] = 0, \qquad\qquad (4.3.3-3)$$

and hence,

$$\hat{Q}^a|\hat{\Phi}\rangle = \hat{Q}^a\hat{\varphi}|\hat{0}\rangle = [\hat{Q}^a,\hat{\varphi}]|\hat{0}\rangle + \hat{\varphi}\hat{Q}^a|\hat{0}\rangle = 0. \qquad (4.3.3-4)$$

$$(\text{q.e.d.})$$

Thus, every localized physical state in QCD (QED) is a color singlet (charge-less) state as long as the global color symmetry (global $U(1)$ symmetry) is not broken spontaneously. Needless to say, the above statement concerns only *localized* physical states ($\in \mathcal{F}(\mathcal{O})|0\rangle \cap \mathcal{V}_{\text{phys}}$) and says nothing about *non-localized* physical states. In fact, if Eq.(2) held for every physical state without any restriction, then the electron in QED could not exist in this world. In the case of QED, what Theorem 4.3.3-1 tells us is that the charged physical state cannot be realized in a finite space-time region because of the long-range *Coulomb tails* [Oji 79].

However, one should note that the distinction between the localized state and the non-localized one is made rather subtle by the Reeh-Schlieder theorem [Ree 61] [Eq.(A.2-32) in Appendix A.2]:

$$\begin{aligned}
\mathcal{V} &= \overline{\mathcal{F}|0\rangle}^{\tau} = \mathcal{F}|0\rangle^{\perp\perp} \\
&= \overline{\mathcal{F}(\mathcal{O})|0\rangle}^{\tau} = \mathcal{F}(\mathcal{O})|0\rangle^{\perp\perp}.
\end{aligned} \qquad (4.3.3-5)$$

Owing to this theorem, *every* state in $\mathcal{V}$ can be approximated by *localized* states as closely as one likes [in the sense of the arbitrary admissible topology τ (see Appendix A.2), the weakest of which is the weak topology (w)]. So, it is true that one can always approximate any non-localized physical state by *localized* states. The only relevant question is whether this approximation can be done using exclusively *localized physical* states or not: If it can, then the **color confinement** is achieved. Namely,

<u>Proposition 4.3.3 − 2</u> [Oji 79]   If the equality

$$\overline{\mathcal{F}(\mathcal{O})|0\rangle \cap \mathcal{V}_{\text{phys}}}^{\tau} = \overline{\mathcal{F}(\mathcal{O})|0\rangle}^{\tau} \cap \overline{\mathcal{V}_{\text{phys}}}^{\tau} (= \mathcal{V}_{\text{phys}}) \qquad (4.3.3-6)$$

holds, every physical state $|\Phi\rangle \in \mathcal{V}_{\text{phys}}$ satisfies Eq.(2) with the global charge Q^a of the unbroken global symmetry:

$$\hat{Q}^a|\hat{\Phi}\rangle = 0. \qquad (4.3.3-7)$$

Remark: Note that, by Eq.(4.3.1-34) of Lemma 4.3.1-5, Eq.(6) is equivalent to

$$\mathcal{V}_{\text{phys}} = \overline{\mathcal{A}(\mathcal{O})|0\rangle}^{\tau} (= \mathcal{A}(\mathcal{O})|0\rangle^{\perp\perp}), \qquad (4.3.3-8)$$

from which the Reeh-Schlieder property in H_{phys} with respect to the algebra, $\hat{\mathcal{A}}(\mathcal{O})$, of smeared local observables follows:

$$H_{\text{phys}} = \overline{\hat{\mathcal{A}}(\mathcal{O})|\hat{0}\rangle}. \qquad (4.3.3-9)$$

The conclusion, Eq.(2), of Proposition 4.3.3-2 can really be obtained from Eq.(9) which is *weaker* than Eq.(6):

Proof of Eq.(9) ⇒ Eq.(2): By the assumption, Eq.(9), for any given $|\Phi\rangle, |\Psi\rangle \in \mathcal{V}_{\text{phys}}$ [11] and $\epsilon > 0$, there exists $\hat{\varphi} \in \hat{\mathcal{A}}(\mathcal{O})$ such that

$$|\langle\hat{\Psi}|\hat{Q}^a(|\hat{\Phi}\rangle - \hat{\varphi}|\hat{0}\rangle)\rangle| < \epsilon. \qquad (4.3.3-10)$$

Since $\hat{Q}^a\hat{\varphi}|\hat{0}\rangle = [\hat{Q}^a, \hat{\varphi}]|\hat{0}\rangle = 0$ by Theorem 4.3.2-1, we obtain

$$|\langle\hat{\Psi}|\hat{Q}^a|\hat{\Phi}\rangle| < \epsilon \qquad \text{for } \forall\epsilon > 0, \qquad (4.3.3-11)$$

which says

$$\langle\hat{\Psi}|\hat{Q}^a|\hat{\Phi}\rangle = 0 \qquad \text{for } \forall|\hat{\Phi}\rangle, |\hat{\Psi}\rangle \in H_{\text{phys}}, \qquad (4.3.3-12)$$

or equivalently Eq.(2). (q.e.d.)

From this result, we know that the Reeh-Schlieder property, Eq.(9), in H_{phys} with respect to the smeared local observable algebra should not hold in the case of QED to secure the existence of the electron in this world. In this connection, it

---

[11]   Precisely speaking, $|\hat{\Psi}\rangle$ should belong to the domain, $\text{Dom}(\hat{Q}^a{}^\dagger)$, of $\hat{Q}^a{}^\dagger$.

may be instructive to remark the role of the Reeh-Schlieder property in the proof of Lemma 4.3.2-2. In fact, if this property in H_{phys} with respect to *any smeared local* field algebra held for QED, then the Maxwell equation, Eq.(4.3.2-3), in H_{phys} would lead us again to the absurd conclusion:

$$\text{electric charge } \hat{Q} = 0 \quad \text{in } H_{\mathrm{phys}}, \qquad (4.3.3 - 13)$$

according to Lemma 4.3.2-2, because the electromagnetic $U(1)$ symmetry should not suffer from spontaneous breakdown. Thus, in order to get rid of this pitfall, any type of the Reeh-Schlieder property with respect to smeared local algebra should be invalidated in H_{phys} of QED. On the contrary, this property, Eq.(9) is desirable in QCD. If it holds, the physical world H_{phys} of hadrons is described according to the principles of the ordinary local quantum field theory completely in terms of the color singlet smeared local observable fields $\mathcal{A}(\mathcal{O})$ identified with the *hadron* fields.

4.3.4 Confining quark-antiquark potential and cluster property

In the intuitive picture of the quark confinement, quark q and antiquark $\bar{q}$ should be intermediated by a string-like object which produces a q-$\bar{q}$ potential not decreasing at infinity to confine quarks inside the hadron. This means the failure of the **cluster property**, while it was proved in [Ara 62] that the cluster property should hold in a Lorentz covariant *local* field theory with a *unique vacuum*. From these circumstances, one might conclude that quark confinement contradicts the usual framework of the local field theory. In such quantum field theory with indefinite inner product as the present case, however, this is not the case as was pointed out by Strocchi [Str 76]. This can be understood in the following generalization to the indefinite metric case of the cluster property theorem of Araki, Hepp and Ruelle [Ara 62].

<u>Theorem 4.3.4 − 1</u> [Str 76] On the assumptions of

(i) covariance under translations,

(ii) local commutativity,

(iii) uniqueness of the vacuum

and

(iv) the spectrum condition

$$\begin{cases} \text{a)} & \text{with a mass gap } (0, \ M) \\ \text{or} \\ \text{b)} & \text{without mass gap,} \end{cases}$$

one obtains an inequality:

$$
|\langle 0|B_1(x_1)B_2(x_2)|0\rangle - \langle 0|B_1(x_1)|0\rangle\langle 0|B_2(x_2)|0\rangle|
$$
$$
\leq \begin{cases} C[\xi]^{-3/2}\exp(-M[\xi])[\xi]^{2N}(1 + |\xi^0|/[\xi]), & \cdots \text{a)} \\ \text{or} \\ C'[\xi]^{-2}[\xi]^{2N}(1 + |\xi^0|/[\xi]^2). & \cdots \text{b)} \end{cases} \qquad (4.3.4-1)
$$

Here

$$
B_\iota(x_i) \equiv \int d^4x_1{}' \cdots d^4x_{r(\iota)}{}'f_\iota(x_1{}', \cdots, x_{r(\iota)}{}')\Phi_{I_1}(x_1{}' + x_\iota) \cdots \Phi_{I_{r(\iota)}}(x_{r(\iota)}{}' + x_\iota),
$$

$f_\iota \in \mathcal{D}(\mathbf{R}^{4r(\iota)})$ $[C^\infty$-functions with compact supports],

$\xi \equiv x_1 - x_2;$

$\Phi_I(x)$'s are generic fields and N is a suitable non-negative integer depending on the B_ι's. The precise definition of $[\xi]$ is rather complicated (see [Ara 62]), but it is enough here to note that the large distance behavior of $[\xi]$ can essentially be represented by $|\xi|$. If the Fourier transform $\tilde{h}_{12}(p)$ of

$$
h_{12}(\xi) \equiv \langle 0|B_1(x_1)B_2(x_2)|0\rangle - \langle 0|B_1(x_1)|0\rangle\langle 0|B_2(x_2)|0\rangle \qquad (4.3.4-2)
$$

is a measure, as in the case with a positive metric [Jos 62], the integer N is 0 and thus the cluster property holds for both cases a) and b).

Proof is given in [Str 78].

The above theorem tells us that the cluster property *may fail* without contradicting the usual axioms of quantum field theory only in the case *with an indefinite metric and without mass gap* $[N \neq 0$ in the case b) of Eq.(1)]. Here note that the mass spectrum in the assumption (iv) refers to the one in the *whole* state vector space $\mathcal{V}$ and that the physical spectrum in H_{phys} may have a mass gap also in the case b), as is expected in the case of quark confinement in QCD as well as in the cases with the Higgs mechanism discussed in Sec.4.1.4 (see also [Nak 73a, Str 77]).

To illustrate Theorem 4.3.4-1, we consider the case of the gauge field A_μ and its field strength $F_{\mu\nu}$ in QED. Setting

$$
\begin{aligned}
&\langle 0|A_\mu(x)A_\nu(y)|0\rangle \\
&= \langle 0|A_\mu(x)A_\nu(y)|0\rangle - \langle 0|A_\mu(x)|0\rangle\langle 0|A_\nu(y)|0\rangle \\
&\equiv \eta_{\mu\nu}F(x-y) + \partial_\mu\partial_\nu G(x-y),
\end{aligned}
\qquad (4.3.4-3)
$$

we obtain

$$
\begin{aligned}
&\langle 0|F_{\mu\nu}(x)F_{\rho\sigma}(y)|0\rangle \\
&= -(\eta_{\mu\rho}\partial_\nu\partial_\sigma - \eta_{\nu\rho}\partial_\mu\partial_\sigma - \eta_{\mu\sigma}\partial_\nu\partial_\rho + \eta_{\nu\sigma}\partial_\mu\partial_\rho)F(x-y).
\end{aligned}
\quad (4.3.4-4)
$$

Since $F_{\mu\nu}$ in QED is an *observable*, we should have

$$
\langle 0|F[f]^\dagger F[f]|0\rangle \geq 0, \qquad (4.3.4-5)
$$

where

$$
F[f] \equiv \int d^4x\, F_{\mu\nu}(x)f^{\mu\nu}(x) \ \text{ with } f^{\mu\nu} \in \mathcal{S}(\mathbf{R}^4). \qquad (4.3.4-6)
$$

Thus, the Fourier transform of $\langle 0|F_{\mu\nu}(x)F_{\rho\sigma}(y)|0\rangle$ is a measure and, by b) of Eq.(1) with $N = 0$, $F(\xi)$ damps at infinity at least as $\sim [\xi]^{-2}$. (Note that $G(x-y)$ in Eq.(3) is an unphysical gauge part.)

On the contrary, $F_{\mu\nu}{}^a$ in QCD is *not* an observable as has been shown in Eq.(4.3.2-11), and such a simple relation between Eq.(3) and Eq.(4) does not hold. Therefore, we can have no restrictions on the function $F^{ab}(x-y)$ in $\langle 0|A_\mu{}^a(x)A_\nu{}^b(y)|0\rangle$ corresponding to $F(x-y)$ in Eq.(3). In the case with $N \geq 1$ for b) of Eq.(1), the q-$\bar q$ potential due to $\langle 0|A_\mu{}^a(x)A_\nu{}^b(y)|0\rangle$ may not decrease at infinity (i.e. confining q-$\bar q$ potential) and then, it could happen that the gluon fields $A_\mu{}^a$ have no asymptotic field (gluon confinement). The following proposition [Oji 80b] suggests more convincingly the failure of the cluster property for unobservable quantities.

Proposition 4.3.4 − 2 Let $\mathcal{R}_1$ and $\mathcal{R}_2$ be smeared local operators belonging, respectively, to $\mathcal{F}(\mathcal{O}_1)$ and $\mathcal{F}(\mathcal{O}_2)$ satisfying the following two conditions:

(i) $\mathcal{R}_1(x)\mathcal{R}_2(y)$ is an observable for any $x, y \in \mathbf{R}^4$, where $\mathcal{R}_,(x) \equiv U(x)\mathcal{R}_,U(x)^{-1}$ with the translation operator $U(x)$,

(ii) $\mathcal{R}_1$ and $\mathcal{R}_2$ satisfy the cluster property in the sense that

$$\lim_{-x^2 \to \infty} |\langle 0|\varphi_1 \mathcal{R}_1 U(x)\mathcal{R}_2 \varphi_2|0\rangle - \langle 0|\varphi_1 \mathcal{R}_1|0\rangle\langle 0|\mathcal{R}_2 \varphi_2|0\rangle| = 0 \qquad (4.3.4-7)$$

holds for any[12] smeared local operators φ_1 and φ_2, then, both $\mathcal{R}_1$ and $\mathcal{R}_2$ are smeared local *observables*.

Proof: Since $\mathcal{R}_1(x)\mathcal{R}_2(y) \in \mathcal{F}((\mathcal{O}_1 + x) \cup (\mathcal{O}_2 + y))$ is a smeared local observable according to (i), Proposition 4.3.1-4 tells us the equality

$$[Q_B, \ \mathcal{R}_1(x)\mathcal{R}_2(y)]_{\mp} = 0 \qquad \text{for } \forall x, y \in \mathbf{R}^4, \qquad (4.3.4-8)$$

which is equivalent to

$$[Q_B, \ \mathcal{R}_1 U(x)\mathcal{R}_2]_{\mp} = 0 \qquad \text{for } \forall x \in \mathbf{R}^4, \qquad (4.3.4-9)$$

because of $[Q_B, \ U(x)] = 0$. Since $[Q_B, \ \varphi_i]_{\mp}$ for *any* smeared local operators φ_i, is also a smeared local operator, we obtain, from Eqs.(9), (7) and the subsidiary condition Eq.(4.1.1-7),

$$\begin{aligned}
0 &= \lim_{-x^2 \to \infty} \langle 0|\varphi_1 [Q_B, \ \mathcal{R}_1 U(x)\mathcal{R}_2]_{\mp} Q_B \varphi_2|0\rangle \\
&= \lim_{-x^2 \to \infty} \langle 0|\varphi_1 Q_B \mathcal{R}_1 U(x)\mathcal{R}_2 Q_B \varphi_2|0\rangle \\
&= \lim_{-x^2 \to \infty} \langle 0|[\varphi_1, \ Q_B]_{\mp} \mathcal{R}_1 U(x)\mathcal{R}_2 [Q_B, \ \varphi_2]_{\mp}|0\rangle \\
&= \langle 0|[\varphi_1, \ Q_B]_{\mp} \mathcal{R}_1|0\rangle\langle 0|\mathcal{R}_2 [Q_B, \ \varphi_2]_{\mp}|0\rangle \\
&= \langle 0|\varphi_1 Q_B \mathcal{R}_1|0\rangle\langle 0|\mathcal{R}_2 Q_B \varphi_2|0\rangle. \qquad (4.3.4-10)
\end{aligned}$$

Hence, at least one of the following equations holds:

$$\langle 0|\varphi_1 Q_B \mathcal{R}_1|0\rangle = 0 \qquad \text{for } \forall \varphi_1 \in \mathcal{F}(\mathcal{O}_1), \qquad (4.3.4-11)$$

or

$$\langle 0|\mathcal{R}_2 Q_B \varphi_2|0\rangle = 0, \qquad \text{for } \forall \varphi_2 \in \mathcal{F}(\mathcal{O}_2). \qquad (4.3.4-12)$$

In the case of Eq.(11), we obtain

$$Q_B \mathcal{R}_1|0\rangle = 0, \qquad (4.3.4-13)$$

---

[12]   It is, in fact, sufficient that Eq.(7) holds *only* for smeared local operators φ_1, φ_2 of the form $\varphi_i = [Q_B, \ \varphi_i']_{\mp}$, as is seen from Eq.(10) in the proof.

because the (anti)local operator annihilating the vacuum is vanishing in itself [Corollary $A.2$-6 in Appendix $A.2$]:

$$(\mathcal{F}(\mathcal{O})|0\rangle)^{\perp} = (\mathcal{F}|0\rangle)^{\perp} = \mathcal{V}^{\perp} = 0. \qquad (4.3.4-14)$$

By Lemma 4.3.1-5, Eq.(13) implies

$$[Q_B, \mathcal{R}_1]_{\mp} = 0, \qquad (4.3.4-15)$$

which yields, by the help of Eq.(8),

$$0 = [Q_B, \mathcal{R}_1]_{\mp}\mathcal{R}_2 \pm \mathcal{R}_1[Q_B, \mathcal{R}_2]_{\mp} = \pm\mathcal{R}_1[Q_B, \mathcal{R}_2]_{\mp}. \qquad (4.3.4-16)$$

Thus we have[13]

$$[Q_B, \mathcal{R}_1]_{\mp} = [Q_B, \mathcal{R}_2]_{\mp} = 0. \qquad (4.3.4-17)$$

The same arguments apply also to the case Eq.(12). (q.e.d.)

From this Proposition 4.3.4-2, we know that, if smeared local operators $\mathcal{R}_1$ and $\mathcal{R}_2$, at least one of which is *not* an observable, define an *observable* of the form $\mathcal{R}_1(x)\mathcal{R}_2(y)$, then the cluster property Eq.(7) for $\mathcal{R}_1$ and $\mathcal{R}_2$ is really broken down. The failure of the cluster property means that the correlation between $\mathcal{R}_1(x)$ and $\mathcal{R}_2(y)$ is not damped, however far they may be separated ($-(x-y)^2 \to +\infty$). This implies that it is impossible for us to detect the quanta generated by $\mathcal{R}_1$ and $\mathcal{R}_2$ separately.

Thus, although the physical states are specified by the subsidiary condition, $Q_B|\text{phys}\rangle = 0$, of non-local nature, in terms of the volume-integrated charge Q_B, such a kind of difficulties as the **"behind-the-moon" problem** [Haa 64, Sha 74] does not occur in the present formulation: Namely, we *need not* worry about the possibility of such a state that it is physical as a whole whereas it can be divided into widely separated unphysical two subsystems, like an FP ghost on the earth and an anti-FP ghost behind the moon. This is because, if such separation can be performed satisfying the cluster property, they should, by the above Proposition,

[13]   Precisely speaking, we should assume that, for any spacelikely separated open domains $\mathcal{O}_1$ and $\mathcal{O}_2$, the product, $\varphi_1\varphi_2$, of any non-vanishing local operators, φ_i, belonging respectively to $\mathcal{F}(\mathcal{O}_i)$ with $i = 1, 2$, does not vanish. Although there is no proof of this statement in the indefinite-metric case, it has been proved in the positive metric case [Schl 69].

be physical in themselves from the beginning. On the other hand, if those subsystems are unphysical, then the failure of the cluster property due to Proposition 4.3.4-2 prevents us from performing the detection of the FP ghost on the earth in a manner independent of the anti-FP ghost behind the moon. Physically this means nothing but the impossibility of the detection of this FP ghost. In this way, the failure of the cluster property in the unphysical world operates to protect unphysical particles from being brought to light. On the other hand, the cluster property for (smeared) local observables is ensured in the physical space $H_{\mathrm{phys}} = \overline{\mathcal{V}_{\mathrm{phys}}/\mathcal{V}_0}$ with the positive definite inner product, where the system of smeared local observable algebras $\hat{\mathcal{A}}(\mathcal{O})$ will satisfy the usual **Araki-Haag-Kastler axioms** [Haa 64, Ara 69].[14] In this situation, if (the representation of) the algebra $\hat{\mathcal{A}}$ of quasi-local observables is *reducible* in H_{phys}, the superselection rule holds [Haa 64, Ara 69, Str 74] and group non-invariant physical states may appear, as expected for the case of QED. On the contrary, the condition, Eq.(4.3.3-9): $H_{\mathrm{phys}} = \overline{\hat{\mathcal{A}}(\mathcal{O})|\hat{0}\rangle}$, from which the color confinement $\hat{Q}^a = 0$ follows (Proposition 4.3.3-2), is equivalent to the condition of the *irreducibility* of (the representation of) $\hat{\mathcal{A}}$ in H_{phys}. In any case, if only we can show the color confinement: $\hat{Q}^a = 0$ in H_{phys}, then the above Proposition 4.3.4-2 guarantees that everything goes well about the quark confinement: The failure of the cluster property in the unphysical colored sectors prevents colored objects from coming out of colorless hadrons to be detected, permitting only the color singlets to appear in the physical world H_{phys}. In this colorless physical world H_{phys}, the validity of the cluster property enables us to perform physical measurements on the earth without worrying about the situations behind the moon. In the next section, we will discuss the problem of color confinement, contrasting it with Higgs mechanism.

---

[14]   Precisely speaking, the Araki-Haag-Kastler axioms are formulated in terms of the *bounded* operators, which should be obtained, in a certain canonical manner, from the unbounded ones treated here. "Quasi-local observables" are defined, in the former version, as the operator-norm limits of smeared local observables.

4.4 STRUCTURE OF GLOBAL GAUGE CURRENTS

On the basis of quantum Maxwell equation, we clarify the charateristic structure of currents and charges of global gauge symmetry. Combining it with the Goldstone theorem relating the spontaneous symmetry breaking with the appearance of massless spectrum, we obtain a converse statement of the Higgs mechanism in the sense that the existence of the massive vector bosons in gauge fields implies the spontaneous breakdown of the corresponding global gauge symmetry except for a certain special case. This special case of exception is found to be relevant to realizing the color confinement.

4.4.1 Mass of gauge bosons and Higgs mechanism

We first note that the quantum Maxwell equation, Eq.(4.3.2-7), i.e.

$$g J_\mu{}^a = \partial^\nu F_{\mu\nu}{}^a + \{Q_B, \, D_\mu \overline{C}^a\}, \qquad (4.4.1-1)$$

can be viewed as an equation giving a decomposition of the conserved current, $g J_\mu{}^a$, of the global gauge symmetry into two *conserved currents* given by

$$\mathcal{G}_\mu{}^a \equiv \partial^\nu F_{\mu\nu}{}^a \qquad (4.4.1-2)$$

and

$$\mathcal{N}_\mu{}^a \equiv \{Q_B, \, D_\mu \overline{C}^a\}. \qquad (4.4.1-3)$$

Because of the conservation law,

$$\partial^\mu \mathcal{G}_\mu{}^a = \partial^\mu \mathcal{N}_\mu{}^a = 0, \qquad (4.4.1-4)$$

these currents yield (formally) conserved charges G^a and N^a,

$$G^a \equiv \int d\mathbf{x} \, \mathcal{G}_0{}^a = \int d\mathbf{x} \, \partial^i F_{0,}{}^a, \qquad (4.4.1-5)$$

$$N^a \equiv \int d\mathbf{x} \, \mathcal{N}_0{}^a = \int d\mathbf{x} \, \{Q_B, \, (D_0 \overline{C})^a\}, \qquad (4.4.1-6)$$

and hence, the global charge, Q^a, of the global gauge symmetry is also given as the sum of these two conserved charges:

$$g Q^a = \int d\mathbf{x} \, g J_0{}^a = G^a + N^a. \qquad (4.4.1-7)$$

It is remarkable that the first term, G^a, is the spatial integral of d-coboundary term, $d * (F_{\mu\nu} dx^\mu \wedge dx^\nu)$, with respect to the spacetime exterior differential d, while the second term, N^a, is the spatial integral of BRS-coboundary term, $-i \int \boldsymbol{\delta}(*(D_\mu \overline{C} dx^\mu))$, where $*$ denotes the Hodge star [see Appendix $A.1$]. These characteristic features are important in revealing some interesting aspects of the global gauge symmetry with the charge Q^a, in the light of the Goldstone theorem [see Appendix $A.3$]. The various versions of this theorem, Theorem $A.3$-1, Corollaries $A.3$-2 and $A.3$-3 in Appendix $A.3$, assert the equivalence of the following conditions on a conserved current J_μ and its global charge Q:

(a) $Q \equiv \int d\mathbf{x} J_0$ does *not* suffer from spontaneous symmetry breakdown;

(b) J_μ contains *no discrete massless spectrum*: $\langle 0|J_\mu|\Psi(p^2 = 0)\rangle = 0$.

Combining this statement with Lemma 4.3.2-7, we obtain the following lemma.

Lemma $4.4.1 - 1$ [Kug unp.b, Kug 79d]   If a linear combination, $\mathcal{G}_\mu{}^A \equiv \alpha^{Aa} \mathcal{G}_\mu{}^a = \alpha^{Aa} \partial^\nu F_{\mu\nu}{}^a$, of $\mathcal{G}_\mu{}^a$'s with some coefficients α^{Aa} contains no discrete massless spectrum,

$$\langle 0|\mathcal{G}_\mu{}^A|\Psi(p^2 = 0)\rangle = 0, \qquad (4.4.1 - 8)$$

or equivalently, if the global charge G^A given by

$$G^A \equiv \int d\mathbf{x} \mathcal{G}_0{}^A = \alpha^{Aa} G^a \qquad (4.4.1 - 9)$$

generates an unbroken symmetry transformation, then G^A vanishes:

$$G^A = 0. \qquad (4.4.1 - 10)$$

Now, we utilize the properties of the massless asymptotic fields constituting the "elementary quartet", namely, the asymptotic conditions shown by Eqs.(4.1.3-22)-(4.1.3-25), their (anti)commutation relations, Eqs.(4.1.3-27)-(4.1.3-28), and their BRS transformation property, Eq.(4.1.3-26). First, note that $\chi^a(x)$ field in $A^{as}{}_\mu{}^a(x)$ makes no contribution to $F^{as}{}_{\mu\nu}{}^a(x)$ owing to the antisymmetry of its indices μ and ν, and that the contributions to $F^{as}{}_{\mu\nu}{}^a(x)$ come from the (massless or massive) genuine *vector* fields $U_\mu{}^A$ which are assumed to be contained in $A_\mu{}^a$ with a weight $\tilde{\alpha}^{aA}$

$$A^{as}{}_\mu{}^a(x) = \partial_\mu \chi^a(x) + \tilde{\alpha}^{aA} U_\mu{}^A(x) + \cdots. \qquad (4.4.1 - 11)$$

The coefficients $\tilde{\alpha}^{aA}$ should properly be taken into account, in the cases with particle mixing, so that the $U_\mu{}^A$'s represent particles with *definite masses*. We suppose that, for the eigenchannel of $U_\mu{}^A$, $F^{\mathrm{as}}{}_{\mu\nu}{}^a(x)$ is given, with some coefficients α^{Aa}, by[1]

$$F^{\mathrm{as}}{}_{\mu\nu}{}^A(x) = \alpha^{Aa} F^{\mathrm{as}}{}_{\mu\nu}{}^a(x) = \partial_\mu U_\nu{}^A(x) - \partial_\nu U_\mu{}^A(x) + \cdots. \qquad (4.4.1-12)$$

Then, from this and Lemma 4.4.1-1, we obtain

<u>Corollary 4.4.1 − 2</u>   If $U_\mu{}^A$ is massive, then $G^A = 0$.

Next, in order to examine the charges N^a, we have to introduce dynamical parameters $u^a{}_b$, defined as the pole residues of $g(A_\mu \times C)^a$ at $p^2 = 0$:

$$g(A_\mu \times C)^{\mathrm{as}\,a}(x) = u^{ab} \partial_\mu \overline{\gamma}^b(x) + \cdots, \qquad (4.4.1-13)$$

which can be estimated by the formula following from Eqs.(4.1.3-24) and (4.1.3-27):

$$\int d^4x\, e^{ip\cdot(x-y)} \langle 0|\mathrm{T}\left[(D_\mu C)^b(x) g(A_\nu \times \overline{C})^a(y)\right]|0\rangle$$
$$= -u^{ab} p_\mu p_\nu / p^2 + \cdots. \qquad (4.4.1-14)$$

Then, we obtain[2]

$$(D_\mu \overline{C})^{\mathrm{as}\,a}(x) = (\delta^{ab} + u^{ab})\partial_\mu \overline{\gamma}^b(x) + \cdots. \qquad (4.4.1-15)$$

<u>Lemma 4.4.1 − 3</u> [Kug unp.b, Kug 79d]   The symmetry transformation generated by the conserved charge of the form $\xi^a N^a = \int d\mathbf{x}\, \xi^a \mathcal{N}_0{}^a$ is *not* broken spontaneously, if and only if ξ^a satisfies

$$(\delta^{ab} + u^{ab})\xi^a = 0 \qquad \text{for all } b = 1, \cdots, n, \qquad (4.4.1-16)$$

---

[1]   Owing to the contributions from the term $gA_\mu \times A_\nu$, the matrix α^{Aa} is, in general, neither the inverse of $\tilde{\alpha}^{aA}$ nor unitary.

[2]   The dots $(\cdots)$ here represent the asymptotic fields decoupled from the field $(D_\mu C)^a$, among which there *may* exist still some *massless* components different from $\overline{\gamma}^a$

or equivalently

$$(\mathbf{1} + u^T) \begin{pmatrix} \xi^1 \\ \vdots \\ \xi^n \end{pmatrix} = 0. \qquad (4.4.1-17)$$

Proof: A necessary and sufficient condition for the conserved current $\xi^a \mathcal{N}^a$ to generate an unbroken symmetry transformation is given by

$$\langle 0 | \xi^a \mathcal{N}_\mu{}^a(x) | \Psi(p^2 = 0) \rangle = 0 \qquad (4.4.1-18)$$

for any massless 1-particle state $|\Psi(p^2 = 0)\rangle$. Using the relation

$$\langle 0 | \Phi(x) | 1\text{-particle} \rangle = \langle 0 | \Phi^{as}(x) | 1\text{-particle} \rangle \qquad (4.4.1-19)$$

which follows from the Yang-Feldman equation, Eq.(1.4.2-2), we can obtain, from Eqs.(3) and (15),

$$\begin{aligned}
\langle 0 | \xi^a \mathcal{N}_\mu{}^a(\dot{x}) | \Psi(p^2 = 0) \rangle &= \xi^a \langle 0 | \{ Q_B, \ (D_\mu \overline{C})^a(x) \} | \Psi(p^2 = 0) \rangle \\
&= \xi^a \langle 0 | (D_\mu \overline{C})^a(x) Q_B | \Psi(p^2 = 0) \rangle \\
&= \xi^a \langle 0 | (D_\mu \overline{C})^{asa}(x) Q_B | \Psi(p^2 = 0) \rangle \\
&= \xi^a (\delta^{ac} + u^{ac}) \partial_\mu \langle 0 | \overline{\gamma}^c(x) Q_B | \Psi(p^2 = 0) \rangle.
\end{aligned} \qquad (4.4.1-20)$$

Since the only non-vanishing matrix element $\langle 0 | \overline{\gamma}^c(x) Q_B | \Psi(p^2 = 0) \rangle$ in Eq.(20) is given by taking $|\Psi(p^2 = 0)\rangle = \chi^b(y)|0\rangle$, with the help of Eqs.(4.1.3-26) and (4.1.3-27), we obtain

$$\begin{aligned}
\langle 0 | \overline{\gamma}^c(x) Q_B (\chi^b(y)|0\rangle) &= \langle 0 | \overline{\gamma}^c(x) [Q_B, \chi^b(y)] | 0 \rangle \\
&= \langle 0 | \overline{\gamma}^c(x)(-i\gamma^b(y)) | 0 \rangle \\
&= -\delta^{cb} D^{(+)}(x - y);
\end{aligned} \qquad (4.4.1-21)$$

Eq.(18) in question turns out to be equivalent to Eq.(16), as is easily seen from Eq.(20):

$$\langle 0 | \xi^a \mathcal{N}_\mu{}^a(x) \chi^b(y) | 0 \rangle = -(\delta^{ab} + u^{ab}) \xi^a \partial_\mu D^{(+)}(x - y). \qquad (4.4.1-22)$$

(q.e.d.)

Corollary $4.4.1 - 4$ [Kug unp.b, Kug 79d]   If $\det(1 + u) \neq 0$, every charge $\xi^a N^a$ is a charge suffering from spontaneous breakdown, except for the trivial case with $\xi^a \equiv 0$. The NG boson responsible for the spontaneous breakdown of N^a is given by χ^b with indices b such that $\delta^{ab} + u^{ab} \neq 0$.

Corollary $4.4.1 - 5$ [Kug unp.b, Kug 79d]   If $u = -\mathbf{1}$, the symmetry transformation generated by $\xi^a N^a$ with arbitrary coefficients ξ^a is not broken spontaneously.

Remark: In the abelian case where the structure constant $f^{abc} \equiv 0$, the term $(g A_\mu \times \overline{C})^a = g f^{abc} A_\mu{}^b \overline{C}^c$ vanishes from the beginning, and hence, u^{ab} in Eq.(13) is identically zero. Thus, the condition of Corollary 4.4.1-4 is always satisfied: $\det(\mathbf{1} + u) = \det \mathbf{1} = 1 \neq 0$, and the charge $N = \int d\mathbf{x}\{Q_B, D_0\overline{C}\} = \int d\mathbf{x}\partial_0 B$ suffers from the spontaneous breakdown. In the case $\det(\mathbf{1} + u) \neq 0$, we obtain the following consequence:

Theorem $4.4.1 - 6$ (Converse of the Higgs theorem) [Kug unp.b, Kug 79d] If $\det(\mathbf{1} + u) \neq 0$, the global gauge symmetries corresponding to the charges $Q^A = \alpha^{Aa} Q^a$ are *broken spontaneously* in *each* eigenchannel of the *massive* gauge bosons $U_\mu{}^a$.

Proof: From Corollary 4.4.1-2, $G^A = 0$ follows for each massive $U_\mu{}^A$, and hence, we obtain

$$gQ^A = G^A + N^A = N^A = \alpha^{Aa}N^a. \qquad (4.4.1 - 23)$$

Since $(\alpha^{A1}, \cdots, \alpha^{An}) \neq 0$ for the very existence of the massive gauge boson $U_\mu{}^A$, Corollary 4.4.1-4 asserts the spontaneous breakdown of the charge Q^A:

$$\begin{aligned} \langle 0|g J_\mu{}^A(x)\chi^b(y)|0\rangle &= \langle 0|\mathcal{N}_\mu{}^A(x)\chi^b(y)|0\rangle \\ &= -\alpha^{Aa}(\delta^{ab} + u^{ab})\partial_\mu D^{(+)}(x - y) \neq 0. \end{aligned} \qquad (4.4.1 - 24)$$

$$(q.e.d.)$$

Here, a few remarks are in order:

Remarks.   i) If $\hat{\alpha} = (\hat{\alpha}^{Ab}) = (\alpha^{Aa}(\delta^{ab} + u^{ab})$ $(A, b = 1, \cdots, n)$ is an invertible (n, n)-matrix [where n is the dimension of the Lie algebra of the gauge group G], we can identify the NG boson responsible for the spontaneous breakdown of Q^A with χ^A defined by

$$\chi^A \equiv (\hat{\alpha}^{-1})^{Aa}\chi^a. \qquad (4.4.1 - 25)$$

In fact, it satisfies

$$\langle 0|gJ_\mu{}^A(x)\chi^B(y)|0\rangle = -\delta^{AB}\partial_\mu D^{(+)}(x-y) \qquad (4.4.1-26)$$

for any channel B and the massive-gauge-boson channel A. In the example of the Higgs-Kibble model discussed in Sec.4.2.1, we already know the explicit form of symmetry breaking:

$$A^{\mathrm{as}}{}_\mu{}^a = \sqrt{K}\partial_\mu\chi^{\mathrm{as}a} + (\sqrt{K}-\alpha N)\partial_\mu B^{\mathrm{as}a} + U^{\mathrm{as}}{}_\mu{}^a, \qquad (4.4.1-27)$$

$$\alpha^{Aa} \propto \tilde\alpha^{Aa} \propto \hat\alpha^{Aa} \propto \delta^{Aa}, \qquad (4.4.1-28)$$

where the $\chi^{\mathrm{as}a}$'s are the asymptotic fields of the "elementary" unphysical NG bosons. Eq.(28) is a consequence of another global $SU(2)$ symmetry remaining after the symmetry breaking. Eq.(27) agrees with Eq.(11), in the general context, by the following identification as is noted in Eq.(4.2.1-76):

$$\chi^a \to \sqrt{Z_3}[\sqrt{K}\chi^{\mathrm{as}a} + (\sqrt{K}-\alpha_r N)B^{\mathrm{as}a}] \qquad (4.4.1-29)$$

since the mixture of $B^{\mathrm{as}a}$ components in Eq.(29) causes no change in Eqs.(4.1.3-26) and (4.1.3-27) by virtue of $[Q_B, B^{\mathrm{as}a}] = [B^{\mathrm{as}a}, B^{\mathrm{as}b}] = 0$.

ii) Note that, in the above Theorem 4.4.1-6, the spontaneous breakdown of a charge Q^A is *in one-to-one correspondence* with the appearance of a *massive* gauge boson $U_\mu{}^A$. This should be compared with the result obtained in [Smi 74, Eic 74, Tye 75]: There, occurrence of the spontaneous breakdown of some global charges has been proved only in the case where the gauge bosons acquire *different masses* within a group multiplet. By such a result, we can neither say anything about the case where the gauge bosons acquire a common mass within the group multiplet, nor even about the abelian case. In contrast to it, the above theorem asserts the spontaneous breakdown of a particular charge Q^A corresponding to each massive gauge boson $U_\mu{}^A$ irrespective of its mass value. It is suitable to comment here on the prevailing misunderstanding about the **"Schwinger mechanism"** by which the gauge bosons are believed to become massive without spontaneous symmetry breaking (see, for instance, [Kog 74]): In fact, the original Schwinger mechanism found in the Schwinger model [Sch 62a, Sch 62b] is nothing but a Higgs mechanism (see Sec.2.5.4). Furthermore, since Theorem 4.4.1-6 is always valid irrespective of the detailed mass generation mechanism, the non-vanishing mass of gauge boson, which is generated by the "Schwinger mechanism" (if any) or by some other ones, necessarily implies the spontaneous breakdown of global symmetries.

Next, we consider the case where the global gauge symmetry of the charge Q^A remains unbroken. According to Corollary $A.3$-3 in Appendix $A.3$, we should have, in this case,

$$0 = \langle 0|gJ_\mu{}^A|\Psi(p^2 = 0)\rangle = \langle 0|\mathcal{G}_\mu{}^A|\Psi(p^2 = 0)\rangle + \langle 0|\mathcal{N}_\mu{}^A|\Psi(p^2 = 0)\rangle \quad (4.4.1-30)$$

or equivalently,

$$\langle 0|\alpha^{Aa}\partial^\nu F_{\mu\nu}{}^a|\Psi(p^2 = 0)\rangle = -\langle 0|\mathcal{N}_\mu{}^A|\Psi(p^2 = 0)\rangle$$
$$= -\alpha^{Aa}\langle 0|(D_\mu\overline{C})^a Q_B|\Psi(p^2 = 0)\rangle. \quad (4.4.1-31)$$

It may happen, however, that Eq.(30) or (31) is not satisfied in spite of the absence of the spontaneous breakdown, as is seen explicitly in the case of QED [Swi 67, Kug 79a]. This is due to the massless contribution from $\alpha^{Aa}\partial_\mu\beta^a$ remaining in $gJ_\mu{}^A$ with some weight ζ. In the case $\det(\mathbf{1} + u) \neq 0$, the problem can be settled [Kug unp.b] by the following modification of the definition of the current and the charge:

$$g\tilde{J}_0{}^A(x) \equiv gJ_0{}^A(x) - \zeta\mathcal{N}_\mu{}^A(x), \quad (4.4.1-32)$$
$$g\tilde{Q}^A \equiv \int d\mathbf{x}\, g\tilde{J}_0{}^A(x) = \int d\mathbf{x}(gJ_0{}^A(x) - \zeta\mathcal{N}_\mu{}^A(x)), \quad (4.4.1-33)$$

where the constant ζ is adjusted so that the massless contribution in $gJ_0{}^A$ is exactly cancelled out by the subtraction of $\zeta\mathcal{N}_0{}^A$. The constant ζ in QED is given by $\zeta = 1 - Z_3$. In this context, the case $\zeta = 1$ corresponds to the spontaneous breakdown, which is equivalent to the appearance of the massive gauge boson $U_\mu{}^A$ as is shown by Theorem 4.4.1-6. Namely, if one wants to extract an unbroken charge annihilating the vacuum by eliminating massless components from the charge gQ^A, then the result is nothing but zero: $gQ^A = \int d\mathbf{x}(gJ_0{}^A - \zeta\mathcal{N}_\mu{}^A) = G^A = 0$, while non-vanishing Q^A requires $\zeta \neq 1$, which makes Q^A suffer from spontaneous breakdown owing to the remaining massless contribution.

Thus, if $\det(\mathbf{1} + u) \neq 0$, the unbroken symmetry should be realized with $\zeta \neq 1$. In this case, $Q^a/(1 - \zeta)$ should be adopted as the generator of the global symmetry, in place of the original Q^a which is spontaneously broken by massless contributions. In fact, one can easily check the following commutation relations [Kug unp.b, Kug 79d]:

$$[g\tilde{Q}^a, \varphi^\alpha(x)] = -g(1-\zeta)(\tau^a)^{\alpha\beta}\varphi^\beta(x), \qquad (4.4.1-34)$$

$$[g\tilde{Q}^a, g\tilde{Q}^b] = ig(1-\zeta)f^{abc}g\tilde{Q}^c. \qquad (4.4.1-35)$$

By this modification, the quantum Maxwell equation, Eq.(3.4.3-14), is merely changed as

$$g\tilde{J}_\mu{}^a/(1-\zeta) = \mathcal{G}_\mu{}^a/(1-\zeta) + \mathcal{N}_\mu{}^a, \qquad (4.4.1-36)$$

which preserves the whole arguments made previously by simply replacing $J_\mu{}^a$ and $\mathcal{G}_\mu{}^a$ by $\tilde{J}_\mu{}^a/(1-\zeta)$ and $\mathcal{G}_\mu{}^a/(1-\zeta)$, respectively.

The condition for the charge $\tilde{Q}^a$ to be unbroken is now given by

$$\langle 0|\partial^\nu F_{\nu\mu}{}^a|\Psi(p^2=0)\rangle/(1-\zeta) = \langle 0|(D_\mu\overline{C})^a Q_B|\Psi(p^2=0)\rangle, \qquad (4.4.1-37)$$

which shows that $F_{\mu\nu}{}^a$ should contain massless component, in order to keep the global gauge symmetry unbroken, as long as $\det(\mathbf{1}+u) \neq 0$ [see Eqs.(18) and (21)]. This can be understood in the following way. According to Theorem 4.4.1-6, every gauge boson $U_\mu{}^A = U_\mu{}^a$ contained in $A_\mu{}^a$ should, in this unbroken case, remain massless and, in fact, we know already in Eq.(4.2.2-20) that the $U_\mu{}^a(= A^{\mathrm{as}}{}_\mu{}^a)$ contains not only the transverse components but also scalar components (certain combination of) β^a.[3] Since $[\beta^a, \chi^b] \neq 0$ by Eq.(4.1.3-27), we obtain

$$[U_\mu{}^a(x), \chi^b(y)] \neq 0, \qquad (4.4.1-38)$$

which explains Eq.(37) for $|\Psi(p^2=0)\rangle = \chi^b(y)|0\rangle$:

$$\langle 0|\partial^\nu F_{\nu\mu}{}^a|\Psi(p^2=0)\rangle \propto \langle 0|\Box U_\mu{}^a - \partial_\mu\partial^\nu U_\nu{}^a|\Psi(p^2=0)\rangle \neq 0. \qquad (4.4.1-39)$$

Thus, the two broken charges $G^a/(1-\zeta)$ and N^a conspire to give an unbroken charge $g\tilde{Q}^a$ in the combination $G^a/(1-\zeta) + N^a$ owing to the cancellation of the massless contributions of β^a. From this, we know that the term $\mathcal{G}_0{}^a = \partial^i F_{0i}{}^a$ in

---

[3]   For massless channel in Eq.(11), precisely speaking, neither $U_\mu{}^A$ is a genuine vector nor $\chi^A \equiv (\tilde{\alpha}^{-1})^{Aa}\chi^a$ is a scalar, and only the sum $a_\mu{}^A \equiv U_\mu{}^A + \partial_\mu\chi^A$ transforms properly as a 4-dimensional vector. In fact, Eq.(4.2.2-20) shows that $\chi^A(x) = \sum_k (a_k{}^L g_k(x)+\text{h.c.})$ is the longitudinal component of $a_\mu{}^A$.

the integrand of $g\tilde{Q}^a$ should not be discarded merely for the reason of its spatial divergence form. The massless particles can contribute on the surface at infinity, making the charge G^a broken spontaneously.

4.4.2 A criterion for color confinement

Now, we discuss the case $\det(\mathbf{1} + u) = 0$, which is possible only in the non-abelian case. From Lemma 4.4.1-3, we obtain

Proposition $4.4.2 - 1$ [Kug unp.b, Kug 79d]   Let $U_\mu{}^A$ be a massive gauge boson. If the coefficients $\hat{\alpha}^{Aa}$ $(a = 1, \cdots, n)$ vanish in this channel A:

$$\hat{\alpha}^{Aa} = \alpha^{Ab}(\delta^{ba} + u^{ba}) = 0 \qquad \text{for } a = 1, \cdots, n, \qquad (4.4.2 - 1)$$

the charge Q^A is *not* broken spontaneously in spite of the massiveness of $U_\mu{}^A$.

Proof: By Corollary 4.4.1-2, $G^A = 0$ follows from the assumption that $U_\mu{}^A$ is massive. Then, Lemma 4.4.1-3 tells us that the charge

$$gQ^A = G^A + N^A = N^A = \alpha^{Aa}N^a \qquad (4.4.2 - 2)$$

generates an unbroken symmetry, if Eq.(1) holds.   (q.e.d.)

In the case of QCD, we can obtain a concise criterion for the color confinement as follows:

Theorem $4.4.2 - 2$ [Kug unp.b, Kug 79d]   If the following two conditions are satisfied in QCD,

(A) $u = -\mathbf{1}$,   namely $u^{ab} = -\delta^{ab}$, $(4.4.2 - 3)$

and

(B) the global color gauge symmetry with charges gQ^a is unbroken,

or equivalently, (A) together with

(B') $\partial^\nu F_{\mu\nu}{}^a$ contains *no massless discrete spectrum*,

then the color confinement is realized in the sense of

$$\hat{Q}^a = 0 \qquad \text{in } H_{\text{phys}}. \qquad (4.4.2 - 4)$$

Proof: By Corollary 4.4.1-5, all the charges N^a are ensured to generate an unbroken symmetry by the condition (A). Then, from the condition (B) and Eq.(4.4.1-7), every charge G^a suffers from no spontaneous symmetry breakdown, and hence, by Lemma 4.4.1-1, it should vanish: $G^a = 0$. Thus, we obtain the equation for the *unbroken charges*

$$gQ^a = N^a = \{Q_B, \int d\mathbf{x}\, (D_0\overline{C})^a(x)\}, \qquad (4.4.2-5)$$

which asserts Eq.(4). Eq.(4.4.1-7) clearly shows the equivalence of the two conditions (B) and (B′) under the condition (A). (q.e.d.)

Corollary 4.4.2 − 3 [Kug unp.b, Kug 79d]   If the condition (A) together with the following one holds,

(B″) all the gauge bosons $A_\mu{}^a$ are massive,

then the color confinement Eq.(4) holds.

Proof: (B″) implies (B′), and hence, the conclusion follows immediately from Theorem 4.4.2-2. (q.e.d.)

In QCD, it is a very important problem whether the global color symmetry remains intact or is broken spontaneously. One usually supposes that this symmetry is not broken spontaneously, because, otherwise, one cannot imagine such simple quark configurations as $q\bar{q}$ and qqq for hadrons, and further, the quark confinement, if any, cannot be assured by the color-singletness of physical states. So, it is natural to assume that the color symmetry is not broken spontaneously.

If we adopt this assumption, we know from Eq.(4.4.1-37) that the *gluons* $U_\mu{}^a$ *should remain massless unless* $\det(\mathbf{1} + u) = 0$. If gluons are massless, their transverse components may not be forbidden to appear as physical particles, *on the mass shell* on which infrared divergences arise inevitably in an uncontrollable manner [Sug 77].[4] Thus, as long as the color symmetry is assumed to be unbroken, the requirement of $\det(\mathbf{1}+u) = 0$ is almost inevitable, which is further strengthened to $\mathbf{1} + u = 0$ in view of the irreducibility of adjoint represenation: The unbroken color symmtery implies that $\mathbf{1} + u$ is a scalar multiple of the unit matrix $\mathbf{1}$ as a matrix commuting with all the matrices giving the adjoint representation of the

---

[4]   In the perturbative approach made in [Sug 77], infrared divergences are shown to survive, in the on-shell renormalization scheme, after the cancellations of those appearing in the off-shell renormalization.

color $SU(3)$, and hence, that $\mathbf{1} + u = 0$ directly follows from $\det(\mathbf{1} + u) = 0$. Thus, we have arrived at the two conditions (A) and (B) of Theorem 4.4.1-6, which lead us to the conclusion of the color confinement.

Here, one may doubt the plausibility of the condition $u = -1$: By small changes of the number and/or interaction type of matter fields, the dynamically determined parameter u would be easily perturbed to shift from -1, even if it was just set on the desired value $-\mathbf{1}$ at the outset. There are, however, some possibilities of its stability against such perturbations as the above, as is seen in the following example [Mas unp]: Suppose that the constant parameter u^{ab} is realized as the value at $p^2 = 0$ of the function $u^{ab}(p^2)$ given by

$$u^{ab}(p^2) = \delta^{ab}\overline{g}^2(p^2)/(r - \overline{g}^2(p^2)), \qquad (4.4.2-6)$$

where $\overline{g}(p^2)$ is the effective coupling at p^2 and r is an unspecified constant. In the weak coupling limit of Eq.(6), it reduces to such a form reasonable in the perturbative sense as

$$u^{ab} \sim \delta^{ab}\overline{g}^2/r. \qquad (4.4.2-7)$$

If the infrared behavior of $\overline{g}(p^2)$ is just in accordance with the one expected in the usual renormalization group approach, i.e., if

$$\overline{g}^2(p^2) \to \infty \quad \text{as} \quad p^2 \to 0, \qquad (4.4.2-8)$$

then Eq.(6) reproduces the desired form Eq.(3), irrespective of the value of the constant r and of the rate of $\overline{g}^2$ approaching infinity. This arbitrariness in the constant r and in the infrared behavior of $\overline{g}^2$ tending to infinity would endow the condition $\mathbf{1} + u = 0$ with a considerable extent of stability. As a matter of course, it is difficult to verify in a non-perturbative sense that such a form as Eq.(6) is actually realized in the standard theory of QCD, but this sort of possibility gives certain reliability to the condition $\mathbf{1} + u = 0$.

It is interesting to note that the same confinement criterion as the above has been formulated in a different way by Nishijima [Nis 82, Nis 85b]: The condition for the absence of massless spectrum in the current $\mathcal{N}_\mu{}^a$ can be related with the condition for the existence of massless bound states in $(C \times C)^a$ and $(\overline{C} \times \overline{C})^a$, etc., by analyzing WT identities involving $\boldsymbol{\delta\overline{\delta}}A_\mu{}^a = \boldsymbol{\delta}(D_\mu\overline{C})^a$ such as

$$\partial^\mu \langle 0|\mathrm{T}\boldsymbol{\delta\overline{\delta}}A_\mu{}^a(x) \cdot q^\alpha(y)\overline{q}^\beta(z)|0\rangle$$

$$= ig(\delta^4(x-y)(\tau^a)^{\alpha\gamma}\langle 0|Tq^\gamma(y)\bar{q}^\beta(z)|0\rangle - \delta^4(x-z)\langle 0|Tq^\alpha(y)\bar{q}^\gamma(z)|0\rangle(\tau^a)^{\gamma\beta}),$$

$$(4.4.2-9)$$

$$\partial^\mu\langle 0|T\boldsymbol{\delta\bar{\delta}}A_\mu{}^a(x)\cdot A_\nu{}^b(y)A_\lambda{}^c(z)|0\rangle$$

$$= g(\delta^4(x-y)f^{abd}\langle 0|TA_\nu{}^d(y)A_\lambda{}^c(z)|0\rangle - \delta^4(x-z)f^{acd}\langle 0|TA_\nu{}^b(y)A_\lambda{}^d(z)|0\rangle).$$

$$(4.4.2-10)$$

In [Nis 86], the restriction on the number of flavors of quarks imposed by the requirement of gluon confinement has also been derived from the renormalization-group-theoretical analysis [Oeh 80a, Oeh 80b] of the asymptotic behavior of the gluon propagator: It should not exceed 9. (This is more stringent than the well-known result ≤ 16 [Gro 73] required by the asymptotic freedom.)

Now, we note that the assumption Eq.(8) crucial for the above example argument of Eq.(6) is the familiar anticipation known as the **"infrared slavery"** [Wei 73, Gro 73]. Since this means a *strong coupling at long distances*, it must have a close connection with the failure of the cluster property discussed in Sec.4.3.4 from the viewpoint of the confining q-$\bar{q}$ potential. For the sake of the failure of the cluster property in $\langle 0|A_\mu{}^a(x)A_\nu{}^b(y)|0\rangle$, it is sufficient that

$$\text{F.T.}(\langle 0|A_\mu{}^a A_\nu{}^b|0\rangle)(p) \sim 1/(p^2)^\alpha \quad \text{with } \alpha \geq 3/2, \qquad (4.4.2-11)$$

and the $\alpha = 2$ case of a massless double pole, $1/(p^2)^2$, corresponds to the "linear potential" case. In view of the condition (B'), of Theorem 4.4.2-2 necessary for protecting the charge $G^a = \int d\boldsymbol{x}\, \partial^i F_{0i}{}^a$ from spontaneous breakdown, we note here the following contrast: The condition (B') requires a *weaker* singularity at $p^2 = 0$ in Eq.(11) (with smaller α, $\alpha < 1$), whereas the *stronger* one (with larger α, $\alpha \geq 3/2$) is more favorable for the failure of the cluster property (and infrared slavery). These two requirements are really *compatible* with each other because the Green's function in Eq.(11) in fact has two independent components, namely, the transverse and longitudinal parts: In this connection, one should first recall that the *isotropic* linear q-$\bar{q}$ potential does *not* realize the *string*-like picture of the confinement which seems to have a directional dependence determined by the q-$\bar{q}$ configuration. Next it may be instructive to refer to the remark made by Frenkel and Taylor [Fre 76] that the components of Yang-Mills field responsible for the peculiar property of asymptotic freedom is only the longitudinal one in Coulomb gauge whereas other transverse ones contribute destructively to the asymptotic freedom similarly to the usual matter fields. Thus, it seems likely that the components responsible for the failure of the cluster property differ from those which could contribute to $\partial^\nu F^{\text{as}}{}_{\mu\nu}{}^a$ or to the charge G^a which should vanish: The former components may be identified with the "longitudinal" one $\chi = A^L$ and the latter would be "transverse", if any. In other words, the failure of the cluster property in the *unphysical* world realizes the condition $u = -\mathbf{1}$ keeping the charge N^a unbroken, while the absence of the infrared tails in the *physical* world protects the charge G^a from being broken spontaneously. Both of them should cooperate to achieve the color confinement. This contrast reminds us of the situation encountered in Sec.4.3.4, where the former operates to confine unphysical particles and the latter guarantees the physical measurement independent of the "behind-the-moon". These observations may give us some clues to the understanding of the naive string picture and of the notion of complete anti-dielectricity of the vacuum.

As for the possibility of (A) and (B″) discussed in Corollary 4.4.2-3, some comments might be necessary. One might suspect that the very existence of massive gluons having colors is contradictory to the conclusion Eq.(4) of color confinement, which is, however, not the case. The colored particles can exist as asymptotic fields, but only in the quartet representations. It is the subject of the next section to discuss this point in some details.

4.4.3 Color confinement from the viewpoint of quartet mechanism

The quartet mechanism, by which the quartet particles essentially decouple from the physical sector as has been discussed in Sec.4.1, can be judged to take place according to a concise criterion. Consider a BRS transformation

$$[iQ_B,\ X(x)] = \mathcal{C}(x), \qquad\qquad (4.4.3-1)$$

where $X(x)$ is an arbitrary Heisenberg field with vanishing FP ghost number ($N_{\text{FP}} = 0$) and $\mathcal{C}(x)$ is its BRS transform having $N_{\text{FP}} = +1$, then the following theorem holds.

Theorem 4.4.3 − 1 [Kug 79c]   If the field operator $\mathcal{C}(x)$ in Eq.(1) has an asymptotic field $\gamma(x)$,

$$\mathcal{C}^{\text{as}}(x) = Z^{1/2}\gamma(x) + \cdots, \qquad\qquad (4.4.3-2)$$

then, the following statements hold:

(i) The field operator $X(x)$ in Eq.(1) also has an asymptotic field, which we denote by $\chi(x)$, and the pair $(\chi(x),\ \gamma(x))$ forms a BRS-doublet:

$$[iQ_B,\ \chi(x)] = \gamma(x). \qquad (4.4.3-3)$$

(ii) There exists another Heisenberg field $\overline{C}(x)$ with the FP ghost number $N_{\mathrm{FP}} = -1$, which has the asymptotic field $\overline{\gamma}(x)$ FP-conjugate to $\gamma(x)$. This $\overline{\gamma}$ also forms another BRS-doublet:

$$\{Q_B,\ \overline{\gamma}(x)\} = \beta(x), \qquad (4.4.3-4)$$

where $\beta(x)$ is supplied as the asymptotic field of the Heisenberg field $\mathcal{B}(x)$ with $N_{\mathrm{FP}} = 0$, defined by

$$\{Q_B,\ \overline{C}(x)\} = \mathcal{B}(x). \qquad (4.4.3-5)$$

(iii) These two BRS-doublets $(\{\chi(x),\ \gamma(x);\ \overline{\gamma}(x),\ \beta(x)\})$ constitute a quartet having a common mass, spin and color indices, and hence these asymptotic fields appear in the physical subspace only in the zero-norm combinations (confinement).

Proof: In view of the BRS transformation, Eq.(1), the existence of asymptotic field γ in Eq.(2) for C necessarily implies the existence of χ also for X which should satisfy Eq.(3), as is explained in Sec.1.4.2. Next, since the assumed asymptotic field γ of C should appear as a pole in a certain two-point Green's function, there should exist (at least one) such Heisenberg field $\overline{C}$ FP-conjugate to C that the propagator $\langle 0|T C(x)\overline{C}(y)|0\rangle$ has a pole at the mass m of γ-field:[5]

$$\mathrm{F.T.}\langle 0|T C(x)\overline{C}(y)|0\rangle = -Z\frac{1}{p^2 - m^2} + \cdots. \qquad (4.4.3-6)$$

This indicates the existence of asymptotic field $\overline{\gamma}$ for $\overline{C}$. Note the following WT identity:

$$0 = \langle 0|\{Q_B,\ T(X(x)\overline{C}(y)\}|0\rangle$$
$$= \langle 0|T X(x)\mathcal{B}(y)|0\rangle - i\langle 0|T C(x)\overline{C}(y)|0\rangle, \qquad (4.4.3-7)$$

where use has been made of Eq.(1), $Q_B|0\rangle = 0$ and the definition, Eq.(5), of $\mathcal{B}$. Eqs.(7) and (6) lead to

$$\mathrm{F.T.}\langle 0|T X(x)\mathcal{B}(y)|0\rangle = -iZ\frac{1}{p^2 - m^2} + \cdots, \qquad (4.4.3-8)$$

---

[5]   Just for simplicity, Eqs.(6), (8) and (10) are explicitly written for the case of scalar fields.

which says the existence of asymptotic field $\beta(x)$ of the field $\mathcal{B}(x)$. Defining these asymptotic fields by

$$X(x) \to Z^{1/2}\chi(x) + \cdots, \quad \mathcal{B}(x) \to Z^{1/2}\beta(x) + \cdots,$$
$$\overline{\mathcal{C}}(x) \to Z^{1/2}\overline{\gamma}(x) + \cdots, \tag{4.4.3 – 9}$$

we can easily see, with the aid of Eqs.(A.4-14) and (A.4-15), that Eq.(4) holds, and can derive

$$[\chi(x),\ \beta(y)] = i\{\gamma(x),\ \overline{\gamma}(y)\} = -i\Delta(x - y; m^2) \tag{4.4.3 – 10}$$

from Eqs.(6) and (8) by virtue of Greenberg-Robinson theorem (Appendix A.4). All other (anti)commutators between χ, β, γ and $\overline{\gamma}$ except for $[\chi(x),\ \chi(y)]$ are found to vanish from the FP ghost number conservation and the nilpotency $Q_B{}^2 = 0$; e.g.,

$$[\beta(x),\ \beta(y)] = \{\gamma(x),\ \gamma(y)\} = \{\overline{\gamma}(x),\ \overline{\gamma}(y)\} = 0. \tag{4.4.3 – 11}$$

$$(\text{q.e.d.})$$

One may notice that the reasoning in the arguments presented here is identical to those utilized in Sec.4.1 in showing the presence of the "elementary" quartet. In fact, the present BRS transformation property, Eqs.(3) and (4), and the structure of (anti)commutation relations, Eqs.(10) and (11), just identical with those [Eqs.(4.1.3-26), (4.1.3-27) and (4.1.3-28)] for the "elementary" quartet, and hence also with those [Eqs.(4.1.3-9)-(4.1.3-12)] for the general quartet.

Theorem 4.4.3 – 2   If the Heisenberg field $\mathcal{C}(x)$ in Eq.(1) has no asymptotic field and if $X(x)$ has its asymptotic field $X^{\text{as}}(x)$, then we have

$$[Q_B,\ X^{\text{as}}(x)] = 0, \tag{4.4.3 – 12}$$

which implies that the particle associated with $X^{\text{as}}(x)$ appears in the physical subspace $\mathcal{V}_{\text{phys}}$, and that $X^{\text{as}}(x)$ becomes
   (i) a BRS-singlet physical particle with positive norm,
or otherwise,
   (ii) a BRS coboundary with zero-norm having its parent field $\overline{\gamma}_X(x)$ with FP-ghost number $N_{\text{FP}} = -1$ such that

$$\{Q_B,\ \overline{\gamma}_X(x)\} = X^{\text{as}}(x). \tag{4.4.3 – 13}$$

Proof: Eq.(12) directly follows from Eq.(1) according to Eqs.(A.4-14) and (A.4-15). As we have seen in Sec.4.1, any asymptotic field belongs to a BRS-singlet or otherwise doublet. Hence $\chi(x)$ should represent either (i) or (ii) in the above.

(q.e.d.)

In the Sec.4.4.2, we have shown in Theorem 4.4.2-2 that the color confinement really takes place in the case with $u = -1$ and with unbroken color symmetry: In fact, the color charges Q^a vanish in H_{phys}, or equivalently

$$\langle \Psi | Q^a | \Phi \rangle = 0 \qquad \text{for } \forall | \Psi \rangle, | \Phi \rangle \in V_{\mathrm{phys}}, \qquad (4.4.3 - 14)$$

owing to the remarkable equality Eq.(4.4.2-5),

$$Q^a = -i\boldsymbol{\delta}(M^a) = \{Q_B, \ M^a\} \ \text{ with } M^a \equiv g^{-1} \int d\mathbf{x} \, (D_0 \overline{C})^a(x). \qquad (4.4.3 - 15)$$

What does this color confinement imply on the structure of asymptotic fields? We know already from the arguments in Sec.4.1 that any asymptotic field is either a BRS-singlet or a quartet member. The color confinement Eq.(14) or (15) means that the *colored asymptotic fields*, if any, should *belong to quartet representations*: In fact, the BRS-singlet (and hence physical) particles, denoted by φ^α, are necessarily color singlets. This is because Eq.(14) with $| \Psi \rangle = \varphi^{\alpha \dagger} | 0 \rangle$ and $| \Phi \rangle = \varphi^{\beta \dagger} | 0 \rangle$ leads to

$$0 = \langle 0 | \varphi^\alpha Q^a \varphi^{\beta \dagger} | 0 \rangle = \langle 0 | \varphi^\alpha [Q^a, \ \varphi^{\beta \dagger}] | 0 \rangle$$
$$= \langle 0 | \varphi^\alpha \varphi^{\gamma \dagger} (\tau^a)^{\gamma\beta} | 0 \rangle = (\tau^a)^{\alpha\beta}, \qquad (4.4.3 - 16)$$

by using the normalization $[\varphi^\alpha, \ \varphi^{\beta \dagger}] = \delta^{\alpha\beta}$. Eq.(16) expresses that the representation matrix $(\tau^a)^{\alpha\beta}$ of color charge Q^a on the particles φ^α should vanish. On the other hand, the quartet particles denoted by $(\chi_\iota, \beta_\iota, \gamma_\iota$ and $\overline{\gamma}_\iota)$, can have color charges in a consistent way with Eq.(15): e.g., $[Q^a, \ \chi_\iota] = -(T^a)_{\iota\jmath}\chi_\jmath$ with $T^a \neq 0$. Indeed, Eq.(15) only requires the following forms for Q^a and M^a:

$$Q^a \sim -\sum_{\iota,\jmath}(\beta_\iota{}^\dagger (T^a)_{\iota\jmath} \chi_\jmath + \chi_\iota{}^\dagger (T^a)_{\iota\jmath} \beta_\jmath + \beta_\iota{}^\dagger (\omega T^a)_{\iota\jmath} \beta_\jmath$$
$$+ i\overline{\gamma}_\iota{}^\dagger (T^a)_{\iota\jmath} \gamma_\jmath - i\gamma_\iota{}^\dagger (T^a)_{\iota\jmath} \overline{\gamma}_\jmath), \qquad (4.4.3 - 17)$$
$$M^a \sim -\sum_{\iota,\jmath}(\overline{\gamma}_\iota{}^\dagger (T^a)_{\iota\jmath} \chi_\jmath + \chi_\iota{}^\dagger (T^a)_{\iota\jmath} \overline{\gamma}_\jmath$$
$$+ \frac{1}{2}[\beta_\iota{}^\dagger (\omega T^a)_{\iota\jmath} \overline{\gamma}_\jmath + \overline{\gamma}_\iota{}^\dagger (T^a \omega)_{\iota\jmath} \beta_\jmath]), \qquad (4.4.3 - 18)$$

where use has been made of the metric matrix, Eq.(4.1.3-12), for the quartets, and the symbol $\sim$ denotes an equality restricted on the Fock space spanned by the asymptotic fields. If the asymptotic completeness holds, both of Eqs.(17) and (18) become exact equalities. Thus, these arguments explicitly show that the color confinement, Eq.(4.4.2-4), namely, $\hat{Q}^a = 0$ in H_{phys}, is really a confinement, the consistency of which is guaranteed by the quartet mechanism found in Sec.4.1. We already know the existence of one color-octet of "elementary" quartet. Further, there may exist asymptotic fields even for the quark fields $q(x)$ and the vector parts of gluon fields $A_\mu{}^a(x)$. In this case, we see, from Theorem 4.4.3-1 and the BRS transformation properties,

$$[iQ_B,\ q(x)] = igC^a(x)\frac{\lambda^a}{2}q(x), \qquad\qquad (4.4.3-19)$$

$$[iQ_B,\ A_\mu{}^a(x)] = \partial_\mu C^a(x) + g(A_\mu(x) \times C(x))^a, \qquad (4.4.3-20)$$

that the quark (and/or gluon) asymptotic fields, if they exist, should necessarily be accompanied by bound states of FP ghosts and quarks (and/or gluons) in the channels of $C^a(x)\lambda^a q(x)$ (and/or $A_\mu(x) \times C(x)$). Note that the formations of such bound-states are possible only in non-abelian cases, because the FP-ghosts are completely free in abelian cases. This point agrees with the previous observation that the confinement condition (A) $u = -1$ in Theorem 4.4.2-2 can be realized only in non-abelian cases.

Some comments are in order:

(i) Since the quartet mechanism takes place in this color confinement, the physical S-matrix unitarity holds on the assumption of asymptotic completeness.

(ii) The confinement by Theorem 4.4.2-2 holds irrespective of the presence or the absence of quark asymptotic fields. One may, however, prefer the case where the quark fields have their own asymptotic fields, because the asymptotic completeness can hold only in such case. In fact, if the quark fields belonging to fundamental representations of $SU(3)$ color group do not have their asymptotic fields, then the Heisenberg fields of quarks cannot be expressed by other asymptotic fields of non-fundamental representations alone.

Finally, we add a few comments on the consistency problem involved in the color-singletness condition on the physical states, which may also open a kinematical approach to the color confinement [Nak 84a, Nak 84b]. Namely, we can obtain the following two statements:

1) Let $\mathcal{O}_1$ and $\mathcal{O}_2$ be two (possibly unbounded) regions localized within two spacelike cones which are spacelikely separated, and $|\Phi\rangle$ be a product state given by $|\Phi\rangle = \varphi_1 \varphi_2 |0\rangle$ with $\varphi_i \in \mathcal{F}(\mathcal{O}_i)$ $(i = 1, 2)$. Then, from the semi-simplicity of $LieG$, it follows that the color-singlet condition, $Q^a |\Phi\rangle = 0$ for the product state implies $Q^a \varphi_i |0\rangle = 0$ for $i = 1, 2$.

2) For a linear-combination state $|\Phi\rangle = \sum_k \varphi^{(k)}{}_1 \varphi^{(k)}{}_2 |0\rangle$, it is impossible to observe the factor states $\varphi^{(k)}{}_1 |0\rangle$ independently of $\varphi^{(k)}{}_2 |0\rangle$, even if the contraction of wave packet by measurement is taken into account.

These results imply that, *even if* a colored BRS-singlet asymptotic state may exist and *even if* a fractional charge due to colored particles may be observed, any localized *particle* with color degrees of freedom cannot be observed in the measurements performed in the color-singlet states. This enables one to adopt the color-singlet condition as a *subsidiary condition* $Q^a |\text{phys}\rangle = 0$ in a consistent manner *without* analyzing the details of the non-perturbative dynamics, although we must then admit the fact that the physical Hilbert space H_{phys} does not have the usual Fock-space structure with respect to asymptotic fields.

In this kinematical approach to color confinement, colored particles are not observable but *states involving colored particles contribute as intermediate states to the unitarity sum.* Hence the total sum of all observable exclusive reaction cross sections do not necessarily saturate the total cross section. Such a possibility does not seem to be excluded experimentally in high energies.

4.4.4 Universality of electric charge in the electroweak theory

The electroweak theory based on the local $SU(2) \times U(1)$ has been established as *the* unified theory of weak and electromagnetic interactions with remarkable agreements with all the experimental results, among which the experimental verifications of W- and Z- bosons (see, for instance, [Lang 86], as a review) have been the most important monuments confirming the success of the principle of local gauge invariance and the Higgs mechanism.

Although a lot of perturbative calculations have been done with some non-perturbative corrections taken into account through the renormalization-group methods, some general important theoretical aspects cannot be fully discussed by such perturbative approach: For instance, in contrast to the case of *abelian QED*, the electromagnetic field in this theory contains some non-abelian components, and hence, its field strength can*not* be an observable quantity according to Theorem 4.3.2-1. How can this be consistent with those results which are derived from the standard abelian QED and which show marvelous agreement with the experimental data? And what about the universality problem of electric charges, and so on?

Somewhat lengthy Lagrangian density of electroweak theory is now well known, and is omitted here. We remark that such asymptotic-field analysis as done in Sec.4.2 explicitly for Yang-Mills theories with broken and unbroken symmetries can be performed also for the electroweak theory. The result is, however, of no novelty, although some complications occur owing to the mixing of two gauge bosons in the neutral channel: For the symmetry-broken parts and unbroken $U(1)$ part of the $SU(2) \times U(1)$ group, the situations are quite similar to the broken $SU(2)$ Higgs-Kibble model (Sec.4.2.1) and the unbroken Yang-Mills model (Sec.4.2.2), respectively.

Here we want to discuss only the problem of *(electric-)charge universality* in the electroweak theory. We mean by *charge universality* that the *on-shell coupling constants of photon* to the charged fields are universal. This should be proved also in the electroweak theory because the absolute values of the charges of electron (or muon) and proton are known to coincide with quite a good accuracy. This problem was trivial in the case of QED. The Ward identity, which is very simple in the abelian case, assures the proportionality of the bare coupling constants e_i for the charged fields φ_i to the on-shell coupling constants $(e_i)_r$:

$$(e_i)_r = \frac{Z_i Z_3^{1/2}}{Z_1} e_i = Z_3^{1/2} e q_i, \qquad (4.4.4-1)$$

where Z_1, Z_3 and Z_i are, respectively, the vertex renormalization constant of $\overline{\varphi}_i \varphi_i A_\mu$, wave-function renormalization constants of A_μ and φ_i. The second equality is due to the Ward identity, $Z_1 = Z_i$. We should recall that quantum numbers q_i of the electric charge operator Q, of course, determine the bare coupling constants e_i as $e_i = e q_i$ and that

$$[Q, \ \varphi_i{}^\dagger] = q_i \varphi_i{}^\dagger. \qquad (4.4.4-2)$$

Since the proportionality constant $Z_3^{1/2}$ or $Z_3^{1/2} e$ is independent of the matter fields φ_i, Eq.(1) proves the charge universality.

In the electroweak theory, however, the WT identities become quite complicated and the proof of such proportionality relation as Eq.(1) is made difficult by the non-abelian character of this theory. In fact, the proof of such proportionality by use of WT identities was accomplished only in the Landau gauge [Tay 71, Sla 72, Aok 82] and in the unitary gauge.[6] In other covariant gauges, only the charge conservation law in arbitrary scattering processes,

$$\sum_{i \in \{outgoing\}} (e_i)_r = \sum_{j \in \{incoming\}} (e_j)_r, \qquad (4.4.4-3)$$

was derived from the WT identity [Aok 82]. This conservation law, Eq.(3), which can be proved also from more general S-matrix theoretical arguments alone as was done by Weinberg [Wei 64], is not sufficient unfortunately for the charge universality proof; for instance, in the case where the muon (μ) and electron (e) numbers are separately conserved, one cannot derive the equality $(e_\mu)_r = (e_e)_r$ from Eq.(3) alone even when $e_\mu = e_e$, because of the possible difference in the radiative corrections.

Now, we give the proof of the proportionality

$$(e_i)_r = \text{constant} \cdot e_i = \text{constant} \cdot eq_i, \qquad (4.4.4-4)$$

by means of the machinery of the present formalism, in the electroweak theory in arbitrary covariant gauges respecting the original global $SU(2) \times U(1)$ symmetry, namely, in such type of gauges as Eq.(3.4.1-3):

$$\mathcal{L}_{\text{GF}} = -\partial^\mu B^a \cdot A_\mu{}^a + \frac{\alpha}{2} B^a B^a, \qquad (4.4.4-5)$$

where the index a runs over 0 (corresponding to $U(1)$) and 1, 2 and 3 (of $SU(2)$). This proof was first given in [Aok 82]. We present it in a more complete form.

Let us recall the quantum Maxwell equations, Eq.(3.4.3-14):

$$\partial^\nu F_{\mu\nu}{}^a = g J_\mu{}^a - \{Q_B, D_\mu \overline{C}^a\} \quad \text{for } a = 1, 2, 3, \qquad (4.4.4-6)$$

$$\partial^\nu F_{\mu\nu}{}^0 = \frac{1}{2} g' J_\mu{}^0 - \{Q_B, \partial_\mu \overline{C}^0\}. \qquad (4.4.4-7)$$

Owing to the spontaneous breakdown of $SU(2) \times U(1)$ to the electromagnetic $U(1)$, only one charge operator, identified with the electric charge operator Q, is unbroken and as usual is written as

$$Q = Q^3 + \frac{1}{2} Q^0. \qquad (4.4.4-8)$$

[6]   There exist, in the unitary gauge, some doubts in the renormalizability, and hence, in the well-definedness of it.

In view of this combination, the 0- and 3-components of quantum Maxwell equations, Eqs.(6) and (7), are combined to yield the equation

$$\partial^\nu F_{\mu\nu} = e(J_\mu{}^3 + \frac{1}{2} J_\mu{}^0 - \mathcal{N}_\mu)$$
(4.4.4 − 9)

where the following definitions are made:

$$F_{\mu\nu} \equiv (g' F_{\mu\nu}{}^3 + g F_{\mu\nu}{}^0)/\sqrt{g^2 + g'^2},$$
(4.4.4 − 10)

$$\mathcal{N}_\mu \equiv \{Q_B, [g'(D_\mu \overline{C})^3 + g \partial_\mu \overline{C}{}^0]/\sqrt{g^2 + g'^2}\},$$
(4.4.4 − 11)

$$e \equiv gg'/\sqrt{g^2 + g'^2}.$$
(4.4.4 − 12)

Both of the formal charges $\int d\mathbf{x}\ (J_0{}^3 + \frac{1}{2}J_0{}^0)$ and $\int d\mathbf{x}\ \mathcal{N}_0/e$ may be identified with Q of Eq.(8), since both of them formally reproduce the commutation relations Eq.(2), as easily assured by using the canonical commutation relations. As has been explained in Sec.4.4.1, however, neither of them is an unbroken charge because of the massless one-particle contribution from the 'elementary' quartet members β^a ($a = 3$ and 0). By the assumption that one-dimensional symmetry transformation corresponding to the combination of Eq.(8) remains unbroken, a certain linear combination of the above formal two charges,

$$Q = \int d\mathbf{x} \left[(J_0{}^3 + \frac{1}{2} J_0{}^0) - \zeta \mathcal{N}_0/e \right]/(1 - \zeta)$$
$$\equiv \int d\mathbf{x}\ J_0{}^{e.m} ,$$
(4.4.4 − 13)

similarly to Eq.(4.4.1-33), must provide an unbroken charge, when the constant $\zeta(\neq 1)$ is suitably adjusted so that the massless contribution in $(J_0{}^3 + J_0{}^0/2)$ is cancelled by that in $\zeta \mathcal{N}_0/e$. Since Eq.(13) clearly reproduces the commutation relations Eq.(2), we see that the unbroken charge in Eq.(13) gives a desired correct expression for the electric charge operator Q formally given by Eq.(8). According to Eq.(13), the quantum Maxwell equation, Eq.(9), is now rewritten as

$$\partial^\nu F_{\mu\nu}/(1 - \zeta) = e J_\mu{}^{e.m.} - \mathcal{N}_\mu.$$
(4.4.4 − 14)

This equation sandwiched by two physical 1-particle states $|i\rangle$ and $|f\rangle(\in \mathcal{V}_{\text{phys}})$ leads to

$$\langle f| \partial^\nu F_{\mu\nu} |i\rangle/(1 - \zeta) = e \langle f| J_\mu{}^{e\ m.} |i\rangle,$$
(4.4.4 − 15)

where use has been made of Eq.(11) and $Q_B|f\rangle = Q_B|i\rangle = 0$

$$\langle f|\mathcal{N}_\mu|i\rangle = \langle f|\{Q_B, *\}|i\rangle = 0. \tag{4.4.4 − 16}$$

Applying $\int d^4x e^{ikx}$ to both sides of Eq.(15), we consider the limit $k_\mu \to 0$. First, similarly to the soft-pion technique, all the contributions to the lhs. of Eq.(15) remaining in this limit come from the massless one-particle in the channel $F_{\mu\nu}$, namely, the photon. The one-photon contribution can be estimated by the asymptotic form of $F_{\mu\nu}$:

$$F^{\mathrm{as}}{}_{\mu\nu} = Y(\partial_\mu A_\nu{}^{\mathrm{ph.}} - \partial_\nu A_\mu{}^{\mathrm{ph.}}) + \cdots, \tag{4.4.4 − 17}$$

where $A_\mu{}^{\mathrm{ph.}}$ represents the renormalized photon asymptotic field and the dots $(\cdots)$ stand for the other massive particles irrelevant here. The constant Y can be evaluated by the pole residue, for example, of

$$\mathrm{F.T.}\langle 0|\mathrm{T}F_{\mu\nu}(x)F_{\rho\sigma}(y)|0\rangle$$
$$= -Y^2(p_\mu p_\rho \eta_{\nu\sigma} + p_\nu p_\sigma \eta_{\mu\rho} - p_\nu p_\rho \eta_{\mu\sigma} - p_\mu p_\sigma \eta_{\nu\rho})/p^2 + \cdots. \tag{4.4.4 − 18}$$

Noting that Eq.(17) yields

$$\mathrm{F.T.}\langle 0|\mathrm{T}F_{\mu\nu}(x)A_\rho{}^{\mathrm{ph.}}(y)|0\rangle = -Y(p_\mu \eta_{\nu\rho} - p_\nu \eta_{\mu\rho})/p^2, \tag{4.4.4 − 19}$$

one can show that the lhs. of Eq.(15) becomes

$$\lim_{k\to 0}\int d^4x e^{ikx}\langle f|\partial^\nu F_{\mu\nu}(x)|i\rangle/(1-\zeta)$$
$$= [Y/(1-\zeta)](2\pi)^4\delta^4(p^f - p^i)\cdot(e_{f_i})_r(p^i + p^f)_\mu, \tag{4.4.4 − 20}$$

where $(e_{f_i})_r$ and $(p^i + p^f)_\mu$ are the renormalized on-shell photon coupling constant and the kinematical factor of the one-particle-irreducible vertex $\langle f|A_\mu{}^{\mathrm{ph.}}|i\rangle_{1\mathrm{PI}}$, respectively. Next, since the unbroken charge operator Q is given by Eq.(13), the time-component ($\mu = 0$) of the rhs. of Eq.(15) produces

$$\lim_{k\to 0}\int d^4x e^{ikx}\langle f|eJ_0{}^{\mathrm{e\ m.}}(x)|i\rangle = eq_i\delta_{f_i}(2\pi)^4\delta^4(p^f - p^i)\cdot 2p_0{}^i, \tag{4.4.4 − 21}$$

where use has been made of Eq.(2). The normalization convention adopted in Eqs.(20) and (21) is $\langle f|i \rangle = (2\pi)^3 2p_0{}' \delta_{f_i} \delta(\boldsymbol{p}^j - \boldsymbol{p}')$. By comparing Eqs.(20) and (21), we obtain

$$(e_{f_i})_r = [(1 - \zeta)/Y]\delta_{f_i} e_{q_i}. \qquad (4.4.4 - 22)$$

This result concludes not only the *diagonality* of the on-shell photon coupling constant with respect to the different types of matter, but also the desired proportionality Eq.(4), since the constants ζ and Y are manifestly independent of i and f. This finishes the proof of the charge universality. The above result Eq.(22) is obtained for arbitrary covariant gauges of the type in Eq.(5). In Landau gauge, only in which such proportionality as Eq.(22) is proved also by the WT identity method, our result Eq.(22) can easily be assured to coincide with that due to the WT method. We should finally note the crucial step Eq.(16) in the above proof. The implication of Eq.(16) is buried in many complicated WT identities in the usual proof based upon the WT identity method. From this example also, we see that the quantum Maxwell equation, Eq.(3.4.3-14), and the subsidiary condition $Q_B|\text{phys}\rangle = 0$, in our covariant operator formalism, are really powerful and useful.

CHAPTER FIVE

QUANTUM THEORY OF THE
GRAVITATIONAL FIELD

We present the manifestly covariant canonical operator formalism of
quantum gravity in detail. Any other formalism of quantum gravity
is not described.

5.1 INTRODUCTORY REMARK

Gravity is the most time-honored force. It is now well-established that the correct
theory of gravity is Einstein's general relativity. It is an extremely beautiful and
successful theory: It is supported by a vast amount of observations and experiments in spite of the fact that it contains no ajustable parameters. As far as
large-scale phenomena are concerned, there is absolutely no need for modifying
general relativity.

On the other hand, particle physics is the basis for all natural phenomena other than gravitational ones. It is well-established that particle physics is
described by quantum field theory. We thus have two fundamental theories, quantum field theory and general relativity. Unfortunately, however, these two theories
have quite different structure. The former is a quantum theory, while the latter is
a classical one; the former is based on the fixed Minkowski spacetime, while in the
latter the spacetime metric $g_{\mu\nu}$ is a dynamical and spacetime-dependent quantity.

It is evidently quite inadequate to regard gravity as an orphan in nature.
Indeed, it is well confirmed experimentally that the wave function of a neutron
is affected by the earth's gravity precisely as predicted by the Schrödinger equation with a gravitational potential (including the inertial force due to the earth's
rotation) [Grer 80]. Particle physics must be enlarged to include gravity.

Since no quantum effects of gravity have ever been observed, there is an opinion that $g_{\mu\nu}$ need not be quantized. One constructs quantum field in a curved spacetime [DeW 75, Bir 82] by simply replacing $\eta_{\mu\nu}$ by $g_{\mu\nu}$. Although one encounters a difficulty in defining the vacuum state in general, this theory is certainly a good *approximation* theory in describing celestial quantum phenomena. Quantum field theory in a cuved spacetime *cannot*, however, be regarded as a fundamental theory [Kib 81, Duf 81], because there is no satisfactory way of determining $g_{\mu\nu}$. Since the matter energy-momentum tensor $T_{\mu\nu}$ is a q-number while $g_{\mu\nu}$ is a c-number, one must take a certain expectation value in the rhs. of the Einstein equation in order to determine $g_{\mu\nu}$, that is, one must choose a particular state, but one has no reasonable principle for determining it. Furthermore, a serious trouble is encountered for the measurement of that state owing to the problem of the contraction of wave packet.[1] We may thus conclude that in order to have a consistent theory, $g_{\mu\nu}$ must be quantized.

It is not difficult to quantize the gravitational field if one regards $g_{\mu\nu} - \eta_{\mu\nu}$ as a quantum field in the Minkowski spacetime. But one then obtains a non-renormalizable theory because its perturbative expansion parameter is the Einstein gravitational constant κ, which has dimension $(\text{mass})^{-2}$ [in natural units, $\sqrt{\kappa} \sim 10^{-33}\text{cm}$ (Planck length), $1/\sqrt{\kappa} \sim 10^{19}\,\text{GeV}$ (Planck mass)]. Accordingly, the divergences encountered in quantum gravity are totally out of control in perturbation theory. On the basis of this fact, there is a wide-spread opinion that quantum gravity is unsatisfactory; some people even say that there is no quantum theory of gravity.

There are two standpoints on which the above opinion stands: The one is a pragmatic one, while the other is a mathematical one. On the pragmatic standpoint, people say that any physical theory should make definite predictions; if not predictable, it cannot be ragrded as a theory of exact science. At first sight, this opinion looks quite reasonable, but a closer examination shows that the meaning of predictability is quite vague. Of course, if the theory were exactly solvable in closed form, everything would be clear. But, except for very special cases, it is too optimistic to expect such a fantastic luck; one must appeal to an approximation method. Then one must show that error is small compared with the approximate value. In quantum field theory, however, *it is impossible to prove that*. What

---

[1]   By using the Einstein-Podolsky-Rosen type setting-up, the quantum state can be changed suddenly without any classical disturbance. For further discussion, see [Pag 81]

physicists really do is to justify the employed approximation method *a posteriori* by comparing the approximate values with experimental results. In quantum gravity, however, there are *absolutely no* experimental results to be compared with. Thus there is no way of confirming the validity of the approximation method in quantum gravity. One might say that since the perturbation theory of quantum electrodynamics is quite successful because of the smallness of the fine structure constant $\alpha \equiv e^2/4\pi$ and since the Planck length $\sqrt{\kappa}$ is extremely small, it is natural to expect that the perturbation theory of quantum gravity should be reliable. This assertion is totally wrong because $\sqrt{\kappa}$ is a dimensionful constant and therefore can be transformed away by taking $x^\mu/\sqrt{\kappa}$ as new spacetime coordinates; that is, the perturbation series is *not* a power series. Of course, it is very desirable (especially for the researchers who wish to write papers) that the theory is calculable, but there is no reason to believe that Nature is always kind enough to prepare a good approximation method.

Quantum gravity is criticized also from the mathematical standpoint: A consistent theory must be finite, that is, the existence of non-removable divergences shows that the theory is inconsistent. Of course, the finiteness requirement should be satisfied, but one must be very careful about the fact that it is required only for the *exact* solution. The *perturbative* infinities are nothing to do with the consistency of the theory. For instance, consider Planck's formula for blackbody radiation; it is completely finite in closed form, but if it is expanded into a series in powers of the Planck constant h then each order is badly ultraviolet divergent. Thus as long as the exact treatment of quantum gravity is not made, the criticism from the mathematical standpoint is totally unacceptable. The perturbative finiteness is neither a necessary condition nor a sufficient one for the consistency of the theory.

Some people wish to compare quantum gravity with the theories of weak interactions. In Fermi's theory of weak interactions, we have a four-fermion interaction term, which evidently makes the theory non-renormalizable. This theory is presently regarded as a low-energy effective theory of the electroweak theory, which is renormalizable. Then analogy suggests that there might exist a correct renormalizable theory of gravity and that the Einstein gravity might be its low-energy effective theory. Such an idea is, however, nothing but a consequence of being too much overconfident of analogy. The Einstein gravity is *totally dissimilar* to Fermi's theory just except for their non-renormalizability. The former is a very deep theory derived uniquely from the first principles, while the latter is a purely phenomenological theory based on Pauli's "deseparate idea". It seems more natural to take the converse standpoint that *renormalizable theories are low-energy*[2] *effective theories*, because renormalizability is nothing but the statement that we can calculate low-energy quantities without knowing the true high-energy behavior. The ultimate theory should not contain infinite renormalization constants, and the natural parameter which cuts off their divergence is believed to be the Planck mass $1/\sqrt{\kappa}$. This belief could not be realized if quantum gravity were a renormalizable theory similar to that in particle physics. Thus it is rather desirable that quantum gravity is a "non-renormalizable" theory *in the terminology of particle physics*.

Yet, those who wish to regard the Einstein gravity as a low-energy effective theory propose the so-called "induced gravity" and "composite gravity" [Adl 82]. They expel the Einstein-Hilbert Lagrangian density from their fundamental Lagrangian density, say $\mathcal{L}_{\text{fund}}$. Of course, they do not wish to violate invariance under general coordinate transformations in order to reproduce the Einstein gravity in low energies. Then their theories necessarily become very artificial in the introduction of the *contravariant* metric tensor $g^{\mu\nu}$. If $g^{\mu\nu}$ is present in $\mathcal{L}_{\text{fund}}$ as an independent quantity, one will have an unwelcome field equation $T_{\mu\nu} = 0$. It is logically inadmissible to take $g^{\mu\nu}$ into account *after* the effective theory is derived. If $g^{\mu\nu}$ is absent in $\mathcal{L}_{\text{fund}}$, it must be constructed from *covariant* quantities through a procedure involving matrix inversion. The covariant metric tensor $g_{\mu\nu}$ is also constructed from covariant quantities. Since those two procedures are logically independent, there is *no reason* why the $g^{\mu\nu}$ satisfying $g^{\mu\nu}g_{\nu\lambda} = \delta^{\mu}{}_{\lambda}$ is the *only* correct one among infinitely many possible candidates.

From the above considerations, we may conclude that the gravitational field $g_{\mu\nu}$ should be a *primary field*[3] present in the fundamental Lagrangian density and *must be quantized without recourse to perturbation theory*. Since the spacetime metric $g_{\mu\nu}$ becomes a q-number, we must recognize the important fact that the principle of Einstein causality is not meaningful at the operator level. Indeed, we should not assume any particular background spacetime at the Lagrangian level

---

[2]   Here "low energy" means the energy region accessible by means of accelerators.

[3]   More precisely, vierbein is a primary field in the vierbein formalism (see Sec.5.7). In this case, local Lorentz invariance eliminates the arbitrariness of defining $g_{\mu\nu}$ and $g^{\mu\nu}$.

because it is absurd that the *classical* equation which determines the backgroumd spacetime logically precedes the *quantum* theory of gravity.

There have appeared many theories of quantum gravity [Arn 62, Ish 75, Kuc 81]. Among them, the most popular one[4] is the path-integral formalism [Fad 74, Fra unp, Haw 79]. But, since the non-perturbative definition of the functional integral over spacetime metrics is quite unclear even in the physicist's standard for rigor, this formalism is essentially formulated in the perturbative way; otherwise one must substitute for it a drastically simplified object, calling the latter "approximation".[5]

On the other hand, the old-fashioned canonical formalism (quantum-gravity counterpart of the Coulomb-gauge quantum electrodynamics) is extremely complicated and badly violates manifest covariance. Furthermore, since one encounters local-operator constraints, which cannot be solved in a non-perturbative way, it is difficult to obtain a consistent quantum theory as discussed in Sec.2.2.3.

The most satisfactory way of quantizing gravity is to extend the covariant operator formalism of the non-abelian gauge field developed in Chapters 3 and 4. We find that the theory becomes remarkably beautiful for a particular choice of gauge-fixing and FP-ghost Lagrangian densities. In the following sections of this chapter, we develop this formalism of quantum gravity in detail. It is a formalism quite satisfactory except for the following two points.

1. We cannot prove finiteness or consistency of the theory.
2. We know no systematic method of approximation which is free of divergence difficulty.

The first point is a prohibitively difficult mathematical problem. One should note that nobody knows even whether or not there exists any non-trivial consistent quantum field theory in the four-dimensional spacetime. Physicists must satisfy themselves by checking that no immediate inconsistency is encountered.[6] The second point is very important practically, but one should note that it is *not the*

---

[4]   Recently, in cosmology, the Wheeler-DeWitt equation also becomes popular, but it is *not* a probabilistically interpretable theory.

[5]   Although analytic continuation to the Euclidean spacetime is often made, one should note that analytic continuation does *not* preserve approximation.

[6]   Divergence problem is discussed in Sec.5.8.2.

problem of the formalism itself but the one which should be studied in the *next* stage. Thus as far as the formulation of the theory is concerned, the following formalism seems to be the most natural and most satisfactory one of quantum gravity.[7]

Further discussion on the fundamental questions concerning quantum gravity is made in Sec.5.8.4.

---

[7]   This theory has been developed by Nakanishi (with various collaborators) in a series of papers entitled "Indefinite-Metric Quantum Field Theory of General Relativity". [The number in this series are shown by Roman numerals in the References.] We make some notational changes; important ones are sign changes in the FP anti-ghost $\bar{c}_r$ and the BRS transformation δ in conformity with a review article [Nak 83d].

5.2 CLASSICAL GENERAL RELATIVITY

In this section, we briefly review the classical theory of general relativity. Since quantum gravity is not necessarily a direct extension of general relativity, classical equations need not remain valid after quantization. But we note that identities are always valid also in quantum gravity.

5.2.1 General coordinate transformation

In general relativity, the **gravitational field** $g_{\mu\nu}$ is identified with the **metric tensor** of a four-dimensional pseudo-Riemannian space. We take its signature as $(+, -, -, -)$, though it loses its meaning *at the operator level* in quantum gravity. The infinitesimal distance squared is given by

$$ds^2 = g_{\mu\nu}dx^\mu dx^\nu, \qquad (5.2.1-1)$$

but this equation is not relevant in quantum gravity.

Since $g_{\mu\nu}$ is a symmetric tensor ($g_{\mu\nu} = g_{\nu\mu}$), it has ten independent components. For the covariant tensor $g_{\mu\nu}$, we define the contravariant metric tensor $g^{\mu\nu}$ by its matrix inverse:

$$g^{\mu\nu}g_{\nu\lambda} = \delta^\mu{}_\lambda. \qquad (5.2.1-2)$$

Here we, of course, assume the non-vanishing of the determinant of $g_{\mu\nu}$. Then the Lorentzian signature implies

$$g \equiv \det g_{\mu\nu} < 0. \qquad (5.2.1-3)$$

Hence we can set

$$h \equiv \sqrt{-g} > 0. \qquad (5.2.1-4)$$

When $g_{\mu\nu}$ is quantized, such an inequality as Eq.(3) may lose its meaning, but the reality (hermiticity) of h can be guaranteed if we start with the vierbein field, as explained in Sec.5.7.1.

General relativity is formulated so as to be invariant under **general coordinate transformations**. Let x^μ and x'^μ be old and new coordinates, respectively, of the same spacetime point. Then

$$dx'^\mu = \frac{\partial x'^\mu}{\partial x^\nu}dx^\nu, \qquad \det\left(\frac{\partial x'^\mu}{\partial x^\nu}\right) \neq 0. \qquad (5.2.1-5)$$

A scalar field $\phi(x)$, a **covariant** vector field $A_\mu(x)$, and a **contravariant** vector field $A^\mu(x)$ transform as

$$\phi'(x') = \phi(x), \qquad\qquad (5.2.1-6)$$

$$A'_\mu(x') = \frac{\partial x^\nu}{\partial x'^\mu} A_\nu(x), \qquad\qquad (5.2.1-7)$$

$$A'^\mu(x') = \frac{\partial x'^\mu}{\partial x^\nu} A^\nu(x), \qquad\qquad (5.2.1-8)$$

respectively. A general **tensor field** transforms like a certain product of vector fields. In particular, $g_{\mu\nu}(x)$ is a rank-2 covariant tensor, $g^{\mu\nu}(x)$ is a rank-2 contravariant tensor, and $\delta^\mu{}_\nu$ is a rank-2 mixed tensor. As noted in Sec.5.1, we emphasize that $g^{\mu\nu}$ is related to $g_{\mu\nu}$ through a very special relation, Eq.(2); such a special feature of the metric tensor could not exist if it (or the vierbein field) were not a primary field.

Tensors have the following three properties:

1. Tensor indices can be raised by $g^{\mu\nu}$ and lowered by $g_{\mu\nu}$; e.g., $A^\mu \equiv g^{\mu\nu} A_\nu$ and $A_\mu \equiv g_{\mu\nu} A^\nu$.

2. A product of a rank-n tensor and a rank-m one is a rank-$(n+m)$ tensor; e.g., for two vectors (i.e., rank-1 tensors) A_μ and B_ν, $A_\mu B_\nu$ is a rank-2 tensor.

3. A rank-n tensor becomes a rank-$(n-2)$ tensor by **contraction**; e.g., a rank-2 tensor $T_{\mu\nu}$ is contracted into a scalar (i.e., rank-0 tensor) $T \equiv T^\mu{}_\mu$.

The quantity which is obtained from a tensor by multiplying by h [defined in Eq.(4)] is called a **tensor density**. In particular, a rank-2 contravariant tensor density

$$\tilde{g}^{\mu\nu} \equiv h g^{\mu\nu} \qquad\qquad (5.2.1-9)$$

plays a very important role. We note that $h d^4 x$ is the invariant volume element.

Let δ be any **derivation**[1]; then from Eqs.(2)-(4) we have

$$\delta g^{\mu\nu} = -g^{\mu\lambda} g^{\nu\rho} \delta g_{\lambda\rho}, \qquad\qquad (5.2.1-10)$$

$$\delta h = \frac{1}{2} h g^{\mu\nu} \delta g_{\mu\nu} = -\frac{1}{2} h g_{\mu\nu} \delta g^{\mu\nu}. \qquad\qquad (5.2.1-11)$$

---

[1]   Derivation means linear operation satisfying the Leibniz rule.

5.2.2 Covariant derivative

Although $\partial_\mu \phi$ is a vector field if ϕ is a scalar, $\partial_\mu A_\nu$ is not a tensor field if A_ν is a vector. To obtain a tensor field, we define **covariant derivative**

$$\nabla_\mu A_\nu \equiv \partial_\mu A_\nu - \Gamma^\lambda{}_{\mu\nu} A_\lambda, \qquad (5.2.2-1)$$

where the **affine connection** $\Gamma^\lambda{}_{\mu\nu}$ should transform as

$$\Gamma^\lambda{}_{\mu\nu}{}'(x') = \frac{\partial x'^\lambda}{\partial x^\rho} \frac{\partial x^\sigma}{\partial x'^\mu} \frac{\partial x^\tau}{\partial x'^\nu} \Gamma^\rho{}_{\sigma\tau}(x) + \frac{\partial x'^\lambda}{\partial x^\rho} \cdot \frac{\partial^2 x^\rho}{\partial x'^\mu \partial x'^\nu}. \qquad (5.2.2-2)$$

For a generic tensor $T_\Sigma \equiv T^{\sigma_1 \cdots \sigma_r}_{\tau_1 \cdots \tau_s}$ (Σ summarizes all tensorial indices), we define

$$\nabla_\mu T_\Sigma \equiv \partial_\mu T_\Sigma + \Gamma^\lambda{}_{\mu\nu}[T_\Sigma]^\nu{}_\lambda, \qquad (5.2.2-3)$$

where

$$[T^{\sigma_1 \cdots \sigma_r}_{\tau_1 \cdots \tau_s}]^\nu{}_\lambda \equiv \sum_{p=1}^r \delta^{\sigma_p}{}_\lambda T^{\sigma_1 \cdots \sigma_{p-1} \nu \sigma_{p+1} \cdots \sigma_r}_{\tau_1 \cdots \tau_s}$$

$$- \sum_{q=1}^s \delta^\nu{}_{\tau_q} T^{\sigma_1 \cdots \sigma_r}_{\tau_1 \cdots \tau_{q-1} \lambda \tau_{q+1} \cdots \tau_s}. \qquad (5.2.2-4)$$

It should be noted that $[T_\Sigma]^\nu{}_\lambda \cdot \varepsilon^\lambda{}_\nu$ is nothing but the change of T_Σ under an infinitesimal **general linear transformation** characterized by $\varepsilon^\lambda{}_\nu$. So, even for a non-tensor Φ, we define $[\Phi]^\nu{}_\lambda$ by its infinitesimal transformation under $GL(4)$.

The covariant differentiation ∇_μ satisfies the Leibniz rule.

It is the characteristic of the (pseudo-)Riemannian geometry that $\Gamma^\lambda{}_{\mu\nu}$ is expressible in terms of the metric tensor. Indeed, from the requirements of metricity,

$$\nabla_\mu g_{\nu\lambda} \equiv \partial_\mu g_{\nu\lambda} - \Gamma^\rho{}_{\mu\nu} g_{\rho\lambda} - \Gamma^\rho{}_{\mu\lambda} g_{\nu\rho} = 0, \qquad (5.2.2-5)$$

and of the torsionless condition, $\Gamma^\rho{}_{\mu\nu} - \Gamma^\rho{}_{\nu\mu} = 0$, we have

$$\Gamma^\lambda{}_{\mu\nu} \equiv \frac{1}{2} g^{\lambda\sigma}(\partial_\mu g_{\nu\sigma} + \partial_\nu g_{\mu\sigma} - \partial_\sigma g_{\mu\nu}). \qquad (5.2.2-6)$$

In particular, Eqs.(6) and (5.2.1-11) imply

$$\Gamma^\lambda{}_{\mu\lambda} = \frac{1}{2} g^{\lambda\sigma} \partial_\mu g_{\lambda\sigma} = h^{-1} \partial_\mu h, \qquad (5.2.2-7)$$

whence, noting $[h]^\nu{}_\lambda = -\delta^\nu{}_\lambda h$, we see

$$\partial_\mu h + \Gamma^\lambda{}_{\mu\nu}[h]^\nu{}_\lambda = 0. \tag{5.2.2 - 8}$$

Accordingly, Eq.(3) holds also for a tensor density, though the rhs. of Eq.(4) then acquires a term $-\delta^\nu{}_\lambda T_\Sigma$. In particular, for a contravariant vector density J^μ, we find

$$\nabla_\mu J^\mu = \partial_\mu J^\mu, \tag{5.2.2 - 9}$$

whence the current conservation law is invariant under general coordinate transformations.

Covariant differentiations are not commutative. Their commutator defines the **Riemann (curvature) tensor**:

$$[\nabla_\mu, \nabla_\nu]A^\lambda = R^\lambda{}_{\rho\mu\nu}A^\rho \tag{5.2.2 - 10}$$

with

$$R^\lambda{}_{\rho\mu\nu} \equiv \partial_\mu\Gamma^\lambda{}_{\nu\rho} - \partial_\nu\Gamma^\lambda{}_{\mu\rho} + \Gamma^\lambda{}_{\mu\sigma}\Gamma^\sigma{}_{\nu\rho} - \Gamma^\lambda{}_{\nu\sigma}\Gamma^\sigma{}_{\mu\rho}. \tag{5.2.2 - 11}$$

The Riemann tensor $R_{\lambda\rho\mu\nu}$ is antisymmetric under $\lambda \leftrightarrow \rho$ and under $\mu \leftrightarrow \nu$ and symmetric under $(\lambda, \rho) \leftrightarrow (\mu, \nu)$. Furthermore, from the Jacobi identity for covariant differentiations, we obtain an identity

$$R_{\lambda\rho\mu\nu} + R_{\lambda\mu\nu\rho} + R_{\lambda\nu\rho\mu} = 0 \tag{5.2.2 - 12}$$

together with the **Bianchi identity**

$$\nabla_\rho R_{\lambda\sigma\mu\nu} + \nabla_\mu R_{\lambda\sigma\nu\rho} + \nabla_\nu R_{\lambda\sigma\rho\mu} = 0. \tag{5.2.2 - 13}$$

From the above symmetry properties, we see that $R_{\lambda\rho\mu\nu}$ has 20 independent components.[2]

The **Ricci tensor** $R_{\mu\nu}$ is defined by

$$R_{\mu\nu} \equiv R^\lambda{}_{\nu\mu\lambda} = \partial_\mu\Gamma^\lambda{}_{\nu\lambda} - \partial_\lambda\Gamma^\lambda{}_{\mu\nu} + \Gamma^\lambda{}_{\mu\sigma}\Gamma^\sigma{}_{\nu\lambda} - \Gamma^\lambda{}_{\sigma\lambda}\Gamma^\sigma{}_{\mu\nu}$$
$$= R_{\nu\mu}, \tag{5.2.2 - 14}$$

and the **scalar curvature** R is its trace $R^\mu{}_\mu$.[3]

---

[2] In D dimensions, the number of its independent components is $D^2(D^2 - 1)/12$.

[3] The Riemann tensor is expressible in terms of $R_{\mu\nu}$ in three dimensions and in terms of R in two dimensions

5.2.3 Einstein theory

The fundamental Lagrangian density of general relativity consists of the **Einstein-Hilbert Lagrangian density**

$$\mathcal{L}_\text{E} \equiv (2\kappa)^{-1} h R \qquad (5.2.3-1)$$

and the matter one $\mathcal{L}_\text{M}$. If one wishes to include the **cosmological term** $\kappa^{-1}\lambda h$, it is convenient to suppose that it is contained in $\mathcal{L}_\text{M}$.

Since $\mathcal{L}_\text{E}$ contains second derivatives of $g_{\mu\nu}$, we rewrite it as

$$\mathcal{L}_\text{E} = \tilde{\mathcal{L}}_\text{E} + \partial_\mu \mathcal{V}^\mu, \qquad (5.2.3-2)$$

where

$$\tilde{\mathcal{L}}_\text{E} \equiv (2\kappa)^{-1} \tilde{g}^{\mu\nu} (\Gamma^\lambda_{\ \sigma\lambda} \Gamma^\sigma_{\ \mu\nu} - \Gamma^\lambda_{\ \mu\sigma} \Gamma^\sigma_{\ \nu\lambda}), \qquad (5.2.3-3)$$

$$\mathcal{V}^\mu \equiv (2\kappa)^{-1} (\tilde{g}^{\mu\lambda} g^{\sigma\tau} - \tilde{g}^{\mu\sigma} g^{\lambda\tau}) \partial_\lambda g_{\sigma\tau}. \qquad (5.2.3-4)$$

Since the second term of the rhs. of Eq.(2) is a total divergence, we can take $\tilde{\mathcal{L}}_\text{E}$ as the starting point of canonical formalism.

The Euler derivatives of $\tilde{\mathcal{L}}_\text{E}$ and $\mathcal{L}_\text{M}$ with respect to $g_{\mu\nu}$ are[4]

$$\delta\tilde{\mathcal{L}}_\text{E}/\delta g_{\mu\nu} = -(2\kappa)^{-1} h \left(R^{\mu\nu} - \frac{1}{2} g^{\mu\nu} R \right), \qquad (5.2.3-5)$$

$$\delta\mathcal{L}_\text{M}/\delta g_{\mu\nu} = -\frac{1}{2} h T^{\mu\nu}, \qquad (5.2.3-6)$$

where $T^{\mu\nu}(= T^{\nu\mu})$ is the matter energy-momentum tensor.[5] We have the Bianchi identity

$$\nabla_\mu \left(R^{\mu\nu} - \frac{1}{2} g^{\mu\nu} R \right) = 0 \qquad (5.2.3-7)$$

from Eq.(5.2.2-13), while $T^{\mu\nu}$ satisfies the *covariant* conservation law

$$\nabla_\mu T^{\mu\nu} = 0. \qquad (5.2.3-8)$$

---

[4]   Here $\delta/\delta g_{\mu\nu}$ is defined by $\frac{1}{2}(\delta/\delta g_{\mu\nu} + \delta/\delta g_{\nu\mu})$, regarding $g_{\mu\nu}$ as if all its components were independent.

[5]   It is the matter symmetric energy-momentum tensor. Therefore one should write it as $\Theta_{\mu\nu}$, but we follow the conventional notation.

The **Einstein equation** for gravity follows from $\tilde{\mathcal{L}}_E + \mathcal{L}_M$:[6]

$$R_{\mu\nu} - \frac{1}{2}g_{\mu\nu}R = -\kappa T_{\mu\nu}. \qquad (5.2.3 - 9)$$

Because of the identities, Eqs.(7) and (8), only six out of ten equations are dynamical ones, and the remaining four ones are constraints. Indeed, the Einstein equation involves $\ddot{g}_{k\ell}$ but not $\ddot{g}_{0\nu}$.

In order to dertermine $g_{\mu\nu}$, it is necessary to introduce a **coordinate condition**, explicitly violating general coordinate invariance. In abuse of language, this procedure is often called "gauge fixing", but one must be careful about the difference between the choice of a coordinate system and gauge fixing. The former has physical meaning; observable effects (i.e., inertial force) actually depend on it. Indeed, the properties of the solution to the Einstein equation generally depend on the choice of a coordinate system; in order to make definite predictions, one should add a coordinate condition as a field equation to the Einstein equation. From the theoretical point of view, the most natural choice of the coordinate condition seems to be the **de Donder condition**,

$$\partial_\mu \tilde{g}^{\mu\nu} = 0, \qquad (5.2.3 - 10)$$

which is called also **harmonic coordinate condition**[7] because it can be rewritten as

$$h^{-1}\partial_\mu(\tilde{g}^{\mu\nu}\partial_\nu x^\lambda) = 0, \qquad (5.2.3 - 11)$$

where $h^{-1}\partial_\mu \tilde{g}^{\mu\nu}\partial_\nu$ is nothing but the Riemannian extension of the d'Alembertian. Owing to $\nabla_\mu \tilde{g}^{\mu\nu} = 0$, Eq.(10) can be rewritten as

$$\Gamma^\nu{}_{\mu\lambda}g^{\mu\lambda} = 0. \qquad (5.2.3 - 12)$$

The de Donder condition is invariant under $GL(4)$.

One should note that Eq.(8) is *not* a genuine conservation law. The Einstein equation, Eq.(9), can be rewritten into a "Maxwell-like" form,

$$\partial_\lambda \mathcal{F}_C{}^{\lambda\mu}{}_\nu + \kappa T_C{}^\mu{}_\nu = 0, \qquad (5.2.3 - 13)$$

---

[6]   In two dimensions, Eq.(9) does not make sense because its lhs. vanishes identically.
In $D(\geq 3)$ dimensions, Eq.(9) can be rewritten as $R_{\mu\nu} + \kappa[T_{\mu\nu} - (D-2)^{-1}g_{\mu\nu}T^\sigma{}_\sigma] = 0$.
[7]   The privileged character of the harmonic coordinate system was emphasized by Fock [Foc 64].

as is shown in the next subsection, where

$$\mathcal{F}_{\mathrm{C}}{}^{\lambda\mu}{}_{\nu} = -\mathcal{F}_{\mathrm{C}}{}^{\mu\lambda}{}_{\nu}, \qquad (5.2.3-14)$$

and

$$\mathcal{T}_{\mathrm{C}}{}^{\mu}{}_{\nu} \equiv h(t^{\mu}{}_{\nu} + T^{\mu}{}_{\nu}). \qquad (5.2.3-15)$$

Here

$$\begin{aligned}
ht^{\mu}{}_{\nu} &\equiv \partial_{\nu}g_{\sigma\tau} \cdot \frac{\partial \tilde{\mathcal{L}}_{\mathrm{E}}}{\partial(\partial_{\mu}g_{\sigma\tau})} - \delta^{\mu}{}_{\nu}\tilde{\mathcal{L}}_{\mathrm{E}} \\
&= (2\kappa)^{-1}[\partial_{\nu}\tilde{g}^{\rho\sigma} \cdot \Gamma^{\mu}{}_{\rho\sigma} - \partial_{\nu}\tilde{g}^{\mu\rho} \cdot \Gamma^{\sigma}{}_{\rho\sigma}] - \delta^{\mu}{}_{\nu}\tilde{\mathcal{L}}_{\mathrm{E}}, \qquad (5.2.3-16)
\end{aligned}$$

is Einstein's **gravitational energy-momentum pseudotensor density**. From Eq.(13) we obtain the genuine energy-momentum conservation law

$$\partial_{\mu}\mathcal{T}_{\mathrm{C}}{}^{\mu}{}_{\nu} = 0, \qquad (5.2.3-17)$$

but $\mathcal{T}_{\mathrm{C}}{}^{\mu}{}_{\nu}$ is not covariant under general coordinate transformations but a tensor density under $GL(4)$.

Since $\mathcal{T}_{\mathrm{C}}{}^{0}{}_{\nu}$ is a three-dimensional divergence as seen from Eqs.(13) and (14), the classical total energy-momentum

$$P_{\mathrm{C}\nu} \equiv \int d\mathbf{x}\, \mathcal{T}_{\mathrm{C}}{}^{0}{}_{\nu} \qquad (5.2.3-18)$$

reduces to a surface integral because of the Gauss theorem. But if $g_{\mu\nu} - \eta_{\mu\nu}$ asymptotically behaves like $1/r$ as the spatial distance r goes to infinity, then $P_{\mathrm{C}\nu}$ gives a definite vector covariant under $GL(4)$. Furthermore, under certain reasonable assumptions for $T^{\mu}{}_{\nu}$, it can be proved that $P_{\mathrm{C}0}$ is positive and vanishes only when the spacetime is flat [Scho 79, Scho 82, Wit 81, Nes 81].

5.2.4 Second Noether theorem

The **second Noether theorem** is the extension of the Noether theorem described in Sec.1.2.4 to the case of a *local* symmetry. Although it has been already discussed in Sec.3.2.2, we here again present it in the form suitable to the gravitational theory.

We denote a primary field by φ_A as a generic symbol. Under a certain local transformation (x^μ is kept unchanged) characterized by a set of infinitesimal functions ε^α, we suppose that φ_A transforms as

$$\delta_*^\varepsilon \varphi_A \equiv G_{A\alpha} \varepsilon^\alpha + T_A{}^\mu{}_\alpha \partial_\mu \varepsilon^\alpha. \qquad (5.2.4-1)$$

Let $\mathcal{F}$ be a certain function of φ_A, which does not explicitly depend on x^μ, and suppose that $\int d^4 x\, \mathcal{F}$ is invariant under the above local transformation, that is, there exist functions $F_\varepsilon{}^\mu$ of φ_A and ε^α such that

$$\delta_*^\varepsilon \mathcal{F} = \partial_\mu F_\varepsilon{}^\mu. \qquad (5.2.4-2)$$

In most cases, it is sufficient to assume that no derivatives of ε^α are contained in $F_\varepsilon{}^\mu$, but in order to apply our result to general relativity, we should admit the case in which the first and second derivatives of ε^α are contained in $F_\varepsilon{}^\mu$, that is, we set

$$F_\varepsilon{}^\mu \equiv F^{(0)\mu}{}_\alpha \varepsilon^\alpha + F^{(1)\mu\lambda}{}_\alpha \partial_\lambda \varepsilon^\alpha + F^{(2)\mu\lambda\rho}{}_\alpha \partial_\lambda \partial_\rho \varepsilon^\alpha \qquad (5.2.4-3)$$

with $F^{(2)\mu\lambda\rho}{}_\alpha = F^{(2)\mu\rho\lambda}{}_\alpha$.

Now, we substitute Eqs.(1) and (3) into Eq.(2). Since ε^α's are arbitrary functions, the coefficients of the zeroth, first, second, and third derivatives of ε^α must coincide *identically* in both sides, that is, we have

$$G_{A\alpha} \frac{\partial \mathcal{F}}{\partial \varphi_A} + \partial_\mu G_{A\alpha} \cdot \frac{\partial \mathcal{F}}{\partial(\partial_\mu \varphi_A)} = \partial_\mu F^{(0)\mu}{}_\alpha, \qquad (5.2.4-4)$$

$$T_A{}^\mu{}_\alpha \frac{\partial \mathcal{F}}{\partial \varphi_A} + G_{A\alpha} \frac{\partial \mathcal{F}}{\partial(\partial_\mu \varphi_A)} + \partial_\nu T_A{}^\mu{}_\alpha \cdot \frac{\partial \mathcal{F}}{\partial(\partial_\nu \varphi_A)} = F^{(0)\mu}{}_\alpha + \partial_\lambda F^{(1)\lambda\mu}{}_\alpha, \quad (5.2.4-5)$$

$$T_A{}^\mu{}_\alpha \frac{\partial \mathcal{F}}{\partial(\partial_\nu \varphi_A)} + (\mu \leftrightarrow \nu) = F^{(1)\mu\nu}{}_\alpha + \partial_\lambda F^{(2)\lambda\mu\nu}{}_\alpha + (\mu \leftrightarrow \nu), \qquad (5.2.4-6)$$

$$0 = F^{(2)\mu\lambda\rho}{}_\alpha + (5 \text{ permutations of } \mu,\ \lambda,\ \rho). \qquad (5.2.4-7)$$

Setting

$$\tilde{F}^{(2)\lambda\mu\nu}{}_\alpha \equiv F^{(2)\lambda\mu\nu}{}_\alpha + (F^{(2)\nu\mu\lambda}{}_\alpha - F^{(2)\mu\nu\lambda}{}_\alpha)/3, \qquad (5.2.4-8)$$

$$\mathcal{K}^{\mu\nu}{}_\alpha \equiv T_A{}^\nu{}_\alpha \frac{\partial \mathcal{F}}{\partial(\partial_\mu \varphi_A)} - F^{(1)\mu\nu}{}_\alpha - \partial_\lambda \tilde{F}^{(2)\lambda\mu\nu}{}_\alpha, \qquad (5.2.4-9)$$

we have

$$\mathcal{K}^{\mu\nu}{}_\alpha = -\mathcal{K}^{\nu\mu}{}_\alpha, \qquad (5.2.4-10)$$

$$\partial_\nu \partial_\lambda \tilde{F}^{(2)\lambda\nu\mu}{}_\alpha = 0. \qquad (5.2.4-11)$$

Hence Eq.(5) can be rewritten as

$$J^\mu{}_\alpha = -\partial_\nu \mathcal{K}^{\nu\mu}{}_\alpha - T_A{}^\mu{}_\alpha \frac{\delta \mathcal{F}}{\delta \varphi_A}, \qquad (5.2.4-12)$$

where

$$J^\mu{}_\alpha \equiv G_{A\alpha} \frac{\partial \mathcal{F}}{\partial(\partial_\mu \varphi_A)} - F^{(0)\mu}{}_\alpha. \qquad (5.2.4-13)$$

Note that if $\mathcal{F}$ is the total Lagrangian density, $J^\mu{}_\alpha$ is nothing but the Noether current. In this case, since $\frac{\delta \mathcal{F}}{\delta \varphi_A} = 0$, $J^\mu{}_\alpha$ becomes a total divergence of an anti-symmetric tensor.

On the other hand, Eq.(4) is rewritten as

$$\partial_\mu J^\mu{}_\alpha = -G_{A\alpha} \frac{\delta \mathcal{F}}{\delta \varphi_A}. \qquad (5.2.4-14)$$

This is nothing but the (first) Noether theorem.

We now apply the above second Noether theorem to general relativity. Under an infinitesimal general coordinate transformation, φ_A transforms as

$$\varphi'_A(x') = \varphi_A(x) + [\varphi_A]^\nu{}_\mu \partial_\nu \varepsilon^\mu, \qquad (5.2.4-15)$$

using the notation of Sec.5.2.2. Therefore, we have

$$\begin{aligned} \delta_*^\varepsilon \varphi_A(x) &\equiv \varphi'_A(x) - \varphi_A(x) \\ &= -\partial_\alpha \varphi_A \cdot \varepsilon^\alpha + [\varphi_A]^\mu{}_\alpha \partial_\mu \varepsilon^\alpha \end{aligned} \qquad (5.2.4-16)$$

as the formula to be identified with Eq.(1).

First, we consider the case $\mathcal{F} = \mathcal{L}_\mathrm{M}$. Since $\mathcal{L}_\mathrm{M}$ is a scalar density, we have

$$\delta_*^\varepsilon \mathcal{L}_\mathrm{M} = -\partial_\mu(\mathcal{L}_\mathrm{M} \varepsilon^\mu), \qquad (5.2.4-17)$$

that is, derivative terms are absent. Then Eq.(10) implies

$$\mathcal{K}_\mathrm{M}{}^{\mu\nu}{}_\alpha \equiv [\varphi_A]^\nu{}_\alpha \frac{\partial \mathcal{L}_\mathrm{M}}{\partial(\partial_\mu \varphi_A)} = -\mathcal{K}_\mathrm{M}{}^{\nu\mu}{}_\alpha. \qquad (5.2.4-18)$$

Furthermore, Eq.(12) becomes[8]

$$\partial_\alpha \varphi_A \frac{\partial \mathcal{L}_\mathrm{M}}{\partial(\partial_\mu \varphi_A)} - \delta^\mu{}_\alpha \mathcal{L}_\mathrm{M} = \partial_\nu \mathcal{K}_\mathrm{M}{}^{\nu\mu}{}_\alpha + hT^\mu{}_\alpha, \qquad (5.2.4-19)$$

---

[8]   Note that the lhs. of Eq.(19) is the matter *canonical* energy-momentum tensor density plus $\partial_\alpha g_{\sigma\tau} \cdot \partial \mathcal{L}_\mathrm{M}/\partial(\partial_\mu g_{\sigma\tau})$.

owing to matter field equations and Eq.(5.2.3-6).

Next, we consider the case in which $\mathcal{F}$ is identified with the total Lagrangian density $\tilde{\mathcal{L}}_{\mathrm{E}} + \mathcal{L}_{\mathrm{M}}$. Because of Eq.(5.2.3-2) we have

$$\tilde{\mathcal{L}}_{\mathrm{E}} + \mathcal{L}_{\mathrm{M}} = (\mathcal{L}_{\mathrm{E}} + \mathcal{L}_{\mathrm{M}}) - \partial_\mu \mathcal{V}^\mu, \qquad (5.2.4-20)$$

$\mathcal{L}_{\mathrm{E}} + \mathcal{L}_{\mathrm{M}}$ being a scalar density. Accordingly, we find

$$F_\varepsilon{}^\mu = -(\mathcal{L}_{\mathrm{E}} + \mathcal{L}_{\mathrm{M}})\varepsilon^\mu - \delta_*{}^\varepsilon \mathcal{V}^\mu, \qquad (5.2.4-21)$$

that is,

$$F^{(0)\mu}{}_\alpha = -\delta^\mu{}_\alpha (\mathcal{L}_{\mathrm{E}} + \mathcal{L}_{\mathrm{M}}) + \partial_\alpha \mathcal{V}^\mu, \qquad (5.2.4-22)$$

$$F^{(1)\mu\nu}{}_\alpha = \delta^\nu{}_\alpha \mathcal{V}^\mu - \delta^\mu{}_\alpha \mathcal{V}^\nu, \qquad (5.2.4-23)$$

$$F^{(2)\mu\nu\lambda}{}_\alpha = -(4\kappa)^{-1}(2\delta^\mu{}_\alpha \tilde{g}^{\nu\lambda} - \delta^\nu{}_\alpha \tilde{g}^{\mu\lambda} - \delta^\lambda{}_\alpha \tilde{g}^{\mu\nu}). \qquad (5.2.4-24)$$

Of course, Eq.(24) satisfies Eq.(7).

We set

$$\mathcal{K}_{\mathrm{E}}{}^{\mu\nu}{}_\alpha \equiv [\varphi_A]^\nu{}_\alpha \frac{\partial \tilde{\mathcal{L}}_{\mathrm{E}}}{\partial(\partial_\mu \varphi_A)} - \delta^\nu{}_\alpha \mathcal{V}^\mu + \delta^\mu{}_\alpha \mathcal{V}^\nu + (2\kappa)^{-1}\partial_\lambda(\delta^\lambda{}_\alpha \tilde{g}^{\nu\mu} - \delta^\mu{}_\alpha \tilde{g}^{\lambda\nu}).$$
$$(5.2.4-25)$$

It is antisymmetric under $\mu \leftrightarrow \nu$, owing to Eq.(10) together with Eq.(18); indeed, explicitly

$$\mathcal{K}_{\mathrm{E}}{}^{\mu\nu}{}_\alpha = -(2\kappa)^{-1}(\tilde{g}^{\mu\sigma} g^{\nu\tau} - \tilde{g}^{\mu\tau} g^{\nu\sigma})\partial_\nu g_{\alpha\tau}. \qquad (5.2.4-26)$$

From Eq.(12) with field equations, we have

$$\partial_\alpha \varphi_A \cdot \frac{\partial(\tilde{\mathcal{L}}_{\mathrm{E}} + \mathcal{L}_{\mathrm{M}})}{\partial(\partial_\mu \varphi_A)} - \delta^\mu{}_\alpha (\tilde{\mathcal{L}}_{\mathrm{E}} + \mathcal{L}_{\mathrm{M}}) + \partial_\alpha \mathcal{V}^\mu = \partial_\nu(\mathcal{K}_{\mathrm{E}}{}^{\nu\mu}{}_\alpha + \mathcal{K}_{\mathrm{M}}{}^{\nu\mu}{}_\alpha). \qquad (5.2.4-27)$$

Then, by using Eqs.(19) and (5.2.3-16), we obtain

$$h(t^\mu{}_\alpha + T^\mu{}_\alpha) = \partial_\nu \mathcal{K}_{\mathrm{E}}{}^{\nu\mu}{}_\alpha - \partial_\nu(\delta^\nu{}_\alpha \mathcal{V}^\mu - \delta^\mu{}_\alpha \mathcal{V}^\nu). \qquad (5.2.4-28)$$

We have thus proved Eq.(5.2.3-13) with Eq.(5.2.3-14), where

$$\begin{aligned}
\mathcal{F}_{\mathrm{C}}{}^{\nu\mu}{}_\alpha &\equiv -\kappa(\mathcal{K}_{\mathrm{E}}{}^{\nu\mu}{}_\alpha - \delta^\nu{}_\alpha \mathcal{V}^\mu + \delta^\mu{}_\alpha \mathcal{V}^\nu) \\
&= -\mathcal{F}_{\mathrm{C}}{}^{\mu\nu}{}_\alpha.
\end{aligned} \qquad (5.2.4-29)$$

It should be noted that, in deriving Eq.(27), *we have used the Einstein equation* $\delta(\tilde{\mathcal{L}}_{\mathrm{E}} + \mathcal{L}_{\mathrm{M}})/\delta g_{\sigma\tau} = 0$, whence Eq.(28) is *valid only in the classical theory*. On the other hand, Eqs.(18) and (19) remain valid also in quantum gravity.

5.3 QUANTUM LAGRANGIAN DENSITY

The quantum Lagrangian density of the gravitational field consists of the classical Lagrangian density $\mathcal{L}_E + \mathcal{L}_M$, the gauge-fixing one, and the FP-ghost one, as in the case of the non-abelian gauge field. The sum of the latter two is, of course, invariant under the BRS transformation, so as to guarantee the unitarity of the physical S-matrix, as discussed in Sec.4.1.4.

5.3.1 Gravitational BRS transformation

Since the invariance under general coordinate transformations is a local symmetry, we should construct the corresponding BRS transformation in order to write down the quantum Lagrangian density of the gravitational field. The recipe of defining the BRS transformation is essentially the same as in the non-abelian gauge theory, but we encounter a new aspect: General coordinate invariance is a *spacetime* symmetry in contrast to gauge invariance.

We consider an infinitesimal general coordinate transformation $x'^\lambda = x^\lambda + \delta x^\lambda$. Let $\Phi(x)$ be a generic field; Φ transforms into $\Phi + \delta\Phi$. For example, if $\Phi(x)$ is a tensor, $\delta\Phi$ is given by

$$\delta\Phi = \partial_\nu \delta x^\lambda \cdot [\Phi]^\nu{}_\lambda, \qquad (5.3.1-1)$$

where $[\Phi]^\nu{}_\lambda$ is defined in Sec.5.2.2. It should be noted, however, that the argument of $\Phi + \delta\Phi$ is not x^λ but $x^\lambda + \delta x^\lambda$. If we want to keep the original coordinates x^λ, we must subtract $\delta x^\lambda \cdot \partial_\lambda \Phi$, the term due to the coordinate change; that is, we define

$$\delta_*\Phi \equiv \delta\Phi - \delta x^\lambda \cdot \partial_\lambda \Phi. \qquad (5.3.1-2)$$

In contrast to δ, δ_* does not change the values of coordinates, whence δ_* commutes with ∂_μ.

We can transcribe the above consideration into the BRS transformation. Corresponding to δ and δ_*, we define *two* kinds of the BRS transformation, denoted by $\boldsymbol{\delta}$ and $\boldsymbol{\delta_*}$. We call $\boldsymbol{\delta}$ the **intrinsic BRS transformation** and $\boldsymbol{\delta_*}$ the **total BRS transformation**. At the operator level, only $\boldsymbol{\delta_*}$ is realizable as an algebraic transformation, but $\boldsymbol{\delta}$ is conceptually simpler and important for the fundamental consideration. They are mutually related through

$$\boldsymbol{\delta_*}(\Phi) = \boldsymbol{\delta}(\Phi) - \boldsymbol{\delta}(x^\lambda)\partial_\lambda \Phi. \qquad (5.3.1-3)$$

Of course, $\boldsymbol{\delta}_*$ commutes with ∂_μ. Furthermore, from Eq.(3), we see $\boldsymbol{\delta}_*(x^\lambda) = 0$. As in Sec.1.2.4, we call the second term of Eq.(3) the **orbital part of the BRS symmetry**.

According to the recipe of defining the BRS transformation presented in Sec.3.3.2, the infinitesimal transformation function δx^λ is replaced by an FP ghost c^λ. In order to make $\boldsymbol{\delta}$ dimension-conserving, it is customary to insert a coefficient κ. We thus have

$$\boldsymbol{\delta}(x^\lambda) = \kappa c^\lambda, \qquad (5.3.1-4)$$

whence Eq.(3) becomes

$$\boldsymbol{\delta}_*(\Phi) = \boldsymbol{\delta}(\Phi) - \kappa c^\lambda \partial_\lambda \Phi. \qquad (5.3.1-5)$$

If Φ is a tensor, $\boldsymbol{\delta}(\Phi)$ equals

$$\boldsymbol{\delta}(\Phi) = \kappa \partial_\nu c^\lambda \cdot [\Phi]^\nu{}_\lambda, \qquad (5.3.1-6)$$

whence

$$\boldsymbol{\delta}_*(\Phi) = -\kappa c^\lambda \partial_\lambda \Phi + \kappa \partial_\nu c^\lambda \cdot [\Phi]^\nu{}_\lambda. \qquad (5.3.1-7)$$

The intrinsic BRS transformation does not commute with ∂_μ; the commutator between them is given by

$$\boldsymbol{\delta}(\partial_\mu \Phi) - \partial_\mu \boldsymbol{\delta}(\Phi) = -\kappa \partial_\mu c^\lambda \cdot \partial_\lambda \Phi, \qquad (5.3.1-8)$$

as is verified by using Eq.(5) together with the commutativity between $\boldsymbol{\delta}_*$ and ∂_μ.

The BRS transformation should be nilpotent. From Eq.(4), we therefore have

$$\boldsymbol{\delta}(c^\lambda) = 0. \qquad (5.3.1-9)$$

Then, extending Eq.(5) to c^ν, we set

$$\boldsymbol{\delta}_*(c^\nu) = -\kappa c^\lambda \partial_\lambda c^\nu. \qquad (5.3.1-10)$$

It is very important to note that Eq.(9) expresses the *abelian nature of the translation group* whose local version is nothing but the group of general coordinate transformations. On the other hand, Eq.(10) is quite akin to the corresponding formula for the non-abelian gauge theory. This non-abelian nature corresponds to the non-commutativity of two local translations in the gauge-theoretical approach to general relativity.

When Φ is a tensor T_Σ, we can directly confirm

$$\boldsymbol{\delta}(\boldsymbol{\delta}(\Phi)) = 0 \qquad (5.3.1 - 11)$$

by using Eqs.(6), (8), (9), and (5.2.2-4), but the manipulation is rather complicated. To prove Eq.(11), it is much simpler to employ mathematical induction with respect to the number of tensorial indices. First, one confirms $\boldsymbol{\delta}(\boldsymbol{\delta}(\phi)) = 0$, $\boldsymbol{\delta}(\boldsymbol{\delta}(A_\lambda)) = 0$, and $\boldsymbol{\delta}(\boldsymbol{\delta}(B^\nu)) = 0$ for a scalar field ϕ and vector fields A_λ and B^ν. Then one construct a new quantity by contracting one of the indices Σ of $\Phi = T_\Sigma$ by either A_λ or B^ν. Finally, one takes the intrinsic BRS transformation of this quantity twice and uses the induction assumption. In this way, Eq.(11) is established.

The nilpotency of the total BRS transformation $\boldsymbol{\delta}_*$ immediately follows from Eqs.(11) and (5) together with Eq.(10). Here the sign choice in Eq.(10), which was not determined by Eq.(5) because of the ordering ambiguity of the FP ghosts, becomes important.

Historically, the gravitational BRS transformation was introduced independently by several authors[1] [Del 76, Ste 77, Tow 77]. They consider the total BRS transformation $\boldsymbol{\delta}_*$, regarding the gravitational field as a gauge field in the Minkowski spacetime. The intrinsic BRS transformation $\boldsymbol{\delta}$ was introduced by Nakanishi [Nak 78b], who emphasized the fundamental importance of the BRS transformation for formulating quantum gravity in the Heisenberg picture. The relation, Eq.(5), was found afterwards.

5.3.2 Gauge fixing and FP-ghost Lagrangian densities

Just as in the gauge theory, general relativity cannot be quantized without gauge fixing, and in order to maintain manifest covariance, we should introduce a gauge-fixing term into the fundamental Lagrangian density. Corresponding to the de Donder condition, Eq.(5.2.3-10), such a gauge-fixing term as

$$(1/2\alpha)\eta_{\mu\nu}\partial_\lambda\tilde{g}^{\lambda\mu} \cdot \partial_\rho\tilde{g}^{\rho\nu} \qquad (5.3.2 - 1)$$

was often employed [Gup 68, Fra 70]. Here $\eta_{\mu\nu}$ is, of course, the Minkowski metric.

---

[1]   It seems that Dixon was the first (according to [Del 76]), but unfortunately his work remains unpublished.

If one requires Poincaré invariance only, the choice of a gauge-fixing term is quite arbitrary, because there exist *two* metric tensors $g_{\mu\nu}$ and $\eta_{\mu\nu}$ and also an invariant quantity h. From the spirit of general relativity, it is quite unnatural to introduce $\eta_{\mu\nu}$ into the fundamental Lagrangian density. Indeed, there is no reason for assuming the *a priori* existence of a particular Cartesian coordinate system. To forbid $\eta_{\mu\nu}$, it is reasonable to require the $GL(4)$ invariance [Nak 78b], which is relevant already at the classical level, as discussed in Sec.5.2.3.

It is very important to note that *gauge fixing can be made in a GL(4)-invariant way if and only if one employs the B-field formalism.* Let $b_\rho(x)$ be the gravitational B-field, which is a $GL(4)$-*covariant* vector field. Then we adopt

$$\mathcal{L}_{\text{GF}} \equiv -\kappa^{-1} \tilde{g}^{\mu\nu} \partial_\mu b_\nu \qquad (5.3.2-2)$$

as a gauge-fixing Lagrangian density [Nak 78b].

From the experience in gauge theories, it might be tempting to add an α-term, i.e., a pure B-field term, to Eq.(2). But it cannot be $GL(4)$-invariant without using $g^{\mu\nu}$. Thus, under the $GL(4)$ invariance, the choice of Eq.(2) is quite distinguished.

The FP-ghost Lagrangian density $\mathcal{L}_{\text{FP}}$ is introduced in such a way that $\int d^4x \, (\mathcal{L}_{\text{GF}} + \mathcal{L}_{\text{FP}})$ is BRS-invariant. There are two ways for doing so, corresponding to $\boldsymbol{\delta}$ and $\boldsymbol{\delta}_*$. Let $c^\sigma(x)$ and $\bar{c}_r(x)$ be the gravitational FP ghost and FP antighost, respectively, which, of course, obey Fermi statistics. From the consideration in Sec.5.3.1, c^σ is a $GL(4)$-*contravariant* vector field, while $\bar{c}_r$ is a $GL(4)$-*covariant* vector field because it is related to b_ρ through the BRS transformation.

On the basis of $\boldsymbol{\delta}$, it is natural to set

$$\boldsymbol{\delta}(\bar{c}_\rho) = ib_\rho, \qquad (5.3.2-3)$$

$$\boldsymbol{\delta}(b_\rho) = 0. \qquad (5.3.2-4)$$

Since hd^4x is the invariant volume, $h^{-1}(\mathcal{L}_{\text{GF}} + \mathcal{L}_{\text{FP}})$ should be BRS-invariant. Then, from Eqs.(2) and (3) and the nilpotency of $\boldsymbol{\delta}$, we should have

$$\mathcal{L}_{\text{GF}} + \mathcal{L}_{\text{FP}} = h\boldsymbol{\delta}(i\kappa^{-1}g^{\mu\nu}\partial_\mu \bar{c}_\nu), \qquad (5.3.2-5)$$

that is[2], we set [Nak 78b]

$$\mathcal{L}_{\text{FP}} \equiv -i\tilde{g}^{\mu\nu}\partial_\mu\bar{c}_\rho \cdot \partial_\nu c^\rho. \qquad (5.3.2-6)$$

Thus the total Lagrangian density is

$$\mathcal{L} \equiv \mathcal{L}_{\text{E}} + \mathcal{L}_{\text{GF}} + \mathcal{L}_{\text{FP}} + \mathcal{L}_{\text{M}}, \qquad (5.3.2-7)$$

where $\mathcal{L}_{\text{E}}$, $\mathcal{L}_{\text{GF}}$, and $\mathcal{L}_{\text{FP}}$ are given by Eqs.(5.2.3-1), (2), and (6), respectively, and $\mathcal{L}_{\text{M}}$ denotes the matter Lagrangian density. Note that $\boldsymbol{\delta}(h^{-1}\mathcal{L}) = 0$.

There is another way of determining the FP-ghost Lagrangian density, based on $\boldsymbol{\delta}_*$ [Nis 78, Kug 78b]. This method is more analogous to the case of gauge theories. On the basis of $\boldsymbol{\delta}_*$, it is natural to set

$$\boldsymbol{\delta}_*(\bar{c}_\rho) = ib_\rho, \qquad (5.3.2-8)$$

$$\boldsymbol{\delta}_*(b_\rho) = 0 \qquad (5.3.2-9)$$

instead of Eqs.(3) and (4). Here, of course, Eq.(5.3.1-5) is violated for b_ρ and $\bar{c}_r$. The total Lagrangian density should be BRS-invariant up to a total divergence. The FP-ghost Lagrangian density $\mathcal{L}_{*\text{FP}}$ is determined by

$$\mathcal{L}_{\text{GF}} + \mathcal{L}_{*\text{FP}} = \boldsymbol{\delta}_*(i\kappa^{-1}\tilde{g}^{\mu\nu}\partial_\mu\bar{c}_\nu), \qquad (5.3.2-10)$$

that is,

$$\mathcal{L}_{*\text{FP}} \equiv -i\partial_\mu\bar{c}_\nu \cdot [\tilde{g}^{\lambda\nu}\partial_\lambda c^\mu + \tilde{g}^{\mu\lambda}\partial_\lambda c^\nu - \partial_\lambda(\tilde{g}^{\mu\nu}c^\lambda)]. \qquad (5.3.2-11)$$

At first sight, it is quite strange that there exist two FP-ghost Lagrangian densities $\mathcal{L}_{\text{FP}}$ and $\mathcal{L}_{*\text{FP}}$ for one and the same $\mathcal{L}_{\text{GF}}$. This puzzle is resolved if we note the fact that there is no reason for assuming that the B-fields in both methods are identical. If we replace b_ρ in Eqs.(8)–(10) by

$$b_{*\rho} \equiv b_\rho + i\kappa c^\lambda\partial_\lambda\bar{c}_\rho, \qquad (5.3.2-12)$$

---

[2]   Bibliographically, we note that an expression like Eq.(6) was once written just in a course of the discussion of the path-integral formalism in 1973 [Fra 73] Of course, no BRS-invariance consideration existed at that time, and the FP ghosts were regarded as world scalars One of the authors (N.N.) would like to thank M Abe for this information

then we can easily see that Eq.(10) reduces to Eq.(5) plus a total divergence [Nak 78c]. Thus both methods are equivalent.

One should note, however, that $\mathcal{L}_{FP}$ is much simpler than $\mathcal{L}_{*FP}$. To employ $\mathcal{L}_{FP}$ makes the theory very beautiful and transparent. This fact implies that to use $\boldsymbol{\delta}$ is more natural than to do $\boldsymbol{\delta}_*$ for determining the Lagrangian density. In other words, it is important to respect the *abelian nature*, Eq.(5.3.1-9), of the translation group, rather than to respect the analogy to the non-abelian gauge theory. In the past, the similarity between general relativity and the non-abelian gauge theory was dogmatically emphasized, but, in the following sections, we shall see that *the abelian nature is essential in our formalism of quantum gravity*. The gauge-theoretical approach spoils the beauty of general relativity.

5.3.3 Field equations

From the total Lagrangian density, Eq.(5.3.2-7), we obtain field equations by taking its Euler derivatives. We have

$$R_{\mu\nu} - \frac{1}{2}g_{\mu\nu}R - E_{\mu\nu} + \frac{1}{2}g_{\mu\nu}E = -\kappa T_{\mu\nu}, \qquad (5.3.3-1)$$

$$\partial_\mu \tilde{g}^{\mu\nu} = 0, \qquad (5.3.3-2)$$

$$\partial_\mu(\tilde{g}^{\mu\nu}\partial_\nu c^\sigma) = 0, \qquad (5.3.3-3)$$

$$\partial_\mu(\tilde{g}^{\mu\nu}\partial_\nu \bar{c}_\tau) = 0, \qquad (5.3.3-4)$$

and the matter field equations. Here,

$$E_{\mu\nu} \equiv \partial_\mu b_\nu + i\kappa\partial_\mu \bar{c}_\rho \cdot \partial_\nu c^\rho + (\mu \leftrightarrow \nu) \qquad (5.3.3-5)$$

and $E \equiv E^\mu{}_\mu$. With this notation, we can write

$$\mathcal{L}_{GF} + \mathcal{L}_{FP} = -(2\kappa)^{-1}hE. \qquad (5.3.3-6)$$

We call Eq.(1) the **quantum Einstein equation**. It contains extra terms $-E_{\mu\nu} + \frac{1}{2}g_{\mu\nu}E$. As in the case of gauge theories, we should take the covariant derivative of Eq.(1) to obtain the field equation for the B-field. We then have

$$\nabla_\mu(E^{\mu\nu} - \frac{1}{2}g^{\mu\nu}E) = 0 \qquad (5.3.3-7)$$

owing to Eqs.(5.2.3-7) and (5.2.3-8). With the aid of Eqs.(2)–(4), we can reduce Eq.(7) into a surprisingly simple equation

$$\partial_\mu(\tilde{g}^{\mu\nu}\partial_\nu b_\rho) = 0. \qquad (5.3.3-8)$$

As noted in Eq.(5.2.3-11), the de Donder condition, Eq.(2), can be rewritten as

$$\partial_\mu(\tilde{g}^{\mu\nu}\partial_\nu x^\lambda) = 0. \qquad (5.3.3-9)$$

We thus see that all of x^λ, b_ρ, c^σ, and $\bar{c}_\tau$ satisfy the *same* d'Alembert equation. Remarkable consequences of this very interesting fact are discussed in Sec.5.5.

5.4 EQUAL-TIME COMMUTATION RELATIONS

After setting up the canonical (anti)commutation relations, we calculate the equal-time ones between the primary fields. It is quite remarkable that all equal-time commutators can be obtained *explicitly in closed form* [Nak 78e]. This fact was never achieved in any preceding attempt at quantizing gravity.

To shorten formulae, we employ the following abbreviations throughout this section:

$$[A, \ B'] \equiv [A(x), \ B(x')]|_{x^0=x'^0}, \qquad (5.4.0-1)$$

$$\delta^3 \equiv \delta(\boldsymbol{x} - \boldsymbol{x}'), \qquad (5.4.0-2)$$

$$\tilde{f} \equiv (\tilde{g}^{00})^{-1}, \qquad (5.4.0-3)$$

$$(\delta^3)_k \equiv \partial_k(\tilde{f}\delta^3). \qquad (5.4.0-4)$$

We here assume that $\tilde{g}^{00}$ is invertible.

5.4.1 Canonical (anti)commutation relations

The primary fields are $g_{\mu\nu}$, b_ρ, c^σ, $\bar{c}_\tau$, and matter fields, but we do not explicitly discuss matter fields and assume that $\mathcal{L}_M$ involves no derivatives of $g_{\mu\nu}$ (The usual scalar-field and vector-field Lagrangian densities are all right.).

Canonical quantization is carried out by taking $g_{\mu\nu}$, c^σ, and $\bar{c}_\tau$ as canonical variables but *not* b_ρ, as in gauge theories. In contrast to the case of gauge theories, however, it is *wrong* to take g_{kl}, b_ρ, c^σ, and $\bar{c}_\tau$ as canonical variables. The reason for this can be clarified in the light of the Dirac method of quantization as described in Addendum 1.A.

Our starting Lagrangian density is

$$\tilde{\mathcal{L}} \equiv \tilde{\mathcal{L}}_{\mathrm{E}} + \tilde{\mathcal{L}}_{\mathrm{GF}} + \mathcal{L}_{\mathrm{FP}} + \mathcal{L}_{\mathrm{M}}, \qquad (5.4.1-1)$$

where $\tilde{\mathcal{L}}_{\mathrm{E}}$ is defined by Eq.(5.2.3-3) and

$$\tilde{\mathcal{L}}_{\mathrm{GF}} \equiv \kappa^{-1} \partial_\mu \tilde{g}^{\mu\nu} \cdot b_\nu. \qquad (5.4.1-2)$$

Hence we have

$$\mathcal{L} - \tilde{\mathcal{L}} = \partial_\mu(\mathcal{V}^\mu - \kappa^{-1}\tilde{g}^{\mu\nu}b_\nu), \qquad (5.4.1-3)$$

where $\mathcal{V}^\mu$ is given by Eq.(5.2.3-4).

The canonical conjugates of $g_{\mu\nu}$, c^σ, and $\bar{c}_\tau$ are, respectively, as follows:[1]

$$\pi_g{}^{\mu\nu} \equiv (\partial/\partial \dot{g}_{\mu\nu})\tilde{\mathcal{L}}$$

$$= -(4\kappa)^{-1} h [g^{0\lambda} g^{\mu\nu} g^{\sigma\tau} + g^{0\tau} g^{\mu\lambda} g^{\nu\sigma} + g^{0\sigma} g^{\mu\tau} g^{\nu\lambda}$$

$$- g^{0\lambda} g^{\mu\tau} g^{\nu\sigma} - g^{0\tau} g^{\mu\nu} g^{\lambda\sigma} - \frac{1}{2}(g^{0\mu} g^{\nu\lambda} + g^{0\nu} g^{\mu\lambda}) g^{\sigma\tau}] \partial_\lambda g_{\sigma\tau}$$

$$- (2\kappa)^{-1} h (g^{0\mu} g^{\nu\rho} + g^{0\nu} g^{\mu\rho} - g^{0\rho} g^{\mu\nu}) b_\rho, \qquad (5.4.1-4)$$

$$\pi_{c\sigma} \equiv (\partial/\partial \dot{c}^\sigma)\tilde{\mathcal{L}} = i\tilde{g}^{\mu 0} \partial_\mu \bar{c}_\sigma, \qquad (5.4.1-5)$$

$$\pi_{\bar{c}}{}^\tau \equiv (\partial/\partial \dot{\bar{c}}_\tau)\tilde{\mathcal{L}} = -i\tilde{g}^{\mu 0} \partial_\mu c^\tau. \qquad (5.4.1-6)$$

The canonical (anti)commutation relations are

$$[\pi_g{}^{\rho\lambda}, g_{\mu\nu}'] = -i\frac{1}{2}(\delta^\lambda{}_\mu \delta^\rho{}_\nu + \delta^\lambda{}_\nu \delta^\rho{}_\mu)\delta^3, \qquad (5.4.1-7)$$

$$\{\pi_{c\lambda}, c^{\sigma\prime}\} = -i\delta^\sigma{}_\lambda \delta^3, \qquad (5.4.1-8)$$

$$\{\pi_{\bar{c}}{}^\lambda, \bar{c}_\tau'\} = -i\delta^\lambda{}_\tau \delta^3; \qquad (5.4.1-9)$$

all other (anti)commutators vanish.

5.4.2 Equal-time commutators involving no time derivative

The equal-time (anti)commutators between two fields vanish if they both are canonical variables. But, since b_ρ is not a canonical field, the commutators involving b_ρ may not vanish.

First, we note that since a commutator is a derivation, Eqs.(5.2.1-10) and (5.2.1-11) imply that

$$[g^{\mu\nu}, \Phi'] = -g^{\mu\lambda} g^{\nu\rho}[g_{\lambda\rho}, \Phi'], \qquad (5.4.2-1)$$

$$[h, \Phi'] = \frac{1}{2} h g^{\mu\nu}[g_{\mu\nu}, \Phi'] \qquad (5.4.2-2)$$

for an arbitrary quantity Φ.

From Eq.(5.4.1-4), we can write

$$\pi_g{}^{\lambda 0} = A^\lambda + B^{\lambda\rho} b_\rho, \qquad (5.4.2-3)$$

---

[1]   Here, as in Sec.5.2.3, $\partial/\partial \dot{g}_{\mu\nu}$ is defined by $\frac{1}{2}(\partial/\partial \dot{g}_{\mu\nu} + \partial/\partial \dot{g}_{\nu\mu})$, regarding $g_{\mu\nu}$ as if all its components were independent.

where A^λ and $B^{\lambda\rho}$ involve no $\dot{g}_{\sigma\tau}$ and hence commute with $g_{\mu\nu}$. Furthermore, $B^{\lambda\rho} \equiv -(2\kappa)^{-1}\tilde{g}^{00}g^{\lambda\rho}$ is invertible as a 4×4 matrix. Accordingly, Eq.(5.4.1-7) (with $\rho = 0$) yields

$$[g_{\mu\nu},\, b_\rho{}'] = -i\kappa\tilde{f}(\delta^0{}_\mu g_{\rho\nu} + \delta^0{}_\nu g_{\mu\rho})\delta^3, \qquad (5.4.2-4)$$

where $\tilde{f}$ is given by Eq.(5.4.0-3).

Next, from $[\pi_g{}^{\mu 0},\, \pi_g{}^{\nu 0'}] = 0$, we can calculate $[b_\rho,\, b_\lambda{}']$ by using Eq.(4). We then find

$$[b_\rho,\, b_\lambda{}'] = 0. \qquad (5.4.2-5)$$

Finally, the commutators $[b_\rho,\, c^{\sigma'}]$ and $[b_\rho,\, \bar{c}_\tau{}']$ evidently vanish.

5.4.3 Equal-time commutators involving one time derivative other than $\dot{b}_\rho$

From Eq.(5.4.1-4), we can write

$$\pi_g{}^{k\ell} = \hat{A}^{k\ell} + \hat{B}^{k\ell\rho}b_\rho + \hat{C}^{k\ell mn}\dot{g}_{mn}, \qquad (5.4.3-1)$$

where $\hat{A}^{k\ell}$, $\hat{B}^{k\ell\rho}$, and $\hat{C}^{k\ell mn}$ commute with $g_{\mu\nu}$ as before. Furthermore, we have

$$\hat{C}^{k\ell mn} = -(4\kappa)^{-1}hK^{k\ell mn}, \qquad (5.4.3-2)$$

where

$$K^{\mu\nu\sigma\tau} \equiv \begin{vmatrix} g^{00} & g^{0\nu} & g^{0\tau} \\ g^{\mu 0} & g^{\mu\nu} & g^{\mu\tau} \\ g^{\sigma 0} & g^{\sigma\nu} & g^{\sigma\tau} \end{vmatrix}. \qquad (5.4.3-3)$$

Evidently, $K^{\mu\nu\sigma\tau}$ is antisymmetric under $\mu \leftrightarrow \sigma$ and under $\nu \leftrightarrow \tau$, and vanishes if any one index is zero, whence

$$K^{\mu\nu\sigma\tau}X_{\sigma\tau} = K^{\mu\nu mn}X_{mn} \qquad (5.4.3-4)$$

for any $X_{\sigma\tau}$. Thanks to Eq.(4), it is easy to see that

$$K^{k\ell mn}\frac{1}{2}(g^{00})^{-1}(g_{ij}g_{mn} - g_{im}g_{jn} - g_{in}g_{jm}) = \frac{1}{2}(\delta^k{}_i\delta^\ell{}_j + \delta^\ell{}_i\delta^k{}_j). \qquad (5.4.3-5)$$

That is, $K^{k\ell mn}$ is invertible as a 6×6 matrix.

By using the above results, it is now possible to calculate $[g_{\mu\nu},\ \dot{g}_{mn}{}']$ from Eq.(5.4.1-7) (with $\lambda = k$, $\rho = \ell$). Then we use the de Donder condition, Eq.(5.3.3-2), to express $\dot{g}_{\lambda 0}$ in terms of $\dot{g}_{mn}$. In this way, we find[2] [Nak 78e]

$$[g_{\mu\nu},\ \dot{g}_{\sigma\tau}{}'] = -2i\kappa\tilde{f}[g_{\mu\nu}g_{\sigma\tau} - g_{\mu\sigma}g_{\nu\tau} - g_{\mu\tau}g_{\nu\sigma}$$
$$+ h\tilde{f}(\delta^0{}_\mu\delta^0{}_\sigma g_{\nu\tau} + \delta^0{}_\mu\delta^0{}_\tau g_{\nu\sigma} + \delta^0{}_\nu\delta^0{}_\sigma g_{\mu\tau} + \delta^0{}_\nu\delta^0{}_\tau g_{\mu\sigma})]\delta^3. \quad (5.4.3-6)$$

We then have

$$[g_{\mu\nu},\ \dot{\tilde{g}}^{\sigma\tau}{}'] = [\tilde{g}^{\sigma\tau},\ \dot{g}_{\mu\nu}{}']$$
$$= -2i\kappa h\tilde{f}^2[(\delta^\sigma{}_\mu\delta^\tau{}_\nu\tilde{g}^{00} - \delta^\sigma{}_\mu\delta^0{}_\nu\tilde{g}^{0\tau}$$
$$- \delta^0{}_\mu\delta^\tau{}_\nu\tilde{g}^{\sigma 0} - \delta^0{}_\mu\delta^0{}_\nu\tilde{g}^{\sigma\tau}) + (\sigma \leftrightarrow \tau)]\delta^3, \quad (5.4.3-7)$$
$$[\tilde{g}^{\mu\nu},\ \dot{\tilde{g}}^{\sigma\tau}{}'] = -2i\kappa h^3\tilde{f}^2(K^{\mu\nu\sigma\tau} + K^{\mu\nu\tau\sigma})\delta^3. \quad (5.4.3-8)$$

Next, we consider $[\dot{g}_{\mu\nu},\ b_\rho{}']$. We calculate $[\pi_g{}^{k\ell},\ b_\rho{}']$ from $[\pi_g{}^{k\ell},\ \pi_g{}^{\lambda 0}{}'] = 0$ with Eq.(5.4.2-3) by using Eq.(5.4.1-7). Then substituting Eq.(1) into $[\pi_g{}^{k\ell},\ b_\rho{}']$, we can calculate $[\dot{g}_{mn},\ b_\rho{}']$ by using Eqs.(5.4.2-4) and (5.4.2-5). To obtain $[\dot{g}_{\lambda 0},\ b_\rho{}']$, we again make use of the de Donder condition, Eq.(5.3.3-2). After very elaborate calculation, we find

$$[\dot{g}_{\mu\nu},\ b_\rho{}'] = -i\kappa\{\tilde{f}(\partial_\rho g_{\mu\nu} + \delta^0{}_\mu\dot{g}_{\rho\nu} + \delta^0{}_\nu\dot{g}_{\mu\rho}) \cdot \delta^3$$
$$+ [(\delta^k{}_\mu - 2\delta^0{}_\mu\tilde{f}\tilde{g}^{0k})g_{\rho\nu} + (\mu \leftrightarrow \nu)](\delta^3)_k\}, \quad (5.4.3-9)$$

where $(\delta^3)_k$ is given by Eq.(5.4.0-4).

Finally, it is very easy to calculate (anti)commutation relations involving c^σ and/or $\bar{c}_\tau$. We have

$$[\dot{c}^\sigma,\ b_\rho{}'] = -i\kappa\tilde{f}\partial_\rho c^\sigma \cdot \delta^3, \quad (5.4.3-10)$$
$$[\dot{\bar{c}}_\tau,\ b_\rho{}'] = -i\kappa\tilde{f}\partial_\rho\bar{c}_\tau \cdot \delta^3, \quad (5.4.3-11)$$
$$\{c^\sigma,\ \dot{\bar{c}}_\tau{}'\} = -\{\bar{c}_\tau,\ \dot{c}^{\sigma}{}'\} = -\delta^\sigma{}_\tau\tilde{f}\delta^3; \quad (5.4.3-12)$$

others vanish.

---

[2]   In D dimensions, the term $g_{\mu\nu}g_{\sigma\tau}$ acquires a coefficient $2/(D-2)$.

5.4.4 Equal-time commutators involving $\dot{b}_\rho$

Since

$$[\Phi, \ \dot{\Psi}'] = \partial_0[\Phi, \ \Psi'] - [\dot{\Phi}, \ \Psi'], \qquad (5.4.4-1)$$

$[\Phi, \ \dot{b}_\rho']$ for $\Phi \neq b_\rho$ is immediately obtained from $[\dot{\Phi}, \ b_\rho']$. Hence we have only to consider $[b_\rho, \ \dot{b}_\lambda']$. To calculate it, we need the quantum Einstein equation, Eq.(5.3.3-1).

First, we solve $\dot{b}_\rho$ from Eq.(5.3.3-5):

$$\dot{b}_0 = \frac{1}{2}E_{00} - i\kappa\dot{\bar{c}}_\sigma\dot{c}^\sigma, \qquad (5.4.4-2)$$

$$\dot{b}_k = E_{0k} - \partial_k b_0 - i\kappa(\dot{\bar{c}}_\sigma\partial_k c^\sigma + \partial_k\bar{c}_\sigma \cdot \dot{c}^\sigma). \qquad (5.4.4-3)$$

Since $R^0{}_\nu - \frac{1}{2}\delta^0{}_\nu R$ contains no $\ddot{g}_{mn}$, we make use of Eq.(5.3.3-1) in the form

$$E^0{}_\nu - \frac{1}{2}\delta^0{}_\nu E = R^0{}_\nu - \frac{1}{2}\delta^0{}_\nu R + \kappa T^0{}_\nu. \qquad (5.4.4-4)$$

Then we obtain

$$\dot{b}_0 = (g^{00})^{-1}(R^0{}_0 - \frac{1}{2}R + \kappa T^0{}_0 + \frac{1}{2}g^{k\ell}E_{k\ell}) - i\kappa\dot{\bar{c}}_\sigma\dot{c}^\sigma, \qquad (5.4.4-5)$$

$$\dot{b}_k = (g^{00})^{-1}(R^0{}_k + \kappa T^0{}_k - g^{0\jmath}E_{\jmath k})$$
$$- \partial_k b_0 - i\kappa(\dot{\bar{c}}_\sigma\partial_k c^\sigma + \partial_k\bar{c}_\sigma \cdot \dot{c}^\sigma). \qquad (5.4.4-6)$$

Assuming a concrete expression for $\mathcal{L}_M$ (e.g., the scalar-field Lagrangian density), we show that

$$[T^\mu{}_\nu, \ b_\rho'] = i\kappa\tilde{f}(\delta^\mu{}_\rho T^0{}_\nu - \delta^0{}_\nu T^\mu{}_\rho)\delta^3. \qquad (5.4.4-7)$$

We shall see later (Sec.5.6) that the validity of Eq.(7) is model independent. By using Eqs.(5.4.2-4) and (5.4.3-9), we can show that

$$[R^0{}_\nu - \frac{1}{2}\delta^0{}_\nu R, \ b_\rho'] = i\kappa\tilde{f}(\delta^0{}_\rho R^0{}_\nu - \delta^0{}_\nu R^0{}_\rho)\delta^3. \qquad (5.4.4-8)$$

This calculation is quite complicated. An elegant proof of Eq.(8) can be given if one makes use of the vierbein formalism presented in Sec.5.7.

We can now calculate $[b_\rho, \ \dot{b}_\lambda']$ with the aid of the above formulae together with Eqs.(5.4.2-4), (5.4.3-10), and (5.4.3-11). Rewriting the resultant by using Eqs.(5) and (6) again, we finally obtain a beautiful formula [Nak 78e]

$$[b_\rho, \ \dot{b}_\lambda'] = i\kappa\tilde{f}(\partial_\rho b_\lambda + \partial_\lambda b_\rho) \cdot \delta^3. \qquad (5.4.4-9)$$

5.5 SYMMETRIES OF QUANTUM GRAVITY

The quantum theory of gravity based on the Lagrangian density given by Eq.(5.3.2-7) is, of course, invariant under $GL(4)$ and BRS transformations. But they are not the only symmetries. Indeed, it turns out that the theory has an unexpectedly beautiful symmetry $IOSp(8,8)$, which is called "choral symmetry". This extremely remarkable result uniquely characterizes the expression for $\mathcal{L}_{\text{GF}} + \mathcal{L}_{\text{FP}}$.

5.5.1 $GL(4)$ and BRS symmetries

The (total) canonical energy-momentum tensor density $\mathcal{T}^\mu{}_\nu$ is defined by[1]

$$\mathcal{T}^\mu{}_\nu \equiv \partial_\nu \varphi_A \frac{\partial \tilde{\mathcal{L}}}{\partial(\partial_\mu \varphi_A)} - \delta^\mu{}_\nu \tilde{\mathcal{L}}. \qquad (5.5.1-1)$$

Owing to Eqs.(5.2.3-16) and (5.2.4-19), Eq.(1) can be rewritten as

$$\mathcal{T}^\mu{}_\nu = h(t^\mu{}_\nu + T^\mu{}_\nu + \tau^\mu{}_\nu) + \partial_\lambda \mathcal{K}_{\text{M}}{}^{\lambda\mu}{}_\nu, \qquad (5.5.1-2)$$

where $\mathcal{K}_{\text{M}}{}^{\lambda\mu}{}_\nu$ is given by Eq.(5.2.4-18) and

$$
\begin{aligned}
h\tau^\mu{}_\nu \equiv & [\partial_\nu g_{\sigma\tau} \cdot \partial/\partial(\partial_\mu g_{\sigma\tau}) - \delta^\mu{}_\nu](\tilde{\mathcal{L}}_{\text{GF}} + \mathcal{L}_{\text{FP}}) \\
& + [\partial_\nu c^\sigma \cdot \partial/\partial(\partial_\mu c^\sigma) + \partial_\nu \bar{c}_\tau \cdot \partial/\partial(\partial_\mu \bar{c}_\tau)]\mathcal{L}_{\text{FP}} \\
= & \kappa^{-1}\partial_\nu \tilde{g}^{\mu\rho} \cdot b_\rho - i\tilde{g}^{\mu\lambda}(\partial_\lambda \bar{c}_\rho \cdot \partial_\nu c^\rho + \partial_\nu \bar{c}_\rho \cdot \partial_\lambda c^\rho) \\
& - \delta^\mu{}_\nu(\tilde{\mathcal{L}}_{\text{GF}} + \mathcal{L}_{\text{FP}}). \qquad (5.5.1-3)
\end{aligned}
$$

We wish to make use of Eq.(5.2.4-28), but, as noted there, it is no longer true because the classical Einstein equation given by Eq.(5.2.3-9) is used in the last term of Eq.(5.2.4-12). When Eq.(5.2.3-9) is replaced by the quantum Einstein equation given by Eq.(5.3.3-1), $h(t^\mu{}_\nu + T^\mu{}_\nu)$ acquires an extra term, that is,

$$h(t^\mu{}_\nu + T^\mu{}_\nu) = -\kappa^{-1}\partial_\lambda \mathcal{F}_{\text{C}}{}^{\lambda\mu}{}_\nu + \kappa^{-1}h(E^\mu{}_\nu - \frac{1}{2}\delta^\mu{}_\nu E), \qquad (5.5.1-4)$$

where $\mathcal{F}_{\text{C}}{}^{\lambda\mu}{}_\nu$ is defined by Eq.(5.2.4-29). Substituting Eqs.(3) and (4) into Eq.(2), we obtain

$$\mathcal{T}^\mu{}_\nu = \kappa^{-1}\tilde{g}^{\mu\lambda}\partial_\lambda b_\nu - \kappa^{-1}\partial_\lambda \mathcal{F}^{\lambda\mu}{}_\nu, \qquad (5.5.1-5)$$

---

[1] φ_A goes over all primary fields including b_ρ, c^σ, and $\bar{c}_\tau$.

where

$$\mathcal{F}^{\lambda\mu}{}_\nu \equiv \mathcal{F}_C{}^{\lambda\mu}{}_\nu - \kappa\mathcal{K}_M{}^{\lambda\mu}{}_\nu + (\delta^\mu{}_\nu\tilde{g}^{\lambda\rho} - \delta^\lambda{}_\nu\tilde{g}^{\mu\rho})b_\rho$$
$$= -\mathcal{F}^{\mu\lambda}{}_\nu. \qquad (5.5.1-6)$$

In contrast to the classical theory, $T^\mu{}_\nu$ does not become a total divergence.

The total energy-momentum P_ν is defined by

$$P_\nu \equiv \int d\mathbf{x}\, T^0{}_\nu, \qquad (5.5.1-7)$$

but as a $translation\ generator$, it is equivalent to [Nak 78f]

$$\hat{P}_\nu \equiv \kappa^{-1}\int d\mathbf{x}\,\tilde{g}^{0\lambda}\partial_\lambda b_\nu, \qquad (5.5.1-8)$$

because the last term of Eq.(5) is a total divergence of an antisymmetric quantity. The above equivalence is completely analogous to that between the electric charge Q and $\int d\mathbf{x}\,\partial_0 B$ as a $U(1)$ generator (see Sec.2.3.3).

The $GL(4)$ angular-momentum tensor density $\mathcal{M}^{\lambda\mu}{}_\nu$ is defined by

$$\mathcal{M}^{\lambda\mu}{}_\nu \equiv x^\mu T^\lambda{}_\nu + \mathcal{S}^{\lambda\mu}{}_\nu, \qquad (5.5.1-9)$$

where the $GL(4)$ intrinsic-spin tensor density $\mathcal{S}^{\lambda\mu}{}_\nu$ is defined by

$$\mathcal{S}^{\lambda\mu}{}_\nu \equiv -[\varphi_A]^\mu{}_\nu\frac{\partial\tilde{\mathcal{L}}}{\partial(\partial_\lambda\varphi_A)}. \qquad (5.5.1-10)$$

With the aid of Eqs.(5.2.4-18), (5.2.4-25), (5.2.4-29), and (6), we rewrite Eq.(10) as

$$\mathcal{S}^{\lambda\mu}{}_\nu = \kappa^{-1}\mathcal{F}^{\lambda\mu}{}_\nu + (2\kappa)^{-1}\partial_\sigma(\delta^\sigma{}_\nu\tilde{g}^{\mu\lambda} - \delta^\lambda{}_\nu\tilde{g}^{\mu\sigma}) - \kappa^{-1}\tilde{g}^{\mu\lambda}b_\nu$$
$$- i\tilde{g}^{\lambda\sigma}(\bar{c}_\nu\partial_\sigma c^\mu - \partial_\sigma\bar{c}_\nu\cdot c^\mu). \qquad (5.5.1-11)$$

We thus have

$$\mathcal{M}^{\lambda\mu}{}_\nu = \kappa^{-1}\tilde{g}^{\lambda\sigma}(x^\mu\partial_\sigma b_\nu - \delta^\mu{}_\sigma b_\nu) - i\tilde{g}^{\lambda\sigma}(\bar{c}_\nu\partial_\sigma c^\mu - \partial_\sigma\bar{c}_\nu\cdot c^\mu)$$
$$- \kappa^{-1}\partial_\sigma(x^\mu\mathcal{F}^{\sigma\lambda}{}_\nu) + (2\kappa)^{-1}\partial_\sigma(\delta^\sigma{}_\nu\tilde{g}^{\mu\lambda} - \delta^\lambda{}_\nu\tilde{g}^{\mu\sigma}). \qquad (5.5.1-12)$$

Accordingly, the $GL(4)$ generators are [Nak 78f]

$$\hat{M}^\mu{}_\nu \equiv \kappa^{-1}\int d\mathbf{x}\,\tilde{g}^{0\sigma}[x^\mu\partial_\sigma b_\nu - \delta^\mu{}_\sigma b_\nu - i\kappa(\bar{c}_\nu\partial_\sigma c^\mu - \partial_\sigma\bar{c}_\nu\cdot c^\mu)]. \qquad (5.5.1-13)$$

It should be noted that the $GL(4)$ generators *cannot* be well-defined operators. This is because if they were all well-defined, we could define *finite* $GL(4)$ transformations; then *any* two field operators of different spacetime points would always (anti)commute, because any two spacetime points can be transformed into equal-time ones by a finite $GL(4)$ transformation.

By using the equal-time (anti)commutation relations, it is straightforward to confirm that $\hat{P}_\nu$ and $\hat{M}^\mu{}_\nu$ are indeed the translation and $GL(4)$ generators, respectively:

$$[i\hat{P}_\nu, \varphi_A] = \partial_\nu \varphi_A, \qquad (5.5.1-14)$$

$$[i\hat{M}^\mu{}_\nu, \varphi_A] = x^\mu \partial_\nu \varphi_A - [\varphi_A]^\mu{}_\nu. \qquad (5.5.1-15)$$

Furthermore, we can explicitly show that they correctly form the **affine algebra**:

$$[\hat{P}_\nu, \hat{P}_\rho] = 0, \qquad (5.5.1-16)$$

$$[\hat{M}^\mu{}_\nu, \hat{P}_\rho] = -i\delta^\mu{}_\rho \hat{P}_\nu, \qquad (5.5.1-17)$$

$$[\hat{M}^\mu{}_\nu, \hat{M}^\lambda{}_\rho] = i\delta^\lambda{}_\nu \hat{M}^\mu{}_\rho - i\delta^\mu{}_\rho \hat{M}^\lambda{}_\nu. \qquad (5.5.1-18)$$

Finally, we discuss the BRS symmetry. Its Noether supercurrent is defined by [Nis 78, Kug 78b]

$$\mathcal{J}_b{}^\mu \equiv \boldsymbol{\delta}_*(\varphi_A)[\partial/\partial(\partial_\mu \varphi_A)]\tilde{\mathcal{L}} + \kappa c^\mu \mathcal{L} + \boldsymbol{\delta}_*(\mathcal{V}^\mu - \kappa^{-1}\tilde{g}^{\mu\nu}b_\nu). \qquad (5.5.1-19)$$

In a way similar to the above, it reduces to

$$\mathcal{J}_b{}^\mu = \tilde{g}^{\mu\nu}(b_\rho \partial_\nu c^\rho - \partial_\nu b_\rho \cdot c^\rho) + \partial_\nu \mathcal{I}_b{}^{\mu\nu}, \qquad (5.5.1-20)$$

where we omit writing the expression for $\mathcal{I}_b{}^{\mu\nu} = -\mathcal{I}_b{}^{\nu\mu}$, because it is a special case of the formula presented later [see Eq.(5.5.3-14)].

The BRS generator Q_b is defined by [Nak 78b]

$$Q_b \equiv \int d\mathbf{x}\, \tilde{g}^{0\nu}(b_\rho \partial_\nu c^\rho - \partial_\nu b_\rho \cdot c^\rho). \qquad (5.5.1-21)$$

Indeed, by using the equal-time (anti)commutation relations presented in Sec.5.4, we can explicitly verify that

$$[iQ_b, \varphi_A]_\mp = \boldsymbol{\delta}_*(\varphi_A). \qquad (5.5.1-22)$$

5.5.2 Sixteen-dimensional Poincaré-like superalgebra

As remarked in Sec.5.3.3, all of x^λ, b_ρ, c^σ, and $\bar{c}_\tau$ satisfy the same d'Alembert equation. It is, therefore, convenient to deal with them in a unified way. We set

$$X = (\hat{x}^\lambda,\ b_\rho,\ c^\sigma,\ \bar{c}_\tau), \qquad \hat{x}^\lambda \equiv x^\lambda/\kappa, \qquad (5.5.2-1)$$

and call X the **sixteen-dimensional supercoordinate** because, as seen later, it naturally behaves like a sixteen-dimensional extension of x^λ, including Grassmann-like dimensions. When we need more than one supercoordinates, we use also the letters Y, Z, U, etc.

Now, Eqs.(5.3.3-9), (5.3.3-8), (5.3.3-3), and (5.3.3-4) are unified into

$$\partial_\mu(\tilde{g}^{\mu\nu}\partial_\nu X) = 0. \qquad (5.5.2-2)$$

Evidently, Eq.(2) is a conservation law for

$$\mathcal{P}^\mu(X) \equiv \tilde{g}^{\mu\nu}\partial_\nu X. \qquad (5.5.2-3)$$

One should note that Eq.(3) is a sixteen-dimensional extension of $\kappa T^\mu{}_\rho$ apart from a total-divergence term [see Eq.(5.5.1-5)]. On the other hand, $\mathcal{M}^{\mu\lambda}{}_\rho$ and $\mathcal{J}_b{}^\mu$ are quadratic in supercoordinates. Generalizing their structure, we find the existence of a conserved supercurrent [Nak 80d]

$$\mathcal{M}^\mu(X,\ Y) \equiv \sqrt{\epsilon(X,\ Y)}\tilde{g}^{\mu\nu}(X\partial_\nu Y - \partial_\nu X \cdot Y), \qquad (5.5.2-4)$$

where a coefficient $\sqrt{\epsilon(X,\ Y)}$, defined below, is inserted for making $\mathcal{M}^\mu(X,\ Y)$ hermitian.

In order to deal with both bosonic and fermionic quantities in a unified way, it is convenient to introduce the following notation. Let

$$\epsilon(A,\ B) = -1 \quad \text{if both } A \text{ and } B \text{ are fermionic}^2,$$
$$= +1 \quad \text{otherwise}; \qquad (5.5.2-5)$$

hence, for instance, we can write $[A,\ B]_\mp = AB - \epsilon(A,\ B)BA$. We set $\epsilon(A) \equiv \epsilon(A,\ A)$, that is, $\epsilon(A) = +1$ for A bosonic and $\epsilon(A) = -1$ for A fermionic; then we have a relation

$$\epsilon(A,\ B) = \frac{1}{2}[1 + \epsilon(A) + \epsilon(B) - \epsilon(A)\epsilon(B)]. \qquad (5.5.2-6)$$

---

2   More precisely, A and B here belong to the same superselection sector.

The following properties are evident:

$$\epsilon(A, \, B) = \epsilon(B, \, A), \qquad\qquad\qquad (5.5.2 - 7)$$

$$\epsilon(AC, \, B) = \epsilon(A, \, B)\epsilon(C, \, B). \qquad\qquad (5.5.2 - 8)$$

Thus the ϵ for a pair of products reduces to a product of the ϵ's for a pair of single fields.

It is convenient to introduce a square root of the ϵ by the artificial convention $\sqrt{+1} = +1$ and $\sqrt{-1} = +i$. There are various identities between products of square roots of the ϵ's [Nak 80e], e.g.,

$$\sqrt{\epsilon(B)}\sqrt{-\epsilon(AB, \, B)} = i\sqrt{\epsilon(A, \, B)}. \qquad (5.5.2 - 9)$$

In order to make the sixteen-dimensional structure of the theory transparent, we introduce the **sixteen-dimensional supermetric** $\eta(X, \, Y)$ by[3]

$$\eta(\hat{x}^{\mu}, \, b_{\nu}) = \eta(b_{\nu}, \, \hat{x}^{\mu}) = \eta(c^{\mu}, \, \bar{c}_{\nu}) = -\eta(\bar{c}_{\nu}, \, c^{\mu}) = \delta^{\mu}{}_{\nu},$$

$$\eta(X, \, Y) = 0 \quad \text{otherwise.} \qquad\qquad (5.5.2 - 10)$$

Here X and Y in $\eta(X, \, Y)$ are regarded as mere indices, just like μ and ν in $\eta_{\mu\nu}$. The 16×16 matrix formed by $\eta(X, \, Y)$ is non-diagonal, but its determinant is equal to $+1$. Hence we can construct its matrix inversion, that is, we define $\tilde{\eta}(X, \, Y)$ by

$$\sum_{Z} \eta(X, \, Z)\tilde{\eta}(Z, \, Y) = \delta(X, \, Y), \qquad (5.5.2 - 11)$$

where $\delta(X, \, Y)$ denotes the sixteen-dimensional Kronecker delta. Explicitly, we have

$$\tilde{\eta}(\hat{x}^{\mu}, \, b_{\nu}) = \tilde{\eta}(b_{\nu}, \, \hat{x}^{\mu}) = -\tilde{\eta}(c^{\mu}, \, \bar{c}_{\nu}) = \tilde{\eta}(\bar{c}_{\nu}, \, c^{\mu}) = \delta^{\nu}{}_{\mu},$$

$$\tilde{\eta}(X, \, Y) = 0 \quad \text{otherwise.} \qquad\qquad (5.5.2 - 12)$$

Now, from Eqs.(3) and (4) we have the following symmetry generators [Nak 80d]:

$$P(X) \equiv \int d\boldsymbol{x}\, \mathcal{P}^{0}(X), \qquad\qquad (5.5.2 - 13)$$

$$M(X, \, Y) \equiv \int d\boldsymbol{x}\, \mathcal{M}^{0}(X, \, Y). \qquad (5.5.2 - 14)$$

---

[3]   Note that Eq.(10) implies $\epsilon(X, \, Y)\eta(X, \, Y) = \epsilon(X)\eta(X, \, Y)$.

Here, X and Y in $P(X)$ and $M(X, Y)$ are regarded as mere indices, just as in $\eta(X, Y)$. From the definition, it is evident that

$$M(Y, X) = -\epsilon(X, Y)M(X, Y); \qquad (5.5.2-15)$$

otherwise its components are independent. Hence, $M(X, Y)$ represents 128 symmetry generators. Since $P(X)$ represents 16, we have 144 symmetry generators in total.

The symmetry generators presented in Sec.5.5.1 are expressed as[4]

$$\hat{P}_\nu = \kappa^{-1} P(b_\nu), \qquad (5.5.2-16)$$

$$\hat{M}^\mu{}_\nu = M(\hat{x}^\mu, b_\nu) - M(c^\mu, \bar{c}_\nu), \qquad (5.5.2-17)$$

$$Q_b = M(b_\rho, c^\rho). \qquad (5.5.2-18)$$

The FP-ghost number operator is given by

$$Q_c \equiv M(\bar{c}_\rho, c^\rho). \qquad (5.5.2-19)$$

The (anti)commutator between $P(X)$ or $M(X, Y)$ and a primary field can be calculated from the results presented in Sec.5.4. [For matter fields, we should assume a concrete expression for $\mathcal{L}_M$.] Let φ be any primary field other than b_ρ, c^σ, and $\bar{c}_\tau$. Then we have

$$[iP(X), \varphi] = \kappa\eta(X, \hat{x}^\nu)\partial_\nu\varphi, \qquad (5.5.2-20)$$

$$[iP(X), Z]_\mp = i\sqrt{-\epsilon(X)}\eta(X, Z) + \kappa\eta(X, \hat{x}^\nu)\partial_\nu Z; \qquad (5.5.2-21)$$

$$[iM(X, Y), \varphi] = \kappa\eta(Y, \hat{x}^\nu)(X\partial_\nu\varphi - \partial_\mu X \cdot [\varphi]^\mu{}_\nu) - (X \leftrightarrow Y), \qquad (5.5.2-22)$$

$$[iM(X, Y), Z]_\mp = \left\{ i\sqrt{-\epsilon(XY, Z)}\eta(Y, Z)X + \kappa\eta(Y, \hat{x}^\nu)X\partial_\nu Z \right\} - \epsilon(X, Y)(X \leftrightarrow Y). \qquad (5.5.2-23)$$

Note that Eqs.(21) and (23) vanish for $Z = \hat{x}^\lambda$, as it should be.

The 144 symmetry generators, $P(X)$ and $M(X, Y)$, form the following superalgebra [Nak 80e]:

$$[P(X), P(Y)]_\mp = 0, \qquad (5.5.2-24)$$

$$[M(X, Y), P(Z)]_\mp = \sqrt{-\epsilon(XY, Z)}\{\eta(Y, Z)P(X) - \epsilon(X, Y)(X \leftrightarrow Y)\}, \qquad (5.5.2-25)$$

$$[M(X, Y), M(U, V)]_\mp = \sqrt{-\epsilon(XY, UV)}\{[\eta(Y, U)M(X, V) - \epsilon(X, Y)(X \leftrightarrow Y)] - \epsilon(U, V)(U \leftrightarrow V)\}. \qquad (5.5.2-26)$$

---

[4] $\hat{P}_\nu$ and $\hat{M}^\mu{}_\nu$ are BRS-invariant because $\hat{P}_\nu = \{Q_b, \kappa^{-1}P(\bar{c}_\nu)\}$ and $\hat{M}^\mu{}_\nu = \{Q_b, M(\hat{x}^\mu, b_\nu)\}$, as seen from Eqs.(25) and (26) below.

Since the above (anti)commutation relations are quite similar to those of the Poincaré algebra (see Sec.1.1.2), this superalgebra is called the **sixteen-dimensional Poincaré-like superalgebra**. Mathematically, it is denoted by $IOSp(8, 8)$ (inhomogeneous ortho-symplectic superalgebra).

Finally, we note that it is often convenient to identify $P(X)$ with $M(1, X) = -M(X, 1)$ under the definition $\eta(1, X) = \eta(X, 1) = \eta(1, 1) = 0$; then any formula involving $P(X)$ can be regarded as a special case of the corresponding one involving $M(X, Y)$ only.

5.5.3 Noether supercurrents for choral symmetry

Usually, given symmetry first, one then constructs a Lagrangian density which has that symmetry. But we are now proceeding in the reversed way: Given a Lagrangian density first, we then find its symmetry.[5]

The (anti)commution relation Eqs.(5.5.2-20)-(5.5.2-23) imply the following symmetry transformations:

$$\boldsymbol{\delta}_{X_*}(\Phi) = \boldsymbol{\delta}_X(\Phi) - \kappa \boldsymbol{\delta}_X(\hat{x}^\nu)\partial_\nu\Phi, \qquad (5.5.3-1)$$

$$\boldsymbol{\delta}_{XY_*}(\Phi) = \boldsymbol{\delta}_{XY}(\Phi) - \kappa \boldsymbol{\delta}_{XY}(\hat{x}^\nu)\partial_\nu\Phi \qquad (5.5.3-2)$$

with

$$\boldsymbol{\delta}_X(\varphi) = 0, \qquad (5.5.3-3)$$

$$\boldsymbol{\delta}_X(Z) = i\sqrt{-\epsilon(X)}\eta(X, Z), \qquad (5.5.3-4)$$

$$\boldsymbol{\delta}_{XY}(\varphi) = -\kappa\left\{\eta(Y, \hat{x}^\nu)\partial_\mu X - \eta(X, \hat{x}^\nu)\partial_\mu Y\right\}[\varphi]^\mu{}_\nu, \qquad (5.5.3-5)$$

$$\boldsymbol{\delta}_{XY}(Z) = i\sqrt{-\epsilon(XY, Z)}\left\{\eta(Y, Z)X - \epsilon(X, Y)\eta(X, Z)Y\right\}. \qquad (5.5.3-6)$$

We see $\boldsymbol{\delta}_X(Z) \neq 0$ for $Z = \hat{x}^\nu$ if and only if $X = b_\rho$, that is, $\boldsymbol{\delta}_{b_\rho}$ is a spacetime symmetry (translation) and $\boldsymbol{\delta}_X$ for $X \neq b_\rho$ is an internal symmetry. Likewise, $\boldsymbol{\delta}_{XY}$ is an internal symmetry if neither X nor Y is b_ρ, and it is a spacetime symmetry otherwise. We call the totality of $\boldsymbol{\delta}_X$ and $\boldsymbol{\delta}_{XY}$ **choral symmetry**. Thus choral symmetry contains both spacetime and internal symmetries in complete harmony.[6]

---

[5]   Note that this way of finding symmetry is impossible in the path-integral approach.

[6]   This beautiful result does not contradict the no-go theorem concerning unification of spacetime and internal symmetries [Haa 75], because it merely refers to the S-matrix symmetries but *not* the Lagrangian ones

We can directly check the invariance of the action under choral-symmetry transformations. We first note Eq.(5.3.3-6) and

$$E_{\mu\nu} = \sum_{X,Y} \sqrt{\epsilon(X)} \tilde{\eta}(X,Y) \partial_\mu X \cdot \partial_\nu Y. \qquad (5.5.3-7)$$

Then we can see that $h^{-1}\mathcal{L}_E$, $h^{-1}(\mathcal{L}_{GF}+\mathcal{L}_{FP})$, and $h^{-1}\mathcal{L}_M$ are separately invariant under $\boldsymbol{\delta}_X$ and under $\boldsymbol{\delta}_{XY}$. The existence of choral symmetry uniquely characterizes our expression for $\mathcal{L}_{GF} + \mathcal{L}_{FP}$.

Since we have found symmetry transformations, it is now possible to construct the Noether supercurrents for choral symmetry. As noted at the end of Sec.5.5.2, it is sufficient to consider $\boldsymbol{\delta}_{XY}$ only. The Noether supercurrent corresponding to $\boldsymbol{\delta}_{XY}$ is given by

$$\hat{\mathcal{M}}^\mu(X,Y) \equiv \boldsymbol{\delta}_{XY*}(\varphi_A)[\partial/\partial(\partial_\mu\varphi_A)]\tilde{\mathcal{L}} + \kappa\boldsymbol{\delta}_{XY}(\hat{x}^\mu)\tilde{\mathcal{L}}$$
$$+ \boldsymbol{\delta}_{XY*}(\mathcal{V}^\mu - \kappa^{-1}\tilde{g}^{\mu\nu}b_\nu). \qquad (5.5.3-8)$$

First, we calculate the contribution from the classical part, for which $\varphi_A \neq b_\rho$, c^σ, $\bar{c}_\tau$. Substituting

$$\boldsymbol{\delta}_{XY}(\varphi) = \kappa\partial_\mu\boldsymbol{\delta}_{XY}(\hat{x}^\lambda) \cdot [\varphi]^\mu{}_\lambda, \qquad (5.5.3-9)$$
$$\boldsymbol{\delta}_{XY*}(\mathcal{V}^\mu) = -\kappa\boldsymbol{\delta}_{XY}(\hat{x}^\lambda)\partial_\lambda\mathcal{V}^\mu - \kappa\partial_\sigma\boldsymbol{\delta}_{XY}(\hat{x}^\lambda) \cdot (\delta^\sigma{}_\lambda\mathcal{V}^\mu - \delta^\mu{}_\lambda\mathcal{V}^\sigma)$$
$$+ \frac{1}{2}\partial_\sigma\partial_\tau\boldsymbol{\delta}_{XY}(\hat{x}^\lambda) \cdot (\delta^\mu{}_\lambda\tilde{g}^{\sigma\tau} - \delta^\sigma{}_\lambda\tilde{g}^{\mu\tau}), \qquad (5.5.3-10)$$

we obtain

$$[\boldsymbol{\delta}_{XY*}(\varphi_A)\partial/\partial(\partial_\mu\varphi_A) + \kappa\boldsymbol{\delta}_{XY}(\hat{x}^\mu)](\tilde{\mathcal{L}}_E + \mathcal{L}_M) + \boldsymbol{\delta}_{XY*}(\mathcal{V}^\mu)$$
$$= -\kappa\partial_\nu[\boldsymbol{\delta}_{XY}(\hat{x}^\lambda)(\mathcal{K}_E{}^{\nu\mu}{}_\lambda + \mathcal{K}_M{}^{\nu\mu}{}_\lambda)] + \frac{1}{2}\partial_\nu[\tilde{g}^{\nu\lambda}\partial_\lambda\boldsymbol{\delta}_{XY}(\hat{x}^\mu) - \tilde{g}^{\mu\lambda}\partial_\lambda\boldsymbol{\delta}_{XY}(\hat{x}^\nu)]$$
$$- \boldsymbol{\delta}_{XY}(\hat{x}^\lambda)h(E^\mu{}_\lambda - \frac{1}{2}\delta^\mu{}_\lambda E), \qquad (5.5.3-11)$$

where use has been made of Eq.(5.2.4-27) (modified by the replacement of the classical Einstein equation by the quantum one) and of Eq.(5.2.4-25).

Next, we calculate the contribution from the gauge-fixing and FP-ghost part, for which φ_A runs over $g_{\mu\nu}$, b_ρ, c^σ, $\bar{c}_\tau$ only. It is straightforward to have

$$[\boldsymbol{\delta}_{XY*}(\varphi_A)\partial/\partial(\partial_\mu\varphi_A) + \kappa\boldsymbol{\delta}_{XY}(\hat{x}^\mu)](\tilde{\mathcal{L}}_{GF} + \mathcal{L}_{FP}) - \kappa^{-1}\boldsymbol{\delta}_{XY*}(\tilde{g}^{\mu\nu}b_\nu)$$
$$= -\kappa^{-1}\tilde{g}^{\mu\nu}\boldsymbol{\delta}_{XY*}(b_\nu) + i\tilde{g}^{\mu\nu}[\boldsymbol{\delta}_{XY*}(c^\rho)\partial_\nu\bar{c}_\rho - \boldsymbol{\delta}_{XY}(\bar{c}_\rho)\partial_\nu c^\rho]$$
$$- \frac{1}{2}\boldsymbol{\delta}_{XY}(\hat{x}^\mu)hE. \qquad (5.5.3-12)$$

Finally, we combine Eqs.(11) and (12). By using Eqs.(2), (5.5.2-10), (6), (5.5.2-11), and (5.5.2-9) in this order, we obtain [Nak 81d]

$$\hat{\mathcal{M}}^\mu(X, Y) = \mathcal{M}^\mu(X, Y) + \partial_\nu \mathcal{I}^{\nu\mu}(X, Y), \qquad (5.5.3-13)$$

where $\mathcal{M}^\mu(X, Y)$ is given by Eq.(5.5.2-4) and [7]

$$
\begin{aligned}
\mathcal{I}^{\nu\mu}(X, Y) &\equiv -\kappa \boldsymbol{\delta}_{XY}(\hat{x}^\lambda)(\mathcal{K}_{\mathrm{E}}{}^{\nu\mu}{}_\lambda + \mathcal{K}_{\mathrm{M}}{}^{\nu\mu}{}_\lambda) \\
&\quad + \frac{1}{2}[\tilde{g}^{\nu\lambda}\partial_\lambda \boldsymbol{\delta}_{XY}(\hat{x}^\mu) - \tilde{g}^{\mu\lambda}\partial_\lambda \boldsymbol{\delta}_{XY}(\hat{x}^\nu)] \\
&= -\mathcal{I}^{\mu\nu}(X, Y). \qquad (5.5.3-14)
\end{aligned}
$$

5.5.4 Spontaneous breakdown of symmetries

We now postulate the existence of a unique vacuum state $|0\rangle$, which has positive norm

$$\langle 0 | 0 \rangle = 1. \qquad (5.5.4-1)$$

Then, from Eqs.(5.5.2-21) and (5.5.2-23), we find

$$\langle 0 | [iP(\hat{x}^\mu), b_\rho] | 0 \rangle = -\delta^\mu{}_\rho, \qquad (5.5.4-2)$$

$$\langle 0 | \{iP(c^\mu), \bar{c}_\rho\} | 0 \rangle = i\delta^\mu{}_\rho, \qquad (5.5.4-3)$$

$$\langle 0 | \{iP(\bar{c}_\mu), c^\rho\} | 0 \rangle = -i\delta^\rho{}_\mu, \qquad (5.5.4-4)$$

$$\langle 0 | [iM(\hat{x}^\mu, \hat{x}^\nu), \tfrac{1}{2}(\partial_\lambda b_\rho - \partial_\rho b_\lambda)] | 0 \rangle = -\kappa^{-1}(\delta^\mu{}_\lambda \delta^\nu{}_\rho - \delta^\nu{}_\lambda \delta^\mu{}_\rho), \qquad (5.5.4-5)$$

$$\langle 0 | \{iM(\hat{x}^\mu, c^\nu), \partial_\lambda \bar{c}_\rho\} | 0 \rangle = i\kappa^{-1}\delta^\mu{}_\lambda \delta^\nu{}_\rho, \qquad (5.5.4-6)$$

$$\langle 0 | \{iM(\hat{x}^\mu, \bar{c}_\nu), \partial_\lambda c^\rho\} | 0 \rangle = -i\kappa^{-1}\delta^\mu{}_\lambda \delta^\rho{}_\nu. \qquad (5.5.4-7)$$

Thus the symmetries generated by $P(\hat{x}^\mu)$, $P(c^\mu)$, $P(\bar{c}_\mu)$, $M(\hat{x}^\mu, \hat{x}^\nu)$, $M(\hat{x}^\mu, c^\nu)$, and $M(\hat{x}^\mu, \bar{c}_\nu)$ are *necessarily* spontaneously broken, and b_ρ, c^σ, and $\bar{c}_\tau$ must contain massless spectrum owing to the Goldstone theorem.

We postulate that the vacuum is invariant under translation:

$$P_\mu | 0 \rangle = 0, \qquad (5.5.4-8)$$

---

[7] For the expressions for $\mathcal{K}_{\mathrm{E}}{}^{\mu\nu}{}_\lambda$ and $\mathcal{K}_{\mathrm{M}}{}^{\mu\nu}{}_\lambda$, see Eqs.(5.2.4-26) and (5.2.4-18).

that is, translational invariance is not spontaneously broken. Then we should have[8]

$$\langle\, 0\,|\, g_{\mu\nu}\,|\, 0\,\rangle = \eta_{\mu\nu}. \qquad (5.5.4-9)$$

In Sec.5.7.6, we show that no generality is lost by setting Eq.(5.5.4-9).

For $\varphi_A = g_{\sigma\tau}$, Eq.(5.5.1-15) becomes

$$[i\hat{M}^{\mu}{}_{\nu},\ g_{\sigma\tau}] = x^{\mu}\partial_{\nu}g_{\sigma\tau} + \delta^{\mu}{}_{\sigma}g_{\nu\tau} + \delta^{\mu}{}_{\tau}g_{\sigma\nu}. \qquad (5.5.4-10)$$

Hence Eq.(9) implies

$$\langle\, 0\,|\,[i\hat{M}^{\mu}{}_{\nu},\ g_{\sigma\tau}]\,|\, 0\,\rangle = \delta^{\mu}{}_{\sigma}\eta_{\nu\tau} + \delta^{\mu}{}_{\tau}\eta_{\sigma\nu}. \qquad (5.5.4-11)$$

From Eq.(11) we see that the symmetric part $\eta_{\mu\lambda}\hat{M}^{\lambda}{}_{\nu} + \eta_{\nu\lambda}\hat{M}^{\lambda}{}_{\mu}$ is spontaneously broken, and the corrensponding NG mode is nothing but $g_{\sigma\tau}$. That is, *gravitons must be exactly massless* [Nak 79b]. On the other hand, the antisymmetric part,

$$\bar{M}_{\mu\nu} \equiv \eta_{\mu\lambda}\hat{M}^{\lambda}{}_{\nu} - \eta_{\nu\lambda}\hat{M}^{\lambda}{}_{\mu}, \qquad (5.5.4-12)$$

has the vanishing Goldstone commutator. We thus see that $GL(4)$ invariance is spontaneously broken up to Lorentz invariance.[9]

The remaining generators $M(b_{\mu},\ b_{\nu})$, $M(b_{\mu},\ c^{\nu})$, $M(b_{\mu},\ \bar{c}_{\nu})$, $M(c^{\mu},\ \bar{c}_{\nu})$, $M(c^{\mu},\ c^{\nu})$, and $M(\bar{c}_{\mu},\ \bar{c}_{\nu})$ have no reason for spontaneous breakdown. Hence it is natural to suppose that they are unbroken. One can indeed confirm that **Ward-Takahashi-type identities**

$$\langle\, 0\,|\,[M(X,Y),\ \mathcal{O}]\,|\, 0\,\rangle = 0 \qquad (5.5.4-13)$$

hold, where $\mathcal{O}$ denotes a T-product of field operators, for the above generators at the tree and one-loop levels in perturbation theory [Nak 81c].

For spontaneously broken generators, the Ward-Takahashi-type identities must be modified by the contribution from the NG mode. By analyzing those modified Ward-Takahashi-type identities, one obtains various consistency conditions [Nak 81a].

---

[8]   The presence of the cosmological term is incompatible with Eq (5.5.4-9).

[9]   The breakdown of $GL(4)$ is previously discussed in the sense of nonlinear realization, but only at the classical level [Bori 74].

5.6 GEOMETRIC COMMUTATION RELATION[*]

The classical general relativity is invariant under general coordinate transformations. But the quantum theory of gravity loses this property because of gauge fixing. Quite remarkably, however, it is found that our theory of quantum gravity has a remnant of general covariance purely at the operator level. The gravitational B-field b_ρ plays the role of the generator of general coordinate transformations, just as the electromagnetic B-field does that of local gauge transformations.

We employ the abbreviated notation of Sec.5.4.

5.6.1 Tensorlike commutation relation

We have calculated the equal-time commutation relations in Sec.5.4. Since both $E_{k\ell}$ and $T_{k\ell}$ commute with b_ρ, the quantum Einstein equation, Eq.(5.3.3-1), implies $[R_{k\ell},\ b_\rho{}'] = 0$. Combining this fact with Eq.(5.4.4-8), we can show that

$$[R_{\mu\nu},\ b_\rho{}'] = -i\kappa\tilde{f}(\delta^0{}_\mu R_{\rho\nu} + \delta^0{}_\nu R_{\mu\rho})\delta^3. \qquad (5.6.1-1)$$

The similarity between Eq.(1) and Eq.(5.4.2-4) is remarkable.

Calculating the equal-time commutation relations between various matter fields and b_ρ, we arrive at the following rule: Let T_Σ be a local quantity which is a tensor (or a tensor density) under general coordinate transformation at the classical level. Then we have the equal-time commutation relation [Nak 80b]

$$[T_\Sigma,\ b_\rho{}'] = i\kappa[T_\Sigma]^0{}_\rho\tilde{f}\delta^3. \qquad (5.6.1-2)$$

We call Eq.(2) the **tensorlike commutation relation**, because it is consistent with tensorial rules stated in Sec.5.2.1.

Since the affine connection $\Gamma^\lambda{}_{\mu\nu}$ is not a tensor, it does not, of course, satisfy Eq.(2). Explicit calculation yields

$$\begin{aligned}[\Gamma^\lambda{}_{\mu\nu},\ b_\rho{}'] =& i\kappa(\delta^\lambda{}_\rho\Gamma^0{}_{\mu\nu} - \delta^0{}_\mu\Gamma^\lambda{}_{\rho\nu} - \delta^0{}_\nu\Gamma^\lambda{}_{\mu\rho})\tilde{f}\delta^3 \\ &+ i\kappa\delta^\lambda{}_\rho[2\delta^0{}_\mu\delta^0{}_\nu\tilde{f}\tilde{g}^{0k} - \delta^0{}_\mu\delta^k{}_\nu - \delta^0{}_\nu\delta^k{}_\mu](\delta^3)_k. \end{aligned} \qquad (5.6.1-3)$$

[*]   This section is of a rather sophisticated topic. One may skip this section for first reading.

Using Eq.(3), we can show that Eq.(2) is also consistent with taking *spatial* covariant derivative. Since Eq.(2) is an equal-time commutation relation, it is impossible to take time covariant derivative. But, if we *require* the validity of Eq.(2) for $\nabla_0 T_\Sigma$, we find what expression $[T_\Sigma, \dot{b}'_\rho]$ must have [Nak 82]:

$$[T_\Sigma, \dot{b}'_\rho] = i\kappa \{\partial_\rho T_\Sigma \cdot \tilde{f}\delta^3 + [T_\Sigma]^0{}_\rho(\partial_0 \tilde{f} \cdot \delta^3$$
$$+ 2\tilde{f}\tilde{g}^{0k}(\delta^3)_k) - [T_\Sigma]^k{}_\rho(\delta^3)_k\}. \qquad (5.6.1-4)$$

This formula is again tensorlike in the sense that it is consistent with tensorial rules.

The main question is whether or not Eqs.(2) and (4) are always valid. The answer is *no*. For example, although $E_{\mu\nu}$ is expressible in terms of classical tensors owing to the quantum Einstein equation, it does not satisfy Eq.(4); $[E_{\mu\nu}, \dot{b}'_\rho]$ acquires an extra term [Nak 82]

$$2i\kappa\partial_k\partial_\ell b_\rho \cdot (\delta^k{}_\mu\delta^\ell{}_\nu\tilde{g}^{00} - \delta^k{}_\mu\delta^0{}_\nu\tilde{g}^{0\ell} - \delta^0{}_\mu\delta^\ell{}_\nu\tilde{g}^{k0} + \delta^0{}_\mu\delta^0{}_\nu\tilde{g}^{k\ell})\tilde{f}^2\delta^3. \qquad (5.6.1-5)$$

Note that this **non-tensorlike term** is proportional to the second derivatives of b_ρ. If b_ρ is replaced by c^σ in Eq.(5), the resultant expression is precisely equal to $[E_{\mu\nu}, \dot{c}^{\sigma\prime}]$.

In general, Eqs.(2) and (4) are broken by non-tensorlike terms if T_Σ contains higher derivatives of primary fields. But non-tensorlike terms are proportional to higher derivatives of b_ρ, and if b_ρ is replaced by c^σ, the non-tensorlike parts of $[T_\Sigma, b_\rho{}']$ and $[T_\Sigma, \dot{b}'_\rho]$ become equal to $[T_\Sigma, c^{\sigma\prime}]$ and $[T_\Sigma, \dot{c}^{\sigma\prime}]$, respectively.

In order to prove the above proposition, however, we need a more general formulation of the problem.

5.6.2 Quantum-gravity D function

In the Minkowski spacetime, the Pauli-Jordan D function $D(x-y)$ plays a fundamental role in the four-dimensional commutator of massless fields. As mentioned in Sec.1.3.1, it can be defined by the following Cauchy problem:

$$\Box^x D(x-y) = 0, \qquad (5.6.2-1)$$

$$D(x-y)|_0 = 0, \qquad (5.6.2-2)$$

$$\partial_0{}^x D(x-y)|_0 = -\delta^3. \qquad (5.6.2-3)$$

Here the symbol $|_0$ means to set $x^0 = y^0$. Of course, we know the explicit expression for $D(x - y)$, but its important properties are: i) It is real, ii) it is odd in $x - y$, iii) it is Poincaré-invariant, and iv) it vanishes for $x - y$ spacelike.

The Pauli-Jordan D function can be extended to the case of a curved space-time in the following way [Nak 79a]. Since the d'Alembertian in a curved spacetime is $h^{-1}\partial_\mu \tilde{g}^{\mu\nu}\partial_\nu$, we define the curved-spacetime D function $\mathcal{D}(x, y)$ by the following Cauchy problem:[1]

$$\partial_\mu{}^x \tilde{g}^{\mu\nu}(x)\partial_\nu{}^x \mathcal{D}(x, y) = 0, \qquad (5.6.2-4)$$

$$\mathcal{D}(x, y)|_0 = 0, \qquad (5.6.2-5)$$

$$\partial_0{}^x \mathcal{D}(x, y)|_0 = -\tilde{f}\delta^3. \qquad (5.6.2-6)$$

Since there is no translational invariance, $\mathcal{D}(x, y)$ is not a function of $x - y$ alone.

It is known how to construct the solution to the differential equation

$$h^{-1}(x)\partial_\mu{}^x \tilde{g}^{\mu\nu}(x)\partial_\nu{}^x \mathcal{G}(x, y) = h^{-1}\delta^4(x - y). \qquad (5.6.2-7)$$

In a geodesically convex domain, there exist two fundamental solutions, $\mathcal{D}_R(x, y)$ and $\mathcal{D}_A(x, y)$, which vanish outside the future cone and outside the past cone, respectively [Fri 75]. Then it can be shown that [Kan 85]

$$\mathcal{D}(x, y) = -\mathcal{D}_R(x, y) + \mathcal{D}_A(x, y). \qquad (5.6.2-8)$$

Since $\mathcal{D}_R$ and $\mathcal{D}_A$ are real and since $\mathcal{D}_R(x, y) = \mathcal{D}_A(y, x)$, $\mathcal{D}(x, y)$ is real and antisymmetric under $x \leftrightarrow y$.

Now, we consider the **quantum-gravity extension of the Pauli-Jordan D function** [Nak 82]. We define it by the Cauchy problem, Eqs.(4)-(6). Although the coefficient functions are now operators, we *assume* that the Cauchy problem is uniquely solvable.

We investigate various properties of the quantum-gravity D function $\mathcal{D}(x, y)$ [Kan 85], which is a bilocal operator.

First, we discuss the equal-time behavior of the derivatives of $\mathcal{D}(x, y)$. For this purpose, it is convenient to transform x^0 and y^0 into $t = (x^0 + y^0)/2$ and $\tau = x^0 - y^0$. The equal time $x^0 = y^0$ becomes $\tau = 0$, while t is arbitrary. Hence

---

[1]   Extension to the massive case is straightforward, but we do not need it.

we can freely differentiate Eqs.(5) and (6) with respect to t. Transforming back, we obtain

$$(\partial_0{}^x + \partial_0{}^y)^n \mathcal{D}(x, y)|_0 = 0, \qquad\qquad (5.6.2-9)$$

$$(\partial_0{}^x + \partial_0{}^y)^n \partial_0{}^x \mathcal{D}(x, y)|_0 = -(\partial_0)^n \tilde{f} \cdot \delta^3 \qquad\qquad (5.6.2-10)$$

for $n = 0, 1, 2, \cdots$.

By using Eqs.(4), (10), and (9), we can calculate $(\partial_0{}^x)^p (\partial_0{}^y)^{m-p} \mathcal{D}(x, y)|_0$ for any $m \geq p \geq 0$. Indeed, for $p \geq 2$, use of Eq.(4) reduces it to those of smaller values of m; for $p = 1$, use of Eq.(10) reduces it to those of the same value of m but of $p \geq 2$; for $p = 0$, use of Eq.(9) reduces it to those of the same value of m but of $p \geq 1$. For example, we have

$$\partial_0{}^y \mathcal{D}(x, y)|_0 = \tilde{f}\delta^3, \qquad\qquad (5.6.2-11)$$

$$(\partial_0{}^x)^2 \mathcal{D}(x, y)|_0 = A - \partial_0 \tilde{f} \cdot \delta^3, \qquad\qquad (5.6.2-12)$$

$$\partial_0{}^x \partial_0{}^y \mathcal{D}(x, y)|_0 = -A, \qquad\qquad (5.6.2-13)$$

$$(\partial_0{}^y)^2 \mathcal{D}(x, y)|_0 = A + \partial_0 \tilde{f} \cdot \delta^3 \qquad\qquad (5.6.2-14)$$

with

$$A \equiv 2\tilde{f}\tilde{g}^{0k}(\delta^3)_k + \tilde{f}^2 \partial_k \tilde{g}^{0k} \cdot \delta^3. \qquad\qquad (5.6.2-15)$$

Next, we prove

$$\mathcal{D}(x, y) \overset{\leftarrow}{\partial}{}_\mu{}^y \tilde{g}^{\mu\nu}(y) \overset{\leftarrow}{\partial}{}_\nu{}^y = 0, \qquad\qquad (5.6.2-16)$$

where an arrow indicates that differentiation acts to the left. Because $\mathcal{D}(x, y)$ is an operator depending on x^λ, the order between $\mathcal{D}(x, y)$ and $\tilde{g}^{\mu\nu}(y)$ cannot be interchanged. We postulate the uniqueness of the solution of the Cauchy problem; then, denoting the lhs. of Eq.(16) by $\mathcal{F}(x, y)$, we have only to show that

$$\partial_\mu{}^x \tilde{g}^{\mu\nu}(x) \partial_\nu{}^x \mathcal{F}(x, y) = 0, \qquad\qquad (5.6.2-17)$$

$$\mathcal{F}(x, y)|_0 = 0, \qquad\qquad (5.6.2-18)$$

$$\partial_0{}^x \mathcal{F}(x, y)|_0 = 0. \qquad\qquad (5.6.2-19)$$

The validity of Eq.(17) is obvious from Eq.(4). It is also straightforward to prove Eq.(18) by using Eqs.(5), (11), and (14). Here we should remember that no operator-ordering problem exists when we calculate the equal-time values (see Sec.1.1). The lhs. of Eq.(19) can be expressed in terms of

$\partial_0{}^x \partial_\mu{}^x \tilde{g}^{\mu\nu}(x) \partial_\nu{}^x \mathcal{D}(x, y)|_0$ $(= 0)$, $(\partial_0{}^x + \partial_0{}^y) \partial_0{}^x \partial_0{}^y \mathcal{D}(x, y)|_0$ $(= -\partial_0 A)$, and $(\partial_0{}^x)^p (\partial_0{}^y)^{m-p} \mathcal{D}(x, y)|_0$ for $m \leq 2$, and then is seen to vanish. Thus Eq.(16) is established.

Evidently, Eq.(16) together with Eqs.(5) and (11) constitutes a Cauchy problem with respect to y. Since its solution is $-\mathcal{D}^\dagger(y, x)$, the uniqueness postulate implies that

$$\mathcal{D}(x, y) = -\mathcal{D}^\dagger(y, x). \qquad (5.6.2 - 20)$$

Owing to Eqs.(16) and (4), the non-local current

$$\partial_\nu{}^z \mathcal{D}(x, z) \cdot \tilde{g}^{\mu\nu}(x) \mathcal{D}(z, y) - \mathcal{D}(x, z) \tilde{g}^{\mu\nu} \partial_\nu{}^z \mathcal{D}(z, y) \qquad (5.6.2 - 21)$$

is conserved with respect to z. Hence the spatial integral of its $\mu = 0$ component is independent of z^0. Setting $z^0 = x^0$ (or $z^0 = y^0$), we see that this integral equals $\mathcal{D}(x, y)$. Thus we obtain an **integral representation** for $\mathcal{D}(x, y)$:

$$\mathcal{D}(x, y) = \int d\mathbf{z} \{\partial_\nu{}^z \mathcal{D}(x, z) \cdot \tilde{g}^{0\nu}(z) \mathcal{D}(z, y) - \mathcal{D}(x, z) \tilde{g}^{0\nu}(z) \partial_\nu{}^z \mathcal{D}(z, y)\}.$$
$$(5.6.2 - 22)$$

Since the supercoordinate X satisfies the d'Alembert equation, Eq.(5.5.2-2), the same reasoning as above yields integral representations for X:

$$X(y) = \int d\mathbf{z} \{\partial_\nu X(z) \cdot \tilde{g}^{0\nu}(z) \mathcal{D}(z, y) - X(z) \tilde{g}^{0\nu}(z) \partial_\nu{}^z \mathcal{D}(z, y)\}, \qquad (5.6.2 - 23)$$

$$X(x) = \int d\mathbf{z} \{\partial_\nu{}^z \mathcal{D}(x, z) \cdot \tilde{g}^{0\nu}(z) X(z) - \mathcal{D}(x, z) \tilde{g}^{0\nu}(z) \partial_\nu X(z)\}. \qquad (5.6.2 - 24)$$

Finally, we show that $\mathcal{D}(x, y)$ is scalar under the choral-symmetry transformations, that is, we have [Kan 85]

$$[iP(X), \mathcal{D}(x, y)] = \kappa \eta(X, \hat{x}^\lambda)(\partial_\lambda{}^x + \partial_\lambda{}^y) \mathcal{D}(x, y), \qquad (5.6.2 - 25)$$

$$[iM(X, Y), \mathcal{D}(x, y)] = \kappa \eta(Y, \hat{x}^\lambda)\{X(x)\partial_\lambda{}^x \mathcal{D}(x, y)$$
$$+ \partial_\lambda{}^y \mathcal{D}(x, y) \cdot X(y)\} - (X \leftrightarrow Y). \quad (5.6.2 - 26)$$

As remarked at the end of Sec.5.5.2, it is sufficient to prove Eq.(26) only. Let $\mathcal{F}(x, y)$ be the difference between both sides of Eq.(26); then the uniqueness of the solution of the Cauchy problem implies the validity of Eq.(26) if we can show Eqs.(17)-(19).

Owing to Eq.(4), we have

$$\partial_\mu{}^x \tilde{g}^{\mu\nu}(x)\partial_\nu{}^x \mathcal{F}(x,y) = -\partial_\mu{}^x [iM(X,Y),\, \tilde{g}^{\mu\nu}(x)]\partial_\nu{}^x \mathcal{D}(x,y)$$
$$- \kappa\{\eta(Y,\hat{x}^\lambda)\partial_\mu{}^x \tilde{g}^{\mu\nu}(x)\partial_\nu{}^x X(x)\partial_\lambda{}^x \mathcal{D}(x,y)$$
$$- (X \leftrightarrow Y)\}, \qquad (5.6.2-27)$$

where differentiations act on all their right quantities. Substituting [see Eq.(5.5.2-22)]

$$[iM(X,Y),\, \tilde{g}^{\mu\nu}] = \kappa\{\eta(Y,\hat{x}^\lambda)\partial_\lambda(\tilde{g}^{\mu\nu}X)$$
$$- \eta(Y,\hat{x}^\nu)\tilde{g}^{\mu\lambda}\partial_\lambda X - \eta(Y,\hat{x}^\mu)\tilde{g}^{\lambda\nu}\partial_\lambda X\}$$
$$- (X \leftrightarrow Y), \qquad (5.6.2-28)$$

into Eq.(27) and using Eq.(4) in the form

$$\partial_\lambda{}^x X(x)\partial_\mu{}^x \tilde{g}^{\mu\nu}(x)\partial_\nu{}^x \mathcal{D}(x,y) = 0, \qquad (5.6.2-29)$$

we obtain Eq.(17).

It is straightforward to prove Eq.(18) by using Eqs.(5), (6), and (11).

With the help of Eqs.(6), (12), and (13), we obtain

$$\partial_0{}^x \mathcal{F}(x,y)|_0 = [iM(X,Y),\, -\tilde{f}\delta^3] - \kappa\{\eta(Y,\hat{x}^\lambda)[\delta^0{}_\lambda(-2\tilde{f}^2\tilde{g}^{0k}\partial_k X - \partial_0\tilde{f}X)$$
$$+ \delta^k{}_\lambda(\tilde{f}\partial_k X - \partial_k\tilde{f}\cdot X)]\cdot\delta^3 - (X \leftrightarrow Y)\}. \qquad (5.6.2-30)$$

From Eq.(28) [remember $\tilde{f} = (\tilde{g}^{00})^{-1}$], we find that both terms of Eq.(30) cancel exactly; this proves Eq.(19).

Thus Eqs.(25) and (26) are established. In particular, $\mathcal{D}(x,y)$ is affine scalar.

5.6.3 Equal-time geometric commutation relation

We extend the tensorlike commutation relation to the case of a general local operator $\Phi(x)$, which can be constructed from the primary fields other than b_ρ, c^σ, and $\bar{c}_r$ by making the following operations finite times:

i) To take a linear combination.

ii) To take a product (and to take the inverse for such an invertible quantity as h and $\tilde{g}^{00}$).

iii) To apply any of differentiations ∂_μ ($\mu = 0, 1, 2, 3$).

The key to this extension is to make use of the quantum-gravity D function introduced in Sec.5.6.2.

Since $\Phi(x)$ has its classical counterpart, its change under an infinitesimal general coordinate transformation is well-defined. Let

$$\delta_*{}^\epsilon \Phi(x) \equiv \Phi'(x) - \Phi(x) \equiv \pounds_\lambda(\Phi)\epsilon^\lambda(x). \qquad (5.6.3-1)$$

Here $\pounds_\lambda$ is a differential operator[2] having the form

$$\pounds_\lambda(\Phi)^x = \sum_{j=0}^{r} L^{\mu_1 \cdots \mu_j}{}_\lambda(x;\, \Phi)\partial_{\mu_1}{}^x \cdots \partial_{\mu_j}{}^x, \qquad (5.6.3-2)$$

where $L^{\mu_1 \cdots \mu_j}{}_\lambda(x;\, \Phi)$ is a local operator determined by $\Phi(x)$ and the integer $r - 1$ does not exceed the highest order p of derivatives of the primary fields involved in $\Phi(x)$. Of course, we know

$$L_\lambda(x;\, \Phi) = -\partial_\lambda \Phi(x), \qquad (5.6.3-3)$$

$$L^\mu{}_\lambda(x;\, \Phi) = [\Phi(x)]^\mu{}_\lambda. \qquad (5.6.3-4)$$

From the definition, Eq.(1), it is obvious that $\pounds_\lambda$ is a derivation *as an operator acting on* Φ. Furthermore, we have

$$\pounds_\lambda(\partial_\nu \Phi) = \partial_\nu \pounds_\lambda(\Phi), \qquad (5.6.3-5)$$

where, of course, ∂_ν in the rhs acts not only on $\pounds_\lambda(\Phi)$ but also on its operand.

As extensions of the tensorlike commutation relations, Eqs.(5.6.1-2) and (5.6.1-4), we propose [Nak 83a]

$$[\Phi(x),\, X(z)]|_0 = -i\kappa\eta(X,\, \hat{x}^\lambda)\pounds_\lambda(\Phi)^x \mathcal{D}(x, z)|_0$$
$$+ \sum_{j=2}^{p} \partial_{\mu_1} \cdots \partial_{\mu_j} X(z) \cdot f^{\mu_1 \cdots \mu_j}(z, x;\, \Phi), \quad (5.6.3-6)$$

$$[\Phi(x),\, \partial_0 X(z)]|_0 = -i\kappa\eta(X,\, \hat{x}^\lambda)\pounds_\lambda(\Phi)^x \partial_0{}^z \mathcal{D}(x, z)|_0$$
$$+ \sum_{j=2}^{p+1} \partial_{\mu_1} \cdots \partial_{\mu_j} X(z) \cdot g^{\mu_1 \cdots \mu_j}(z, x;\, \Phi). \quad (5.6.3-7)$$

---

[2]   For a tensor, $-\pounds_\lambda$ is the Lie derivative.

Here $f^{\mu_1 \cdots \mu_j}(z, x; \Phi)$ and $g^{\mu_1 \cdots \mu_j}(z, x; \Phi)$ are some operators *independent* of X, which are defined only at $z^0 = x^0$ and have the form

$$\sum_{s=0}^{n} \omega^{\mu_1 \cdots \mu_j k_1 \cdots k_s}(z) \partial_{k_1}{}^z \cdots \partial_{k_s}{}^z \delta(\boldsymbol{z} - \boldsymbol{x}), \qquad (5.6.3-8)$$

where n is an unspecified integer and $\omega^{\mu_1 \cdots \mu_j k_1 \cdots k_s}(z)$ is a local operator, symmetric in $\mu_1, \cdots, \mu_j$ and in $k_1, \cdots, k_s$.

It is easy to confirm that Eqs.(6) and (7) reduce to Eqs.(5.6.1-2) and (5.6.1-4), respectively, apart from the non-tensorlike term, when $\Phi = T_\Sigma$ and $X = b_\rho$.

Now, we prove Eqs.(6) and (7) by mathematical induction with respect to Φ [Nak 84c].

As the first step, we prove Eqs.(6) and (7) for the case in which Φ is a primary field φ. Their validity has been checked for $\varphi = g_{\mu\nu}$. For matter fields, it is easily checked if $\mathcal{L}_M$ is explicitly given. But without knowing $\mathcal{L}_M$, it is impossible to calculate commutation relations. We therefore present a general proof on the assumption that the BRS invariance of the theory is correctly incorporated at the quantum level.

Since the validity of Eqs.(6) and (7) is trivial for $X = \hat{x}^\lambda$, c^σ, and $\bar{c}_\tau$, we consider $X = b_\rho$ only. The BRS invariance implies

$$\boldsymbol{\delta}_*([\varphi, \bar{c}_\rho{}']) = \{\boldsymbol{\delta}_*(\varphi), \bar{c}_\rho{}'\} + [\varphi, (ib_\rho - \kappa c^\lambda \partial_\lambda \bar{c}_\rho)'] \qquad (5.6.3-9)$$

with

$$\boldsymbol{\delta}_*(\varphi) = \kappa \partial_\mu c^\nu \cdot [\varphi]^\mu{}_\nu - \kappa c^\mu \partial_\mu \varphi. \qquad (5.6.3-10)$$

Because of canonical commutation relations, φ commutes with c^σ, $\dot{c}^\sigma$, $\bar{c}_\tau$, and $\dot{\bar{c}}_\tau$. Hence the lhs. of Eq.(9) vanishes and the rhs. is simplified. The first term of the rhs. is calculated by means of Eq.(5.4.3-12); we find

$$\{\boldsymbol{\delta}_*(\varphi), \bar{c}_\rho{}'\} = \kappa [\varphi]^0{}_\rho \tilde{f} \delta^3. \qquad (5.6.3-11)$$

Hence we obtain the tensorlike commutation relation

$$[\varphi, b_\rho{}'] = i\kappa [\varphi]^0{}_\rho \tilde{f} \delta^3. \qquad (5.6.3-12)$$

Likewise, we calculate

$$\boldsymbol{\delta}_*([\varphi, \dot{\bar{c}}_\rho{}']) = \{\boldsymbol{\delta}_*(\varphi), \dot{\bar{c}}_\rho{}'\} + [\varphi, (i\dot{b}_\rho - \kappa \dot{c}^\lambda \partial_\lambda \bar{c}_\rho - \kappa c^\lambda \partial_\lambda \dot{\bar{c}}_\rho)']. \qquad (5.6.3-13)$$

Since $[\varphi,\ \tilde{g}^{\mu\nu\prime}] = 0$, the field equation, Eq.(5.3.3-4), implies that $[\varphi, \ddot{\tilde{c}}_\rho{}'] = 0$, whence $[\dot\varphi,\ \dot{\tilde{c}}_\rho{}'] = 0$. We therefore obtain

$$[\varphi,\ \dot{b}_\rho{}'] = i\kappa([\varphi]^\mu{}_\nu\{\partial_\mu c^\nu,\ \dot{\tilde{c}}_\rho{}'\} - \partial_\mu\varphi\cdot\{c^\mu,\ \dot{\tilde{c}}_\rho{}'\}). \qquad (5.6.3-14)$$

The rhs. can easily be calculated by means of

$$\{\dot{c}^\nu,\ \dot{\tilde{c}}_\rho{}'\} = \delta^\nu{}_\rho(\partial_0\tilde{f}\cdot\delta^3 + 2\tilde{f}\tilde{g}^{0k}(\delta^3)_k) \qquad (5.6.3-15)$$

and Eq.(5.4.3-12). Thus the tensorlike commutation relation for $[\varphi,\ \dot{b}_\rho{}']$ [see Eq.(5.6.1-4)] is proved.

We proceed to the case of general Φ. Suppose that we have proved Eqs.(6) and (7) for any derivative of φ; then they are valid for Φ owing to the fact that $\pounds_\lambda$ is a derivation on Φ. We therefore prove our proposition by mathematical induction with respect to the order of the derivative of φ. As for spatial differentiation, induction is trivial because of Eq.(5). Thus what we must prove is this: Under the assumption of Eqs.(6) and (7), the corresponding formulae for $[\partial_0\Phi(x),\ X(z)]|_0$ and $[\partial_0\Phi(x),\ \partial_0 X(z)]|_0$ are valid.

First, by using Eqs.(6) and (7), we have

$$\begin{aligned}
[\partial_0\Phi(x),\ X(z)]|_0 &= \partial_0([\Phi(x),\ X(z)]|_0) - [\Phi(x),\ \partial_0 X(z)]|_0 \\
&= -i\kappa\eta(X,\ \hat{x}^\lambda)\{(\partial_0{}^x + \partial_0{}^z)\pounds_\lambda(\Phi)^x\mathcal{D}(x,\ z)|_0 \\
&\quad - \pounds_\lambda(\Phi)^x\partial_0{}^z\mathcal{D}(x,\ z)|_0\} + \partial_0\{\sum_{j=2}^{p}\partial_{\mu_1}\cdots\partial_{\mu_j}X(z)\cdot f^{\mu_1\cdots\mu_j}(z,\ x;\ \Phi)\} \\
&\quad - \sum_{j=2}^{p+1}\partial_{\mu_1}\cdots\partial_{\mu_j}X(z)\cdot g^{\mu_1\cdots\mu_j}(z,\ x;\ \Phi). \qquad (5.6.3-16)
\end{aligned}$$

Hence, with the aid of Eq.(5), we obtain

$$\begin{aligned}
[\partial_0\Phi(x),\ X(z)]|_0 &= -i\kappa\eta(X,\ \hat{x}^\lambda)\pounds_\lambda(\partial_0\Phi)^x\mathcal{D}(x,\ z)|_0 \\
&\quad + \sum_{j=2}^{p+1}\partial_{\mu_1}\cdots\partial_{\mu_j}X(z)\cdot f^{\mu_1\cdots\mu_j}(z,\ x;\ \partial_0\Phi), \qquad (5.6.3-17)
\end{aligned}$$

where

$$\begin{aligned}
f^{\mu_1\cdots\mu_j}(z,\ x;\ \partial_0\Phi) &\equiv j^{-1}\sum_{\imath=1}^{\jmath}\delta^{\mu_\imath}{}_0 f^{\mu_1\cdots\mu_{\imath-1}\mu_{\imath+1}\cdots\mu_j}(z,\ x;\ \Phi) \\
&\quad + \partial_0 f^{\mu_1\cdots\mu_j}(z,\ x;\ \Phi) - g^{\mu_1\cdots\mu_j}(z,\ x;\ \Phi) \qquad (5.6.3-18)
\end{aligned}$$

with the understanding $f^{\mu_1 \cdots \mu_j}(z, x; \Phi) = 0$ for $j = 1$ and $j = p + 1$.

The same reasoning can be applied to $[\partial_0 \Phi(x), \ \partial_0 X(z)]|_0$ if a formula for $[\Phi(x), \ (\partial_0)^2 X(z)]|_0$ is available, but we do not have it as an induction assumption. We therefore make use of the field equation, Eq.(5.5.2-2), together with Eq.(5.3.3-2):

$$(\partial_0)^2 X = -\tilde{f} \tilde{g}^{k\ell} \partial_k \partial_\ell X - 2\tilde{f} \tilde{g}^{0k} \partial_k \partial_0 X, \qquad (5.6.3 - 19)$$

where $\tilde{f} \equiv (\tilde{g}^{00})^{-1}$ as before. Then we can substitute Eqs.(6) and (7), obtaining

$$
\begin{aligned}
&[\Phi(x), (\partial_0)^2 X(z)]|_0 \\
&= -i\kappa\eta(X, \hat{x}^\lambda)\mathcal{L}_\lambda(\Phi)^x \mathcal{D}(x, z)\{- \overleftarrow{\partial}_k{}^z \overleftarrow{\partial}_\ell{}^z \tilde{f}(z)\tilde{g}^{k\ell}(z) \\
&\quad - 2\overleftarrow{\partial}_0{}^z \overleftarrow{\partial}_k{}^z \tilde{f}(z)\tilde{g}^{0k}(z)\}|_0 \\
&\quad - \Big\{ \sum_{j=2}^{p} \partial_{\mu_1} \cdots \partial_{\mu_j} X(z) \cdot f^{\mu_1\cdots\mu_j}(z, x; \Phi)\Big\} \overleftarrow{\partial}_k{}^z \overleftarrow{\partial}_\ell{}^z \tilde{f}(z)\tilde{g}^{k\ell}(z) \\
&\quad - 2\Big\{ \sum_{j=2}^{p+1} \partial_{\mu_1} \cdots \partial_{\mu_j} X(z) \cdot g^{\mu_1\cdots\mu_j}(z, x; \Phi)\Big\} \overleftarrow{\partial}_k{}^z \tilde{f}(z)\tilde{g}^{k\ell}(z) \\
&\quad - \partial_k \partial_\ell X(z) \cdot [\Phi(x), \ \tilde{f}(z)\tilde{g}^{k\ell}(z)]|_0 \\
&\quad - 2\partial_k \partial_0 X(z) \cdot [\Phi(x), \ \tilde{f}(z)\tilde{g}^{0k}(z)]|_0. \qquad (5.6.3 - 20)
\end{aligned}
$$

The first term is reduced to

$$-i\kappa\eta(X, \hat{x}^\lambda)\mathcal{L}_\lambda(\Phi)^x (\partial_0{}^z)^2 \mathcal{D}(x, z), \qquad (5.6.3 - 21)$$

owing to Eq.(5.6.2-16) together with Eq.(5.3.3-2). The remainder can be written as

$$\sum_{j=2}^{p+2} \partial_{\mu_1} \cdots \partial_{\mu_j} X(z) \cdot h^{\mu_1 \cdots \mu_j}(z, x; \Phi), \qquad (5.6.3 - 22)$$

where $h^{\mu_1 \cdots \mu_j}(z, x; \Phi)$ has the form of Eq.(8).

In this way, we complete mathematical induction.

We call Eqs.(6) and (7) the **equal-time geometric commutation relations**, because $\mathcal{L}(\Phi)$ represents the geometric nature of the classical general relativity. Of course, such a geometric property cannot be completely valid in quantum gravity because spacetime structure itself has quantum fluctuation. This breaking is represented by the second terms, which we call **non-geometric terms**. As is seen from Eq.(20), the source of their appearance is $[\Phi(x), \ \tilde{g}^{\mu\nu}(z)]|_0$, that is, *the operator nature of the spacetime metric is the origin of non-geometric terms.*

5.6.4 Four-dimensional geometric commutation relation

We extend the equal-time geometric commutation relation, Eqs.(5.6.3-6) and (5.6.3-7), to the four-dimensional form with the help of the integral representations presented in Sec.5.6.2.

Our aim of this subsection is to prove the following **four-dimensional geometric commutation relation** [Nak 83a]:

$$[\Phi(x),\ X(y)] = -i\kappa\eta(X,\ \hat{x}^\lambda)\mathcal{L}_\lambda(\Phi)^x \mathcal{D}(x,\ y)$$
$$+ \int dz\, \partial_\mu\partial_\nu X(z) \cdot K^{\mu\nu}(z,\ x,\ y;\ \Phi), \qquad (5.6.4-1)$$

where $K^{\mu\nu}(z,\ x,\ y;\ \Phi)$, defined only at $x^0 = z^0$, is an unknown non-local operator *independent of X*. Since the first term vanishes for $X \neq b_\rho$, we see that $b_\rho(y)$ indeed plays the role of the generator of general coordinate transformations and that the structure of the non-geometric term is common for b_ρ, c^σ, and $\bar{c}_r$. [It vanishes for $X(z) = \hat{z}^\lambda$, as it should be.]

Now, we prove Eq.(1). Owing to Eq.(5.6.2-23) with $z^0 = x^0$, we have

$$[\Phi(x),\ X(y)]$$
$$= \int dz\,\{[\Phi(x),\ \partial_\nu X(z)] \cdot \tilde{g}^{0\nu}(z)\mathcal{D}(z,\ y)$$
$$- [\Phi(x),\ X(z)]\tilde{g}^{0\nu}(z)\partial_\nu{}^z\mathcal{D}(z,\ y)\}|_{z^0=x^0}$$
$$+ \int dz\,\{\partial_\nu X(z) \cdot [\Phi(x),\ \tilde{g}^{0\nu}(z)\mathcal{D}(z,\ y)]$$
$$- X(z)[\Phi(x),\ \tilde{g}^{0\nu}(z)\partial_\nu{}^z\mathcal{D}(z,\ y)]\}|_{z^0=x^0}. \qquad (5.6.4-2)$$

The first integral can be calculated by substituting Eqs.(5.6.3-6) and (5.6.3-7). It becomes

$$-i\kappa\eta(X,\ \hat{x}^\lambda) \int dz\,\{\mathcal{L}_\lambda(\Phi)^x\partial_\nu{}^z\mathcal{D}(x,\ z) \cdot \tilde{g}^{0\nu}(z)\mathcal{D}(z,\ y)$$
$$- \mathcal{L}_\lambda(\Phi)^x\mathcal{D}(x,\ z)\tilde{g}^{0\nu}(z)\partial_\nu{}^z\mathcal{D}(z,\ y)\}|_{z^0=x^0} \qquad (5.6.4-3)$$

apart from the contribution from the non-geometric terms. Since z^0 is arbitrary in the integral representation, Eq.(5.6.2-22), we can freely differentiate it with respect to x^μ. Hence Eq.(3) exactly produces the first term of the rhs. of Eq.(1). The remainder of the first integral of Eq.(2) can easily be rewritten in the form of the second term of Eq.(1); reduction to the second order derivatives of X is achieved by means of the field equation, Eq.(5.6.3-19), and integrations by parts.

Our remaining task is to show that the second integral of Eq.(2) can be rewritten in the form of the second term of Eq.(1). This can be done by some manipulation based on the fact that Eq.(2) holds also for $X(y) = 1$ and for $X(y) = \hat{y}^\lambda$ [Nak 83a].

It is quite remarkable that we have obtained a four-dimensional commutation relation in quantum gravity, though we have almost no information about $K^{\mu\nu}$.

5.6.5 Non-geometric terms

As seen in the proof [see Eq.(5.6.3-20)] of the equal-time geometric commutation relations, the "seed" of non-geometric terms is $[\Phi, \tilde{g}^{\mu\nu\prime}]$. Therefore, by examining this commutator, we can see in what order of time derivative of a primary field φ a non-geometric term arises for the first time [Nak 83b].

Suppose $[\Phi, \tilde{g}^{\mu\nu\prime}] = 0$; then the de Donder condition, Eq.(5.3.3-2), implies that

$$[\partial_0 \Phi, \tilde{g}^{0\nu\prime}] = -[\Phi, \partial_0 \tilde{g}^{0\nu\prime}]$$
$$= \partial_k{}'[\Phi, \tilde{g}^{k\nu\prime}] = 0. \qquad (5.6.5-1)$$

This fact means that in order to be non-vanishing, the last term of Eq.(5.6.3-20) requires one more ∂_0 than the second last term. Thus the first-appearing non-geometric term is always proportional to $\partial_k \partial_l X$ but not to $\partial_k \partial_0 X$.

Let r be the least integer such that $[(\partial_0)^r \varphi, \tilde{g}^{kl\prime}] \neq 0$. Then Eq.(5.6.3-20) implies that a non-geometric term in $[(\partial_0)^n \varphi, X']$ arises first at $n = r + 2$, that is,

$$\text{N.G.}[(\partial_0)^n \varphi, X'] = 0 \quad \text{for} \quad n \leq r+1, \qquad (5.6.5-2)$$
$$\text{N.G.}[(\partial_0)^{r+2} \varphi, X'] = -[(\partial_0)^r \varphi, \tilde{g}^{kl\prime}](\tilde{f}\partial_k\partial_l X)', \qquad (5.6.5-3)$$

where N.G. denotes to take the non-geometric term.

For example, since $r = 1$ for $g_{\mu\nu}$, we have

$$\text{N.G.}[(\partial_0)^n g_{\mu\nu}, X'] = 0 \qquad \text{for} \quad n \leq 2. \qquad (5.6.5-4)$$

From Eq.(4), we can immediately infer that the Riemann tensor $R^\lambda{}_{\rho\mu\nu}$ satisfies the tensorlike commutation relation, Eq.(5.6.1-2). [Direct check of this fact is extremely laborious.] From Eqs.(3) and (5.4.3-7), we have

$$N.G.[(\partial_0)^3 g_{\mu\nu}, \ X'] = -4i\kappa h \tilde{f}^3 (\delta^k{}_\mu \delta^\ell{}_\nu \tilde{g}^{00} - \delta^k{}_\mu \delta^0{}_\nu \tilde{g}^{0\ell}$$
$$- \delta^0{}_\mu \delta^\ell{}_\nu \tilde{g}^{k0} - \delta^0{}_\mu \delta^0{}_\nu \tilde{g}^{k\ell}) \partial_k \partial_\ell X \cdot \delta^3. \quad (5.6.5-5)$$

5.6.6 Commutation relation between supercoordinates

The quantum-gravity D function $\mathcal{D}(x, y)$ is useful also for the (anti)commutation relations between two sixteen-dimensional supercoodinates (see Sec.5.5.2). The equal-time (anti)commutation relations concerning b_ρ, c^σ, and $\bar{c}_\tau$, which are given in Sec.5.4, are summarized as follows:

$$[X, Y']_{\mp} = 0, \quad (5.6.6-1)$$
$$[X, \partial_0 Y']_{\mp} = -\sqrt{-\epsilon(X)}\eta(X, Y)\tilde{f}\delta^3$$
$$+ i\kappa\{\eta(Y, \hat{x}^\lambda)\partial_\lambda X + \eta(X, \hat{x}^\lambda)\partial_\lambda Y\} \cdot \tilde{f}\delta^3. \quad (5.6.6-2)$$

We now propose the four-dimensional (anti)commutation relation between X and Y [Nak 83b]:

$$[X(x), Y(y)]_{\mp} \cong -\sqrt{-\epsilon(X)}\eta(X, Y)\mathcal{D}(x, y)$$
$$+ i\kappa\{\eta(Y, \hat{x}^\lambda)\partial_\lambda X(x) + \eta(X, \hat{x}^\lambda)\partial_\lambda Y(y)\} \cdot \mathcal{D}(x, y)$$
$$+ \mathcal{H}(x, y; X, Y). \quad (5.6.6-3)$$

Here $\mathcal{H}(x, y; X, Y)$ is a solution to the following Cauchy problem:

$$\tilde{g}^{\sigma\tau}(x)\partial_\sigma{}^x \partial_\tau{}^x \mathcal{H}(x, y; X, Y)$$
$$= -\partial_\sigma \partial_\tau X(x) \cdot \int dz \, \partial_\mu \partial_\nu Y(z) \cdot K^{\mu\nu}(z, x, y; \tilde{g}^{\sigma\tau}) \quad (5.6.6-4)$$

with

$$\mathcal{H}(x, y; X, Y)|_0 = \partial_0{}^x \mathcal{H}(x, y; X, Y)|_0 = 0, \quad (5.6.6-5)$$

where $K^{\mu\nu}(z, x, y; \Phi)$ is the quantity appearing in the four-dimensional geometric commutation relation, Eq.(5.6.4-1). The symbol $\cong$ means that Eq.(3) holds not in the sense of the genuine four-dimensional formula but in the sense of the formal equal-time expansion, that is, it yields the correct formula for

$$F_{m, p}(x, x') \equiv [(\partial_0)^p X, (\partial_0)^{m-p} Y']_{\mp} \quad (5.6.6-6)$$

for *any* $m \geq p \geq 0$.

To prove Eq.(3) in the sense of Eq.(6), we employ mathematical induction with respect to m. For $m = 0$, 1, it is valid because of Eqs.(1) and (2) together with Eq.(5). Hence we set $m \geq 2$.

First, we note that for fixed m it is sufficient to prove the proposition for a particular value of p, because

$$F_{m,\,p\pm1} = \partial_0 F_{m-1,\,p-(1\mp1)/2} - F_{m,\,p}. \qquad (5.6.6-7)$$

Hence we consider the $p = 2$ case only.

From Eq.(5.6.3-19), we have

$$\begin{aligned}
F_{m,\,2} = & -[\tilde{f}(2\tilde{g}^{0k}\partial_k\partial_0 X + \tilde{g}^{kl}\partial_k\partial_l X),\ (\partial_0)^{m-2}Y']_{\mp} \\
= & -\tilde{f}\partial_\sigma\partial_\tau X \cdot [\tilde{g}^{\sigma\tau},\ (\partial_0)^{m-2}Y'] \\
& -2\tilde{f}\tilde{g}^{0k}\partial_k F_{m-1,\,1} - \tilde{f}\tilde{g}^{kl}\partial_k\partial_l F_{m-2,\,0}. \qquad (5.6.6-8)
\end{aligned}$$

Here, from Eq.(5.6.4-1), we know that

$$\begin{aligned}
[\tilde{g}^{\sigma\tau},\ (\partial_0)^{m-2}Y'] = & i\kappa\eta(Y,\ \hat{x}^\lambda)\{\partial_\lambda\tilde{g}^{\sigma\tau} \cdot (\partial_0{}^y)^{m-2}\mathcal{D}(x,\ y)|_0 \\
& -(\delta^\sigma{}_\lambda\tilde{g}^{\mu\tau} + \delta^\tau{}_\lambda\tilde{g}^{\sigma\mu} - \delta^\mu{}_\lambda\tilde{g}^{\sigma\tau})\partial_\mu{}^x(\partial_0{}^y)^{m-2}\mathcal{D}(x,\ y)|_0\} \\
& + \int d\mathbf{z}\,\partial_\mu\partial_\nu Y(z) \cdot (\partial_0{}^y)^{m-2}K^{\mu\nu}(z,\ x,\ y;\ \tilde{g}^{\sigma\tau}). \quad (5.6.6-9)
\end{aligned}$$

The last two terms of Eq.(8) can be calculated by using Eq.(3) because of the induction assumption. With the aid of Eq.(5.6.2-4), we rewrite their sum as

$$\begin{aligned}
& -\sqrt{-\epsilon(X)}\eta(X,\ Y)(\partial_0{}^x)^2(\partial_0{}^y)^{m-2}\mathcal{D}(x,\ y)|_0 \\
& +i\kappa\eta(Y,\ \hat{x}^\lambda)\partial_\lambda X \cdot (\partial_0{}^x)^2(\partial_0{}^y)^{m-2}\mathcal{D}(x,\ y)|_0 \\
& -i\kappa\eta(Y,\ \hat{x}^\lambda)\tilde{f}\{2\tilde{g}^{0k}\partial_k\partial_\lambda X \cdot \partial_0{}^x(\partial_0{}^y)^{m-2}\mathcal{D}(x,\ y)|_0 \\
& +2\tilde{g}^{0k}\partial_k{}^x(\partial_0\partial_\lambda X \cdot (\partial_0{}^y)^{m-2}\mathcal{D}(x,\ y)|_0) \\
& +\tilde{g}^{kl}\partial_k\partial_l\partial_\lambda X \cdot (\partial_0{}^y)^{m-2}\mathcal{D}(x,\ y)|_0 \\
& +2\tilde{g}^{kl}\partial_l\partial_\lambda X \cdot \partial_k{}^x(\partial_0{}^y)^{m-2}\mathcal{D}(x,\ y)|_0\} \\
& +i\kappa\eta(X,\ \hat{x}^\lambda)\sum_{j=0}^{m-2}\binom{m-2}{j}(\partial_0)^j\partial_\lambda Y(y) \cdot (\partial_0{}^x)^2(\partial_0{}^y)^{m-2-j}\mathcal{D}(x,\ y)|_0 \\
& -2\tilde{f}\tilde{g}^{0k}\partial_k{}^x\partial_0{}^x(\partial_0{}^y)^{m-2}\mathcal{H}(x,\ y;\ X,\ Y)|_0 \\
& -\tilde{f}\tilde{g}^{kl}\partial_k{}^x\partial_l{}^x(\partial_0{}^y)^{m-2}\mathcal{H}(x,\ y;\ X,\ Y)|_0. \qquad (5.6.6-10)
\end{aligned}$$

Substituting Eqs.(9) and (10) into Eq.(8) and making use of the field equation, Eq.(5.6.3-19), and the equation of $\mathcal{H}(x, y; X, Y)$, Eq.(4), we obtain

$$
\begin{aligned}
f_{m,2} = & - \sqrt{-\epsilon(X)}\eta(X, Y)(\partial_0{}^x)^2(\partial_0{}^y)^{m-2}\mathcal{D}(x, y)|_0 \\
& + i\kappa\eta(Y, \hat{x}^\lambda)\{\partial_\lambda X \cdot (\partial_0{}^x)^2(\partial_0{}^y)^{m-2}\mathcal{D}(x, y)|_0 \\
& + 2\partial_0\partial_\lambda X \cdot \partial_0{}^x(\partial_0{}^y)^{m-2}\mathcal{D}(x, y)|_0 + \partial_0{}^2\partial_\lambda X \cdot (\partial_0{}^y)^{m-2}\mathcal{D}(x, y)|_0\} \\
& + i\kappa\eta(X, \hat{x}^\lambda)\sum_{j=0}^{m-2}\binom{m-2}{j}(\partial_0)^j\partial_\lambda Y(y) \cdot (\partial_0{}^x)^2(\partial_0{}^y)^{m-2-j}\mathcal{D}(x, y)|_0 \\
& + (\partial_0{}^x)^2(\partial_0{}^y)^{m-2}\mathcal{H}(x, y; X, Y)|_0. \qquad (5.6.6-11)
\end{aligned}
$$

The rhs. of Eq.(11) is precisely equal to the expression for $(\partial_0{}^x)^2(\partial_0{}^y)^{m-2}$ [rhs. of Eq.(3)]$|_0$. Thus mathematical induction is completed.

Although we do not know the explicit expression for $\mathcal{H}(x, y; X, Y)$, Eq.(3) unifies quite a large number of equal-time (anti)commutation relations, because

$$
(\partial_0{}^x)^p(\partial_0{}^y)^{m-p}\mathcal{H}(x, y; X, Y)|_0 = 0 \quad \text{for} \quad p \le m \le 4 \qquad (5.6.6-12)
$$

and

$$
\begin{aligned}
(\partial_0{}^x)^p(\partial_0{}^y)^{5-p}\mathcal{H}(x, y; X, Y)|_0 \\
= (-1)^{p+1}4i\kappa h^3 \tilde{f}^4 K^{ijkl}\partial_i\partial_j X \cdot \partial_k\partial_l Y \cdot \delta^3, \qquad (5.6.6-13)
\end{aligned}
$$

where K^{ijkl} is defined by Eq.(5.4.3-3). For instance, Eq.(5.6.3-15) immediately follows from Eq.(3) with Eq.(5.6.2-13).

Eq.(12) is shown as follows. The $m \le 1$ case is nothing but Eq.(5). The $2 \le m \le 4 \, (= 2 + 2)$ case follows from Eqs.(4) and (5.6.5-2) for $\varphi = \tilde{g}^{\sigma\tau}$ [cf. Eq.(5.6.5-4)]. Finally, Eq.(13) follows from Eqs.(4) and (5.6.5-3) for $\varphi = \tilde{g}^{\sigma\tau}$ together with Eq.(5.4.3-8).

5.7 VIERBEIN FORMALISM

The existence of Dirac fields in the real world is firmly confirmed in particle physics. But the spinor representation of Lorentz group cannot be linearly extended to that of such a large group as the general -coordinate-transformation group or $GL(4)$. In the framework of general relativity, therefore, the Dirac field ψ must be regarded as a *scalar* field. On the other hand, the Dirac gamma matrices γ^μ must be regarded as a matrix vector field because their anticommutator is $2g^{\mu\nu}$. In order not to introduce too much redundancy, it is most convenient to express γ^μ in terms of the vierbein field,[1] $h_\mu{}^a$, a set of four vector fields. Thus the theory of quantum gravity must be extended to the case of the vierbein formalism.

5.7.1 Classical vierbein formalism

In this subsection, we briefly review the classical theory of the vierbein formalism.

Since the spacetime of the classical general relativity is locally Minkowskian, we can set up a local orthogonal coordinate system at each point x^μ. The four basis vectors $h_\mu{}^a(x)$ ($a = 0$, 1, 2, 3) are called the **vierbein** (or **tetrad**). They are related to the metric tensor $g_{\mu\nu}(x)$ through

$$g_{\mu\nu} = \eta_{ab} h_\mu{}^a h_\nu{}^b, \qquad (5.7.1-1)$$

where η_{ab} is the Minkowski metric in the tangent space.

Since $h_\mu{}^a$ has 16 independent components while $g_{\mu\nu}$ has 10, the vierbein contains six extra degrees of freedom. They are nothing but the degrees of freedom of the Lorentz transformations in the tangent space, which, of course, leave Eq.(1) invariant. This Lorentz transformation is called the **local Lorentz transformation**, because it depends on x^μ. It is *not* a spacetime transformation but a local gauge transformation.

We can regard Eq.(1) as a matrix relation. Taking determinants, therefore, we have

$$g = -h^2, \qquad (5.7.1-2)$$

where

$$h \equiv \det h_\mu{}^a. \qquad (5.7.1-3)$$

---

[1]   It is more popular to denote it by $e_\mu{}^a$.

It is important to note that Eq.(2) implies the reality of $\sqrt{-g}$. Since $g \neq 0$, we have $h \neq 0$, that is, the matrix $(h_\mu{}^a)$ is invertible. Let $h^\mu{}_c{}' \equiv g^{\mu\nu} h_{\nu c}$, and we multiply Eq.(1) by it; then owing to the above invertibility we obtain

$$h^\mu{}_c h_\mu{}^a = \delta^a{}_c, \qquad (5.7.1-4)$$

that is, $(h^\mu{}_c)$ is the (transposed) inverse matrix of $(h_\mu{}^a)$.

In the vierbein formalism, the primary field of gravity is $h_\mu{}^a$, and $g_{\mu\nu}$ is a composite (fusion) field.

Extending Eq.(5.2.2-5), we require

$$\nabla_\mu h_\nu{}^a \equiv \partial_\mu h_\nu{}^a - \Gamma^\lambda{}_{\mu\nu} h_\lambda{}^a + \omega_\mu{}^a{}_b h_\nu{}^b = 0, \qquad (5.7.1-5)$$

where $\omega_\mu{}^a{}_b$ denotes the **spin connection**, which is the connection concerning the local gauge transformation, that is, $\omega_\mu{}^a{}_b$ is nothing but the gauge field for the Lorentz group. From Eq.(5), it should be defined by

$$\omega_\mu{}^a{}_b \equiv (\Gamma^\lambda{}_{\mu\nu} h_\lambda{}^a - \partial_\mu h_\nu{}^a) \cdot h^\nu{}_b, \qquad (5.7.1-6)$$

that is, $-h_\nu{}^b \omega_\mu{}^a{}_b$ is covariant derivative of $h_\nu{}^a$ *with respect to spacetime only*. By direct calculation or by taking covariant derivative of Eq.(4) with respect to spacetime only, we find

$$\omega_\mu{}^{ba} = -\omega_\mu{}^{ab}. \qquad (5.7.1-7)$$

When it is necessary to make the distinction clear, we call a tensor with respect to spacetime transformations a **world tensor** and a tensor with respect to local-Lorentz transformations a **Lorentz tensor**. Any world tensor can be transcribed into a local-Lorentz tensor by means of vierbeins, and *vice versa*; for instance, $A^a = h_\lambda{}^a A^\lambda$. Applying $[\nabla_\mu, \nabla_\nu]$ to it and using Eqs.(5) and (5.2.2-10), we find

$$R^\lambda{}_{\rho\mu\nu} = h^\lambda{}_a h_\rho{}^b \hat{R}_{\mu\nu}{}^a{}_b, \qquad (5.7.1-8)$$

where $\hat{R}_{\mu\nu}{}^a{}_b$ is the curvature tensor concerning the spin connection, that is,

$$\hat{R}_{\mu\nu}{}^a{}_b \equiv \partial_\mu \omega_\nu{}^a{}_b - \partial_\nu \omega_\mu{}^a{}_b + \omega_\mu{}^a{}_c \omega_\nu{}^c{}_b - \omega_\nu{}^a{}_c \omega_\mu{}^c{}_b$$
$$= -\hat{R}_{\nu\mu}{}^a{}_b = -\hat{R}_{\mu\nu b}{}^a. \qquad (5.7.1-9)$$

Since $\omega_\mu{}^a{}_b$ is a world vector, Eq.(9) is a manifestly covariant relation in contrast to Eq.(5.2.2-11).

Now, we formulate the **generally covariant Dirac theory** on the basis of the vierbein formalism.

Let $\overset{\circ}{\gamma}{}^a$ be the flat-space gamma matrices:

$$\{\overset{\circ}{\gamma}{}^a, \overset{\circ}{\gamma}{}^b\} = 2\eta^{ab}. \qquad (5.7.1-10)$$

We set

$$\overset{\circ}{\sigma}{}^{ab} \equiv [\overset{\circ}{\gamma}{}^a, \overset{\circ}{\gamma}{}^b]/4 = -\overset{\circ}{\sigma}{}^{ba}. \qquad (5.7.1-11)$$

The generally covariant gamma matrices γ^μ are defined by

$$\gamma^\mu \equiv h^\mu{}_a \overset{\circ}{\gamma}{}^a. \qquad (5.7.1-12)$$

Then Eqs.(1) and (4) imply

$$\{\gamma^\mu, \gamma^\nu\} = 2g^{\mu\nu}. \qquad (5.7.1-13)$$

The Dirac field ψ is a world scalar and a local-Lorentz spinor. Hence its covariant derivative is defined by

$$\nabla_\mu \psi \equiv (\partial_\mu + \omega_\mu)\psi, \qquad (5.7.1-14)$$

where

$$\omega_\mu \equiv \frac{1}{2}\overset{\circ}{\sigma}{}^{ab}\omega_{\mu ab}. \qquad (5.7.1-15)$$

The Dirac conjugate, $\bar{\psi}$, of ψ is defined by

$$\bar{\psi} \equiv \psi^\dagger \overset{\circ}{\gamma}{}^0. \qquad (5.7.1-16)$$

Hence, from Eq.(14), the covariant derivative of $\bar{\psi}$ is given by

$$\bar{\psi}\overleftarrow{\nabla}_\mu = \bar{\psi}(\overleftarrow{\partial}_\mu - \omega_\mu). \qquad (5.7.1-17)$$

Let m be the mass of the Dirac field. Then the Lagrangian density of the Dirac field ψ is given by

$$\mathcal{L}_{\mathrm{D}} \equiv \frac{1}{2}ih\bar{\psi}\gamma^\mu\nabla_\mu\psi - \frac{1}{2}i\bar{\psi}\overleftarrow{\nabla}_\mu\gamma^\mu\psi h - mh\bar{\psi}\psi. \qquad (5.7.1-18)$$

From Eq.(18), we obtain the Dirac equation

$$(\gamma^\mu\nabla_\mu + im)\psi = 0. \qquad (5.7.1-19)$$

In the vierbein formalism, the matter energy-momentum tensor is defined by

$$hT^{\mu\nu} \equiv -h^{\nu a}\delta\mathcal{L}_{\mathrm{M}}/\delta h_{\mu}{}^{a}, \qquad (5.7.1-20)$$

which, of course, reduces to Eq.(5.2.3-6) if $\mathcal{L}_{\mathrm{M}}$ involves the gravitational field only through $g_{\mu\nu}$. When Eq.(20) is applied to $\mathcal{L}_{\mathrm{D}}$, we obtain an expression for $hT_{\mathrm{D}}{}^{\mu\nu}$ which is not manifestly symmetric under $\mu \leftrightarrow \nu$, but we can rewrite it as

$$T_{\mathrm{D}}{}^{\mu\nu} = \frac{1}{4}i\bar{\psi}(\gamma^{\mu}\nabla^{\nu} - \overleftarrow{\nabla}{}^{\nu}\gamma^{\mu})\psi + (\mu \leftrightarrow \nu) \qquad (5.7.1-21)$$

by using Eq.(19) and its conjugate. In general, $T^{\mu\nu}$ is symmetric under $\mu \leftrightarrow \nu$ if $\mathcal{L}_{\mathrm{M}}$ is local-Lorentz invariant; this fact follows from the second Noether theorem in the following way.

For the local-Lorentz transformations, we have $T_A{}^{\mu}{}_{\alpha} = 0$ in Eq.(5.2.4-1) and, setting $\mathcal{F} = \mathcal{L}_{\mathrm{M}}$, $F_{\epsilon}{}^{\mu} = 0$ in Eq.(5.2.4-2). Hence Eqs.(5.2.4-4) and (5.2.4-5) reduce to

$$G_{A\alpha}\frac{\partial\mathcal{L}_{\mathrm{M}}}{\partial\varphi_A} + \partial_{\mu}G_{A\alpha}\frac{\partial\mathcal{L}_{\mathrm{M}}}{\partial(\partial_{\mu}\varphi_A)} = 0, \qquad (5.7.1-22)$$

$$G_{A\alpha}\frac{\partial\mathcal{L}_{\mathrm{M}}}{\partial(\partial_{\mu}\varphi_A)} = 0, \qquad (5.7.1-23)$$

respectively. We therefore obtain

$$G_{A\alpha}\frac{\delta\mathcal{L}_{\mathrm{M}}}{\delta\varphi_A} = 0. \qquad (5.7.1-24)$$

With the aid of matter field equations, Eq.(24) becomes

$$(\eta^{ac}h_{\lambda}{}^{b} - \eta^{cb}h_{\lambda}{}^{a})\frac{\delta\mathcal{L}_{\mathrm{M}}}{\delta h_{\lambda}{}^{c}} = 0. \qquad (5.7.1-25)$$

Then multiplication by $h^{\mu}{}_{a}h^{\nu}{}_{b}$ yields $T^{\nu\mu} = T^{\mu\nu}$.

5.7.2 Local-Lorentz BRS transformation

We now construct the quantum theory of the vierbein formalism [Nak 79c].

The six degrees of freedom of $h_{\mu}{}^{a}$ corresponding to the local Lorentz transformation is totally non-dynamical, that is, there is no Lagrangian density which characterizes them as dynamical freedom. Therefore, one might feel that it is natural to eliminate these degrees of freedom. But such a procedure inevitably violates the beautiful symmetry properties of the theory. The most natural way of quantizing $h_\mu{}^a$ is again to make use of the BRS symmetry.

For this purpose, we introduce three world-scalar, antisymmetric local-Lorentz-tensor fields s_{ab}, t_{ab}, and $\bar{t}_{ab}$. Here s_{ab} is the local-Lorentz B-field (bosonic), and t_{ab} and $\bar{t}_{ab}$ are the local-Lorentz FP ghost and FP antighost, respectively (fermionic).

We denote the local-Lorentz BRS transformation by $\boldsymbol{\delta}_\mathrm{L}$. From the expression for the changes of fields under the infinitesimal local Lorentz transformation, we set

$$\boldsymbol{\delta}_\mathrm{L}(h_\mu{}^a) = -t^a{}_b h_\mu{}^b, \qquad (5.7.2-1)$$

$$\boldsymbol{\delta}_\mathrm{L}(\psi) = -\frac{1}{2} t_{ab} \overset{\circ}{\sigma}{}^{ab} \psi. \qquad (5.7.2-2)$$

For any local-Lorentz scalar Φ, we have

$$\boldsymbol{\delta}_\mathrm{L}(\Phi) = 0. \qquad (5.7.2-3)$$

Since the commutation relation between two Lorentz generators can be written as

$$\left[\frac{1}{2} M_{ab}, \frac{1}{2} M_{cd}\right] = i f_{ab,cd}{}^{ef} \frac{1}{2} M_{ef} \qquad (5.7.2-4)$$

with

$$f_{ab,cd}{}^{ef} \equiv \frac{1}{4}[(\eta_{bc} \delta^e{}_a \delta^f{}_d - \eta_{ac} \delta^e{}_b \delta^f{}_d + \eta_{ad} \delta^e{}_b \delta^f{}_c \\ - \eta_{bd} \delta^e{}_a \delta^f{}_c) - (e \leftrightarrow f)], \qquad (5.7.2-5)$$

we set

$$\boldsymbol{\delta}_\mathrm{L}(t^{ab}) = -\frac{1}{2} f_{cd,ef}{}^{ab} t^{cd} t^{ef} \\ = -t^{ac} t_c{}^b. \qquad (5.7.2-6)$$

Furthermore, as usual, we set

$$\boldsymbol{\delta}_\mathrm{L}(\bar{t}_{ab}) = i s_{ab}, \qquad (5.7.2-7)$$

$$\boldsymbol{\delta}_\mathrm{L}(s_{ab}) = 0. \qquad (5.7.2-8)$$

Then, of course, $\boldsymbol{\delta}_{\mathrm{L}}$ is a nilpotent, hermiticity-preserving anti-derivation, as it should be. Since $\boldsymbol{\delta}_{\mathrm{L}}$ is independent of $\boldsymbol{\delta}$, we assume that they commute. [Alternatively, we may assume that they anticommute.] Since $\boldsymbol{\delta}_{\mathrm{L}}$ is an internal symmetry, it commutes with ∂_μ.

From Eqs.(5.7.1-6), (1), and (3), we obtain

$$\boldsymbol{\delta}_{\mathrm{L}}(\omega_\mu{}^{ab}) = -t^a{}_c\omega_\mu{}^{cb} + t^b{}_c\omega_\mu{}^{ca} + \partial_\mu t^{ab}$$
$$= (D_\mu t)^{ab}, \qquad\qquad (5.7.2-9)$$

where we set

$$(D_\mu\varphi)^{ab} \equiv D_\mu{}^{ab}{}_{cd}\varphi^{cd}$$
$$\equiv \partial_\mu\varphi^{ab} + f_{ef,cd}{}^{ab}\omega_\mu{}^{ef}\varphi^{cd} \qquad (5.7.2-10)$$

for any antisymmetric local-Lorentz tensor φ^{ab}.

5.7.3 Local-Lorentz gauge fixing

In order to determine the local-Lorentz gauge-fixing Lagrangian density $\mathcal{L}_{\mathrm{LGF}}$, we require it to satisfy the following five conditions:

1. Although it must violate the local-Lorentz invariance, it should not violate the global internal Lorentz invariance.
2. It should be written in terms of the vierbein and the local-Lorentz B-field only.
3. The number of time differentiations involved in its term should be at most two.
4. It should contain the time derivatives of $h_\mu{}^a$ corresponding to the six degrees of freedom of the local-Lorentz transformation.
5. It should be a scalar density under general coordinate transformations.

The first and second requirements are natural ones for the gauge-fixing term in the B-field formalism. The third and fourth requirements are necessary for canonical quantization. What we want to emphasize is to impose the fifth requirement. Since the invariance under general coordinate transformations is already broken by $\mathcal{L}_{\mathrm{GF}} + \mathcal{L}_{\mathrm{FP}}$, most physicists do not want to respect it, and they adopt a seemingly simpler gauge-fixing term. Such a choice may be sufficient for the perturbative investigation, but it makes the theory in the Heisenberg picture complicated and extremely ugly. It will turn out that *without the fifth requirement, the vierbein formalism does not reproduce the theory formulated on the basis of $g_{\mu\nu}$*.

The simplest choice satisfying all the above requirements is [Nak 79c]

$$\mathcal{L}_{\text{LGF}} = -\tilde{g}^{\mu\nu}\omega_\mu{}^{ab}\partial_\nu s_{ab}. \qquad (5.7.3-1)$$

We may add to Eq.(1) any linear combination of such terms as $hh^{\nu a}\partial_\nu h^{\mu b}\cdot\partial_\mu s_{ab}$ and $hs^{ab}s_{ab}$, which satisfy the requirements other than the fourth. But, for simplicity, we do not take account of such gauge terms.

Clearly, Eq.(1) has close analogy to the Landau-gauge gauge-fixing Lagrangian density of the non-abelian gauge field (see Sec.3.4.1). One must note, however, that $\omega_\mu{}^a{}_b$ is *not* a primary field but it contains time derivatives of a primary field $h_\nu{}^c$. Accordingly, we should not integrate by parts in Eq.(1), that is, the B-field s_{ab} is *not a Lagrange multiplier*. Thus $\mathcal{L}_{\text{LGF}}$ makes the local-Lorentz degrees of freedom dynamical. This resolves the difficulty pointed out at the beginning of Sec.5.7.2.

The local-Lorentz FP-ghost Lagrangian density $\mathcal{L}_{\text{LFP}}$ is determined by the requirement [Nak 79c]

$$\mathcal{L}_{\text{LGF}} + \mathcal{L}_{\text{LFP}} = \boldsymbol{\delta}_{\text{L}}(i\tilde{g}^{\mu\nu}\omega_\mu{}^{ab}\partial_\nu\bar{t}_{ab}) \qquad (5.7.3-2)$$

as usual. We obtain

$$\mathcal{L}_{\text{LFP}} = -i\tilde{g}^{\mu\nu}\partial_\mu\bar{t}_{ab}\cdot(D_\nu t)^{ab}. \qquad (5.7.3-3)$$

The total Lagrangian density of the quantum theory of the vierbein formalism is thus given by

$$\boldsymbol{\mathcal{L}} \equiv \mathcal{L} + \mathcal{L}_{\text{LGF}} + \mathcal{L}_{\text{LFP}}, \qquad (5.7.3-4)$$

where $\mathcal{L}$ is given by Eq.(5.3.2-7) and $\mathcal{L}_{\text{M}}$ may include $\mathcal{L}_{\text{D}}$. Field equations are as follows.

Firstly, the quantum Einstein equation, Eq.(5.3.3-1), acquires a contribution from $\mathcal{L}_{\text{LGF}} + \mathcal{L}_{\text{LFP}}$, that is, we have

$$R_{\mu\nu} - \frac{1}{2}g_{\mu\nu}R - E_{\mu\nu} + \frac{1}{2}g_{\mu\nu}E = -\kappa(T_{\mu\nu} + V_{\mu\nu}), \qquad (5.7.3-5)$$

where we define

$$hV^{\mu\nu} \equiv -h^{\nu a}(\delta/\delta h_\mu{}^a)(\mathcal{L}_{\text{LGF}} + \mathcal{L}_{\text{LFP}}) \qquad (5.7.3-6)$$

in analogy to Eq.(5.7.1-20). In contrast to $T_{\mu\nu}$, however, $V_{\mu\nu}$ is *not* a symmetric quantity under $\mu \leftrightarrow \nu$ because $\mathcal{L}_{\mathrm{LGF}} + \mathcal{L}_{\mathrm{LFP}}$ is not invariant under local Lorentz transformations. Since all other terms in Eq.(5) are symmetric under $\mu \leftrightarrow \nu$, $V_{\mu\nu}$ satisfies

$$V_{\mu\nu} - V_{\nu\mu} = 0 \qquad (5.7.3-7)$$

as a field equation.

Secondly, Eqs.(5.3.3-2)-(5.3.3-4) evidently remain unchanged. Since $\mathcal{L}_{\mathrm{LGF}}$ and $\mathcal{L}_{\mathrm{LFP}}$ are scalar densities under general coordinate transformations, we have $\nabla_\mu V^{\mu\nu} = 0$, whence Eq.(5.3.3-8) also remains unchanged. This fact is a very important consequence of the fifth requirement for choosing $\mathcal{L}_{\mathrm{LGF}}$.

Thirdly, we obtain the following new field equations:

$$\partial_\mu(\tilde{g}^{\mu\nu}\omega_\nu{}^{ab}) = 0, \qquad (5.7.3-8)$$

$$\partial_\mu[\tilde{g}^{\mu\nu}(D_\nu t)^{ab}] = 0, \qquad (5.7.3-9)$$

$$\partial_\mu[\tilde{g}^{\mu\nu}(D_\nu \bar{t})^{ab}] = 0. \qquad (5.7.3-10)$$

As is seen from Eq.(5.7.2-9), the $\boldsymbol{\delta}_{\mathrm{L}}$-transform of Eq.(8) is nothing but Eq.(9). On the other hand, the $\boldsymbol{\delta}_{\mathrm{L}}$-transform of Eq.(10) becomes

$$\partial_\mu\{\tilde{g}^{\mu\nu}(D_\nu s)^{ab} - i\tilde{g}^{\mu\nu}[\bar{t}^a{}_c(D_\nu t)^{cb} - (a \leftrightarrow b)]\} = 0. \qquad (5.7.3-11)$$

By explicit calculation, we can see that Eq.(7) is nothing but Eq.(11) [Nak 79c].

5.7.4 Equal-time commutation relations

For simplicity, we assume that matter is a Dirac field only. Then canonical quantization is based on the Lagrangian density

$$\tilde{\boldsymbol{\mathcal{L}}} \equiv \tilde{\mathcal{L}}_{\mathrm{E}} + \tilde{\mathcal{L}}_{\mathrm{GF}} + \mathcal{L}_{\mathrm{FP}} + \mathcal{L}_{\mathrm{LGF}} + \mathcal{L}_{\mathrm{LFP}} + \tilde{\mathcal{L}}_{\mathrm{D}}, \qquad (5.7.4-1)$$

where

$$\tilde{\mathcal{L}}_{\mathrm{D}} \equiv h\bar{\psi}(i\gamma^\mu\nabla_\mu - m)\psi \qquad (5.7.4-2)$$

differs form $\mathcal{L}_{\mathrm{D}}$ by a total divergence only.

The canonical variables are $h_\mu{}^a$, c^σ, $\bar{c}_\tau$, s_{ab}, t_{ab}, $\bar{t}_{ab}$, and ψ. It should be noted that the local-Lorentz B-field s_{ab} is canonical. The canonical conjugates are as follows:

$$
\begin{aligned}
\pi_h{}^\mu{}_a &\equiv (\partial/\partial\dot{h}_\mu{}^a)\tilde{\boldsymbol{L}} \\
&= 2h_{\nu a}\pi_g{}^{\mu\nu} + \tilde{g}^{\nu\lambda}\frac{\partial\omega_\nu{}^{cd}}{\partial\dot{h}_\mu{}^a}(\partial_\lambda s_{cd} - 2i\partial_\lambda\bar{t}_{ed}\cdot t^e{}_c) \\
&\quad - ih\bar{\psi}\gamma^\nu\frac{\partial\omega_\nu}{\partial\dot{h}_\mu{}^a}\psi,
\end{aligned} \tag{5.7.4 - 3}
$$

where $\pi_g{}^{\mu\nu}$ is given by Eq.(5.4.1-4); $\pi_{c\sigma}$ and $\pi_{\bar{c}}{}^\tau$ are the same as Eqs.(5.4.1-5) and (5.4.1-6), respectively; furthermore,[2]

$$
\pi_s{}^{ab} \equiv (\partial/\partial\dot{s}_{ab})\tilde{\boldsymbol{L}} = \tilde{g}^{0\nu}\omega_\nu{}^{ab}, \tag{5.7.4 - 4}
$$

$$
\pi_t{}^{ab} \equiv (\partial/\partial\dot{t}_{ab})\tilde{\boldsymbol{L}} = -i\tilde{g}^{0\nu}\partial_\nu\bar{t}^{ab}, \tag{5.7.4 - 5}
$$

$$
\pi_{\bar{t}}{}^{ab} \equiv (\partial/\partial\dot{\bar{t}}_{ab})\tilde{\boldsymbol{L}} = i\tilde{g}^{0\nu}(D_\nu t)^{ab}, \tag{5.7.4 - 6}
$$

$$
\pi_\psi \equiv (\partial/\partial\dot{\psi})\tilde{\boldsymbol{L}} = ih\bar{\psi}\gamma^0. \tag{5.7.4 - 7}
$$

Setting up the canonical (anti)commutation relations, we can calculate all the equal-time (anti)commutation relations concerning the primary fields in closed form [Nak 79d]. We here present the final results only, employing the notation of Sec.5.4.

We have

$$
[h_\mu{}^a, \ b_\rho'] = -i\kappa\delta^0{}_\mu h_\rho{}^a \tilde{f}\delta^3, \tag{5.7.4 - 8}
$$

$$
\begin{aligned}
[h_\mu{}^a, \ \dot{b}_\rho'] = i\kappa\{\partial_\rho h_\mu{}^a \cdot \tilde{f}\delta^3 &- \delta^0{}_\mu h_\rho{}^a(\partial_0\tilde{f}\cdot\delta^3 + 2\tilde{f}\tilde{g}^{0k}(\delta^3)_k) \\
&+ \delta^k{}_\mu h_\rho{}^a(\delta^3)_k\}
\end{aligned} \tag{5.7.4 - 9}
$$

in conformity with Eqs.(5.6.1-2) and (5.6.1-4), respectively;

$$
\begin{aligned}
[h_\mu{}^a, \ \dot{h}_\nu{}^{b\prime}] = -\frac{1}{2}i\kappa\tilde{f}[h_\mu{}^a h_\nu{}^b &- h_\mu{}^b h_\nu{}^a - g_{\mu\nu}\eta^{ab} \\
+ h\tilde{f}(\delta^0{}_\mu\delta^0{}_\nu\eta^{ab} &+ \delta^0{}_\mu h^{0b}h_\nu{}^a + \delta^0{}_\nu h^{0a}h_\mu{}^b + h^{oa}h^{0b}g_{\mu\nu})]\delta^3, \tag{5.7.4 - 10}[3]
\end{aligned}
$$

---

[2]   In Eq.(4), $\partial/\partial\dot{s}_{ab}$ is defined by $\frac{1}{2}(\partial/\partial\dot{s}_{ab} - \partial/\partial\dot{s}_{ba})$, regarding s_{ab} as if all its components were independent. Likewise for Eqs.(5) and (6).

[3]   In D dimensions, the term $h_\mu{}^a h_\nu{}^b$ is multiplied by $2/(D-2)$

$$[h_\mu{}^a, \ \dot{s}_{cd}{}'] = \frac{1}{2}i(\delta^a{}_c h_{\mu d} - \delta^a{}_d h_{\mu c})\tilde{f}\delta^3, \tag{5.7.4 - 11}$$

$$[\varphi_{cd}, \ \dot{b}_\rho{}'] = i\partial_\rho \varphi_{cd} \cdot \tilde{f}\delta^3 \quad \text{for} \quad \varphi_{cd} = s_{cd}, \ t_{cd}, \ \bar{t}_{cd}, \tag{5.7.4 - 12}$$

$$[s_{ab}, \ \dot{t}_{cd}{}'] = \frac{1}{2}i[(\eta_{bd}t_{ac} - \eta_{ad}t_{bc}) - (c \leftrightarrow d)]\tilde{f}\delta^3, \tag{5.7.4 - 13}$$

$$\{t_{ab}, \ \dot{\bar{t}}_{cd}{}'\} = -\frac{1}{2}(\eta_{ac}\eta_{bd} - \eta_{ad}\eta_{bc})\tilde{f}\delta^3. \tag{5.7.4 - 14}$$

The (anti)commutation relations, Eqs.(8)-(14), together with Eqs.(5.4.3-10)-(5.4.3-12) and (5.4.4-9) and those which can be obtained from them by means of the Leibniz rule, Eq.(5.4.4-1), exhaust all non-vanishing (anti)commutators between one or two of $h_\mu{}^a$, b_ρ, c^σ, s_{ab}, t_{ab}, $\bar{t}_{ab}$ and at most one of their first time derivatives. The (anti)commutation relations involving ψ are as follows:

$$\{\psi_\alpha, \ \bar{\psi}'_\beta\} = (\gamma^0)_{\alpha\beta}\tilde{f}\delta^3, \tag{5.7.4 - 15}$$

$$[\psi, \ \dot{h}_\mu{}^a{}'] = \frac{1}{4}\kappa\tilde{f}(h_\mu{}^a + 2h\tilde{f}\delta^0{}_\mu h^{0a})\psi\delta^3, \tag{5.7.4 - 16}$$

$$[\psi, \ \dot{b}_\rho{}'] = i\kappa\partial_\rho\psi \cdot \tilde{f}\delta^3, \tag{5.7.4 - 17}$$

$$[\psi, \ \dot{s}_{ab}{}'] = \frac{1}{2}i\tilde{f}\overset{\circ}{\sigma}_{ab}\psi\delta^3, \tag{5.7.4 - 18}$$

$$[\psi, \ \dot{t}_{ab}{}'] = [\psi, \ \dot{\bar{t}}_{ab}{}'] = 0. \tag{5.7.4 - 19}$$

Some remarks are in order.

It is an extremely important consequence of the fifth requirement for choosing $\mathcal{L}_{\text{LGF}}$ that *all equal-time (anti)commutators concerning $g_{\mu\nu}$, b_ρ, c^σ, and $\bar{c}_\tau$ remain unchanged*. Accordingly, all results related to the sixteen-dimensional Poincaré-like superalgebra and the geometric commutation relation remain valid (at the operator level).

By direct computation, it is straightforward to show

$$[\omega_\mu{}^{ab}, \ b_\rho{}'] = -i\kappa\delta^0{}_\mu \omega_\rho{}^{ab}\tilde{f}\delta^3. \tag{5.7.4 - 20}$$

Combining it with Eq.(5.7.1-9), we can easily prove Eq.(5.4.4-8), as noted there [Nak 80b].

As is seen from Eq.(3), $\pi_h{}^\mu{}_a$ depends on s_{ab}, t_{ab}, and $\bar{t}_{ab}$ and even on ψ. Nevertheless, Eqs.(8)-(10) have no explicit dependence on them. The cancellation is quite noteworthy. Such cancellation occurs also in the calculation of the energy-momentum tensor density, the BRS Noether current, etc.

5.7.5 Symmetries in the local-Lorentz part

In the vierbein formalism, we have some extra symmetries corresponding to the degrees of freedom of the local-Lorentz transformation.

First, we note that $\tilde{\mathcal{L}}$ is invariant under global internal Lorentz transformations. The corresponding Noether current is given by

$$\mathcal{M}_{\mathrm{L}}{}^{\mu ab} \equiv [\varphi_A]_{\mathrm{L}}{}^{ab}[\partial/\partial(\partial_\mu\varphi_A)]\tilde{\mathcal{L}}, \qquad (5.7.5-1)$$

where $[\,\cdot\,]_{\mathrm{L}}{}^{ab}$ denotes the infinitesimal internal Lorentz transformation; for example [cf. Eq.(5.7.1-25)],

$$[h_\mu{}^c]_{\mathrm{L}}{}^{ab} = \eta^{ac}h_\mu{}^b - \eta^{bc}h_\mu{}^a, \qquad (5.7.5-2)$$

$$[\psi]_{\mathrm{L}}{}^{ab} = \overset{\circ}{\sigma}{}^{ab}\psi. \qquad (5.7.5-3)$$

Calculating Eq.(1), we obtain

$$\mathcal{M}_{\mathrm{L}}{}^{\mu ab} = 2\{\tilde{g}^{\mu\nu}(D_\nu s)^{ab} - i\tilde{g}^{\mu\nu}[\bar{t}^a{}_c(D_\nu t)^{cb} - (a \leftrightarrow b)]\}. \qquad (5.7.5-4)$$

The conservation law of $\mathcal{M}_{\mathrm{L}}{}^{\mu ab}$ is nothing but Eq.(5.7.3-11).

Owing to Eqs.(5.7.3-8)-(5.7.3-10), we have three similar conserved currents. The Noether current of the local-Lorentz BRS transformation is given by

$$\begin{aligned} J_{\mathrm{s}}{}^\mu &\equiv \boldsymbol{\delta}_{\mathrm{L}}(\varphi_A)[\partial/\partial(\partial_\mu\varphi_A)]\tilde{\mathcal{L}} \\ &= \tilde{g}^{\mu\nu}[s_{ab}(D_\nu t)^{ab} - \partial_\nu s_{ab}\cdot t^{ab} + i\partial_\nu\bar{t}_{ab}\cdot t^a{}_c t^{cb}]. \end{aligned} \qquad (5.7.5-5)$$

We can construct four analogous conserved currents.

From the above consideration, we have 29 symmetry generators [Nak 80f]:

$$P(\omega^{ab}) \equiv \int d\mathbf{x}\,\tilde{g}^{0\nu}\omega_\nu{}^{ab}, \qquad (5.7.5-6)$$

$$P(s^{ab}) \equiv \frac{1}{2}M_{\mathrm{L}}{}^{ab} \equiv \frac{1}{2}\int d\mathbf{x}\,\mathcal{M}_{\mathrm{L}}{}^{0ab}, \qquad (5.7.5-7)$$

$$P(t^{ab}) \equiv \int d\mathbf{x}\,\tilde{g}^{0\nu}(D_\nu t)^{ab}, \qquad (5.7.5-8)$$

$$P(\bar{t}^{ab}) \equiv \int d\mathbf{x}\,\tilde{g}^{0\nu}(D_\nu\bar{t})^{ab}; \qquad (5.7.5-9)$$

$$Q(s,t) \equiv Q_{\mathrm{s}} \equiv \int d\mathbf{x}\,J_{\mathrm{s}}{}^0, \qquad (5.7.5-10)$$

$$Q(s, \bar{t}) \equiv \bar{Q}_s \equiv \int d\mathbf{x}\, \tilde{g}^{0\nu} [s_{ab}(D_\nu \bar{t})^{ab} - \partial_\nu s_{ab} \cdot \bar{t}^{ab} + i\bar{t}_{ac}\bar{t}^c{}_b (D_\nu t)^{ab}], \quad (5.7.5-11)$$

$$Q(t, \bar{t}) \equiv Q_t \equiv i \int d\mathbf{x}\, \tilde{g}^{0\nu}[\bar{t}_{ab}(D_\nu t)^{ab} - \partial_\nu \bar{t}_{ab} \cdot t^{ab}], \qquad (5.7.5-12)$$

$$Q(t, t) \equiv i \int d\mathbf{x}\, \tilde{g}^{0\nu}[t_{ab}(D_\nu t)^{ab} - \partial_\nu t_{ab} \cdot t^{ab}], \qquad (5.7.5-13)$$

$$Q(\bar{t}, \bar{t}) \equiv i \int d\mathbf{x}\, \tilde{g}^{0\nu}[\bar{t}_{ab}(D_\nu \bar{t})^{ab} - \partial_\nu \bar{t}_{ab} \cdot \bar{t}^{ab}]. \qquad (5.7.5-14)$$

Here, $M_L{}^{ab}$ ($\equiv 2P(s^{ab})$) is the global internal Lorentz generator, Q_s ($\equiv Q(s, t)$) the local-Lorentz BRS generator, $\bar{Q}_s$ ($\equiv Q(s, \bar{t})$) the local-Lorentz anti-BRS generator, and Q_t ($\equiv Q(t, \bar{t})$) the local-Lorentz FP-ghost number operator. For example, by using the equal-time commutation relations, we can show that

$$[iM_L{}^{ab}, \varphi_A] = [\varphi_A]_L{}^{ab}, \qquad (5.7.5-15)$$

$$[iQ_s, \varphi_A]_{\mp} = \boldsymbol{\delta}_L(\varphi_A). \qquad (5.7.5-16)$$

All the above 29 symmetry generators commute with all ones of the sixteen-dimensional Poincaré-like superalgebra, and form a superalgebra whose structure is completely the same as that of the Landau-gauge non-abelian gauge theory having $SO(3, 1)$ as its gauge group (see Sec.3.4.6).

As in Sec.5.5.4, the symmetries generated by $P(\omega^{ab})$, $P(t^{ab})$, and $P(\bar{t}^{ab})$ are necessarily spontaneously broken, and the corresponding NG modes are s_{ab}, $\bar{t}_{ab}$, and t_{ab}, respectively. The spontaneous breakdown of $P(s^{ab}) = \frac{1}{2}M_L{}^{ab}$ has an important implication, which is discussed in the next subsection. The other five symmetries are expected to be unbroken.

5.7.6 Meaning of the Lorentz invariance of particle physics

In the vierbein formalism, the primary gravitational field is $h_\mu{}^a$. Since we postulate that the vacuum is invariant under translations, the vacuum expectation value

$$\langle 0 | h_\mu{}^a | 0 \rangle \equiv \overset{\circ}{h}_\mu{}^a \qquad (5.7.6-1)$$

is *independent of* x^λ. We assume that the matrix $(\overset{\circ}{h}_\mu{}^a)$ is invertible, because otherwise $\langle 0 | h^\nu{}_b | 0 \rangle$ would not exist. [We do not necessarily require that $\overset{\circ}{h}_\mu{}^a \overset{\circ}{h}^\mu{}_b = \delta^a{}_b$ holds strictly, but our requirement is that the qualitative feature remains unchanged by quantum corrections just as in the case of the Higgs potential.] Then the matrix $(\mathring{h}_\mu{}^a)$ defines a general linear transformation. For the sake of convenience, we denote this transformation by $\hat{u}^\nu{}_\sigma$ and its inverse by $u_\lambda{}^\sigma$ $(\hat{u}^\nu{}_\sigma u_\lambda{}^\sigma = \delta^\nu{}_\lambda)$. Then Eq.(1) is rewritten as

$$\langle\, 0\,|\, u_\mu{}^\sigma h_\sigma{}^a\,|\, 0\,\rangle = \delta_\mu{}^a. \qquad (5.7.6-2)$$

Now, we consider the $GL(4)$ generators $\hat{M}^\mu{}_\nu$. We can explicitly show, by using the equal-time commutation relations, that

$$[i\hat{M}^\mu{}_\nu,\ h_\lambda{}^c] = x^\mu \partial_\nu h_\lambda{}^c + \delta^\mu{}_\lambda h_\nu{}^c, \qquad (5.7.6-3)$$

as it should be. Hence Eq.(1) implies that

$$\langle\, 0\,|\,[i\hat{M}^\mu{}_\nu,\ h_\lambda{}^c]\,|\, 0\,\rangle = \delta^\mu{}_\lambda \mathring{h}_\nu{}^c. \qquad (5.7.6-4)$$

Thus the $GL(4)$ invariance is *totally* spontaneously broken in the vierbein formalism.

In order to make the situation clearer, we perform the $GL(4)$ transformation defined by $u_\lambda{}^\sigma$ for all world tensors.[4] In particular, we set

$$\hat{M}^{\#\mu}{}_\nu \equiv \hat{u}^\mu{}_\sigma u_\nu{}^\tau \hat{M}^\sigma{}_\tau, \qquad (5.7.6-5)$$

$$h^\#{}_\lambda{}^c \equiv u_\lambda{}^\sigma h_\sigma{}^c. \qquad (5.7.6-6)$$

Then, according to Eq.(2), Eq.(4) is rewritten as

$$\langle\, 0\,|\,[i\hat{M}^{\#\mu}{}_\nu,\ h^\#{}_\lambda{}^c]\,|\, 0\,\rangle = \delta^\mu{}_\lambda \delta_\nu{}^c. \qquad (5.7.6-7)$$

Next, from Eq.(5.7.5-15) together with Eqs.(5.7.5-2) and (1), we have

$$\langle\, 0\,|\,[iM_L{}^{ab},\ h_\lambda{}^c]\,|\, 0\,\rangle = \eta^{ac}\mathring{h}_\lambda{}^b - \eta^{bc}\mathring{h}_\lambda{}^a; \qquad (5.7.6-8)$$

or, equivalently,

$$\langle\, 0\,|\,[iM_L{}^{ab},\ h^\#{}_\lambda{}^c]\,|\, 0\,\rangle = \eta^{ac}\delta_\lambda{}^b - \eta^{bc}\delta_\lambda{}^a. \qquad (5.7.6-9)$$

---

[4]   This procedure is analogous to that by which $\langle\, 0\,|\,\phi\,|\, 0\,\rangle$ is made real positive in the Goldstone model (see Sec.1.5.2.)

Accordingly, the global internal Lorentz invariance is also spontaneously broken.

It should be noted, however, that if we consider

$$M_{\alpha\beta} \equiv \bar{M}^{\#}{}_{\alpha\beta} + \eta_{\alpha a}\eta_{\beta b}M_{\mathrm{L}}{}^{ab}, \qquad (5.7.6-10)$$

where

$$\bar{M}^{\#}{}_{\alpha\beta} \equiv \eta_{\alpha\sigma}\hat{M}^{\#\sigma}{}_\beta - \eta_{\beta\sigma}\hat{M}^{\#\sigma}{}_\alpha, \qquad (5.7.6-11)$$

then Eqs.(7) and (9) imply that

$$\langle 0 \,|\, [iM_{\alpha\beta},\, h^{\#}{}_\lambda{}^c] \,|\, 0 \rangle = 0. \qquad (5.7.6-12)$$

Thus *the spacetime symmetry generated by $M_{\alpha\beta}$ is not spontaneously broken* [Nak 79e, Nak 81d].

The generators $M_{\alpha\beta}$ are nothing but the Lorentz generators of particle physics. Indeed, for instance, it is easy to show that

$$[iM_{\alpha\beta},\, \psi] = (x^{\#}{}_\alpha\partial^{\#}{}_\beta - x^{\#}{}_\beta\partial^{\#}{}_\alpha + \overset{\circ}{\sigma}_{\alpha\beta})\psi, \qquad (5.7.6-13)$$

where

$$x^{\#}{}_\alpha \equiv \eta_{\alpha\mu}\hat{u}^\mu{}_\sigma x^\sigma, \qquad (5.7.6-14)$$

$$\partial^{\#}{}_\beta \equiv \eta_{\alpha\beta}(\partial/\partial x^{\#}{}_\alpha) = u_\beta{}^\tau\partial_\tau. \qquad (5.7.6-15)$$

That is, the Dirac field ψ behaves as a *world spinor* under $M_{\alpha\beta}$. It should be emphasized that there is no reason why ψ is regarded as a world spinor in the classical generally-covariant Dirac theory.

Thus *the Lorentz generators of particle physics are what are uniquely characterized by Eq.(12).* This extremely important, fundamental result can be established only in the framework of the manifestly covariant canonical operator formalism of quantum gravity.

The fact that the Lorentz generators are characterized at the level of the representation of field operators has an important implication: *The Lorentz invariance of particle physics cannot be the first principle of the fundamental theory including gravity.* The Einstein causality also *cannot* be the most fundamental postulate. Indeed, spacetime coordinates x^μ must be redefined, as shown in Eq.(14).

5.8 UNITARITY AND THE DIVERGENCE PROBLEM

The divergence problem is the most serious problem of quantum gravity, as discussed in Sec.5.1. Although a pessimistic standpoint prevails concerning the Einstein gravity on the basis of the perturbative approach, we here present an optimistic standpoint.

It is well known that the divergence problem is easily resolved if one trades in the unitarity. This is a very bad idea, however, because the unitarity is the *exact* property which must be satisfied by any physically sensible quantum theory, while the divergence problem which we usually consider is based on *approximation*. The divergence problem must be discussed without endangering the unitarity.

5.8.1 Unitarity – Free-Lagrangian method –

Our theory of quantum gravity, of course, has the BRS charge Q_{b} ($\equiv M(b_\rho, c^\rho)$) and the FP-ghost number operator Q_c ($\equiv M(\bar{c}_\rho, c^\rho)$); in addition to them there are the local-Lorentz BRS charge Q_{s} ($\equiv Q(s, t)$) and the local-Lorentz FP-ghost number operator Q_t ($\equiv Q(\bar{t}, t)$) in the vierbein formalism.

As in Sec.3.4.3, we define the physical state $|f\rangle$ by setting up the subsidiary conditions[1]

$$Q_{\mathrm{b}}|f\rangle = 0, \qquad\qquad (5.8.1-1)$$

$$Q_{\mathrm{s}}|f\rangle = 0. \qquad\qquad (5.8.1-2)$$

Then, as proved in Sec.4.1, the physical S-matrix is unitary under the assumption that all the BRS-singlet states have positive norm.

Although logically there is no need for further consideration, we here demonstrate the unitarity of the physical S-matrix on the basis of the free-Lagrangian method (see Sec.2.4.3) in order to see explicitly that the above assumption is satisfied. We assume the following situations:

1. All primary fields have their own (*in* and *out*) asymptotic fields.
2. No other asymptotic fields exist.

---

[1]   In contrast to the Maxwell equation, the classical Einstein equation is *not* reproduced by taking the physical-state expectation value of Eq.(5.3.3-1)

3. All asymptotic fields are governed by the quadratic part of the total Lagrangian density apart from possible renormalization.

Only in this subsection, spacetime metric is Minkowskian.

We first discuss the case in which $g_{\mu\nu}$ is a primary field [Nak 78b]. Let $\varphi_{\mu\nu}$, β_ρ, γ^σ, and $\bar\gamma_\tau$ be the renormalized (*in* and *out*) asymptotic fields of $(g_{\mu\nu} - \eta_{\mu\nu})/2\sqrt{\kappa}$, $b_\rho/\sqrt{\kappa}$, c^σ and $\bar c_\tau$, respectively. Then they are governed by

$$\mathcal{L}^{\mathrm{as}} \equiv \frac{1}{2}\partial_\rho\varphi^{\mu\nu}\cdot\partial^\rho\varphi_{\mu\nu} - \frac{1}{2}\partial_\rho\varphi\cdot\partial^\rho\varphi - \partial_\mu\varphi^{\mu\rho}\cdot\partial^\nu\varphi_{\nu\rho}$$
$$+ \partial^\mu\varphi_{\mu\rho}\cdot\partial^\rho\varphi - (2\partial_\mu\varphi^{\mu\rho} - \partial^\rho\varphi)\cdot\beta_\rho$$
$$- i\partial^\mu\bar\gamma_\rho\cdot\partial_\mu\gamma^\rho + \mathcal{L}_{\mathrm{M}}{}^{\mathrm{as}}, \qquad (5.8.1-3)$$

where $\varphi \equiv \varphi^\lambda{}_\lambda$, and $\mathcal{L}_{\mathrm{M}}{}^{\mathrm{as}}$ denotes the asymptotic Lagrangian density of matter fields. Of course, Eq.(3) is a special case of the linearized quantum Einstein theory [Kim 76]. Since $\mathcal{L}_{\mathrm{M}}{}^{\mathrm{as}}$ decouples, we consider $\varphi_{\mu\nu}$, β_ρ, γ^σ, and $\bar\gamma_\tau$ only. They satisfy

$$\Box\varphi_{\mu\nu} - \partial_\mu\beta_\nu - \partial_\nu\beta_\mu = 0, \qquad (5.8.1-4)$$

$$\partial^\mu\varphi_{\mu\nu} - \frac{1}{2}\partial_\nu\varphi = 0, \qquad (5.8.1-5)$$

$$\Box\beta_\rho = 0, \qquad \Box\gamma^\sigma = 0, \qquad \Box\bar\gamma_\tau = 0. \qquad (5.8.1-6)$$

The four-dimensional (anti)commutation relations between asymptotic fields are as follows:[2]

$$[\varphi_{\mu\nu}(x),\ \varphi_{\sigma\tau}(y)] = -\frac{1}{2}i(\eta_{\mu\nu}\eta_{\sigma\tau} - \eta_{\mu\sigma}\eta_{\nu\tau} - \eta_{\mu\tau}\eta_{\nu\sigma})D(x-y)$$
$$- \frac{1}{2}i(\eta_{\nu\tau}\partial_\mu\partial_\sigma + \eta_{\nu\sigma}\partial_\mu\partial_\tau + \eta_{\mu\tau}\partial_\nu\partial_\sigma + \eta_{\mu\sigma}\partial_\nu\partial_\tau)E(x-y), \ (5.8.1-7)$$

$$[\varphi_{\mu\nu}(x),\ \beta_\rho(y)] = \frac{1}{2}i(\eta_{\mu\rho}\partial_\nu + \eta_{\rho\nu}\partial_\mu)D(x-y), \qquad (5.8.1-8)$$

$$\{\gamma^\sigma(x),\ \bar\gamma_\tau(y)\} = -\delta^\sigma{}_\tau D(x-y), \qquad (5.8.1-9)$$

and other (anti)commutators vanish.[3] We encounter dipole ghosts but not tripole ghosts.[4]

---

[2]   In D dimensions, the term $\eta_{\mu\nu}\eta_{\sigma\tau}$ acquires a coefficient $2/(D-2)$.

[3]   Note that the equal-time versions of those relations coincide with the free-field versions of the equal-time (anti)commutation relations presented in Sec.5.4.

[4]   If we do not adhere to Eq.(3), we encounter tripole ghosts in general, but they cause no trouble [Kug 78b].

The transformation law under the symmetries which are not spontaneously broken (see Sec.5.5.2) are, apart from those under the Poincaré group, as follows:

$$[iM(b_\sigma, b_\tau), \varphi_{\mu\nu}] = \frac{1}{2}\kappa(\eta_{\tau\nu}\partial_\mu + \eta_{\mu\tau}\partial_\nu)\beta_\sigma - (\sigma \leftrightarrow \tau); \quad (5.8.1-10)$$

$$[iM(b_\sigma, c^\tau), \varphi_{\mu\nu}] = -\frac{1}{2}\sqrt{\kappa}(\eta_{\sigma\nu}\partial_\mu + \eta_{\mu\sigma}\partial_\nu)\gamma^\tau, \quad (5.8.1-11)$$

$$\{iM(b_\sigma, c^\tau), \bar{\gamma}_\lambda\} = i\sqrt{\kappa}\delta^\tau{}_\lambda\beta_\sigma; \quad (5.8.1-12)$$

$$[iM(b_\sigma, \bar{c}_\tau), \varphi_{\mu\nu}] = -\frac{1}{2}\sqrt{\kappa}(\eta_{\sigma\nu}\partial_\mu + \eta_{\mu\sigma}\partial_\nu)\bar{\gamma}_\tau, \quad (5.8.1-13)$$

$$\{iM(b_\sigma, \bar{c}_\tau), \gamma^\lambda\} = -i\sqrt{\kappa}\delta^\lambda{}_\tau\beta_\sigma; \quad (5.8.1-14)$$

$$[iM(\bar{c}_\sigma, c^\tau), \gamma^\lambda] = \delta^\lambda{}_\sigma\gamma^\tau, \quad (5.8.1-15)$$

$$[iM(\bar{c}_\sigma, c^\tau), \bar{\gamma}_\lambda] = -\delta^\tau{}_\lambda\bar{\gamma}_\sigma; \quad (5.8.1-16)$$

$$[iM(c^\sigma, c^\tau), \bar{\gamma}_\lambda] = -(\delta^\sigma{}_\lambda\gamma^\sigma + \delta^\sigma{}_\lambda\gamma^\tau); \quad (5.8.1-17)$$

$$[iM(\bar{c}_\sigma, \bar{c}_\tau), \gamma^\lambda] = \delta^\lambda{}_\tau\bar{\gamma}_\sigma + \delta^\lambda{}_\sigma\bar{\gamma}_\tau; \quad (5.8.1-18)$$

other (anti)commute. Of course, Eq.(3) is invariant under $M(b_\sigma, b_\tau)$, $M(b_\sigma, c^\tau)$, $M(b_\sigma, \bar{c}_\tau)$, $M(\bar{c}_\sigma, c_\tau)$, $M(c^\sigma, c^\tau)$ and $M(\bar{c}_\sigma, \bar{c}_\tau)$. Conversely, this fact uniquely characterizes Eq.(3). For example, a BRS-invariant quantity $-\varphi\partial^\nu\beta_\nu + i\partial^\nu\bar{\gamma}_\nu\cdot\partial_\mu\gamma^\mu$ is *not* invariant under any of them [Nak 80d].

Now, we discuss the problem of unitarity. The Fourier transforms of Eqs.(7)-(9) are as follows (Fourier transform is denoted by tilde):

$$[\tilde{\varphi}_{\mu\nu}(p), \tilde{\varphi}_{\sigma\tau}{}^\dagger(q)] = \frac{1}{2}(\eta_{\mu\sigma}\eta_{\nu\tau} + \eta_{\mu\tau}\eta_{\nu\sigma} - \eta_{\mu\nu}\eta_{\sigma\tau})\theta(p_0)\delta(p^2)\delta^4(p-q)$$

$$+ \frac{1}{2}(\eta_{\nu\tau}p_\mu p_\sigma + \eta_{\nu\sigma}p_\mu p_\tau + \eta_{\mu\tau}p_\nu p_\sigma + \eta_{\mu\sigma}p_\nu p_\tau)\theta(p_0)\delta'(p^2)\delta^4(p-q), (5.8.1-19)$$

$$[\tilde{\varphi}_{\mu\nu}(p), \tilde{\beta}_\rho{}^\dagger(q)] = -\frac{1}{2}i(\eta_{\mu\rho}p_\nu + \eta_{\rho\nu}p_\mu)\theta(p_0)\delta(p^2)\delta^4(p-q), \quad (5.8.1-20)$$

$$\{\tilde{\gamma}^\sigma(p), \tilde{\bar{\gamma}}_\tau{}^\dagger(q)\} = i\delta^\sigma{}_\tau\theta(p_0)\delta(p^2)\delta^4(p-q). \quad (5.8.1-21)$$

In order to take out six independent components of $\tilde{\varphi}_{\mu\nu}(p)$, it is convenient to consider it in a Lorentz frame such that $p_1 = p_2 = 0$ and $p_3 > 0$. We see [Nak 79e]

$$\tilde{\varphi}_1 \equiv \tilde{\varphi}_{11}, \qquad \tilde{\varphi}_2 \equiv \tilde{\varphi}_{22}; \quad (5.8.1-22)$$

$$\tilde{\chi}_0 \equiv \frac{1}{2}(\tilde{\varphi}_{00} - \tilde{\varphi}_{11}), \quad \tilde{\chi}_1 \equiv \tilde{\varphi}_{01}, \quad \tilde{\chi}_2 \equiv \tilde{\varphi}_{02}, \quad \tilde{\chi}_3 \equiv \frac{1}{2}(\tilde{\varphi}_{33} + \tilde{\varphi}_{11}). (5.8.1-23)$$

The other four components are expressible in terms of the above six ones owing to Eq.(5). Then Eqs.(19) and (20) become

$$[\tilde{\varphi}_i(p), \tilde{\varphi}_j{}^\dagger(q)] = \frac{1}{2}\delta_{ij}\theta(p_0)\delta(p^2)\delta^4(p-q), \quad (5.8.1-24)$$

$$[\tilde{\varphi}_i(p),\ \tilde{\chi}_\mu{}^\dagger(q)] = 0, \qquad\qquad (5.8.1-25)$$

$$[\tilde{\varphi}_i(p),\ \tilde{\beta}_\rho{}^\dagger(q)] = 0, \qquad\qquad (5.8.1-26)$$

$$[\tilde{\chi}_\mu(p),\ \tilde{\beta}_\rho{}^\dagger(q)] = -\frac{1}{2}i\eta_{\mu\rho}p_0\theta(p_0)\delta(p^2)\delta^4(p-q), \qquad (5.8.1-27)$$

where $[\tilde{\chi}_\mu,\ \tilde{\chi}_\nu{}^\dagger]$ is omitted since it is unnecessary for our purpose. Furthermore, from Eqs.(11) and (12) for $\sigma = \tau$ together with $p_1 = p_2 = 0$, we can easily see that $\tilde{\varphi}_i{}^\dagger$, $\tilde{\beta}_\rho{}^\dagger$, and $\tilde{\gamma}^{\sigma\dagger}$ are BRS-invariant and that the BRS transforms of $\tilde{\chi}_\mu{}^\dagger$ and $\tilde{\bar{\gamma}}_\tau{}^\dagger$ are $\tilde{\gamma}_\mu{}^\dagger$ and $\tilde{\beta}_\tau{}^\dagger$, respectively, up to a multiplicative c-number. Thus we see that $\tilde{\varphi}_1{}^\dagger$ and $\tilde{\varphi}_2{}^\dagger$ are *transverse gravitons of positive norm* and that $\tilde{\chi}_\mu{}^\dagger$, $\tilde{\beta}_\rho{}^\dagger$, $\tilde{\gamma}_\sigma{}^\dagger$, and $\tilde{\bar{\gamma}}_\tau{}^\dagger$ form *four quartets*.[5] We can then prove the unitarity of the physical S-matrix in the same way as in Sec.4.1.4.

In the vierbein case [Nak 79c], we have $\chi_{\mu a}$ as the asymptotic fields of $(h_{\mu a} - \eta_{\mu a})/2\sqrt{\kappa}$ instead of $\varphi_{\mu\nu}$. [Precisely speaking, we should consider $h^\#{}_\mu{}^a$, $x_\alpha{}^\#$, etc. as in Sec.5.7.6, but we omit $\#$ for simplicity.] Moreover, we have σ_{ab}, τ_{ab}, and $\bar{\tau}_{ab}$ as the asymptotic fields of $\sqrt{\kappa}s_{ab}$, t_{ab}, and $\bar{t}_{ab}$, respectively, in addition to β_ρ, γ^σ, and $\bar{\gamma}_\tau$. It is convenient to set

$$\varphi_{\mu\nu} \equiv \chi_{\mu\nu} + \chi_{\nu\mu} = \varphi_{\nu\mu}, \qquad\qquad (5.8.1-28)$$

$$\psi_{\mu\nu} \equiv \chi_{\mu\nu} - \chi_{\nu\mu} = -\psi_{\nu\mu}; \qquad\qquad (5.8.1-29)$$

then $\varphi_{\mu\nu}$ is essentially the previous one.

The asymptotic-field Lagrangian density is given by

$$\boldsymbol{\mathcal{L}}^{\mathrm{as}} \equiv \mathcal{L}^{\mathrm{as}} - \partial^\lambda\psi^{\mu\nu}\cdot\partial_\lambda\sigma_{\mu\nu} - 2\partial^\nu\varphi^{\mu\lambda}\cdot\partial_\lambda\sigma_{\mu\nu}$$
$$- i\partial_\lambda\bar{\tau}_{\mu\nu}\cdot\partial^\lambda\tau^{\mu\nu}. \qquad\qquad (5.8.1-30)$$

The field equations are as follows: Eq.(4) is modified into

$$\Box\varphi_{\mu\nu} - \partial_\mu\beta_\nu - \partial_\nu\beta_\mu - \partial_\nu\partial^\lambda\sigma_{\mu\lambda} - \partial_\mu\partial^\lambda\sigma_{\nu\lambda} = 0; \qquad (5.8.1-31)$$

Eqs.(5) and (6) remain unchanged; the field equation for $\psi_{\mu\nu}$ reduces to the d'Alembert one if we make use of Eq.(5); the field equations for $\sigma_{\mu\nu}$, $\tau_{\mu\nu}$, and $\bar{\tau}_{\mu\nu}$ all are the d'Alembert ones.

---

[5] Their freedom corresponds to the Newtonian force.

The four-dimensional (anti)commutation relations are as follows: Eqs.(7)-(9) remain unchanged;

$$[\psi_{\mu\nu}(x), \ \sigma_{\sigma\tau}(y)] = -\frac{1}{2}i(\eta_{\mu\sigma}\eta_{\nu\tau} - \eta_{\mu\tau}\eta_{\nu\sigma})D(x-y), \qquad (5.8.1-32)$$

$$\{\tau_{\mu\nu}(x), \ \bar{\tau}_{\sigma\tau}(y)\} = -\frac{1}{2}(\eta_{\mu\sigma}\eta_{\nu\tau} - \eta_{\mu\tau}\eta_{\nu\sigma})D(x-y); \qquad (5.8.1-33)$$

all other (anti)commutators vanish. Furthermore, we have

$$[iQ_{\mathbf{s}}, \ \psi_{\mu\nu}] = \tau_{\mu\nu}, \qquad \{iQ_{\mathbf{s}}, \ \tau_{\mu\nu}\} = 0, \qquad (5.8.1-34)$$

$$\{iQ_{\mathbf{s}}, \ \bar{\tau}_{\mu\nu}\} = i\sigma_{\mu\nu}, \qquad [iQ_{\mathbf{s}}, \ \sigma_{\mu\nu}] = 0. \qquad (5.8.1-35)$$

Then, evidently, $\tilde{\psi}_{\mu\nu}{}^{\dagger}$, $\tilde{\sigma}_{\mu\nu}{}^{\dagger}$, $\tilde{\tau}_{\mu\nu}{}^{\dagger}$, and $\tilde{\bar{\tau}}_{\mu\nu}{}^{\dagger}$ form six quartets. The unitarity of the physical S-matrix is shown in the same way as before.

5.8.2 Possibility of resolving the divergence problem

The Einstein gravity is non-renormalizable in the usual power-counting sense. In the standard perturbative treatment, all one-loop divergences cannot be absorbed into the counterterms invariant under general coordinate transformations [Lei 75, Cap 78].

In the background-field approach, the (on-shell) one-loop divergences disappear in pure gravity, but they remain if coupled to a matter field [tHo 74]. Even for pure gravity, the two-loop graviton self-energy is on-shell divergent [Cap 85, Gor 85, Gor 86].

As discussed in Sec.5.1, the perturbative treatment of quantum gravity is meaningless if we encounter the energies larger than the Planck mass $1/\sqrt{\kappa}$. It is quite nonsensical to discuss such a mathematical problem as the divergence problem on the basis of an approximation method whose reliability is totally doubtful. We should rely on the exact results only. The fact that our theory of quantum gravity has such a huge symmetry as the choral symmetry might be regarded as an indication of the divergence-free nature of the theory. We should not give up this beautiful theory unless the exact treatment establishes its divergence difficulty.

As noted at the introductory remark of this section, the most curcial point of the divergence problem is the *compatibility* between ultraviolet convergence and the unitarity of the physical S-matrix, or more precisely, the *positivity* of probability. One should note that the positivity is spoiled not only by ghosts but also by *analytic continuation*. Thus the removal of divergence by means of analytic regularization cannot be regarded as the resolution of the divergence problem.

In contrast to the perturbative divergence difficulty, there is a time-honored expectation [Lan 55, Ber 56] that quantum gravity may eliminate all divergences encountered in the Minkowskian quantum field theory. This expectation is based on the possible disappearance of the light-cone singularity due to quantized metric. Since quantum gravity involves a fundamental constant κ^{-1} having the dimension of mass squared, it is natural to expect that it may play the role of a natural cutoff of ultra high energy. Furthermore, since the gravitational field couples with any field universally, such an expectation is not unreasonable.

If the above expectation is true, that is, if quantum field theory including gravity is finite without renormalization, one must explain the reason why the renormalizability requirement in particle physics has so far been so much successful. This question can be answered in the following way.

The statement that κ^{-1} is a natural cutoff implies that all particle-physics calculations, in which quantum gravity is neglected, are made by taking the improper limit $\kappa^{-1} \to \infty$, which causes divergences. If there exists a quantity which diverges in a power of mass dimension in particle physics, it must be extraordinarily huge, albeit mathematically finite, in the accelerator-energy region when quantum gravity is taken into account. If such a situation is encountered, then we cannot make any reliable calculation in particle physics unless that quantity is irrelevant. Thus we conclude that *all divergences should be logarithmic* in order for particle physics to be meaningful approximation in the accelerator-energy region. This requirement is somewhat stronger than the renormalizability requirement, but it is satisfied by gauge theories except for the Higgs part, which is irrelevant to the present-day experimental physics.[6]

Now, we turn to the central question: How can quantum gravity remove divegences? Of course, this question is too difficult to be answered, but we can point out a fact favorable to the resolution of the divergence difficulty.

It is usually supposed that the perturbative results on the divergence problem are not improved by the exact treatment because of the **Lehmann theorem**,

---

[6]   The quadratic divergence of the Higgs-scalar self-energy may be removed by supersymmetry. Otherwise, the Higgs scalar is an effective field; it does not exist in a more fundamental theory

which states that *the high-energy asymptotic behavior of the exact two-point function cannot be milder than that of the corresponding free one (Feynman propagator), provided that the state-vector space is of positive metric.* For example, for a scalar field $\phi(x)$, we have a spectral representation (see Sec.1.4.1):

$$i\,\text{F.T.}\,\langle\,0\,|\,\text{T}\,\phi(x)\phi(0)\,|\,0\,\rangle = \int_0^\infty ds\,\frac{\rho(s)}{s - p^2 - i0}, \qquad (5.8.2-1)$$

where F.T. indicates Fourier transform and T denotes time-ordered product. The spectral function $\rho(s)$ is essentially the Fourier transform of $\langle\,0\,|\,\phi(x)\phi(0)\,|\,0\,\rangle$. If the state-vector space is of positive metric, inserting a complete set $1 = \sum_n |\,n\,\rangle\langle\,n\,|$, we see that $\rho(p^2)$ is essentially a sum of $|\langle\,n\,|\phi(0)\,|\,0\,\rangle|^2$ over $|\,n\,\rangle$ whose momentum equals p_μ, whence

$$\rho(s) \geq 0. \qquad (5.8.2-2)$$

Then from Eq.(1) we find that $\text{F.T.}\langle\,0\,|\,\text{T}\phi(x)\phi(0)\,|\,0\,\rangle$ cannot behave more mildly than $1/p^2$ asymptotically.

This result is, of course, crucially based on the positivity of metric. Accordingly, the Lehmann theorem does not hold for a gauge field. Unfortunately, however, the Lehmann theorem does hold for any gauge-invariant, or more precisely, BRS-invariant quantity. We remember the fundamental theorem (see Sec.4.1.4) which is crucial for the proof of the unitarity of the physical S-matrix: *For any two physical states $|\,f\,\rangle$ and $|\,g\,\rangle$, we have*

$$\langle\,f\,|\,g\,\rangle = \langle\,f\,|\,P^{(0)}\,|\,g\,\rangle, \qquad (5.8.2-3)$$

where $P^{(0)}$ denotes the projection operator to a positive-norm subspace of the physical subspace V_{phys}. Hence if $\phi(x)$ is BRS-invariant, we have

$$\langle\,0\,|\,\phi(x)\phi(y)\,|\,0\,\rangle = \langle\,0\,|\,\phi(x)P^{(0)}\phi(y)\,|\,0\,\rangle, \qquad (5.8.2-4)$$

whence we again obtain Eq.(2).

The above situation drastically changes when quantum gravity is taken into account [Nak 80c]. The reason originates from the fact that general covariance is a spacetime symmetry. Indeed, as emphasized in Sec.5.3.1, the total gravitational BRS transformation $\boldsymbol{\delta}_*$ consists of the intrinsic one $\boldsymbol{\delta}$ and the orbital part $-\kappa c^\lambda \partial_\lambda$. Let $\Phi(x)$ be any local quantity constructible from the primary fields other than

b_ρ, c^σ, and $\bar{c}_\tau$ as done in Sec.5.6.3. Then, since $\boldsymbol{\delta}(\Phi)$ involves no non-differentiated FP ghost, the orbital part $-\kappa c^\lambda \partial_\lambda \Phi$ cannot be cancelled. Thus we see[7]

$$[iQ_b, \ \Phi(x)] \neq 0, \qquad (5.8.2-5)$$

whence the insertion of $P^{(0)}$ as in Eq.(4) is no longer possible. Thus *the Lehmann theorem is evaded for any two-point function, without spoiling the unitarity, if quantum gravity is taken into account.* It should be noted that the Lehmann theorem is based on the BRS invariance of *off-shell* states, while the unitarity of the physical S-matrix is based on that of *on-shell* states.

From the above consideration, we find that the spectral function $\rho(s)$ can be negative if $s \varsigma \kappa^{-1}$. Suppose that the kinematic term of the Lagrangian density of a scalar field ϕ is

$$\frac{1}{2}\tilde{g}^{\mu\nu}\partial_\mu\phi \cdot \partial_\nu\phi. \qquad (5.8.2-6)$$

Then the canonical commutation relations imply

$$[\phi(x), \ \dot{\phi}(y)]|_0 = i[\tilde{g}^{00}(x)]^{-1}\delta(\boldsymbol{x}-\boldsymbol{y}). \qquad (5.8.2-7)$$

Combining Eq.(7) with

$$\langle 0 \,|\, [\phi(x), \ \phi(y)] \,|\, 0 \rangle = i \int_0^\infty ds \, \rho(s)\Delta(x-y; s), \qquad (5.8.2-8)$$

we find that the sum rule

$$\int_0^\infty ds \, \rho(s) = 1 \qquad (5.8.2-9)$$

still remains to hold (at least approximately), owing to Eq.(5.5.4-9). Therefore, the two-point function, Eq.(1), does not damp faster than $1/p^2$ in its absolute value. We thus expect that $\rho(s)$ is infinitely oscillatory like $(2/\pi s)\sin \kappa s$. The oscillatory behavior can easily remove logarithmic divergence.

5.8.3 Elimination of the Schwinger term

As discussed in Sec.1.4.1, it is one of the most pathological features of the Minkowskian quantum field theory that we must assume the existence of the

---

[7]   More generally, Eq.(5) is valid as long as Φ is not expressible as $\boldsymbol{\delta}_*(\Psi)$.

Schwinger term. We reconsider this problem by taking quantum gravity into account.

The Dirac current j^μ is defined by

$$j^\mu \equiv h\bar{\psi}\gamma^\mu\psi \qquad (5.8.3-1)$$

in the framework of the vierbein formalism (see Sec.5.7). From the equal-time anticommutation relation, Eq.(5.7.4-15), we can easily show that

$$[j^0(x),\, j^k(y)]|_0 = 0. \qquad (5.8.3-2)$$

On the other hand, from the conservation law

$$\partial_\mu j^\mu = 0 \qquad (5.8.3-3)$$

together with Poincaré invariance and Einstein causality, we have

$$e^2\langle 0\,|\,[j^\mu(x),\, j^\nu(y)]\,|\,0\rangle = i\int_0^\infty ds\,\pi(s)(-s\eta^{\mu\nu} - \partial^\mu\partial^\nu)\Delta(x-y;\,s). \qquad (5.8.3-4)$$

with $\pi(s) \not\equiv 0$. Hence we obtain

$$e^2\langle 0\,|\,[j^0(x),\, j^k(y)]|_0\,|\,0\rangle = i\int_0^\infty ds\,\pi(s)\cdot\partial^k\delta(\mathbf{x}-\mathbf{y}). \qquad (5.8.3-5)$$

In the Minkowskian quantum field theory, the Dirac field is quantized in the positive-metric Hilbert space. Then, as in the proof of the Lehmann theorem, it is straightforward to show

$$\pi(s) \geq 0. \qquad (5.8.3-6)$$

Accordingly, Eq.(5) is non-vanishing in contradiction to Eq.(2). We are thus forced to deny the validity of Eq.(2), that is, we must assume the existence of the Schwinger term as the consequence of calculating singular expressions by means of some limiting procedure. This standpoint is inconsistent with the postulate that products of field operators at the same spacetime point are independent of their ordering [see Sec.1.1.1].

The above trouble can be resolved if we take account of quantum gravity [Nak 80c]. Since the Minkowskian quantum field theory is an approximate theory valid for the energy region much below the Planck mass, it is quite unreasonable to extrapolate Eq.(6) to $s \gtrsim \kappa^{-1}$. Indeed, since $j^\mu(x)$ is *not* invariant under the gravitational BRS transformation as shown in Sec.5.8.2, *there is no reason for claiming the validity of Eq.(6) for $s \gtrsim \kappa^{-1}$*. Hence it is quite possible to have

$$\int_0^\infty ds\, \pi(s) = 0. \qquad (5.8.3-7)$$

Thus *we no longer need to assume the existence of the Schwinger term*. We should rather regard Eq.(7) as a sum rule just like Eq.(5.8.2-9). The appearance of perturbative divergence is, of course, consistent with Eq.(7). For example, a model function

$$\pi(s) = (1 - \kappa s) \exp(-\kappa s) \qquad (5.8.3-8)$$

satisfies Eq.(7) and is completely well-behaved, but if it is expanded into a power series of κ, integrals over s are very badly divergent.

Finally, we note that the two-dimensional quantum field theory is quite exceptional in this connection. Since there is no Einstein gravity in two dimensions, we have no possibility of eliminating the Schwinger term. Indeed, as noted in Sec.2.5.3, the presence of the Schwinger term is indispensable for formulating the two-dimensional models consistently.

5.8.4 Remarks and comments

Quantum gravity has been formulated quite satisfactorily in the framework of canonical operator formalism. We have encoutered no difficulty as long as we work in the Heisenberg picture.

Mathematically inclined physicists, however, criticize the Lagrangian formalism as non-rigorous and wish to employ the axiomatic approach. But we emphasize that the axiomatic approach is quite unsuitable to quantum gravity.

Firstly, indefinite metric is indispensable in quantum gravity. It is extremely difficult to construct the theory of indefinite metric in a mathematically *and* physically satisfactory way. In gauge theories, one can partially bypass this difficulty by considering gauge-invariant quantities only in the physical subspace. In the gravitational theory, however, any spacetime-dependent quantities are *non-invariant* under general coordinate transformations.

Secondly, the postulate that product of field operators at the same spacetime point are independent of their operator ordering is very important even if one forgets field equations and canonical commutation relations. Such quantities as $g^{\mu\nu}(x)$ (matrix inverse of $g_{\mu\nu}(x)$) and $h(x)$ cannot be defined without this postulate.[8] And, as shown in Sec.5.8.3, the validity of this postulate is crucially dependent on the presence of indefinite metric.

Thirdly, in the axiomatic approach, one adopts Poincaré invariance and Einstein causality (local commutativity) as fundamental axioms. They are, however, the concepts characteristic to the Minkowski spacetime; therefore, it is quite unnatural to regard them as the basic principles in the gravitational theory. Indeed, as shown in Sec.5.7.6, they are what are derived at the level of representation.

In quantum gravity, x^μ should not be directly identified with the spacetime coordinates which we observe. Since the invariant distance squared,

$$ds^2 \equiv g_{\mu\nu}(x)dx^\mu dx^\nu, \qquad\qquad (5.8.4-1)$$

is an operator, the geometrical interpretation of x^μ is impossible at this stage. Since there are no lightcone singularities, such an approach as the Wilson operator product expansion is inappropriate in quantum gravity.

Planck length is an extraordinarily small scale too far beyond the particle-physics scale. It is quite dangerous to extrapolate particle-physics results naively to quantum gravity. For example, if we first consider the Higgs mechanism and then introduce gravity, we encounter a ridiculously large cosmological term. This is evidently a wrong way of thinking. We should first consider the spontaneous breakdown of $GL(4)$, as discussed in Sec.5.7.6. Then we should make some kind of "coarse graining" in order to define state vectors of particle physics by eliminating the degrees of freedom of gravitons. *After that*, we should discuss the Higgs mechanism.

Although anomaly is one of current topics, we can say nothing about it from quantum gravity. All symmetries found in our theory of quantum gravity are independent of the spacetime dimensions and therefore free of anomaly. Gravitational anomaly is caused by very peculiar matter fields in the unphysical spacetime dimensions $4k + 2$ $(k = 0, 1, 2, \ldots)$ [Alv 84]. Those theories should be excluded as unphysical. Anyway, gravitational anomaly should not be considered before the above-mentioned "coarse graining" is taken into account.

---

[8]   One can avoid the use of $g^{\mu\nu}$ and h only in the three-dimensional pure gravity [Wit 88], but it is a trivial and totally unphysical situation.

Quantum field theory in curved spacetime [Bir 82] deals with the phenomena in the scale very much larger than that of particle physics. It is, therefore, probably absurd to try to derive it directly from quantum gravity. Both theories might be directly connected at the beginning of the universe, but the extrapolation of the Big-Bang cosmology to the scale of Planck time is too much speculative to be considered seriously as physics.

5.9 MISCELLANEOUS TOPICS

In this section, we briefly survey some related topics. Since the literature of quantum gravity is enormous, we restrict our attention to the non-perturbative approaches to quantum gravity in the operator formalism based on the BRS symmetry.

5.9.1 Coupling with gauge field

It is of particular interest to consider the case in which the matter Lagrangian density $\mathcal{L}_M$ contains gauge fields.

The Lagrangian density of an abelian gauge field (electromagnetic field) which couples with the gravitational field is given by

$$\mathcal{L}_{EM} = -\frac{1}{4}\tilde{g}^{\mu\sigma}g^{\nu\tau}F_{\mu\nu}F_{\sigma\tau} - \tilde{g}^{\mu\nu}A_\nu\partial_\mu B$$
$$- i\tilde{g}^{\mu\nu}\partial_\mu\bar{C}\cdot\partial_\nu C \qquad (5.9.1-1)$$

with

$$F_{\mu\nu} \equiv \partial_\mu A_\nu - \partial_\nu A_\mu = -F_{\nu\mu}, \qquad (5.9.1-2)$$

where A_μ, B, C, and $\bar{C}$ are the electromagnetic field, its B-field, its FP ghost, and its FP anti-ghost, respectively. We restrict our consideration to the Landau gauge, because then we have the largest symmetry.

The similarity between the gravitational FP-ghost Lagrangian density $\mathcal{L}_{FP}$ and the electromagnetic one in Eq.(1) is remarkable. This similarity is, of course, the manifestation of the abelian nature of the translation group and $U(1)$. By making use of this fact, we can extend the sixteen-dimensional Poincaré-like superalgebra to the nineteen-dimensional one [Nak 81b]; the supercoordinate X acquires three extra components B, C, and $\bar{C}$. The gravi-electromagnetic theory has 201 symmetry generators (without counting the local-Lorentz part).

As is well known, the Einstein-Hilbert Lagrangian density and the Maxwell one can be combined into the five-dimensional Kaluza-Klein form. Correspondingly, $\hat{x}^\lambda$, b_ρ, c^σ, and $\bar{c}_\tau$ may be supposed to acquire their fifth components $1/\sqrt{\kappa}$, $\sqrt{\kappa}B$, C, and $\bar{C}$, respectively ($\partial_5 \equiv 0$). Then the above-mentioned extended superalgebra can be regarded as a 5×4-dimensional *Lorentz-like* superalgebra. Furthermore, it is possible to extend the geometric commutation relation into the five-dimensional form. Of course, this five-dimensional description is incomplete becuase x^5 is regarded as a constant throughout. It is unrealistic and artificial to regard x^5 as a true coordinate.

When a non-abelian gauge field couples with gravity, it is diffucult to extend the sixteen-dimensonal Poincaré-like superalgebra into a larger indecomposable superalgebra because of the non-abelian nature of the Lie algebra. But, in the Cartan subalgebra (i.e., maximal abelian subalgebra), whose dimension r is (by definition) the rank of the original Lie algebra, we have the abelian nature, and therefore it is possible to extend the sixteen-dimensional Poincaré-like superalgebra in a non-trivial way [Abe 85]. The extended superalgebra is indecomposable and have $146 + 36r + 2r^2$ symmetry generators. Although this formalism is unsatisfactory in the sense that the original non-abelian symmetry is sacrificed, it is remarkable that spacetime symmetry and physical internal symmetry are unified without introducing any *ad hoc* ansatz by hand.

5.9.2 Various extensions

The manifestly covariant canonical operator formalism of quantum gravity presented here can be extended in various ways.

Many physicists wish to introduce the so-called R^2 terms (i.e., the terms proportional to hR^2 and $hR_{\mu\nu}R^{\mu\nu}$) into the gravitational Lagrangian density in order to secure perturbative renormalizability. Then, of course, one gets into the ghost trouble; there are many unsatisfactory proposals for the possible way-out of the unitarity trouble. Instead of criticizing each of them concretely, we here note that nobody has invented a way-out which is applicable solely to the R^2 theory but *not* to any higher-derivative theories in particle physics. This fact nullifies the effectiveness of the renormalizability requirement.

The R^2 theory is, however, still interesting as a model theory. We should introduce some auxiliary fields in order to rewrite the Lagrangian density into the form which apparently contains no higher derivative terms. Adopting $\mathcal{L}_{GF} + \mathcal{L}_{FP}$ given in Sec.5.3.2, we can construct a manifestly covariant canonical operator formalism of the R^2 quantum gravity [Kaw 81, Kaw 82, Kaw 83]. Although equal-time commutation relations concerning $g_{\mu\nu}$ and its time derivatives are drastically modified, choral symmetry and tensorlike commutation relations can be explicitly shown to remain unchanged, as is expected.

An extension to the Poincaré gauge theory has been made without yet carrying out canonical quantization explicitly [Ham 83].

A similar extension to supergravity is not made (See Sec.5.9.4). Only discussed is the unitarity proof in the framework of the canonical operator formalism [Hat 79]. The crucial observations made there are as follows: The supersymmetry FP ghosts, which are bosonic Majorana spinors, should satisfy a *second-order* differential equation in order for the unitarity proof to work, and then the supersymmetry B-field, which is a fermionic Majorana spinor, is indispensable in order to have the (not only off-shell but also *on-shell*) nilpotency of the BRS transformation.

One may also consider various extensions of the gauge-fixing Lagrangian density, sacrificing the beauty of the theory. From the viewpoint of the B-field formalism, the general gauge-fixing Lagrangian density should be of the form $b_\rho F^\rho + G$ apart from a total divergence, where F^ρ is an arbitrary function of the gravitational field and matter fields together with their first derivative and G is a function of b_ρ alone involving no derivative. Then the corresponding FP-ghost Lagrangian density is bilinear with respect to c^σ and $\bar{c}_\tau$. In this case, it is possible to extend the four-dimensional geometric commutation relation, Eq.(5.6.4-1), by using a *4×4 matrix extension* of the quantum gravity D function [Nak 85].

In two-dimensional spacetime, the Einstein gravity does not exist because hR is a total divergence. Hence sometimes two-dimensional gravity is formulated by introducing a scalar field ϕ by hand and then by replacing hR by $hR\phi$. Evidently, such a procedure is too artificial. The right way of formulating the two-dimensional quantum gravity is to take account of the **Weyl invariance**, i.e., the local scale invariance, of $\int d^2x\, hR$. Let b be the B-field of the Weyl invariance; then its gauge fixing term is hRb, precisely the expression which one wants. Then one can extend the manifestly covariant canonical formalism of quantum gravity quite beautifully though one encounters unitarity trouble [Sat unp].

5.9.3 Extension of BRS symmetry

As discussed in Sec.3.5.2, the **double (or extended) BRS symmetry** is of some interest, and it is tempting to extend it to quantum gravity. This extension is, of course, possible, as the anti-BRS generator $\eta^{\mu\nu}M(\bar{c}_\mu, b_\nu)$ does exist [Nak 80d], though it violates $GL(4)$ invariance. In order to have the double BRS algebra, therefore, we must forget about $GL(4)$, introducing $\eta_{\mu\nu}$ from the outset.

The anti-BRS transformation $\bar{\boldsymbol{\delta}}$ can be defined as follows: In all defining formulae of $\boldsymbol{\delta}$ (see Secs.5.3.1 and 5.3.2), we replace c^σ by $\eta^{\sigma\tau}\bar{c}_\tau$ and $\bar{c}_\tau$ by $-\eta_{\tau\sigma}c^\sigma$.[1] Then, for example, we have

$$\bar{\boldsymbol{\delta}}(g^{\mu\nu}) = \eta^{\mu\rho}\partial_\lambda\bar{c}_\rho \cdot g^{\lambda\nu} + \eta^{\nu\rho}\partial_\lambda\bar{c}_\rho \cdot g^{\mu\lambda}, \qquad (5.9.3-1)$$

$$\bar{\boldsymbol{\delta}}(\bar{c}_\tau) = 0, \qquad \bar{\boldsymbol{\delta}}(c^\lambda) = -i\eta^{\lambda\rho}b_\rho. \qquad (5.9.3-2)$$

Of course, $\bar{\boldsymbol{\delta}}$ is nilpotent and anticommutes with $\boldsymbol{\delta}$.

In order to find candidates for the gauge-fixing plus FP-ghost action invariant under both BRS and anti-BRS transformations, it is useful to list up the quantities of the form $\boldsymbol{\delta}\bar{\boldsymbol{\delta}}(\Phi)$ [2] [Pas 82]. For example, one obtains

$$\boldsymbol{\delta}\bar{\boldsymbol{\delta}}(\eta_{\mu\nu}g^{\mu\nu}) = iE = -2i\kappa h^{-1}(\mathcal{L}_{\text{GF}} + \mathcal{L}_{\text{FP}}), \qquad (5.9.3-3)$$

$$\boldsymbol{\delta}\bar{\boldsymbol{\delta}}(c^\lambda\bar{c}_\lambda) = \eta^{\lambda\rho}b_\lambda b_\rho. \qquad (5.9.3-4)$$

Now, we introduce two Grassmann numbers (i.e., anticommuting c-numbers), θ and $\bar{\theta}$, satisfying $\theta^* = -\theta$, $\bar{\theta}^* = -\bar{\theta}$, $\theta^2 = \bar{\theta}^2 = \{\theta, \bar{\theta}\} = 0$ and anticommuting with c^λ and $\bar{c}_\tau$. We then consider a six-dimensional superspace whose coordinates are $(x^\mu, \theta, \bar{\theta})$. Because of the Grassmann nature of θ and $\bar{\theta}$, any six-dimensional function $\Phi(x, \theta, \bar{\theta})$ is expanded into a finite Taylor series

$$\Phi = \Phi_0 + \theta\left(\frac{\partial\Phi}{\partial\theta}\right)_0 + \bar{\theta}\left(\frac{\partial\Phi}{\partial\bar{\theta}}\right)_0 + \theta\bar{\theta}\left(\frac{\partial\Phi}{\partial\bar{\theta}\partial\theta}\right)_0, \qquad (5.9.3-5)$$

where a subscript 0 indicates to set $\theta = \bar{\theta} = 0$. If we identify $\bar{\theta}$ and θ with the transformation parameters of BRS and anti-BRS symmetries, respectively, we should have

$$\left(\frac{\partial\Phi}{\partial\theta}\right)_0 = \bar{\boldsymbol{\delta}}(\Phi_0), \qquad (5.9.3-6)$$

$$\left(\frac{\partial\Phi}{\partial\bar{\theta}}\right)_0 = \boldsymbol{\delta}(\Phi_0), \qquad (5.9.3-7)$$

$$\left(\frac{\partial\Phi}{\partial\bar{\theta}\partial\theta}\right)_0 = \boldsymbol{\delta}\bar{\boldsymbol{\delta}}(\Phi_0). \qquad (5.9.3-8)$$

---

[1] $\mathcal{L}$ is invariant under this replacement, but it is not a discrete symmetry, but generated by $\frac{1}{2}[\eta^{\mu\nu}M(\bar{c}_\mu, \bar{c}_\nu) + \eta_{\mu\nu}M(c^\mu, c^\nu)]$.

[2] In quantum gravity, it is conjectured that $\boldsymbol{\delta}\bar{\boldsymbol{\delta}}(\Phi)$ exhausts BRS-anti-BRS invariant quantities apart from world scalars.

The **superfield formalism** aims at the geometrical description in the six-dimensional superspace; starting from six-dimensionally covariant quantities, one wishes to derive Eqs.(6)-(8). Since the complete six-dimensional symmetry is *explicitly* broken, one must introduce some constraints, which are known to be "soul flatness" conditions (flatness in the Grassmann directions).

A straightforward transcription of the gauge-field case can be made by extending the connection to a six-dimensional supervector [Hir 82, Ham 82, Kik 83a, Kik 83b]. But, since the connection is not a primary field, the spacetime symmetry of gravity is not properly extended. As a gravitational theory, it seems more natural to start with defining a six-dimensional extension of metric tensor [Pas 82, Del 82b, Del 82c, Fal 84]. The extended metric can be constructed from a metric in a preferred system by means of a certain six-dimensional coordinate supertransformation.

The superfield formalism, however, has several unsatisfactory features. For example, "soul flatness" conditions are usually introduced by hand, and there is no way of incorporating matter fields. Moreover, from the outset, we cannot find any appealing reason for considering an *explicitly broken* six-dimensional symmetry at the sacrifice of the perfect sixteen-dimensional one.

The superfield formalism can be extended so as to incorporate the 32 symmetries generated by $M(b_\mu, c^\nu)$ and $M(b_\mu, \bar{c}_\nu)$ [Pas 82], but in order to reproduce the sixteen-dimensional Poincaré-like superalgebra, one must take over $P(\hat{x}^\mu)$ and $M(\hat{x}^\mu, \hat{x}^\nu)$ from the original formalism.

5.9.4 Supersymmetric extension of the vierbein formalism

In this section, we outline a newly proposed supersymmetric extension of the local Lorentz symmetry of the vierbein formalism different from supergravity. This is quite a natural extension of our theory of quantum gravity towards a unified theory.

In the Minkowskian quantum field theory, the **supersymmetry**[3] proposed by Wess and Zumino [Wes 74] is quite remarkable in the sense that it is a unique non-trivial extension of the Poincaré symmetry [Haa 75]. Unfortunately, however, this Poincaré supersymmetry is badly violated in the real world. The breakdown

---

[3]   We here assume that the reader has some knowledge about supersymmetry.

cannot be spontaneous because no Goldstone fermion is observed. The most natural way-out is to promote the supersymmetry to a local one so as to make use of the (super) Higgs mechanism. In this way, one is led to **supergravity** [Wes 83, Gat 83].

It should be noted, however, that there is no Poincaré structure in the Einstein gravity because local Lorentz transformations totally commute with general coordinate transformations. In order to recover the Poincaré structure at the Lagrangian level, one must introduce the notion of the local *internal* translation. Then supergravity is formulated as a *gauge theory* of the supersymmetric extension of the *internal* Poincaré algebra. Thus supergravity is a direct extension of *neither* the Einstein gravity *nor* the particle-physics Poincaré supersymmetry. Indeed, in supergravity, the *global* spacetime translation generator P_μ cannot be expressed in terms of an anticommutator of the supertransformation generators because there is no global internal translation. This fact also implies that it is impossible to construct a globally supersymmetric gauge-fixing term in supergravity, because if it were possible then the non-existent global internal translation generator could be constructed.

Now, as stressed in Sec.5.7.6, the particle-physics Poincaré symmetry is not a fundamental symmetry. It is more natural, therefore, to supersymmetrize the *local Lorentz symmetry only* rather than the artificially constructed local Poincaré symmetry. Indeed, we can quite satisfactorily construct a **supersymmetric extension of the vierbein formalism** by taking a complex orthosymplectic superalgebra $OSp(N, 2; C)$ as a local supersymmetry in such a way that it is totally independent of the general coordinate transformation [Nak 87, Abe 87, Abe 88a, Abe 88b, Abe 88c, Abe 89b, Abe 88d, Abe 88e] (for a review, see [Abe 90]). The bosonic part of $OSp(N, 2; , C)$ is $SL(2, C) \oplus SO(N, C)$, where, of course, $SL(2, C)$ is identified with the Lorentz algebra. New degrees of freedom are $2 \cdot 2N$ fermionic ones and $2 \cdot N(N-1)/2$ bosonic ones.

In contrast to the fact the supertransformation generator Q has mass dimension 1/2 in supergravity, Q is *dimensionless* in this new theory because the anticommutator of Q's is essentially the Lorentz generator. Hence Q cannot induce the transformation between a Klein-Gordon field and a Dirac field. This difficulty is resolved by introducing a new nonlinear realization, called the **ζ-field realiza-**

tion[4]. That is, the superpartners of the vierbein field and any matter field are given not by primary fields but by products of them and a special spinor field, ξ, obeying fermi statistics. Mathematically, the ξ-field is nothing but a frame matrix of a **super-Grassmann manifold** $SGM(N + 2, 2)$ whose bosonic submatrix is normalized to an $N \times N$ unit matrix [Kan 89]. Hence if the bosonic submatrix is left arbitrary, $SO(N, C)$ is naturally extended to $GL(N, C)$. Even without knowing such a mathematical basis, one can see that matter-field representations are systematically constructed by using $GL(N, C)$ tensors [Abe 87]. But, of course, the original ξ-field itself is not a representation of $GL(N, C)$. It is interesting, however, that the metric tensor of $GL(N, C)$ can be expressed by the ξ-field. This metric is naturally decomposed into a product of $GL(N, C)$ vielbeins. By using the $GL(N, C)$ vielbein, one can construct the $OSp(N, 2; C)$ supersymmetric Dirac theory. Here it turns out that the real part of the extra $N(N - 1)/2$ complex degrees of freedom of the $GL(N, C)$ vielbein correspond to the chiral gauge symmetry. In this way, one achieves a unification of the Einstein's theory and the $SO(N)$ chiral non-abelian gauge theory.

The characteristics and the main results of this theory are as follows:

1. In contrast to supergravity, torsion remains zero and gravitino is unnecessary.

2. All symmetries of $OSp(N, 2; C)$ are spontaneously broken.

3. No new elementary particles are predicted to be observed.

4. The gravitational field $g_{\mu\nu}$ and therefore the Einstein-Hilbert action are invariant under $OSp(N, 2; C)$.

5. Matter-field Lagrangian densities are supersymmetrized.

6. The chiral gauge group must be $SO(N)$ [Abe 87]. Hence if one believes in the **grand unification theories**, one should take $SO(10)$ but neither $SU(5)$ nor E_6.

7. The superconnection can be expressed in terms of the spin connection, the ξ-field, and the $GL(N, C)$ vielbein explicitly [Abe 88a, Abe 89b].

8. The supercurvature essentially reduces to the Riemann curvature (plus the chiral gauge field strength if chiral gauge symmetry is taken into account) [Abe 88d].

9. In contrast to supergravity, the gauge-fixing term and the FP-ghost one can

---

[4]   As an example of the ξ-field realization, the BRS transformation can be regarded as that of the BRS algebra by setting $\xi = T^a c^a$ [Abe 89c].

be chosen to be globally supersymmetric [Abe 88c, Abe 89b].

10. The unitarity proof given in Sec.5.8.1 can be extended [Abe 88b, Abe 88c].

11. The extended BRS symmetries stated in Sec.5.7.5 remain unchanged [Abe 88c].

12. Canonical quantization can be carried out, so that our theory of quantum gravity presented in this chapter is beautifully extended. (Only the $N = 1$ case is explicitly worked out [Abe 88c]).

Finally, we note that it is also possible to construct the $SL(N, 2; C)$ supersymmetric extension instead of the $OSp(N, 2; C)$ one [Abe 89a]. But both theories are not so much different.

MATHEMATICAL SUPPLEMENTS

In this appendix, we collect some useful notions and consequences in differential geometry and in the axiomatic theory of relativistic quantum fields extended to the situation with indefinite metric.

A.1 Some basic facts about Grassmann algebras, differential forms and cohomology

a) Grassmann algebra

A **Grassmann algebra** $\mathcal{A}$ is, in short, an associative algebra equipped with an *anticommutative* multiplication, which we denote by $a \wedge b$ ($a, b \in \mathcal{A}$) and which is sometimes called the **exterior product**: To be more precise, it is convenient to take a set $\{\xi_i\}$ of *generators*[1] of $\mathcal{A}$ which are to satisfy

$$\xi_i \wedge \xi_j = -\xi_j \wedge \xi_i, \qquad (A.1-1)$$

$$(\xi_i \wedge \xi_j) \wedge \xi_k = \xi_i \wedge (\xi_j \wedge \xi_k). \qquad (A.1-2)$$

Then, $\mathcal{A}$ is the totality of linear combinations of such monomials as $a \equiv \xi_{i_1} \wedge \xi_{i_2} \wedge \cdots \wedge \xi_{i_r}$.

The number r of the factors ξ_i in this product a is called its **degree** and is denoted by $p(a)$. The degree 0 is allowed by assigning it to the c-numbers $\in K$, where K is the *field* (*not* Feld but Körper in German) of coefficients, e.g., $K = \mathbf{R}$ (totality of real numbers), $\mathbf{C}$ (totality of complex numbers), etc. If the number n of generators $\{\xi_i\}$ is finite, the maximum degree of non-vanishing monomials is clearly n, because of the antisymmetry Eq.(1). The number of independent

---

[1]   For the sake of the intrinsic mathematical definition, it is *not* necessary to choose a set of generators, but only for convenience' sake.

monomials with a degree r is given by $\binom{n}{r} = n!/r!(n-r)!$, and hence, the dimension of the Grassmann algebra $\mathcal{A}$ is $\sum_{r=0}^{n} \binom{n}{r} = 2^n$. $\mathcal{A}$ satisfies the **graded commutativity**; i.e. for two monomials a and b, we have

$$a \wedge b = (-1)^{p(a)p(b)} b \wedge a. \qquad (A.1-3)$$

Let V denote the vector space of linear combinations of the generators $\{\xi_i\}$. For a subspace W of V, linear combinations of the products $\eta_1 \wedge \eta_2$ with $\eta_i \in W$ is denoted by $W \wedge W = \wedge^2 W$, and similar definitions can be given for $W \wedge W \wedge W = \wedge^3 W$, etc. Then, the Grassmann algebra $\mathcal{A}$ is given as the direct sum of the exterior products, $\wedge^r V$, with degree r:

$$\mathcal{A} = \overset{n=\dim V}{\underset{r=0}{\oplus}} \wedge^r V = K \oplus V \oplus \wedge^2 V \oplus \cdots \oplus \wedge^n V \equiv \wedge V. \qquad (A.1-4)$$

Next, $(W \wedge W)^*$, the dual space of $W \wedge W$, consisting of linear functionals α on $W \wedge W$ (taking values in K), can be identified with the space of antisymmetric bilinear forms $\tilde{\alpha}$ on W through the relation

$$\alpha(a \wedge b) = \tilde{\alpha}(a, b) \quad \text{for } \forall a, b \in W. \qquad (A.1-5)$$

Introducing the dual base $\{\eta'\}$ defined by the relation

$$\eta'(\xi_j) = \delta'_j, \qquad (A.1-6)$$

the antisymmetric bilinear form α can also be expressed as an exterior product given by

$$\alpha = \frac{1}{2} \sum_{i,j} \alpha(\xi_i \wedge \xi_j) \eta' \wedge \eta^j, \qquad (A.1-7)$$

which gives the isomorphism between $(W \wedge W)^*$ and $W^* \wedge W^*$:

$$(W \wedge W)^* \cong W^* \wedge W^*. \qquad (A.1-8)$$

Here the exterior product $\varphi \wedge \psi$ of linear functionals φ and ψ on W is defined as an antisymmetric bilinear form on W by

$$(\varphi \wedge \psi)(a, b) \equiv \varphi(a)\psi(b) - \varphi(b)\psi(a). \qquad (A.1-9)$$

The most important example of the Grassmann algebra is provided by the exterior algebra of differential forms on a manifold M, as explained in d).

b) Tangent bundles

Here we assume that the notion of differential manifolds defined by *'patching together local charts smoothly'* is known (see, for instance, [Kob 63]). At a point $x \in M$, a tangent vector X_x is defined as a **derivation**, a linear map acting on functions f given in some neighborhood of x satisfying the following Leibniz rule:

$$X_x(fg) = X_x(f)g(x) + f(x)X_x(g), \qquad (A.1-10)$$

where g is another arbitrary function given in some neighborhood of x. The vector space of all tangent vectors at $x \in M$ is denoted by $T_x(M)$ and, putting together all such vector spaces over M, one can obtain a **tangent bundle** $T(M)$ over the manifold M as a differential manifold with the dimension $2n$, twice of that of M, $n \equiv \dim M$. A **cross section** X assigning smoothly to each $x \in M$ a tangent vector X_x belonging to $T_x(M)$ is called a tangent vector field, or simply, a **vector field** on M. The totality, $\Gamma(T(M))$, of vector fields on M is usually denoted by $\mathcal{X}(M)$ and constitutes a **module** over the algebra $C^\infty(M)$ of smooth functions. (Namely, we can multiply a vector field X by a smooth function $f \in C^\infty(M)$ to obtain another vector field fX, and, aside from the possible non-invertibility of $f \neq 0$, the space $\mathcal{X}(M)$ enjoys almost all the properties of vector spaces taking smooth functions $f \in C^\infty(M)$ as scalar coefficients.) It can be seen to constitute a(n infinite-dimensional) Lie algebra (over $\mathbf{R}$) with respect to the Lie bracket $[X, Y]$ defined for $X, Y \in \mathcal{X}(M)$ by

$$([X, Y]f)(x) = X(Yf)(x) - Y(Xf)(x) \qquad \text{for } f \in C^\infty(M). \qquad (A.1-11)$$

c) Differential forms

Next, taking the dual of tangent bundle $T(M)$, we obtain the **cotangent bundle** $T^*(M)$ over M: At each point $x \in M$, $T_x^*(M)$ is the vector space of all linear functionals on the vector space $T_x(M)$, which are sometimes called **covectors**. A cross section η assigning smoothly to each $x \in M$ a covector $\eta_x \in T^*(M)$ is called a **differential 1-form**, or simply, a 1-form. Similarly to the case of vector fields, the totality of 1-forms, $\Gamma(T^*(M))$, constitutes a vector space, or more properly, a module over the algebra $C^\infty(M)$, which is just in the dual relation with the space $\mathcal{X}(M)$ of vector fields:

$$\Gamma(T^*(M)) \simeq \Gamma(T(M))^* = \mathcal{X}(M)^*. \qquad (A.1-12)$$

A (differential) r-**form** is defined as a linear combination of r-fold exterior products, $\varphi_1 \wedge \cdots \wedge \varphi_r$'s, of 1-forms, $\varphi_1, \cdots, \varphi_r \in \Gamma(T^*(M))$ (with $C^\infty(M)$ as coefficients), the totality of which constitutes $\wedge^r\Gamma(T^*(M)) \simeq \Gamma(\wedge^r T^*(M))$. In terms of a local chart $(x^1 \cdots x^n)$ in some neighborhood in M, a more intuitive expression for an r-form φ can be given by

$$\varphi_x = \sum_{\iota_1 < \cdots < \iota_r} f_{\iota_1 \cdots \iota_r}(x) dx^{\iota_1} \wedge \cdots \wedge dx^{\iota_r}, \qquad (A.1-13)$$

with $f_{\iota_1 \cdots \iota_r} \in C^\infty(M)$. In this context, smooth functions in $C^\infty(M)$ are regarded as 0-forms. Then, taking the direct sum of $\wedge^r\Gamma(T^*(M))$ for $r = 0, \cdots, n$, we obtain the Grassmann algebra of differential forms, $\wedge\Gamma(T^*(M))$.

d) Exterior differential

In view of the isomorphism, Eq.(12), we note that an r-form in $\wedge^r\Gamma(T^*(M))$ can be identified with an antisymmetric (r-)multilinear functions on $\mathcal{X}(M)$ (multilinear with respect to the coefficients in $C^\infty(M)$): By using this identification, the exterior differential d mapping an r-form φ into an $(r + 1)$-form $d\varphi$ can be defined globally by

$$d\varphi(X_0, \cdots, X_r) = \sum_{\iota=0}^{r} (-1)^\iota X_\iota(\varphi(X_0, \cdots, \hat{X}_\iota, \cdots, X_r))$$
$$+ \sum_{\iota<\jmath}^{r} (-)^{\iota+\jmath} \varphi([X_\iota, X_\jmath], X_0, \cdots, \hat{X}_\iota, \cdots, \hat{X}_\jmath, \cdots, X_r),$$
$$(A.1-14)$$

where the symbol $\hat{\ }$ indicates the omitted term. The local expression for $d\varphi$ corresponding to Eq.(13) is given by

$$d\varphi = \sum_{\iota_1 < \cdots < \iota_r} df_{\iota_1 \cdots \iota_r} \wedge dx^{\iota_1} \wedge \cdots \wedge dx^{\iota_r}. \qquad (A.1-15)$$

Just in a parallel with the Lie-algebra cohomology discussed in Sec.3.3.2 (here with $\mathcal{X}(M)$ as the relevant Lie algebra), d is an **anti-derivation**,

$$d(\varphi_1 \wedge \varphi_2) = d\varphi_1 \wedge \varphi_2 + (-1)^{r_1} \varphi_1 \wedge d\varphi_2, \quad \text{for } r_\iota\text{-forms } \varphi_\iota, \ (i = 1, 2), \quad (A.1-16)$$

satisfying the **nilpotency** property :

$$d^2 = 0. \qquad (A.1-17)$$

The family $\wedge^r \Gamma(T^*(M))$ graded by the degree r equipped with the exterior differentials, $d = d_r : \wedge^r \Gamma(T^*(M)) \rightarrow \wedge^{r+1}\Gamma(T^*(M))$, constitutes the **de Rham complex** associated with the manifold M:

$$0 \rightarrow C^\infty(M) \xrightarrow{d} \Gamma(T^*(M)) \xrightarrow{d} \wedge^2\Gamma(T^*(M)) \xrightarrow{d} \cdots \xrightarrow{d} \wedge^n\Gamma(T^*(M)) \xrightarrow{d} 0.$$
$$(A.1-18)$$

From the general viewpoint of cohomology, the de Rham complex is a special case of **cochain complexes** (or **differential complexes**),

$$0 \rightarrow C^0 \xrightarrow{d} C^1 \xrightarrow{d} C^2 \xrightarrow{d} \cdots \xrightarrow{d} C^n \xrightarrow{d} 0, \qquad (A.1-19)$$

with $C^r \equiv \wedge^r\Gamma(T^*(M))$. In this general context, r-forms φ are called r-**cochains** and the exterior differential d a **coboundary operator**. On the other hand, as a *Grassmann algebra equipped with a nilpotent anti-derivation d*, $\wedge\Gamma(T^*(M))$ provides a typical example of **graded differential algebra** (see, e.g., [Sul 77]).

When an r-form or r-cochain φ is annihilated by d,

$$d\varphi = 0, \qquad (A.1-20)$$

it is called an r-**cocycle**, and when given as an image of some $(r-1)$-cochain ψ,

$$\varphi = d\psi, \qquad (A.1-21)$$

it is called an r-**coboundary**, the totality of which are denoted, respectively, by $Z^r(= \mathrm{Ker}\ d_r)$ and by $B^r(= \mathrm{Im}\ d_{r-1})$. The nilpotency, Eq.(17), implies the inclusion relation $B^r \subseteq Z^r$. The r-th **cohomology group** H^r is defined by $H^r \equiv Z^r/B^r$. (To stress the relevance of de Rham complex, it is called **de Rham cohomology**.) Roughly speaking, this is to measure the deviation of the cochain complex, Eq.(19), from being an **exact sequence** characterized by the equalities $Z^r = \mathrm{Ker}\ d_r = \mathrm{Im}\ d_{r-1} = B^r$. Since it is unrealistic for $H^0 = Z^0 = \mathrm{Ker}\ d_0$ to vanish, however, the *most trivial* cohomological situation is that the complex Eq.(19) is **acyclic**, i.e., it becomes an exact sequence if one term $\mathrm{Ker}\ d_0 \equiv A$ is inserted in between 0 and C^0 (with ϵ called an *augmentation*):

$$0 \xrightarrow{\epsilon} A \rightarrow C^0 \xrightarrow{d} C^1 \xrightarrow{d} C^2 \xrightarrow{d} \cdots \xrightarrow{d} C^n \xrightarrow{d} 0. \qquad (A.1-22)$$

This is equivalent to the condition for all the cohomology groups H^r to vanish for $r \geq 1$ and also to the validity of the **Poincaré lemma**[2], which holds, for instance, for the de Rham complex on the flat Minkowski space $M = \mathbf{R}^4$. Beside the exterior differential d, there are two other important operations acting on differential forms, the **interior product** i_X and the **Lie derivative** $\mathcal{L}_X$, which are defined for an r-form φ by

$$i_X \varphi(X_0, \cdots, X_{r-2}) = \varphi(X, X_0, \cdots, X_{r-2}), \qquad (A.1-23)$$

$$\mathcal{L}_X \varphi(X_0, \cdots, X_{r-1}) = X(\varphi(X_0, \cdots, X_{r-1}))$$
$$- \sum_{i=0}^{r-1} \varphi(X_0, \cdots, [X, X_i], \cdots, X_{r-1}). \qquad (A.1-24)$$

i_X maps an r-form to an $(r-1)$-form and $\mathcal{L}_X$ does not change the degree of a form. Although the definition of $\mathcal{L}_X$ looks a little complicated, it can be naturally understood from the general definition of Lie derivative on arbitrary tensors given, through the *Leibniz rule* with respect to the tensor product $\otimes$,

$$\mathcal{L}_X(S \otimes T) = (\mathcal{L}_X S) \otimes T + S \otimes \mathcal{L}_X T, \qquad (A.1-25)$$

by extending the following starting definitions:[3]

$$\mathcal{L}_X f = X f = df(X) \qquad \text{for } f \in C^\infty(M), \qquad (A.1-26)$$

$$\mathcal{L}_X Y = [X, Y] \qquad \text{for } Y \in \mathcal{X}(M). \qquad (A.1-27)$$

From the definitions, one can verify the relations among them given by

$$\mathcal{L}_X = d\, i_X + i_X\, d, \qquad (A.1-28)$$

$$\mathcal{L}_{[X,Y]} = [\mathcal{L}_X, \mathcal{L}_Y], \qquad (A.1-29)$$

$$i_{[X,Y]} = [\mathcal{L}_X, i_Y]. \qquad (A.1-30)$$

e) Hodge star operation

Although we omit its precise definition here, the notion of the integral of differential forms (see, e.g., [deR 60]) is important in its relations, for instance,

---

[2]   Some physicists call this property *inverse of Poincaré lemma* in a confusing way.

[3]   A more intrinsic definition can be given by introducing the notion of (one-parameter family of) *diffeomorphisms generated by a vector field* X, which we omit here, however.

with **Stokes theorem** equating the integration, $\int_N d\eta$, of an r-coboundary $d\eta$ over an r-dimensional submanifold N with $\int_{\partial N} \eta$ over a **boundary** ∂N of N defined in the context of **homology theory**:

$$\int_N d\eta = \int_{\partial N} \eta. \qquad (A.1-31)$$

On the basis of these, **de Rham theorem** exhibits the dual relation between the homology and the cohomology. To facilitate further the analysis of cohomology, the inner product structure[4] of the space of differential forms is important; an inner product can be defined for r-forms α and β by

$$\langle \alpha, \beta \rangle \equiv \int_M *\alpha \wedge \beta = \int_M \alpha \wedge *\beta. \qquad (A.1-32)$$

Here $*$ is the **Hodge star operation** transforming r-forms into $(n-r)$-forms through the formula

$$
\begin{aligned}
&* \left(dx^{\mu_1} \wedge \cdots \wedge dx^{\mu_r} \right) \\
&= \frac{\sqrt{|g|}}{(n-r)!} \epsilon^{\mu_1 \cdots \mu_r, \mu_{r+1} \cdots \mu_n} g_{\mu_{r+1}\nu_{r+1}} \cdots g_{\mu_n \nu_n} dx^{\nu_1} \wedge \cdots \wedge dx^{\nu_n},
\end{aligned}
\qquad (A.1-33)
$$

where $g_{\mu\nu}$ is a (pseudo-)Riemannian metric and g is its determinant: $g = \det g_{\mu\nu}$. The totally antisymmetric unit symbol $\epsilon_{\mu_1 \cdots \mu_r \mu_{r+1} \cdots \mu_n}$ with *lower indices* is defined as the signature of a permutation, and the one with *upper indices* is found to be

$$\epsilon^{\mu_1 \cdot \ \mu_r \mu_{r+1} \cdots \mu_n} = g^{-1} \mathrm{sgn} \begin{pmatrix} 1 & \cdots & r & r+1 & \cdots & n \\ \mu_1 & \cdots & \mu_r & \mu_{r+1} & \cdots & \mu_n \end{pmatrix}. \qquad (A.1-34)$$

Operating the Hodge star twice on an r-form φ, we find

$$*(*\varphi) = (|g|/g)(-1)^{r(n-r)}\varphi = \mathrm{sgn}(g)(-1)^{r(n-r)}\varphi. \qquad (A.1-35)$$

On the basis of the inner product given by Eq.(32) and of the Hodge star, such important notions as the Laplace-Beltrami operator and harmonic forms can be defined. If M is a *compact* manifold without boundary and with (*positive definite*)

---

[4]   We first consider it in the sense of a symmetric bilinear form $\langle \cdot, \cdot \rangle$ in a *real* vector space By extending the coefficients from **R** to **C** and by inserting suitable complex conjugates, it can easily be extended to the *complex* vector space equipped with a *sesquilinear* form.

Riemannian metric, the inner product, Eq.(32), defines a (*positive definite*) Hilbert space structure, which leads to the Hodge theorem giving the analytic expression to cocycles in terms of harmonic forms. However, since the restrictions of *compactness* and *positivity* of a Riemannian metric are too stringent for us, we refrain from pursuing this topic further.

f) Double complex

Our final topic is the notion of a **double complex** equipped with *two* coboundary operators. Let $\{C^{r,s}\}$ be a cochain complex with two gradations, r and s, corresponding to mutually anticommuting coboundary operators, d_1 and d_2:

$$d_1^2 = d_2^2 = d_1 d_2 + d_2 d_1 = 0. \qquad (A.1-36)$$

They act on $C^{r,s}$ in such a way that d_1 increases r by 1 and d_2 does s by 1:

$$d_1 : C^{r,s} \to C^{r+1,s}, \qquad (A.1-37)$$

$$d_2 : C^{r,s} \to C^{r,s+1}. \qquad (A.1-38)$$

One encounters a typical example of double complexes in complex geometry, with $d_1 \equiv d'$ and $d_2 \equiv d''$ given by

$$d'(f_{i_1\cdots i_r \bar{j}_1\cdots\bar{j}_s} dz^{i_1} \wedge \cdots \wedge dz^{i_r} \wedge d\bar{z}^{j_1} \wedge \cdots \wedge d\bar{z}^{j_s})$$
$$= \frac{\partial f_{i_1\cdots i_r \bar{j}_1\cdots\bar{j}_s}}{\partial z^i} dz^i \wedge dz^{i_1} \wedge \cdots \wedge dz^{i_r} \wedge d\bar{z}^{j_1} \wedge \cdots \wedge d\bar{z}^{j_s}, \quad (A.1-39)$$
$$d''(f_{i_1\cdots i_r \bar{j}_1\cdots\bar{j}_s} dz^{i_1} \wedge \cdots \wedge dz^{i_r} \wedge d\bar{z}^{j_1} \wedge \cdots \wedge d\bar{z}^{j_s})$$
$$= \frac{\partial f_{i_1\cdots i_r \bar{j}_1\cdots\bar{j}_s}}{\partial \bar{z}^j} d\bar{z}^j \wedge dz^{i_1} \wedge \cdots \wedge dz^{i_r} \wedge d\bar{z}^{j_1} \wedge \cdots \wedge d\bar{z}^{j_s}, \quad (A.1-40)$$

Note here $d''f = 0$ for a function f is just the Cauchy-Riemann equation as the condition for f to be *holomorphic*.

From a given double complex, several differential complexes can be derived, such as $\{H^r(C^{\bullet,s}, d_1), d_2\}$ and $\{H^s(C^{r,\bullet}, d_2), d_1\}$, the mutual relationships of which provide interesting cohomological information. Here, we discuss the notion of **total complex** $\{D^t\}$ of this double complex $\{C^{r,s}\}$: It is defined by

$$D^t \equiv \bigoplus_{r,s:r+s=t} C^{r,s}, \qquad (A.1-41)$$

which constitutes also a differential complex equipped with a coboundary operator Δ acting on a t-form $\varphi = (\varphi^{t,0}, \varphi^{t-1,1}, \cdots, \varphi^{0,t}) \in D^t$ by the formula

$$\Delta\varphi \equiv (d_1\varphi^{t,0},\ d_1\varphi^{t-1,1} + d_2\varphi^{t,0}, \cdots,\ d_1\varphi^{0,t} + d_2\varphi^{1,t-1},\ d_2\varphi^{0,t}). \qquad (A.1-42)$$

If the complexes

$$0 \to C^{0,s} \xrightarrow{d_1} C^{1,s} \xrightarrow{d_1} C^{2,s} \xrightarrow{d_1} \cdots \xrightarrow{d_1} C^{n,s} \xrightarrow{d_1} 0, \qquad (A.1-43)$$

are *acyclic* for *all* s, with augmentations $\epsilon_s : A^s \longrightarrow C^{0,s}$,

$$0 \to A^s \xrightarrow{\epsilon_s} C^{0,s} \xrightarrow{d_1} C^{1,s} \xrightarrow{d_1} C^{2,s} \xrightarrow{d_1} \cdots \xrightarrow{d_1} C^{n,s} \xrightarrow{d_1} 0 : \text{ exact}, \qquad (A.1-44)$$

and if $\{A^s\}$ constitutes also a differential complex with coboundary operators δ (called a *bordered cochain complex*),

$$0 \to A^0 \xrightarrow{\delta} A^1 \xrightarrow{\delta} A^2 \xrightarrow{\delta} A^3 \xrightarrow{\delta} \cdots \xrightarrow{\delta} A^s \xrightarrow{\delta} \cdots : \text{ exact}, \qquad (A.1-44')$$

then the following isomorphism between the cohomology groups can be shown to hold:

$$H^*(\{D\}, \Delta) \simeq H^*(\{A\}, \delta). \qquad (A.1-45)$$

(We omit the proof which is not so difficult).

A.2 General postulates of relativistic quantum field theory with an indefinite metric and their consequences

In the following three sections, $A.2$, $A.3$ and $A.4$, of this Appendix, we summarize some useful consequences obtained in the axiomatic approach to relativistic quantum field theory [Jos 63, Stre 64, Bog 75, Völ 77], and examine how and to what extent they can be generalized to the cases with indefinite inner product.

The state vector space $\mathcal{V}$ of the indefinite-metric quantum field theory (except for quantum gravity) is required to satisfy the standard postulates of quantum field theory apart from the positivity assumption of the inner product, namely,

(0) Principles of the quantum theory (*apart from the positivity*),

(i) Poincaré covariance,

(ii) Spectrum condition,

(iii) Local (anti)commutativity.

In connection with the postulate (0), the space $\mathcal{V}$ is required to be a *topological vector space* with an *(indefinite) inner product* $\langle \cdot | \cdot \rangle$ *separately continuous* with respect to its topology. If this inner product $\langle \cdot | \cdot \rangle$ is degenerate, namely, if there exists some non-zero vector $|\omega\rangle \in \mathcal{V}$ *orthogonal* to the whole $\mathcal{V}$,

$$\langle \Psi | \omega \rangle = 0 \qquad \text{for } \forall |\Psi\rangle \in \mathcal{V}, \qquad (A.2-1)$$

such a vector $|\omega\rangle$ as the above has no physical effects (which are to be described in terms of inner product), and hence is an irrelevant object according to the principles of the quantum theory. So, we can assume, without loss of generality, the inner product $\langle \cdot | \cdot \rangle$ is **non-degenerate**, namely, a vector $|\omega\rangle$ satisfying the condition, Eq.(1), is nothing but the zero vector:

$$\langle \Psi | \omega \rangle = 0 \quad \text{for } \forall |\Psi\rangle \in \mathcal{V} \quad \Rightarrow \quad |\omega\rangle = 0. \qquad (A.2-2)$$

On this assumption, Eq.(2), of non-degeneracy, any zero-norm vector ('neutral vector' in the mathematical terminology [Bogn 74]) $|\chi\rangle$, orthogonal to itself

$$\langle \chi | \chi \rangle = 0, \qquad (A.2-3)$$

should have some vector $|\Psi\rangle$ *not* orthogonal to it:

$$\exists |\Psi\rangle \in \mathcal{V} \text{ such that } \langle \Psi | \chi \rangle \neq 0. \qquad (A.2-4)$$

It may be instructive to note that the coexistence of the above two conditions, Eqs.(3) and (4), necessarily implies the indefiniteness of the inner product $\langle \cdot | \cdot \rangle$ [Bogn 74]:

$$\langle \Phi | \Phi \rangle \gtreqless 0 \quad \text{in } \mathcal{V}.$$

While the inner product $\langle \cdot | \cdot \rangle$ is assumed to be non-degenerate in the *whole* $\mathcal{V}$, it *may be degenerate in subspaces* of $\mathcal{V}$. Indeed, we have the following result.

Lemma $A.2 - 1$ Let $\mathcal{W}$ be a *positive semi-definite* subspace of $\mathcal{V}$. Then, every zero-norm vector in $\mathcal{W}$ is orthogonal to $\mathcal{W}$:

$$\mathcal{W}_0 \equiv \{ |\chi\rangle \in \mathcal{W}; \ \langle \chi | \chi \rangle = 0 \ \} \quad \perp \ \mathcal{W}. \qquad (A.2-5)$$

The same conclusion as Eq.(5) holds when $\mathcal{W}$ is a negative semi-definite subspace.

Proof: By the semi-definiteness of the inner product in $\mathcal{W}$, the Cauchy-Schwarz inequality holds in $\mathcal{W}$:

$$|\langle \Phi | \Psi \rangle| \leq |\langle \Phi | \Phi \rangle|^{1/2} |\langle \Psi | \Psi \rangle|^{1/2}, \qquad (A.2-6)$$

from which Eq.(5) follows.   (q.e.d.)

Thus, the zero-norm subspace $\mathcal{V}_0$ of the physical subspace $\mathcal{V}_{\text{phys}}$, Eq.(3.4.3-2), is orthogonal to the whole $\mathcal{V}_{\text{phys}}$:

$$\mathcal{V}_0 \ \perp \ \mathcal{V}_{\text{phys}}, \qquad (A.2-7)$$

because $\mathcal{V}_{\text{phys}}$ is positive semi-definite as is proved in Sec.4.1. The above Eqs.(3) and (4) tell us that for any zero-norm physical state $|\chi\rangle \in \mathcal{V}_0$ there is some unphysical state $|\Psi\rangle \notin \mathcal{V}_{\text{phys}}$ not orthogonal to it:

$$\langle \Psi | \chi \rangle \neq 0 \qquad \text{for } \forall |\chi\rangle \in \mathcal{V}_0 \text{ and } \exists |\Psi\rangle \notin \mathcal{V}_{\text{phys}}. \qquad (A.2-8)$$

By Eq.(8), we know that the subspaces $\mathcal{V}_0$ and $\mathcal{V}_{\text{phys}}$ have *no orthogonal projections* onto themselves.

For ensuring the consistency of physical interpretations in scattering processes, the following theorem plays an indispensable role.

Theorem $A.2 - 2$ If the following three conditions are satisfied for the Hamiltonian H and the physical subspace $\mathcal{V}_{\text{phys}}$ of the total state vector space $\mathcal{V}$ with an indefinite inner product $\langle \cdot | \cdot \rangle$:

(0) the hermiticity of Hamiltonian H, $H^\dagger = H$, is satisfied so that the total S-matrix S exists in $\mathcal{V}$ satisfying the (pseudo)unitarity with respect to the indefinite inner product $\langle \cdot | \cdot \rangle$,

$$\langle S\Phi | S\Psi \rangle = \langle \Phi | \Psi \rangle, \qquad (A.2-9)$$

(i) $\mathcal{V}_{phys}$ is invariant under the time development, [or (i') $S\mathcal{V}_{phys} = S^{-1}\mathcal{V}_{phys} = \mathcal{V}_{phys}$]

(ii) the inner product is positive semi-definiteness in $\mathcal{V}_{phys}$,

then the physical S-matrix S_{phys} can be defined consistently in the (completed)[1] quotient space given by

$$H_{phys} \equiv \overline{\mathcal{V}_{phys}/\mathcal{V}_0}. \qquad (A.2-10)$$

H_{phys} is a *Hilbert space* with *positive definite* inner product, and S_{phys} is unitary with respect to this Hilbert space structure:

$$S_{phys}{}^\dagger S_{phys} = S_{phys} S_{phys}{}^\dagger = \mathbf{1}. \qquad (A.2-11)$$

Proof: Because of the orthogonality, Eq.(7), of the zero-norm subspace $\mathcal{V}_0$ to every vector in $\mathcal{V}_{phys}$, two state vectors in $\mathcal{V}_{phys}$, $|\Phi\rangle$ and $|\Phi\rangle + |\chi\rangle$, with $|\chi\rangle \in \mathcal{V}_0$, cannot be distinguished physically; the difference $|\chi\rangle$ of them has no effect on any inner product in $\mathcal{V}_{phys}$. Then, by virtue of (ii), the (completion of) quotient space Eq.(10), $H_{phys} = \overline{\mathcal{V}_{phys}/\mathcal{V}_0}$, of $\mathcal{V}_{phys}$ with respect to $\mathcal{V}_0$ becomes a Hilbert space, equipped with positive definite inner product defined by

$$\langle \hat{\Phi} | \hat{\Psi} \rangle = \langle \Phi | \Psi \rangle \qquad (A.2-12)$$

for $|\hat{\Phi}\rangle = |\Phi\rangle + \mathcal{V}_0, |\hat{\Psi}\rangle = |\Psi\rangle + \mathcal{V}_0$. Next, the condition (i') [which should be derivable from (i)] allows us to define the physical S-matrix S_{phys} in H_{phys} by the equation:

$$S_{phys}|\hat{\Phi}\rangle \equiv (S|\Phi\rangle)\hat{} \qquad \text{for} \qquad |\hat{\Phi}\rangle = |\Phi\rangle + \mathcal{V}_0 \in \mathcal{V}_{phys}/\mathcal{V}_0. \qquad (A.2-13)$$

One can easily check the unitarity of S_{phys}, Eq.(11), with the aid of the condition (0). In fact, e.g.,

$$\langle \hat{\Psi} | S_{phys}{}^\dagger S_{phys} | \hat{\Phi} \rangle = \langle \hat{S\Psi} | \hat{S\Phi} \rangle = \langle S\Psi | S\Phi \rangle = \langle \Psi | S^\dagger S | \Phi \rangle = \langle \Psi | \Phi \rangle = \langle \hat{\Psi} | \hat{\Phi} \rangle.$$

$$\text{(q.e.d.)}$$

---

[1] $\overline{V}$ in Eq.(10) denotes the completion of space V to incorporate all the limiting states of Cauchy sequences in V.

The next problem is the topology of $\mathcal{V}$, which has been assumed to make the inner product $\langle \cdot | \cdot \rangle$ separately continuous. Such a topology is called a *partial majorant* [Bogn 74]. From various points of view, it is natural and convenient to impose additionally on the topology τ the requirement that it should be an **admissible topology** [Bogn 74]: A topology τ is said to be *admissible* if the condition for a linear functional φ on $\mathcal{V}$ to be continuous with respect to the topology τ is equivalent to the existence of such a vector $|\Phi\rangle \in \mathcal{V}$ that

$$\varphi(\Psi) = \langle \Phi | \Psi \rangle. \qquad (A.2-14)$$

The weakest of the admissible topologies is known to be the weak topology (w) [Bogn 74]. As is familiar in the cases of the Hilbert space, the following lemma holds for the admissible topology:

Lemma $A.2-3$ Let $\mathcal{W}$ be an arbitrary subspace of $\mathcal{V}$. Then, the equality

$$\mathcal{W}^{\perp\perp} = \overline{\mathcal{W}}^{\tau} \qquad (A.2-15)$$

holds for any admissible topology τ of $\mathcal{V}$, where $\mathcal{W}^{\perp}$ is defined by

$$\mathcal{W}^{\perp} \equiv \{ |\Phi\rangle \in \mathcal{V}; \ \langle \Phi | \Psi \rangle = 0 \ \text{ for } \forall |\Psi\rangle \in \mathcal{W} \} \qquad (A.2-16)$$

and $\overline{\mathcal{W}}^{\tau}$ is the closure of $\mathcal{W}$ with respect to the topology τ.

(For the proof, see [Bogn 74].)

Corollary $A.2-4$ Let $\mathcal{W}$ be any dense subspace of $\mathcal{V}$ with respect to an admissible topology τ:

$$\overline{\mathcal{W}}^{\tau} = \mathcal{V}. \qquad (A.2-17)$$

Then, there exists no non-trivial vector orthogonal to $\mathcal{W}$:

$$\mathcal{W}^{\perp} = (\overline{\mathcal{W}}^{\tau})^{\perp} = \mathcal{V}^{\perp} = 0. \qquad (A.2-18)$$

Proof: The first equality is due to the equality

$$\mathcal{W}^{\perp} = \mathcal{W}^{\perp\perp\perp} \qquad (A.2-19)$$

and Eq.(15). The second one is due to the assumption, Eq.(17).   (q.e.d.)

As for the postulate (i) of the Poincaré covariance, we assume such standard ingredients as the unitary representation $U(a, L)$ of the Poincaré group and as the fields $\varphi_i(x)$ covariant under the Poincaré transformations, and so on [Jos 63, Stre 64, Bog 75]. In some respects, however, the precise formulation is possible only in certain formal sense, for lack of positivity (at the present stage of development of the mathematical theory of indefinite inner product spaces). For example, the unitarity of $U(a, L)$ means the *unitarity with respect to the indefinite inner product*, which, contrary to the unitarity with respect to the positive definite inner product, does *not* necessarily imply that the operator $U(a, L)$ is bounded, *nor* that $U(a, \mathbf{1})$ can be written in the form

$$U(a, \mathbf{1}) \equiv U(a) = \exp(iP_\mu a^\mu) = \int e^{ip_\mu a^\mu} dE(p). \qquad (A.2-20)$$

Since no correspondent of the spectral resolution theorem,[2] valid in the Hilbert space, has been proved yet in the general context of indefinite inner product spaces, the precise meaning of such an expression as $\exp(iP_\mu a^\mu)$ is not so clear. Thus, the relation between the energy-momentum operator P_μ and the translation operator $U(a)$ is a rather symbolic one. These situations may seem to endanger the postulate (ii) of the spectrum condition, which can, however, be formulated in the following form without any difficulties.

First, let us recall that the fields $\varphi_i(x)$ are not operators by themselves but operator-valued (tempered) distributions which become operators by smearing with test functions:

$$\varphi_i(f) = \int d^4x \varphi_i(x) f(x). \qquad (A.2-21)$$

Precisely speaking, we should have a *common dense domain* Ω in $\mathcal{V}$ of any operator of such a form as in Eq.(21) with $f \in \mathcal{S}(\mathbf{R}^4)$ [3] and of any $U(a, L)$, stable under these operators $\varphi_i(f)$ and $U(a, L)$, where the term "dense" means "dense in any admissible topology τ of $\mathcal{V}$", namely,

$$\overline{\Omega}^\tau = \Omega^{\perp\perp} = \mathcal{V}. \qquad (A.2-22)$$

---

[2]   See SNAG theorem in [Stre 64]

[3] $\mathcal{S}$ here and $\mathcal{D}$ below represent the spaces of test functions decreasing rapidly in $\mathbf{R}^4$ and of those having compact supports, respectively. See a textbook of distributions or [Stre 64].

Further, the linear functional

$$f \;\mapsto\; \langle \Phi | \varphi_\iota(f) | \Psi \rangle \qquad\qquad (A.2-23)$$

should be continuous for any $|\Phi\rangle, |\Psi\rangle \in \mathcal{V}$ with respect to the topology of $\mathcal{S}(\mathbf{R}^4)$. We denote, by $\mathcal{F}$ and $\mathcal{F}(\mathcal{O})$ with $\mathcal{O}$ being a (bounded) open set $\subseteq \mathbf{R}^4$, the polynomial algebras generated by the operators of the form

$$\int d^4 x_1 \cdots d^4 x_r \varphi_{\iota_1}(x_1) \cdots \varphi_{\iota_r}(x_r) f(x_1, \cdots, x_r) \qquad\qquad (A.2-24)$$

with $f \in \mathcal{S}(\mathbf{R}^{4r})$ and with $f \in \mathcal{D}(\overbrace{\mathcal{O} \times \cdots \times \mathcal{O}}^{r})$ $(r = 0, 1, 2, \cdots)$, respectively. In the case that $\mathcal{O} \subset \mathbf{R}^4$ is a finite spacetime region, we call an element of $\mathcal{F}(\mathcal{O})$ a **smeared local operator** taking account of the postulate (iii) of local (anti)commutativity.

Here we add a further postulate (iv).

(iv) Existence of the unique vacuum and its cyclicity : There exists a *unique* state vector $|0\rangle$ (**vacuum**) invariant under any translations,

$$U(a)|0\rangle = |0\rangle \qquad\qquad (A.2-25)$$

or equivalently

$$P_\mu |0\rangle = 0, \qquad\qquad (A.2-26)$$

which is *cyclic* with respect to $\mathcal{F}$:

$$\mathcal{V} = \overline{\mathcal{F}|0\rangle}^{\,r}. \qquad\qquad (A.2-27)$$

Now, the spectrum condition is postulated in the form [Str 77] as

$$\int d^4 a\, e^{-\imath p_\mu a^\mu} \langle \Phi | U(a) | \Psi \rangle = 0, \qquad \text{if } p \notin \overline{V}_+ \equiv \{q \in \mathbf{R}^4;\; q_0 \geq 0,\; q^2 \geq 0\} \qquad (A.2-28)$$

for any $|\Phi\rangle, |\Psi\rangle \in \mathcal{F}|0\rangle$, or equivalently as

$$\int d^4 \xi_1 \cdots d^4 \xi_{r-1}\, e^{\imath(q_1 \xi_1 + \cdots + q_{r-1}\xi_{r-1})} W_{\iota_1 \cdots \iota_r}(\xi_1, \cdots, \xi_{r-1})$$

$$\equiv \tilde{W}_{\iota_1 \cdots \iota_r}(q_1, \cdots, q_{r-1}) = 0 \qquad \text{if } \exists q_j \notin \overline{V}_+, \qquad (A.2-29)$$

where $W_{i_1 \cdots i_r}(\xi_1, \cdots, \xi_{r-1})$ is defined by

$$\langle 0| \varphi_{i_1}(x_1) \cdots \varphi_{i_r}(x_r) |0\rangle \equiv W_{i_1 \cdots i_r}(x_1 - x_2, \cdots, x_{r-1} - x_r), \qquad (A.2-30)$$

on the basis of the translational invariance of the vacuum, Eq.(25). According to the well-known techniques [Jos 63, Stre 64, Bog 75], Eq.(29) combined with the postulates (i) and (iii) allows us to continue $W_{i_1 \cdots i_r}(\xi_1, \cdots, \xi_{r-1})$ analytically to the complex *analytic function* $W_{i_1 \cdots i_r}(\zeta_1, \cdots, \zeta_{r-1})$ in the permuted extended tube, in much the same way as the cases with positive definite inner product. This analyticity property furnishes us with powerful techniques, the well-known one of which is the following theorem [Ree 61]:

<u>Theorem $A.2-5$</u> (**Reeh-Schlieder theorem**)   For any open set $\mathcal{O}$ in spacetime, the equality

$$\overline{\mathcal{F}(\mathcal{O})|0\rangle}^\tau = \overline{\mathcal{F}|0\rangle}^\tau \qquad (A.2-31)$$

holds for any admissible topology τ. On the assumption (iv) of the cyclicity of the vacuum, Eq.(27), we obtain

$$\overline{\mathcal{F}(\mathcal{O})|0\rangle}^\tau = \mathcal{V} \qquad (A.2-32)$$

or

$$(\mathcal{F}(\mathcal{O})|0\rangle)^\perp = (\mathcal{F}|0\rangle)^\perp = 0. \qquad (A.2-33)$$

Proof: The "edge-of-the-wedge" theorem (see, e.g. [Stre 64]) implies that the equality

$$\langle \Psi | \varphi_{i_1}(x_1) \cdots \varphi_{i_r}(x_r) |0\rangle = 0 \qquad (A.2-34)$$

holds if

$$\langle \Psi | \int d^4 x_1 \cdots d^4 x_r f(x_1, \cdots, x_r) \varphi_{i_1}(x_1) \cdots \varphi_{i_r}(x_r) |0\rangle = 0 \qquad (A.2-35)$$

for $f \in \mathcal{D}(\overbrace{\mathcal{O} \times \cdots \times \mathcal{O}}^{r})$; namely,

$$(\mathcal{F}(\mathcal{O})|0\rangle)^\perp = (\mathcal{F}|0\rangle)^\perp. \qquad (A.2-36)$$

By virtue of Lemma $A.2$-3, Eq.(36) tells us

$$\overline{\mathcal{F}(\mathcal{O})|0\rangle}^\tau = (\mathcal{F}(\mathcal{O})|0\rangle)^{\perp\perp} = (\mathcal{F}|0\rangle)^{\perp\perp} = \overline{\mathcal{F}|0\rangle}^\tau. \qquad \text{(q.e.d.)}$$

Combining the above theorem with the postulate (iii) of the *local (anti)commutativity*, we obtain the following corollary [**separating property of the vacuum** with respect to (smeared) local fields].

Corollary $A.2 - 6$ If $\mathcal{O}$ is an open set of spacetime whose causal complement $\mathcal{O}'$, defined by

$$\mathcal{O}' \equiv \text{the interior of the set } \{x \in \mathbf{R}^4 \ ; \ (x-y)^2 < 0 \ \text{ for } \forall y \in \mathcal{O}\}, \quad (A.2 - 37)$$

is not empty, and if $\varphi \in \mathcal{F}(\mathcal{O})$, then

$$\varphi|0\rangle = 0 \qquad\qquad\qquad (A.2 - 38)$$

implies $\varphi = 0$ on the assumption of (iv). In particular, since $\mathcal{O}'$ for a bounded open set $\mathcal{O}$ is not empty, any (smeared) *local* operator $\varphi \in \mathcal{F}(\mathcal{O})$ annihilating the vacuum is vanishing in itself.

Proof: It is sufficient to consider the case with fields satisfying the local commutativity, since the cases containing both local commutative and anticommutative fields can be treated in a similar manner with slight modifications. From Eq.(38) and the local commutativity, we obtain

$$0 = \langle\Psi|\psi\varphi|0\rangle = \langle\Psi|\varphi\psi|0\rangle \qquad\qquad (A.2 - 39)$$

for any $|\Psi\rangle \in \Omega$ and any $\psi \in \mathcal{F}(\mathcal{O}')$. Thus we have

$$\varphi^\dagger|\Psi\rangle \in \mathcal{F}(\mathcal{O}')|0\rangle^\perp = 0, \qquad\qquad (A.2 - 40)$$

by virtue of Theorem A.2-5:

$$\overline{\mathcal{F}(\mathcal{O}')^\tau} = \mathcal{V}, \qquad\qquad (A.2 - 41)$$

and of Corollary A.2-4. Then, Eq.(40) implies

$$\langle\Psi|\varphi|\Phi\rangle = \langle\varphi^\dagger\Psi|\Phi\rangle = 0 \qquad \text{for } \forall|\Phi\rangle \in \Omega, \qquad (A.2 - 42)$$

which says

$$\varphi|\Phi\rangle = 0 \qquad \text{for } \forall|\Phi\rangle \in \Omega \qquad\qquad (A.2 - 43)$$

or

$$\varphi = 0, \qquad\qquad\qquad (A.2 - 44)$$

because of Ω being dense in $\mathcal{V}$, Eq.(22), and of Corollary $A.2$-4.   (q.e.d.)

Next, we comment on the postulate (iv) of the cyclicity of the vacuum, which is nothing but a natural requirement that every state in a field theory should be described in terms of fields. In the cases with *positive-definite* inner product, the following three conditions are known to be equivalent with one another in relativistic quantum field theory satisfying the assumptions (i), (ii) and (iii) [Rue 62, Ara 62] (see also [Jos 63, Stre 64, Bog 75]) :

(iv) uniqueness and cyclicity of the vacuum,

(iv a) **irreducibility** of the field algebra $\mathcal{F}$,

(iv b) **cluster property**.

The relation between the cluster property (iv b) and the uniqueness of the vacuum can be understood more properly in the context of (non-commutative) ergodic theory (see, for instance, [Brat 79]), with the aid of the equality

$$(w\text{-}) \lim_{|\boldsymbol{X}|\to\infty} U(x^0, \boldsymbol{x}) = (w\text{-}) \lim_{|\boldsymbol{X}|\to\infty} \int_{p^2\geq 0, p_0\geq 0} e^{\imath(p_0 x^0 - \boldsymbol{p}\cdot\boldsymbol{x})} dE(p_0, \boldsymbol{p}) = E(p_\mu = 0),$$
$$(A.2-45)$$

following from the Riemann-Lebesgue lemma. Here $E(p_\mu = 0)$ is the projection onto the subspace of *vacua* or vectors invariant under the spacetime translations, in terms of which the uniqueness condition of the vacuum can be expressed as

$$\text{rank } E(p_\mu = 0) = 1 \quad \text{or} \quad E(p_\mu = 0) = |0\rangle\langle 0|. \qquad (A.2-46)$$

As seen in Sec.4.3.4 in connection with the quark confinement, the cluster property does *not* follow from the uniqueness of the vacuum in the indefinite-metric theory.

To see how the irreducibility (iv a) follows from (iv), we take an operator C *commuting* with all the field operators $\mathcal{R}$,

$$[U(x)CU(x)^\dagger,\ \mathcal{R}] = U(x)[C,\ U(-x)\mathcal{R}U(-x)^\dagger]U(x)^\dagger = 0, \qquad (A.2-47)$$

from which the following equality holds:

$$\begin{aligned}
\langle 0|CU(-x)\mathcal{R}|0\rangle &= \langle 0|(U(x)CU(x)^\dagger)\mathcal{R}|0\rangle \\
&= \langle 0|\mathcal{R}(U(x)CU(x)^\dagger)|0\rangle \\
&= \langle 0|\mathcal{R}U(x)C|0\rangle.
\end{aligned} \qquad (A.2-48)$$

The Fourier transform of Eq.(48) and the spectrum condition tell us that its support in p-space is concentrated at the point $p_\mu = 0$. Therefore, the Fourier transform of the operator $U(x)CU(x)^\dagger$ should be proportional to $\delta(p)$ without involving any derivatives of $\delta(p)$, *if C is a bounded operator in a positive-definite* Hilbert space. Thus, the state $C|0\rangle$ is invariant under the translations, and hence, it is required to be proportional to the vacuum $|0\rangle$ by the uniqueness condition of the vacuum: $C|0\rangle = \langle 0|C|0\rangle|0\rangle$. Because of the commutativity, Eq.(47), the cyclicity of the vacuum with respect to the field operators concludes the equality $C = \langle 0|C|0\rangle \mathbf{1}$. Namely, the irreducibility of the totality of field operators follows from the cyclicity and the uniqueness of the vacuum. On the other hand, if the inner product is *indefinite*, some derivatives of $\delta(p)$ can appear in the above argument, and hence, we cannot derive the irreducibility (iv a) from (iv) in this case.

Furthermore, the notion of the irreducibility in the indefinite-metric theory involves some subtle points: Namely, in the indefinite inner product spaces, it splits into the two *different* notions of "**subspace irreducibility**" and "**operator irreducibility**" (see [Gel 66]). The former means the absence of non-trivial (closed) subspace invariant under the operation of field operators, and the latter means the triviality of operators commuting with all the field operators. In the positive-definite Hilbert space, these two notions become equivalent to each other via **Schur lemma** with the aid of the spectral decomposition and of the one-to-one correspondence between (closed) subspaces and *orthogonal projections*, which do not necessarily hold in the indefinite-metric cases. While the notion of subspace irreducibility is more appropriate from the viewpoint of *representation* theory, that of operator irreducibility is more convenient for treating field operators in an algebraic way. This can be particularly said in the context of dealing with field operators in terms of *asymptotic fields* on the assumption of asymptotic completeness. In Appendix A.4, we will see that the cyclicity of the vacuum and the asymptotic completeness imply the validity of operator irreducibility, which, in turn, plays an important role in determining the field equations for asymptotic fields as is seen in Secs.2.4 and 4.2.

Now, we discuss the implication, (iv a) irreducibility + (iii) (weak) locality $\Rightarrow$ (iv) uniqueness of the vacuum. Again, the more general and appropriate setting for discussing this problem is the (non-commutative) ergodic theory (supplemented with *asymptotic abelianness* which generalizes the notion of local commutativity;

see [Brat 79]): Namely, the **ergodicity** and the uniqueness of the invariant vector are equivalent under the condition of the asymptotic abelianness guaranteeing the commutativity between 'infinitely separated' operators. While the ergodicity means in short the irreducibility of the algebra generated by field operators *and* $U(x)$'s, it reduces to the irreducibility of the algebra of field operators in the usual sense, on the assumption of the spectrum condition which makes $U(x)$'s belong to the algebra of field operators [*Arveson-Borchers theorem* (see [Brat 79])]. However, the simplest version of proving the uniqueness of the vacuum from the irreducibility in relativistic quantum field theory may be the proof given by Borchers [Bor 62], **Borchers theorem**. It makes use of the *(weak) locality* property in an algebraically convenient form of PCT symmetry based upon the consequence of a profound theorem, the **PCT theorem** [Jos 57; Jos 63, Stre 64, Bog 75]): From the analyticity of Wightman functions combined with the Lorentz invariance, the PCT invariance of the theory is shown to be equivalent to the *weak local commutativity* which is a weaker condition than the local commutativity,

$$\langle 0|\varphi_{\alpha_1}(x_1)\cdots\varphi_{\alpha_r}(x_r)|0\rangle = (\pm)\langle 0|\varphi_{\alpha_r}(x_r)\cdots\varphi_{\alpha_1}(x_1)|0\rangle,$$
$$\text{if } \left(\sum_{i=1}^{r-1} \lambda_i(x_i - x_{i+1})\right)^2 < 0 \text{ for } \forall\lambda_j \geq 0 \text{ not vanishing simultaneously.}$$

$$(A.2-49)$$

We now examine how this Borchers theorem can be extended to the indefinite-metric cases. However, we note that the PCT theorem presupposes the **spin-statistics theorem** [Jos 63, Stre 64, Bog 75] and that the latter theorem is *not* valid in the general situation with an indefinite inner product which admits the existence of such *scalar fermions* as the FP ghosts, for instance. In case of the Yang-Mills theory, the PCT symmetry holds with a slight modification as has been shown in Sec.3.4.3. Namely, the invalidity of the spin-statistics theorem due to the FP ghosts is harmless except the minor change of their PCT transformation law as in Eq.(3.4.3-16). From this fact and the reconstruction theorem [Wig 56, Yng 77] valid also in cases with indefinite inner product, we can safely assert the existence of the antiunitary PCT operator Θ defined (at least in Ω) by

$$\Theta|0\rangle = |0\rangle, \qquad\qquad\qquad (A.2-50)$$
$$\Theta\Phi_{i_1}(x_1)\cdots\Phi_{i_r}(x_r)|0\rangle = \Phi^{PCT}{}_{i_1}(x_1)\cdots\Phi^{PCT}{}_{i_r}(x_r)|0\rangle, \quad (A.2-51)$$

and satisfying

$$\Theta^2 = 1 \quad (\text{in } \Omega), \qquad\qquad (A.2-52)$$

$$\Theta\Phi_{i}(x)\Theta = \Phi^{PCT}{}_{i}(x). \qquad\qquad (A.2-53)$$

Using this fact, we obtain the following *modified* Borchers theorem concluding the uniqueness of the vacuum (in a rather restricted sense) from the irreducibility. Here we need to mean by "irreducibility" the validity of *both* of the above two kinds of irreducibilities, i.e., subspace and operator irreducibilities.

Proposition $A.2-7$ [Kug 79d]   If the field algebra $\mathcal{F}$ is both subspace and operator irreducible, there exists no such other vacuum[4] $|0'\rangle$ (in Ω) linearly independent of $|0\rangle$ that

$$U(x)|0'\rangle = |0'\rangle \quad \text{for } \forall x \in \mathbf{R}^4 \qquad\qquad (A.2-54)$$

$$\Theta|0'\rangle = e^{i\omega}|0'\rangle \quad (\omega \in \mathbf{R}^4). \qquad\qquad (A.2-55)$$

Proof: If there exists a vacuum $|0'\rangle$ not proportional to $|0\rangle$, then we can construct another vacuum $|0\rangle_{\alpha\beta}$

$$|0\rangle_{\alpha\beta} \equiv \alpha|0\rangle + \beta|0'\rangle, \qquad\qquad (A.2-56)$$

which is normalized by a suitable choice of complex numbers α, β, and is cyclic on the assumption of the subspace irreducibility. Then, according to the (modified) PCT theorem in the Yang-Mills theory, the locality and the spectrum condition of theory with the vacuum $|0\rangle_{\alpha\beta}$ allow us to construct the PCT operator $\Theta_{\alpha\beta}$ referring to this vacuum $|0\rangle_{\alpha\beta}$:

$$\Theta_{\alpha\beta}|0\rangle_{\alpha\beta} = |0\rangle_{\alpha\beta}, \qquad\qquad (A.2-57)$$

$$\Theta_{\alpha\beta}\Phi_{i_1}(x_1)\cdots\Phi_{i_r}(x_r)|0\rangle_{\alpha\beta} = \Phi^{PCT}{}_{i_1}(x_1)\cdots\Phi^{PCT}{}_{i_r}(x_r)|0\rangle_{\alpha\beta},$$
$$(A.2-58)$$

$$\Theta_{\alpha\beta}(\lambda|\Psi\rangle + \mu|\Phi\rangle) = \lambda^*\Theta_{\alpha\beta}|\Psi\rangle + \mu^*\Theta_{\alpha\beta}|\Phi\rangle, \qquad\qquad (A.2-59)$$

$$\Theta_{\alpha\beta}{}^2 = 1 \quad (\text{in } \Omega), \qquad\qquad (A.2-60)$$

$$\Theta_{\alpha\beta}\Phi_{i}(x)\Theta_{\alpha\beta} = \Phi^{PCT}{}_{i}(x). \qquad\qquad (A.2-61)$$

---

[4]   In this context, "vacuum" means just a translationally invariant (normalizable) state. In the cases with positive inner product, we need not require the condition in Eq.(55), since the condition $\Theta^2 = 1$ enables one to find mutually orthogonal vectors invariant under Θ. In the case of Yang-Mills theory, it is satisfied by the states $|0'\rangle \equiv Q|0\rangle$ with $Q = gQ^a, G^a, N^a$, etc., discussed in Sec.4.4, because of Eq.(3.4 3-22).

By Eqs.(53) and (61), one can easily check the commutativity of $\Theta\Theta_{\alpha\beta}$ with every $\Phi_i(x)$. On the other hand, $\Theta\Theta_{\alpha\beta}|0\rangle$, i.e.,

$$\Theta\Theta_{\alpha\beta}|0\rangle_{\alpha\beta} = \Theta|0\rangle_{\alpha\beta} = \alpha^*|0\rangle + \beta^* e^{i\omega}|0'\rangle, \qquad (A.2-62)$$

can be made not proportional to $|0\rangle_{\alpha\beta}$, given by Eq.(56), by choosing α and β such that

$$\alpha^*/\alpha \neq \beta^* e^{i\omega}/\beta. \qquad (A.2-63)$$

Then, $\Theta\Theta_{\alpha\beta}$ is not a c-number operator, whereas it commutes with every $\Phi_i(x)$. This contradicts the assumption of the operator irreducibility. (q.e.d.)

A.3 Symmetries, currents and charges –Goldstone theorem–

As is well known, the **spontaneous symmetry breaking (SSB)** is closely related with the problem of uniqueness of the vacuum. The occurrence of SSB means the following situation: If we choose a formulation with a unique vacuum (i.e., choice of an irreducible representation of the field algebra), then the symmetry cannot be implemented by a unitary transformation in the state vector space with the symmetry generator being *ill defined*; on the other hand, if one wants to respect the symmetry, then the representation of field algebra becomes *reducible* with *degenerate vacua*. In the indefinite-metric theory, however, the relation among SSB, vacuum uniqueness and irreducibility is not so simple: First, in spite of its conceptual naturality, the validity of subspace irreducibility (see Appendix A.2) is very hard to be checked, without which, however, the modified Borchers theorem in Appendix A.2 cannot guarantee the uniqueness of the vacuum. Second, as remarked in Sec.2.4.3, the argument concluding the *ill-definedness* of the spontaneously broken symmetry generator in Sec.1.5.1 may be invalidated by the indefinte-metric: In Secs.2.4.3 and 2.5.2, the possibility of the non-uniqueness of the vacuum is pointed out by utilizing the freedom to create translationally invariant states belonging to the *zero-norm* physical subspace $\mathcal{V}_0$. We note here the simple fact that what is physically meaningful is only the uniqueness of the vacuum *in the physical Hilbert space* H_{phys}, i.e.,

$$\hat{P}_\mu|\hat{\Phi}\rangle = 0 \iff \exists \alpha \in \mathbf{C} \text{ such that } |\hat{\Phi}\rangle = \alpha|\hat{0}\rangle, \qquad (A.3-1)$$

which is equivalently written for $|\Phi\rangle \in \mathcal{V}_{\text{phys}}$ as

$$P_\mu|\Phi\rangle \in \mathcal{V}_0 \iff \exists \alpha \in \mathbf{C} \; \exists|\chi\rangle \in \mathcal{V}_0 \text{ such that } |\Phi\rangle = \alpha|0\rangle + |\chi\rangle. \qquad (A.3-2)$$

Therefore, it is necessary to relax the meaning of the *uniqueness* of the vacuum (iv) into

$$P_\mu|\Phi\rangle = 0 \Rightarrow \exists \alpha \in \mathbf{C} \; \exists|\chi\rangle \in \mathcal{V}_0 \text{ such that } |\Phi\rangle = \alpha|0\rangle + |\chi\rangle. \qquad (A.3-3)$$

From this point of view, it is not sufficient to characterize the SSB by a condition $Q|0\rangle \neq 0$, because the spontaneously broken charge Q may *or* may not be a well-defined quantity. We need to distinguish SSB into two or three classes according to its effect on the physical world H_{phys}:

i) hidden SSB

  i a) SSB of an *unphysical* symmetry transforming a physical state into an unphysical one [irrelevant to H_{phys}],

  i b) SSB with a well-defined charge transforming a physical state$\in \mathcal{V}_{\text{phys}}$ into a *zero-norm physical* state$\in \mathcal{V}_0$ [inducing identity transformation in H_{phys}],

ii) manifest SSB: SSB of a *physical* symmetry transforming a physical observable acting in $\mathcal{V}_{\text{phys}}$ [see Sec.4.3] into another observable one; it induces the usual type SSB in H_{phys} having inevitably an *ill-defined* charge according to the argument in Sec.1.5.1.

For the unified treatment of these different cases, we need to introduce the **Goldstone commutator** and to examine whether it vanishes or not:

$$\langle 0|[iQ, \varphi(x)]|0\rangle \neq 0. \qquad (A.3-4)$$

The existence of a field operator $\varphi(x)$ with non-vanishing Goldstone commutator just indicates the non-invariance of the vacuum, $Q|0\rangle \neq 0$. A more appropriate formulation can be attained by introducing a 'local charge' Q_R, or the algebraic transformation generated by it on the local field algebras which is meaningful *independently* of the action of the symmetry transformation on a state vector space (the latter being dependent on the *choice of a specific representation* of local field algebras). While the *global* charge Q given as the total volume integral, $Q = \int d\mathbf{x} j_0(x)$, of the density current $j_\mu(x)$, may not be well-defined, a 'local charge' Q_R always exists and generates the transformation 'locally' [Kas 66]:

$$Q_R \equiv \int d^4x j_0(x_0, \mathbf{x})\alpha_T(x_0)f_R(\mathbf{x}), \qquad (A.3-5)$$

$$\frac{d}{ds}\langle \Phi|\tau_s(\varphi)|\Psi\rangle|_{s=0} = \langle \Phi|[iQ_R, \varphi]|\Psi\rangle \qquad \text{for } \varphi \in \mathcal{F}(\mathcal{O}). \qquad (A.3-6)$$

In the above, τ_s is a (one-parameter group of the) symmetry transformation of field operators and $\alpha_T \in \mathcal{D}(\mathbf{R}), f_R \in \mathcal{D}(\mathbf{R}^3)$ are such test functions that

$$\int dx_0 \alpha_T(x_0) = 1, \qquad (A.3-7)$$

$$f_R(\mathbf{x}) = \begin{cases} 1 & (|\mathbf{x}| \leq R), \\ 0 & (|\mathbf{x}| > 2R). \end{cases} \qquad (A.3-8)$$

Eq.(6) holds for any sufficiently large $R > 0$ (depending on $\mathcal{O}$) and independently of the choice of α_T as a consequence of the locality and of the conservation law:

$\partial^\mu j_\mu = 0$ (see [Kas 66, Orz 70]). Now, the intuitive expression Eq.(4) for the spontaneous breakdown of the symmetry τ_s can properly be replaced by the condition

$$\lim_{R\to\infty} \langle 0|[iQ_R,\varphi]|0\rangle = \frac{d}{ds}\langle 0|\tau_s(\varphi)|0\rangle|_{s=0} \neq 0 \qquad \exists \varphi \in \mathcal{F}(\mathcal{O}). \qquad (A.3-9)$$

If the local charge Q_R or its limit, $\lim_{R\to\infty} Q_R$, cannot be so chosen that they belong to the local *observable algebra* $\mathcal{A}(\mathcal{O})$ or the global one $\mathcal{A}$ mapping a physical state to another physical one, the non-vanishing Goldstone commutator Eq.(9) corresponds to the case i a) in the above classification. If $Q_R \in \mathcal{A}(\mathcal{O})$ and if any possible field operator $\varphi(x)$ in Eq.(9) with non-vanishing Goldstone commutator does not belong to $\mathcal{A}$, the case i b) is realized. The examples discussed in Secs. 2.4.3 and 2.5.2 belong to this case. The case ii) is realized in the case where both Q_R and $\varphi(x)$ can be made to be a (smeared) local observable.

Now, the implication of Eq.(9) is well known as the **Goldstone theorem** [Gol 62, Kas 66, Eza 67, Ree 68] which asserts the existence of a massless NG boson. In the neatest form, this theorem can be stated as a corollary of the following equation [Mai 72, Völ 74, Völ 77]

$$\begin{aligned}
\lim_{R\to\infty} \langle 0|[Q_R,\varphi]|0\rangle &= \lim_{R\to\infty} \left(\langle 0|Q_R E_1\varphi|0\rangle - \langle 0|\varphi E_1 Q_R|0\rangle\right) \\
&= 2\lim_{R\to\infty} \langle 0|Q_R E_1\varphi|0\rangle \\
&= 2\lim_{R\to\infty} \langle 0|Q_R\varphi|0\rangle,
\end{aligned} \qquad (A.3-10)$$

where E_1 is the projection onto the states $p^2 = 0$. Eq.(10) has been proved only for the positive metric cases, and it is not known that it can be extended to the indefinite metric cases. To our knowledge, there exists the following statement for these cases made by Strocchi [Str 77]:

Theorem $A.3-1$ (Goldstone theorem)   Spontaneous breakdown of the symmetry, Eq.(9), occurs if and only if the Fourier transform of $\langle 0|[j_\mu(x),\varphi]|0\rangle$ contains a $\delta(p^2)$ singularity.

(For the proof, see [Str 77].)

Because of this theorem, it is quite reasonable that if the projection E_1 of the states of mass zero in the indefinite-metric cases can be defined in such a form as

$$E_1 = \sum_{\imath,\jmath} a_{\imath}{}^\dagger|0\rangle\eta^{-1}{}_{\imath\jmath}\langle 0|a_{\jmath}, \qquad (A.3-11)$$

Eq.(10) will be verified in these cases. Here, $a_\imath{}^\dagger$ and $a_\jmath$ denote the creation and annihilation operators of massless asymptotic fields and $\eta^{-1}{}_{\imath\jmath}$ is the inverse of the metric matrix $\eta_{\imath\jmath} = [a_\imath, \ a_\jmath{}^\dagger]_\mp$. In any case, we obtain the following criterion:

Corollary $A.3 - 2$ The criterion for the spontaneous breakdown of the symmetry is given by

$$\langle 0|j_\mu|\Psi\rangle \neq 0 \qquad \text{for } \exists|\Psi\rangle : \text{ massless one-particle state.} \qquad (A.3 - 12)$$

Taking account of the fact that the weak topology is the (weakest) admissible topology, we can verify the following:

Corollary $A.3 - 3$ The necessary and sufficient condition for the global charge Q given by

$$\langle \Phi|Q|\Psi\rangle = \lim_{R\to\infty} \langle \Phi|Q_R|\Psi\rangle \qquad (A.3 - 13)$$

to be a charge without suffering from any spontaneous symmetry breaking defined on the dense domain $\mathcal{F}(\mathcal{O})|0\rangle$ is that one of the following conditions is valid:

(i) $\displaystyle\lim_{R\to\infty} \langle 0|[Q_R, \ \varphi]|0\rangle = 0 \qquad \text{for } \forall\varphi \in \mathcal{F},$ $\qquad (A.3 - 14)$

(ii) $\displaystyle\lim_{R\to\infty} \langle 0|\varphi Q_R|0\rangle = 0 \qquad \text{for } \forall\varphi \in \mathcal{F}(\mathcal{O}),$ $\qquad (A.3 - 15)$

(iii) $\langle 0|j_\mu|\Psi(p^2 = 0)\rangle = \langle 0|j^{as}{}_\mu|\Psi(p^2 = 0)\rangle = 0,$ $\qquad (A.3 - 16)$

(iv) $Q|0\rangle = 0,$ $\qquad (A.3 - 17)$

(v) $Q\varphi|0\rangle = [Q_R, \ \varphi]|0\rangle \qquad \text{for } \forall\varphi \in \mathcal{F}, \ \exists R_0 > 0, \forall R > R_0.$ $(A.3 - 18)$

In Eq.(16) where the Yang-Feldman equation [Eq.$(A.4\text{-}2)$] is used, $j^{as}{}_\mu$ is the asymptotic field corresponding to j_μ.

(Proof is omitted.)

A.4 Asymptotic fields, asymptotic states and their behavior under the symmetry transformation

We have discussed asymptotic fields and asymptotic states for an ordinary scalar field in Secs.1.4.2-1.4.4. In this appendix, we discuss these notions in the general framework including gauge theories.

In the theory with a positive definite inner product and with mass gap, the notions of asymptotic states and asymptotic fields have their sound basis in the **Haag-Ruelle scattering theory** [Haa 58, Rue 62]. Since the existence of a massless field[1] and of an indefinite inner product obstructs the extension of this theory to the general case, we cannot but take a naive attitude to interpret the asymptotic fields and states as the representatives of the poles in p^2 of time-ordered Green's functions in p-space. Namely, the second postulate (ii) in Appendix A.2 is formulated here in the following form.

(ii′) Characterization of asymptotic fields and spectrum condition for them: Corresponding to each discrete spectrum of $P^\mu P_\mu$ appearing as a pole at $p^2 = m_\iota{}^2 (\geq 0)$ of time-orderd Green's functions in momentum space, an asymptotic field $\varphi^{\mathrm{as}}{}_\iota$ (as=in or out) satisfying

$$(\Box + m_\iota{}^2)^{r_\iota} \varphi^{\mathrm{as}}{}_\iota(x) = 0 \qquad (A.4-1)$$

with a positive integer r_ι is assumed to exist.

As for the relation between the asymptotic fields and the original Heisenberg fields, we assume the validity of the **Yang-Feldman equation** [Yan 50], which gives an expression for the asymptotic fields in terms of Heisenberg fields: For instance, in the case with $r_\iota = 1$, it is written as

$$\varphi^{\mathrm{in/out}}{}_\iota(x) = \Phi_\iota(x) - \int d^4 y \Delta_{\mathrm{R/A}}(x - y; m_\iota{}^2) j_\iota(y) \qquad (A.4-2)$$

with a Heisenberg field $\Phi_\iota(x)$ and with its source $j_\iota(x) = (\Box + m_\iota{}^2) \Phi_\iota(x)$ containing no discrete spectrum at $p^2 = m_\iota{}^2$. From this, we obtain a convenient equation

$$\langle 0 | \Phi_\iota(x) | \Psi(p^2 = m_\iota{}^2) \rangle = \langle 0 | \varphi^{\mathrm{as}}{}_\iota(x) | \Psi(p^2 = m_\iota{}^2) \rangle. \qquad (A.4-3)$$

---

[1]   As for an extension of the Haag-Ruelle theory to a certain type of massless theory with positive inner product, see [Buc 75, Buc 77].

In the cases with bound states and with multipole-ghosts ($r_i \geq 2$), due modifications to Eq.(2) are necessary [Zim 58, Nak 78a]. The Haag-Ruelle scattering theory valid in the cases with positive definite inner product and with mass gap tells us that all asymptotic fields are mutually (anti)local [Hep 66]. In the present case with indefinite inner product and without mass gap, such proof is not available. So, we need to assume

(iii′) Locality of asymptotic fields: All asymptotic fields are mutually (anti)local.

[Note that, according to the general results on *Borchers class* [Bor 60], the asymptotic fields which are *free*, can*not*, of course, be (anti)local with the *interacting* Heisenberg fields, unless the S-matrix is trivial.]

The correspondent of the postulate (iv) of the cyclicity of the vacuum in this case means nothing but the assumption of asymptotic completeness.

(iv′) Asymptotic completeness: The vacuum $|0\rangle$ is cyclic with respect to the totality of asymptotic fields: $\mathcal{V} = \mathcal{V}^{\mathrm{in}} = \mathcal{V}^{\mathrm{out}}$.

As an important general consequence of the assumptions (ii′) and of (iii′) combined with the Poincaré covariance (valid also for the asymptotic fields owing to Eq.(2)), the following **Greenberg-Robinson theorem** holds including the indefinite-metric cases.

Theorem $A.4 - 1$ [Dell 61, Gre 62, Rob 62]   If a local field $\varphi(x)$ in the spacetime dimensions $D \geq 3$ satisfying the Poincaré covariance, spectrum condition, locality and cyclicity of the vacuum has a Fourier transform $\tilde{\varphi}(p)$ vanishing in a spacelike open domain in p-space, then $\varphi(x)$ is a **generalized free field**, namely, the commutator $[\varphi(x), \varphi(y)]$ is a c-number in the whole spacetime.

The essence of the proof of this theorem consists in applying to a Wightman function in p-space Gårding's technique for analyzing the support property of its Fourier transform by means of the control over the wave propagation of a solution to a wave equation in $(D + 1)$-dimensional p-space [Wig 67]. Although the assumption of *cyclicity* of the vacuum plays a crucial role, the positivity of the Hilbert-space metric does not enter here, and hence, the theorem holds even in the indefinite-metric quantum field theory. (For the detail of the proof, see the references [Gre 62, Rob 62, Nak 78a].)

Remark: As shown in [Nak 78a], this theorem *breaks down* in the *two-dimensional* spacetime *if* the translation generator P_μ contains the massless spectrum; in this case, a counter-example can be found explicitly, which is closely related to the current algebra models.

Applying the Greenberg-Robinson theorem to the asymptotic fields characterized by (ii'), (iii') and (iv'), we obtain the following corollary.

Corollary $A.4-2$ The (anti)commutator $[\varphi^{as}{}_i(x),\ \varphi^{as}{}_j(y)]_{\mp}$ is a c-number.

Thus, the space of asymptotic states is a *Fock space* of asymptotic fields, and hence, for the asymptotic fields, the cyclicity (iv') of the vacuum implies the *irreducibility* of the asymptotic fields. This is an immediate consequence of the following generalization to the indefinite-metric case of the expansion formula [Haa 55, Gla 57] of arbitrary operator $\mathcal{R}$ in terms of the asymptotic fields, which is valid on the assumption (iv'):

$$\mathcal{R} = \sum_{n=0}^{\infty} \frac{i^n}{n!} \sum_{i_1 \cdots i_n} \sum_{j_1 \cdots j_n} \int (\prod_{a=1}^{n} d\mathbf{x}_a) \langle 0|[\cdots[L,\ \varphi^{as}{}_{i_1}(x_1)],\cdots,\varphi^{as}{}_{i_n}(x_n)]|0\rangle$$

$$\times (\eta^{-1})_{i_1 j_1} \cdots (\eta^{-1})_{i_n j_n} \overset{\leftrightarrow}{\partial}_0{}^{x_1} \cdots \overset{\leftrightarrow}{\partial}_0{}^{x_n} : \varphi^{as}{}_{j_n}(x_n) \cdots, \varphi^{as}{}_{j_1}(x_1) : .$$

$$(A.4-4)$$

Here, $: \cdots :$ means the Wick normal product and $\overset{\leftrightarrow}{\partial}_0 \equiv \overset{\rightarrow}{\partial}_0 - \overset{\leftarrow}{\partial}_0$. For simplicity, we have indicated in Eq.(4) the generalization to the situation with indefinite inner product only in such a case that the indefiniteness is due to that of the (non-singular) matrix η_{ij} appearing in such a commutation relation of $\varphi^{as}{}_i$ as

$$[\varphi^{as}{}_i(x),\ \varphi^{as}{}_j(y)] = i\eta_{ij} \Delta(x-y; m_i{}^2), \qquad (A.4-5)$$

where η_{ij} is supposed to vanish for $m_i \neq m_j$. From Eq.(4), it follows trivially that any operator $\mathcal{R}$ commuting with all the asymptotic fields $\varphi^{as}{}_i$ is nothing but a c-number: $\mathcal{R} = \langle 0|\mathcal{R}|0\rangle \mathbf{1}$, namely the totality of asymptotic fields is irreducible. This fact implies further the following

Proposition $A.4-3$ The assumption (iv') of asymptotic completeness implies the irreducibility of the Heisenberg field algebra $\mathcal{F}$.

Proof: Let $\mathcal{R}$ be any operator commuting with all the Heisenberg fields, then the Yang-Feldman equation Eq.(2) tells us that $\mathcal{R}$ commutes with all the asymptotic fields. Since the totality of asymptotic fields is irreducible on the assumption (iv') of asymptotic completeness, $\mathcal{R}$ is nothing but a c-number. (q.e.d.)

If we take $\mathcal{R}$ as a Heisenberg field $\Phi_\iota(x)$, the expansion formula, Eq.(4), gives us an expression for the Heisenberg fields in terms of the asymptotic fields. Taking account of the LSZ reduction formula [Leh 55] (see Sec.1.4.4), we can rewrite the above formula as follows:

$$S\mathcal{R} = \sum_{n=0}^{\infty} \frac{i^n}{n!} \sum_{i_1 \cdots i_n} \sum_{J_1 \cdots J_n} \int (\prod_{a=1}^{n} d^4 x_a) : \varphi^{\text{in}}{}_{J_n}(x_n) \cdots , \varphi^{\text{in}}{}_{J_1}(x_1) :$$

$$\times \left(\prod_{b=1}^{n} (\eta^{-1})_{i_b J_b} (\Box^{x_b} + m_{J_b}{}^2) \right) \langle 0 | T \mathcal{R} \Phi_{J_n}(x_n) \cdots \Phi_{J_1}(x_1) | 0 \rangle$$

$$\equiv: \exp((\varphi^{\text{in}})^T \eta^{-1} K \delta/\delta J) : \langle 0 | T(\mathcal{R} \exp i J^T \Phi) | 0 \rangle|_{J=0}$$

$$\equiv: \mathcal{K} : \langle 0 | T(\mathcal{R} \exp i J^T \Phi) | 0 \rangle; \qquad (A.4-6)$$

$$\mathcal{R}S = S^{-1}(S\mathcal{R})S = S^{-1} : \exp((\varphi^{\text{in}})^T \eta^{-1} K \delta/\delta J) : \langle 0 | T(\mathcal{R} \exp i J^T \Phi) | 0 \rangle|_{J=0} S$$

$$=: \exp((\varphi^{\text{out}})^T \eta^{-1} K \delta/\delta J) : \langle 0 | T(\mathcal{R} \exp i J^T \Phi) | 0 \rangle|_{J=0}. \qquad (A.4-7)$$

where K is a matrix consisting of Klein-Gordon operators: $K \equiv (K_{\iota J}) = (\delta_{\iota J}(\Box + m_\iota{}^2))$. For φ^{in} and J viewed as column vectors (matrices with *one column*), the superfix T in $(\varphi^{\text{in}})^T$ and J^T means the *transpose*, i.e., to replace them with the corresponding row vectors. We call this formula, Eq.(6), the (generalized) **Glaser-Lehmann-Zimmermann (GLZ) formula**, which holds for any polynomial $\mathcal{R}(\in \mathcal{F})$ of local (Heisenberg) operators. By setting $\mathcal{R} = \mathbf{1}$ in Eq.(6), the S-matrix S is given by

$$S =: \exp((\varphi^{\text{in}})^T \eta^{-1} K \delta/\delta J) : \langle 0 | T(\exp i J^T \Phi) | 0 \rangle|_{J=0}$$

$$=: \mathcal{K} : \langle 0 | T \exp i J^T \Phi | 0 \rangle. \qquad (A.4-8)$$

[See Eq.(1.4.4-14).] These GLZ formulae are useful in the discussion about the behavior of the asymptotic fields under the symmetry transformation.

First, we note that every asymptotic field is transformed *linearly* under *any* unbroken (nonlinear in general) transformation of an internal symmetry.

Theorem $A.4-4$ [Sen 67, Nak 78a, Kug 79a] Let $\delta\varphi^{\text{as}}{}_\iota$ be the infinitesimal transform of the asymptotic field $\varphi^{\text{as}}{}_\iota$, by the symmetry transformation generated by an unbroken charge Q:

$$[iQ, \varphi^{\text{as}}{}_\iota(x)]_{\mp} = \delta\varphi^{\text{as}}{}_\iota(x). \qquad (A.4-9)$$

Then, $\delta\varphi^{\text{as}}{}_\iota(x)$ depends *linearly* upon the asymptotic fields $\varphi^{\text{as}}{}_J(x)$:

$$\delta\varphi^{\text{as}}{}_\iota(x) = a_{\iota J} \varphi^{\text{as}}{}_J(x). \qquad (A.4-10)$$

The coefficients a_{ij} may contain finite-order differential operators in ∂_μ.

Proof: We first note that the WT identity

$$\langle 0|\mathrm{T}J^T\lambda\delta\Phi\exp iJ^T\Phi|0\rangle = \langle 0|[i\lambda Q, \mathrm{T}\exp iJ^T\Phi]|0\rangle = 0, \qquad (A.4-11)$$

follows from the unbroken symmetry generated by Q:

$$[i\lambda Q, \Phi_i] = \lambda[iQ, \Phi_i]_\mp = \lambda\delta\Phi_i. \qquad (A.4-12)$$

Here we have inserted a parameter λ of "infinitesimal transformation" for the sake of notational convenience to treat the ordinary and "super-type" charges on the same footing as done in Sec.3.4.2. Differentiating Eq.(11) with respect to $J_k(x)$, we obtain

$$\langle 0|\mathrm{T}(\delta\Phi_k(x) + i\Phi_k(x)J^T\lambda\delta\Phi)\exp iJ^T\Phi|0\rangle = 0. \qquad (A.4-13)$$

Next, these WT identities, Eqs.(11) and (13), with the operation $: \mathcal{K} :$ defined in Eq.(6) can be simplified as follows. When the operator $: (\varphi^{\mathrm{in}})^T\eta^{-1}K\delta/\delta J :$ is applied to Eqs.(11) and (13), the external sources J_i are replaced by the Klein-Gordon operator K with a coefficient of an on-shell quantity φ^{in}: $J_i \to (\varphi^{\mathrm{in}\,T}\eta^{-1}K)_i$. Because of the presence of the operator $(\varphi^{\mathrm{in}\,T}\eta^{-1}K)_i$ instead of $J_i(x)$, the fields $\delta\Phi_i(x)$, which generally contain non-linear terms of fields also, can be replaced by the *linear* combinations of fields having the *same mass* m_i:

$$\delta\Phi_i(x) \underset{\mathrm{on-shell}}{\Longrightarrow} \lambda a_{ij}\Phi_j(x), \qquad (A.4-14)$$

$$a_{ij} = \int d^4z\langle 0|\mathrm{T}\delta\Phi_i(x)\Phi_k(z)|0\rangle\langle 0|\mathrm{T}\Phi_k(z)\Phi_j(y)|0\rangle^{-1}|_{\mathrm{on-shell}}$$

$$= i\langle 0|\mathrm{T}\delta\Phi_i(x)\Phi_k(y)|0\rangle \overleftarrow{K}_{m_k}(y)(\eta^{-1})_{kj}|_{\mathrm{on-shell}}. \qquad (A.4-15)$$

By this replacement, Eqs.(11) and (13) lead to the "on-shell WT identities":

$$: \mathcal{K} : \langle 0|\mathrm{T}J^T\lambda a\Phi\exp iJ^T\Phi|0\rangle = 0, \qquad (A.4-16)$$

$$: \mathcal{K} : \langle 0|\mathrm{T}(\delta\Phi_k(x) + iJ^T\lambda a\Phi \cdot \Phi_k(x))\exp iJ^T\Phi|0\rangle = 0. \qquad (A.4-17)$$

where a denotes the matrix (a_{ij}) commuting with $K = (K_{ij})$:

$$Ka = aK. \qquad (A.4-18)$$

Eqs.(16) and (17) are further rewritten as

$$: \mathcal{K} \cdot (\varphi^{\text{in}})^T : \eta^{-1} K \lambda a \langle 0|T\Phi \text{exp} i J^T \Phi|0\rangle = 0, \qquad (A.4-19)$$

$$: \mathcal{K} : \langle 0|T\lambda \delta \Phi_k(x) \text{exp} i J^T \Phi|0\rangle = -i : \mathcal{K} \cdot (\varphi^{\text{in}})^T : \eta^{-1} K \lambda a$$
$$\times \langle 0|T\Phi \cdot \Phi_k(x) \text{exp} i J^T \Phi|0\rangle, \qquad (A.4-20)$$

Now, we can determine the form of $\delta\varphi^{\text{in}}{}_{\iota}(x)$. Since Q is the unbroken conserved charge, we should have

$$0 = [i\lambda Q, \ S] =: [i\lambda Q, \ (\varphi^{\text{in}})^T] \eta^{-1} K (\delta/\delta J) \mathcal{K} : \langle 0|T \text{ exp} i J^T \Phi|0\rangle$$
$$=: \mathcal{K} \cdot \lambda \delta(\varphi^{\text{in}})^T : \eta^{-1} K \langle 0|Ti\Phi \text{exp} i J^T \Phi|0\rangle \qquad (A.4-21)$$

by using Eq.(8). In view of Eq.(18) and of the on-shell WT identity, Eq.(19), we find it sufficient for the validity of Eq.(21) to take $\lambda(\delta\varphi^{\text{in}})^T = -(\varphi^{\text{in}})^T \eta^{-1} a \eta \lambda$, i.e.,

$$\lambda\delta\varphi^{\text{in}}{}_{\jmath}(x) = -\eta_{\jmath\imath} a_{k\jmath} (\eta^{-1})_{lk} \varphi^{\text{in}}{}_{\imath}(x) \lambda. \qquad (A.4-22)$$

Using Eq.(18) and the second on-shell WT identity, Eq.(20), we can further verify that Eq.(22) really reproduces the original transformation law, Eq.(12), of the Heisenberg fields:

$$[i\lambda Q, \ S\Phi_k(x)] = -.: \mathcal{K} \cdot \lambda(\delta\varphi^{\text{in}})^T : \eta^{-1} K \langle 0|Ti\Phi \cdot \Phi_k(x) \text{exp} i J^T \Phi|0\rangle$$
$$=: \mathcal{K} : \langle 0|T\lambda\delta\Phi_k(x) \text{exp} i J^T \Phi|0\rangle$$
$$= S\lambda\delta\Phi_k(x). \qquad (A.4-23)$$

In the above, we have used the GLZ formula, Eq.(6), for $\mathcal{R} = \Phi_k(x)$ and $\mathcal{R} = \lambda\delta\Phi_k(x)$. The commutativity, Eq.(21), of Q and S ensures the equivalence of Eq.(23) with Eq.(12). Finally, to verify Eq.(10), we need to show the equality

$$\lambda\delta\varphi^{\text{in}}{}_{\iota}(x) = -\eta_{\jmath\imath} a_{k\jmath} (\eta^{-1})_{lk} \varphi^{\text{in}}{}_{\imath}(x) \lambda = \lambda a_{\imath\jmath} \varphi^{\text{in}}{}_{\jmath}(x). \qquad (A.4-24)$$

In view of irreducibility of asymptotic fields, it is sufficient to check the validity of this equality in the commutator with an arbitrary asymptotic field $\varphi^{\text{in}}{}_m(y)$. Substituting Eq.(22) into the following Jacobi identity involving $i\lambda Q$, $\varphi^{\text{in}}{}_{\iota}(x)$ and $\varphi^{\text{in}}{}_m(y)$,

$$[\lambda\delta\varphi^{\text{in}}{}_{\iota}(x), \varphi^{\text{in}}{}_m(y)]_{\mp} = [[i\lambda Q, \ \varphi^{\text{in}}{}_{\iota}(x)], \ \varphi^{\text{in}}{}_m(y)]_{\mp}$$
$$= -[\varphi^{\text{in}}{}_{\iota}(x), \ [i\lambda Q, \ \varphi^{\text{in}}{}_m(y)]]_{\mp} - [[\varphi^{\text{in}}{}_{\iota}(x), \ \varphi^{\text{in}}{}_m(y)]_{\mp}, i\lambda Q] = 0$$
$$= -[\varphi^{\text{in}}{}_{\iota}(x), \ [i\lambda Q, \ \varphi^{\text{in}}{}_m(y)]]_{\mp}, \qquad (A.4-25)$$

we can check the desired equality.   (q.e.d.)

At the end of this section, we note that, if the 'infinitesimal' transform $\delta\Phi_,(x) = [iQ,\ \Phi_i(x)]_{\mp}$ with a hermitian charge Q is shown to generate in Green's functions a pole at $p^2 = m^2$ represented by an asymptotic field $(\delta\Phi_,)^{\mathrm{as}}$, then the original Heisenberg field $\Phi_,(x)$ should necessarily have a pole of the *same mass and spin* as $\delta\Phi_,(x)$. Namely, the existence of the discrete spectrum of $\delta\Phi_,(x)$ means the following:

$$0 \neq \langle 0|\delta\Phi_,(x)|\Psi(p^2 = m^2)\rangle = \langle 0|[iQ,\ \Phi_,(x)]_{\mp}|\Psi(p^2 = m^2)\rangle$$
$$= \mp i\langle 0|\Phi_,(x)(Q|\Psi(p^2 = m^2)\rangle),\qquad\qquad (A.4-26)$$

which asserts the existence of the asymptotic field $\varphi^{\mathrm{as}}_,(x)$ of $\Phi_,(x)$ satisfying

$$[iQ,\ \varphi^{\mathrm{as}}_,(x)]_{\mp} = (\delta\Phi)^{\mathrm{as}}_,(x).\qquad\qquad (A.4-27)$$

Since Q is usually Lorentz scalar, we know from Eq.(27) that the mass and spin of $\varphi^{\mathrm{as}}_,$ coincide with those of $(\delta\Phi)^{\mathrm{as}}_,$.

A.5 Properties of dipole functions and wave packet systems

Here some invariant delta functions and wave packets related with the massless dipole ghost are collected.

We begin with the definitions of E-functions (see Eqs.(2.3.4-7) and (2.3.4-15)):

$$E^{(\cdot)}(x) \equiv -\frac{\partial}{\partial m^2}\Delta^{(\cdot)}(x; m^2)|_{m^2=0} \qquad (A.5-1)$$

$$= \frac{1}{2}(\partial^2)^{-1}(x_0\partial_0 - 1)D^{(\cdot)}(x), \qquad (A.5-2)$$

where $E^{(\cdot)}$ denotes E, $E^{(1)}$, $E^{(\pm)}$ and E_F corresponding to $D^{(\cdot)} = D$, $D^{(1)}$, $D^{(\pm)}$ and D_F, respectively. These E-functions, in fact, suffer from infrared divergences except for $E(x)$, and hence one should adopt a suitable infrared cutoff procedure to define them properly [Nak 74a]. The cutoff dependence is, however, harmless in gauge theories as shown in Eq.(2.3.4-16).

First note that the integro-differential operator $\mathcal{E}^{(\omega)}$ defined by

$$\mathcal{E}_x^{(\omega)} = (1/2)(\partial^2)^{-1}(x_0\partial_0 - \omega) \qquad (A.5-3)$$

for arbitrary constant ω works as a non-covariant "inverse" of d'Alembertian $\Box$ in front of any solution $f(x)$ of d'Alembert equation:

$$\Box\mathcal{E}^{(\omega)}f(x) = f(x) \quad \text{if } \Box f(x) = 0. \qquad (A.5-4)$$

Hence, from Eq.(2), the equations

$$\Box E^{(\cdot)}(x) = \Box\mathcal{E}^{(\omega)}D^{(\cdot)}(x) = D^{(\cdot)}(x) \qquad (A.5-5)$$

hold except for $E^{(\cdot)} = E_F$, for which Eq.(5) is valid only when $\omega = 1$. It is an easy task to prove the following useful formulae also from Eq.(2):

$$E^{(\cdot)}(x - y) = \mathcal{E}_{x-y}^{(1)}D^{(\cdot)}(x - y) = (\mathcal{E}_x^{(1/2)} + \mathcal{E}_y^{(1/2)})D^{(\cdot)}(x - y), \quad (A.5-6)$$

$$\partial_\mu{}^x E^{(\cdot)}(x - y) = (\partial_\mu{}^x\mathcal{E}_x^{(1/2)} - \mathcal{E}_y^{(1/2)}\partial_\mu{}^y)D^{(\cdot)}(x - y), \qquad (A.5-7)$$

$$\partial_\mu{}^x\partial_\nu{}^x E^{(\cdot)}(x - y) = (\partial_\mu{}^x\mathcal{E}_x^{(1/2)}\partial_\nu{}^x + \partial_\nu{}^y\mathcal{E}_y^{(1/2)}\partial_\mu{}^y)D^{(\cdot)}(x - y). (A.5-8)$$

Next, we introduce wave packet systems for massless scalar and vector fields. Let $\{g_k\}$ be a complete set of positive frequency solutions of the d'Alembert equation:

$$g_k(x) = \int \frac{d\boldsymbol{p}}{\sqrt{(2\pi)^3 2\hat{p}_0}}\varphi_k(\boldsymbol{p})e^{-i\hat{p}x}, \quad \hat{p}_\mu \equiv (|\boldsymbol{p}|, \boldsymbol{p}), \qquad (A.5-9)$$

where $\varphi_k(\boldsymbol{p})$ should satisfy the following conditions

$$\sum_k \varphi_k(\boldsymbol{p})\varphi_k{}^*(\boldsymbol{q}) = \delta(\boldsymbol{p} - \boldsymbol{q}), \qquad (A.5-10)$$

$$\int d\boldsymbol{p}\varphi_k{}^*(\boldsymbol{p})\varphi_l(\boldsymbol{p}) = \delta_{kl}. \qquad (A.5-11)$$

Then the g_k's satisfy

$$\sum_k g_k(x)g_k^*(y) = D^{(+)}(x - y), \qquad (A.5-12)$$

$$i\int d\boldsymbol{x}g_k^*(x)\overleftrightarrow{\partial}_0 g_l(x) \equiv (g_k, g_l) = \delta_{kl}, \qquad (A.5-13)$$

where

$$f\overleftrightarrow{\partial}_0 g \equiv (f\partial_0 g) - (\partial_0 f)g. \qquad (A.5-14)$$

Using the same $\{\varphi_k(\boldsymbol{p})\}$, we define wave packet system $\{f_{k\sigma}{}^\mu\}$ for the massless vector fields:

$$f_{k,\sigma}{}^\mu(x) = \int \frac{d\boldsymbol{p}}{\sqrt{(2\pi)^3 2\hat{p}_0}}\varphi_k(\boldsymbol{p})\epsilon_\sigma{}^\mu(\boldsymbol{p})e^{-i\hat{p}x}, \qquad (A.5-15)$$

where the polarization vectors $\epsilon_\sigma{}^\mu(\boldsymbol{p})$ $(\sigma = 1, 2, L, S)$ are defined by

$$\left.\begin{array}{l} \boldsymbol{p} \cdot \boldsymbol{\epsilon}_\sigma(\boldsymbol{p}) = 0 \\ \boldsymbol{\epsilon}_\sigma(\boldsymbol{p}) \cdot \boldsymbol{\epsilon}_\tau(\boldsymbol{p}) = \delta_{\sigma\tau} \end{array}\right\} \quad \text{for } \sigma, \tau = 1, 2, \qquad (A.5-16)$$

$$\epsilon_L{}^\mu(\boldsymbol{p}) = -i\hat{p}^\mu = -i(|\boldsymbol{p}|, \boldsymbol{p}), \qquad (A.5-17)$$

$$\epsilon_S{}^\mu(\boldsymbol{p}) = -i\bar{p}^\mu/2|\boldsymbol{p}|^2 \equiv -i(|\boldsymbol{p}|, -\boldsymbol{p})/2|\boldsymbol{p}|^2. \qquad (A.5-18)$$

Defining a 'metric' $\tilde{\eta}^{\sigma\tau}$ by

$$\tilde{\eta} = (\tilde{\eta}^{\sigma\tau}) \equiv \begin{array}{c} \\ 1 \\ 2 \\ L \\ S \end{array}\begin{array}{cccc} 1 & 2 & L & S \\ \begin{pmatrix} -1 & 0 & 0 & 0 \\ 0 & -1 & 0 & 0 \\ 0 & 0 & 0 & 1 \\ 0 & 0 & 1 & 0 \end{pmatrix} \end{array}, \qquad (A.5-19)$$

we introduce 'contravariant' polarization vector $\epsilon^{\sigma,\mu}(\boldsymbol{p})$:

$$\epsilon^{\sigma,\mu}(\boldsymbol{p}) = \sum_{\tau=1,2,L,S} \tilde{\eta}^{\sigma\tau}\epsilon_\tau{}^\mu(\boldsymbol{p}). \qquad (A.5-20)$$

Then one can easily check the relations

$$\sum_\sigma \epsilon_\sigma{}^\mu(\boldsymbol{p})\epsilon^{\sigma,\nu}{}^*(\boldsymbol{p}) = \eta^{\mu\nu}, \qquad (A.5-21)$$

$$\epsilon_\sigma{}^\mu(\boldsymbol{p})\epsilon_{\tau,\mu}{}^*(\boldsymbol{p}) = \tilde{\eta}_{\sigma\tau} \qquad (A.5-22)$$

Then $\{f_{k,\sigma}{}^\mu\}$ satisfy

$$\sum_{k,\sigma} f_{k,\sigma}{}^\mu(x)f_k{}^{\sigma,\nu}{}^*(y) = \eta^{\mu\nu}D^{(+)}(x-y), \qquad (A.5-23)$$

$$(f_{k,\sigma}{}^\mu, f_l{}^\tau{}_\mu) = \delta_\sigma{}^\tau\delta_{kl}. \qquad (A.5-24)$$

By virtue of the common use of $\{\varphi_k(\boldsymbol{p})\}$ both in $\{g_k(x)\}$ and $\{f_{k,\sigma}{}^\mu(x)\}$, we have the following useful relations:

$$f_{k,L}{}^\mu(x) = \partial^\mu g_k(x), \qquad (A.5-25)$$

$$\partial_\mu f_{k,S}{}^\mu(x) = -g_k(x), \qquad (A.5-26)$$

$$\partial_\mu f_{k,\sigma}{}^\mu(x) = 0 \qquad \text{for } \sigma = 1, 2, \text{and } L. \qquad (A.5-27)$$

Now the dipole wave packet system $\{h_k(x)\}$ is introduced with the definition given by

$$h_k(x) = \mathcal{E}_x{}^{(1/2)}g_k(x) = (1/2)(\partial^2)^{-1}(x_0\partial_0 - 1/2)g_k(x). \qquad (A.5-28)$$

This $\{h_k(x)\}$ satisfies

$$\sum_k (h_k(x)g_k{}^*(y) + g_k(x)h_k{}^*(y)) = E^{(+)}(x-y), \qquad (A.5-29)$$

$$(g_k, h_l) + (h_k, g_l) = 0. \qquad (A.5-30)$$

Here Eq.(29) follows directly from Eqs.(6) and (12). Eq.(30) can be proved directly by using the definition Eq.(28), but it would be easier to utilize the identity

$$E^{(+)}(x-y) = i\int dz[D^{(+)}(x-z)\overleftrightarrow{\partial_0}{}^z E^{(+)}(z-y) + E^{(+)}(x-z)\overleftrightarrow{\partial_0}{}^z D^{(+)}(z-y)] \qquad (A.5-31)$$

and the completeness relations Eqs.(12) and (29).

REFERENCES

[Abe 78a]   Abe, M. and Miyake, S., Phys. Lett. **76B**, 76 (1978)   [Sec.2.5.4]

[Abe 78b]   Abe, M. and Miyake, S., Prog. Theor. Phys. **59**, 2037 (1978)
[Sec.2.5.4]

[Abe 85]*   Abe, M. and Nakanishi, N., Prog. Theor. Phys. **74**, 881 (1985) [XX]
[Sec.5.9.1]

[Abe 87]   Abe, M. and Nakanishi, N., Prog. Theor. Phys. **78**, 704 (1987)
[Sec.5.9.4]

[Abe 88a]   Abe, M. and Nakanishi, N., Prog. Theor. Phys. **79**, 227 (1988)
[Sec.5.9.4]

[Abe 88b]   Abe, M. and Nakanishi, N., Prog. Theor. Phys. **79**, 240 (1988)
[Sec.5.9.4]

[Abe 88c]   Abe, M., Zeits. f. Physik C, Particles and Fields, **40**, 403 (1988)
[Sec.5.9.4]

[Abe 88d]   Abe, M. and Nakanishi, N., Prog. Theor. Phys. **80**, 731 (1988)
[Sec.5.9.4]

[Abe 88e]   Abe, M. and Nakanishi, N., Prog. Theor. Phys. **80**, 913 (1988)
[Sec.5.9.4]

[Abe 89a]   Abe, M., Zeits. f. Physik C, Particles and Fields, **41**, 637 (1989)
[Sec.5.9.4]

[Abe 89b]   Abe, M. and Nakanishi, N., Int. J. Mod. Phys. **A4**, 2837 (1989)
[Sec.5.9.4]

[Abe 89c]   Abe, M. and Nakanishi, N., Prog. Theor. Phys. **82**, 420(1989)
[Sec.5.9.4]

[Abe 90]   Abe, M., Int. J. Mod. Phys. **A5**, 3277 (1990)   [Sec.5.9.4]

[Aber 73]   Abers, E.S. and Lee, B.W., Phys. Reports **9**, 1 (1973)   [Sec.3.4.4]

[Adl 82]   Adler, S.L., Rev. Mod. Phys. **54**, 729 (1982)   [Sec.5.1]

*   From here, Abe, M(itsuo) is a person different from Abe, M(itue) in [Abe 78a, b]

[Aha 59]   Aharonov, Y. and Bohm, D., Phys. Rev. **115**, 495 (1959) [Sec.2.1.2]

[Alv 84]   Alvarez-Gaumé, L. and Witten, E., Nucl. Phys. **B234**, 269 (1984)
                [Sec.5.8.4]

[And 63]   Anderson, P.W., Phys. Rev. **130**, 433 (1963)   [Sec.2.4.3]

[Aok 82]   Aoki, K.-I., Hioki, Z., Kawabe, R., Konuma, M. and Muta, T., Prog.
                Theor. Phys. Suppl. No.**73**, 1(1982)   [Sec.4.4.4]

[Ara 62]   Araki, H., Hepp, K. and Ruelle, D., Helv. Phys. Acta **35**, 164 (1962)
                [Sec.4.3.4]

[Ara 69]   Araki, H., in "Local Quantum Theory", ed. by Jost, R. (Academic
                Press, 1969), p.65   [Sec.4.3.4]

[Arat 81]   Aratyn, H., J. Phys. A: Math. Gen. **14**, 1313 (1981)   [Sec.2.5.2]

[Arn 62]   Arnowitt, R., Deser, S. and Misner, C.W., in "Gravitation, An In-
                troduction to Current Research" ed. by Witten, L. (John Wiley and
                Sons, 1962), p.227   [Sec.5.1]

[Aur 72]   Aurilia, A., Takahashi, Y. and Umezawa, H., Prog. Theor. Phys. **48**,
                290 (1972)   [Sec.2.4.5]

[Bau 82]   Baulieu, L. and Thierry-Mieg, J., Nucl. Phys. **B197**, 477 (1982).
                [Sec.3.5.2]

[Bau 84]   Baulieu, L., Nucl. Phys. **B241**, 557 (1984).   [Sec.3.4.3]

[Bau 85]   Baulieu, L., in "Perspectives in Particles and Fields, Cargèse 1983",
                ed. by Lévy, M., et al. (Plenum, 1985), p.1   [Sec.3.4.3]

[Bec 76a]   Becchi, C., Rouet, A. and Stora, R., Ann. Phys. **98**, 287 (1976)
                [Secs.3.1, 3.4.4]

[Bec 76b]   Becchi, C., Rouet, A. and Stora, R., in "Renormalization", NATO
                Advanced Study Institute, 1975, ed. by Velo, G. and Wightman, A.S.
                (D.Reidel Publ. Co., 1976), p.299   [Sec.3.4.1]

[Ber 56]   Bergmann, P.G., Helv. Phys. Acta Suppl. **4**, 79 (1956)   [Sec.5.8.2]

[Bir 82]   Birrell, N.D. and Davies, P.C., "Quantum Fields in Curved Space"
                (Cambridge Univ. Press, 1982)   [Secs.5.1, 5.8.4]

[Ble 50]   Bleuler, K., Helv. Phys. Acta **23**, 567 (1950)   [Sec.2.3]

[Bog 59]   Bogoliubov, N.N. and Shirkov, D.V., "Introduction to the Theory of
                Quantized Fields" (Interscience, 1959), Sec.41-Sec.44   [Add.2.A]

[Bog 75]   Bogoliubov, N.N., Logunov, A.A. and Todorov, L.T., "Introduction
             to Axiomatic Quantum Field Theory" (Benjamin, 1975)   [App.A.2]

[Bogn 74]   Bognár, J., "Indefinite Inner Product Space" (Springer-Verlag, 1974)
                                                                  [App.A.2]

[Bon 81a]   Bonora, L., and Tonin, M., Phys. Lett. **98B**, 48 (1981)   [Sec.4.3.2]

[Bon 81b]   Bonora, L., Pasti, P. and Tonin, M., Nuov. Cim. **64A**, 307 (1981)
                                                          [Secs.3.5.1, 4.1.2]

[Bon 83]   Bonora, L., and Cotta-Ramusino, P., Comm. Math. Phys. **87**, 589
             (1983)   [Secs.3.3.2, 4.3.2]

[Bor 60]   Borchers, H.J., Nuovo Cim. **15**, 784 (1960)   [App.A.4]

[Bor 62]   Borchers, H.J., Nuovo Cim. **24**, 214 (1962)   [App.A.2]

[Bori 74]   Borisov, A.B. and Ogievetskii, V.I., Teor. Mat. Fiz. **21**, 329
             (1974)[transl. Theor. Math. Phys. **21**, 1179 (1974)]   [Sec.5.5.4]

[Bra 74]   Brandt, R.A. and Ng, W.-C., Phys. Rev. **D10**, 4198 (1974)
                                                                  [Sec.2.3.3]

[Brat 79]   Bratteli, O, and Robinson, D.W., "Operator Algebras and Quan-
             tum Statistical Mechanics, I" (Springer-Verlag, 1979), Chapter 4
                                                                  [App.A.2]

[Buc 75]   Buchholz, D., Comm. Math. Phys. **42**, 269 (1975)
                                                        [Add.2.A, App.A.4]

[Buc 77]   Buchholz, D., Comm. Math. Phys. **52**, 147 (1977)
                                                        [Add.2.A, App.A.4]

[Cap 78]   Capper, D.M. and Namazie, M.A., Nucl. Phys. **B142**, 535 (1978)
                                                                  [Sec.5.8.2]

[Cap 85]   Capper, D.M., Dulwich, J.J. and Ramón Medrano, M., Nucl. Phys.
             B254, 737 (1985)   [Sec.5.8.2]

[Chu 65]   Chung, V., Phys. Rev. **140**, B1110 (1965)   [Add.2.A]

[Col 73]   Coleman, S., Comm. Math. Phys. **31**, 259 (1973)   [Sec.2.5.3]

[Col 75]   Coleman, S., Phys. Rev. **D11**, 2088 (1975)   [Sec.2.5.3]

[Cur 76a]   Curci, G. and Ferrari, R., Phys. Lett. **63B**, 91 (1976)
                                                          [Secs.3.5.1, 3.5.2]

[Cur 76b]   Curci, G. and Ferrari, R., Nuovo Cim. **32A**, 151 (1976)
[Secs.3.4.1, 3.5.2]

[Cur 76c]   Curci, G. and Ferrari, R., Nuovo Cim. **35A**, 273 (1976)   [Sec.3.4.3]

[Del 76]   Delbourgo, R. and Ramón Medrano, M., Nucl. Phys. **B110**, 467 (1976)   [Sec.5.3.1]

[Del 82a]   Delbourgo, R. and Jarvis, P.D., J. Phys. A: Math. Gen. **15**, 611 (1982)   [Sec.3.5.2]

[Del 82b]   Delbourgo, R., Jarvis, P.D. and Thompson, G., Phys. Lett. **109B**, 25 (1982)   [Sec.5.9.3]

[Del 82c]   Delbourgo, R., Jarvis, P.D. and Thompson, G., Phys. Rev. **D26**, 775 (1982)   [Sec.5.9.3]

[Dell 61]   Dell'Antonio, G.F., J. Math. Phys. **2**, 759 (1961)   [Sec.3.1, App.*A*.4]

[deR 60]   de Rham, G., "Variétés Différentiables" (Hermann, 1960)   [App.*A*.1]

[DeW 67]   DeWitt, B.S., Phys. Rev. **162**, 1195; 1239(1967)   [Secs.3.1, 3.4.1]

[DeW 75]   DeWitt, B.S., Physics Reports **19**, 295 (1975)   [Sec.5.1]

[Dir 64]   Dirac, P.A.M., "Lectures on Quantum Mechanics" (Belfer Graduate School of Science, Yeshiva Univ., 1964)   [Add.1*A*]

[Dub 67]   Dubin, D.A. and Tarski, J., Ann. Phys. **43**, 263 (1967)   [Sec.2.5.5]

[Duf 81]   Duff, M.J., in "Quantum Gravity 2, A Second Oxford Symposium" ed. by Isham, C.J. et al. (Oxford Univ. Press, 1981), p.81   [Sec.5.1]

[Eic 74]   Eichten, E.J. and Feinberg, F.L., Phys. Rev. **D10**, 3254 (1974)
[Sec.4.4.1]

[Eng 64]   Englert, F. and Brout, R., Phys. Rev. Lett. **13**, 321 (1964)   [Sec.2.4.3]

[Eza 67]   Ezawa, H. and Swieca, J.A., Comm. Math. Phys. **5**, 330 (1967)
[Sec.1.5.1, App.*A*.3]

[Fad 67]   Faddeev, L.D. and Popov, V.N., Phys. Lett. **25B**, 29 (1967)
[Secs.3.1, 3.3.1]

[Fad 69]   Faddeev, L.D., Teor. Mat. Fiz. **1**, 3(1969) [transl. Theor. Math. Phys. **1**, 1 (1970)]   [Secs.3.1, 3.3.1]

[Fad 74]   Faddeev, L.D. and Popov, V.N., Sov. Phys. Usp. **16**, 777 (1974)
[Sec.5.1]

[Fal 84]. Falck, N.K. and Hirschfeld, A.C., Phys. Lett. **138B**, 52 (1984)
[Sec.5.9.3]

[Fer 71] Ferrari, R. and Picasso, L.E., Nucl. Phys. **B31**, 316 (1971) [Sec.2.3.3]

[Fer 77] Ferrari, R., Picasso, L.E. and Strocchi, F., Nuovo Cim. **39A**, 1 (1977)
[Sec.4.3.3]

[Fey 63] Feynman, R.P., Acta Phys. Polon. **24**, 697 (1963)   [Secs.3.1, 3.4.1]

[Foc 64] Fock, V., "The Theory of Space, Time and Gravitation", 2nd
Revised Edition, transl. by Kemmer, N. (Pergamon Press, 1964)
[Sec.5.2.3]

[Fra 70] Fradkin, E.S. and Tyutin, I.V., Phys. Rev. **D2**, 2841 (1970)
[Sec.5.3.2]

[Fra 73] Fradkin, E.S. and Vilkovisky, G.A., Phys. Rev. **D8**, 4241 (1973)
[Sec.5.3.2]

[Fra unp] Fradkin, E.S. and Vilkovisky, G.A., CERN preprint TH. 2332 (1977)
[Sec.5.1]

[Fre 76] Frenkel, J. and Taylor, J.C., Nucl. Phys. **B109**, 439 (1976) [Sec.4.4.2]

[Freu 70] Freundlich, Y. and Lurié, D., Nucl. Phys. **B19**, 557 (1970) [Sec.2.4.5]

[Fri 75] Friedlander, F.G., "The Wave Equation on a Curved Space-Time"
(Cambridge Univ. Press, 1975)   [Sec.5.6.2]

[Fuj 64] Fujii, Y. and Kamefuchi, S., Nuovo Cim. **33**, 1639 (1964) [Sec.2.4.1]

[Fujk 72] Fujikawa, K., Lee, B.W. and Sanda, A.I., Phys. Rev. **D6**, 2923 (1972)
[Sec.3.4.3]

[Fujk 78] Fujikawa, K., Prog. Theor. Phys. **59**, 2045 (1978)   [Sec.4.1.4]

[Gat 83] Gates, S.J., Grisaru, M.T., Roček, M. and Siegel, W., "Superspace
or One Thousand and One Lessons in Supersymmetry" (Benjamin/
Cummings, 1983)   [Sec.5.9.4]

[Gel 66] Gel'fand, I.M. Graev, M.I. and Vilenkin, N.Ya., "Generalized Functions," Vol.5 (Academic Press, 1966) pp.148-150   [App.A.2]

[Gho 72] Ghose, P. and Das, A., Nucl. Phys. **B41**, 299 (1972)   [Sec.2.4.2]

[Gla 57] Glaser, V., Lehmann, H. and Zimmermann, W., Nuovo Cim. **6**, 1122
(1957)   [App.A.4]

[Gol 61] Goldstone, J., Nuovo Cim. **19**, 154 (1961)   [Secs.1.5, 1.5.2]

[Gol 62]   Goldstone, J., Salam, A. and Weinberg, S., Phys. Rev. **127**, 965
             (1962)   [Sec.1.5.1, App.*A*.3]

[Gor 85]   Goroff, M.H. and Sagnotti, A., Phys. Lett. **150B**, 81 (1985)
                                                                    [Sec.5.8.2]

[Gor 86]   Goroff, M.H. and Sagnotti, A., Nucl. Phys. **B266**, 709 (1986)
                                                                    [Sec.5.8.2]

[Got 55]   Goto, T. and Imamura, T., Prog. Theor. Phys. **14**, 396 (1955)
                                                                    [Sec.1.4.1]

[Got 67]   Goto, T. and Obara, T., Prog. Theor. Phys. **38**, 871 (1967) [Sec.2.3]

[Gre 62]   Greenberg, O.W., J. Math. Phys. **3**, 859 (1962)
                                                    [Secs.1.4.2, 3.1, App.*A*.4]

[Grer 80]   Greenberger, D.M. and Overhauser, A.W., Scientific American,
             242(May), 54 (1980)   [Sec.5.1]

[Gro 73]   Gross, D.J. and Wilczek, F., Phys. Rev. **D8**, 3633 (1973) [Sec.4.4.2]

[Gup 50]   Gupta, S.N., Proc. Phys. Soc. **A63**, 681 (1950)   [Sec.2.3]

[Gup 68]   Gupta, S.N., Phys. Rev. **172**, 1303 (1968)   [Sec.5.3.2]

[Gur 64]   Guralnik, G.S., Hagen, C.R. and Kibble, T.W.B., Phys. Rev. Lett.
             13, 585 (1964)   [Sec.2.4.3]

[Haa 55]   Haag, R., Mat. Fys. Medd. Dan. Vid. Selsk. **29**, No.12 (1955)
                                                                    [App.*A*.4]

[Haa 58]   Haag, R., Phys. Rev. **112**, 669 (1958)   [Sec.1.4.3, App.*A*.4]

[Haa 59]   Haag, R., Nuovo Cim. Suppl. **14**, 131 (1959)   [Sec.1.4.3]

[Haa 64]   Haag, R. and Kastler, D., J. Math. Phys. **5**, 848 (1964)   [Sec.4.3.4]

[Haa 75]   Haag, R., Łopuszański, J.T. and Sohnius, M., Nucl. Phys. **B88**, 257
             (1975)   [Secs.5.5.3, 5.9.4]

[Had 79a]   Hadjiivanov, L. and Stoyanov, D.T., Rep. Math. Phys. **15**, 361 (1979)
                                                                    [Sec.2.5.2]

[Had 79b]   Hadjiivanov, L., Mikhov, S.G. and Stoyanov, D.T., J. Phys. A: Math.
             Gen. **12**, 119 (1979)   [Sec.2.5.2]

[Ham 82]   Hamamoto, S., Phys. Lett. **119B**, 331 (1982)   [Sec.5.9.3]

[Ham 83]   Hamamoto, S., Zeits. f. Physik C, Particles and Fields, **19**, 353 (1983)
                                                              [Sec.5.9.2]

[Hat 79]   Hata, H. and Kugo, T., Nucl. Phys. **B158**, 357 (1979)   [Sec.5.9.2]

[Haw 79]   Hawking, S.W., in "General Relativity" ed. by Hawking, S.W. and
             Israel, W. (Cambridge Univ. Press, 1979), p.746   [Sec.5.1]

[Hep 66]   Hepp, K., in "Brandeis University Summer Institute in Theoretical
             Physics 1965", ed. by Cretien, M. and Deser, S. (Gordon and Breach,
             1966), p.1   [App.A.4]

[Hig 64a]   Higgs, P.W., Phys. Lett. **12**, 132 (1964)   [Sec.2.4.3]

[Hig 64b]   Higgs, P.W., Phys. Rev. Lett. **13**, 508 (1964)   [Sec.2.4.3]

[Hig 66]   Higgs, P.W., Phys. Rev. **145**, 1156 (1966)   [Sec.2.4.3]

[Hir 82]   Hirayama, M. and Hirano, N., Phys. Lett. **115B**, 107 (1982)
                                                              [Sec.5.9.3]

[Hirs 81]   Hirschfeld, A.C. and Leschke, H. Phys. Lett. **101B**, 48 (1981)
                                                              [Sec.4.3.2]

[Hoy 83]   Hoyos, J., Quirós, M., Ramirez Mittelbrunn J. and de Urries, F.J.,
             Nucl. Phys. **B218**, 159 (1983)   [Sec.3.5.1]

[Ish 75]   Isham, C.J., in "Quantum Gravity, An Oxford Symposium" ed. by
             Isham, C.J. et al.(Clarendon Press, 1975), p.1   [Sec.5.1]

[Ito 75]   Ito, K.R., Prog. Theor. Phys. **53**, 817 (1975)   [Sec.2.5.4]

[Jac 74]   Jackiw, R., Phys. Rev. **D9**, 1686 (1974)   [Sec.3.4.4]

[Joh 61a]   Johnson, K., Nucl. Phys. **25**, 435 (1961)   [Sec.2.4.2]

[Joh 61b]   Johnson, K., Nuovo Cim. **20**, 773 (1961)   [Sec.2.5.3]

[Joh 66]   Johnson, K. and Low, F.E., Prog. Theor. Phys. Suppl. No.**37/38**, 74
             (1966)   [Sec.1.4.1]

[Jos 57]   Jost, R., Helv. Phys. Acta **39**, 409 (1957)   [App.A.2]

[Jos 62]   Jost, R. and Hepp, K., Helv. Phys. Acta **35**, 34 (1962)   [Sec.4.3.4]

[Jos 63]   Jost, R., "General Theory of Quantized Fields" (Amer. Math. Soc.
             Publications, 1963)   [App.A.2]

[Käl 52]   Källén, G., Helv. Phys. Acta **25**, 417 (1952)   [Sec.2.3.5]

[Kan 85]   Kanno, H. and Nakanishi, N., Prog. Theor. Phys. **73**, 496 (1985)
            [XIX]   [Sec.5.6.2]

[Kan 89]   Kanno, H., Int. J. Mod. Phys. **A4**, 2765 (1989)   [Sec.5.9.4]

[Kas 66]   Kastler, D., Robinson, D.W. and Swieca, J.A., Comm. Math. Phys.
            2, 108 (1966)   [Sec.1.5.1, App.A.3]

[Kaw 81]   Kawasaki, S., Kimura, T. and Kitago, K., Prog. Theor. Phys. **66**,
            2085 (1981)   [Sec.5.9.2]

[Kaw 82]   Kawasaki, S. and Kimura, T., Prog. Theor. Phys. **68**, 1749 (1982)
                                                               [Sec.5.9.2]

[Kaw 83]   Kawasaki, S. and Kimura, T., Prog. Theor. Phys. **69**, 1015 (1983)
                                                               [Sec.5.9.2]

[Kib 81]   Kibble, T.W.B., in "Quantum Gravity 2, A Second Oxford Sym-
            posium" ed. by Isham, C.J. et al. (Oxford Univ. Press, 1981), p.63
                                                                 [Sec.5.1]

[Kik 83a]   Kikugawa, H. and Kikugawa, J., Prog. Theor. Phys. **69**, 1835 (1983)
                                                               [Sec.5.9.3]

[Kik 83b]   Kikugawa, H. and Kikugawa, J., Prog. Theor. Phys. **70**, 625 (1983)
                                                               [Sec.5.9.3]

[Kim 76]   Kimura, T., Prog. Theor. Phys. **55**, 1259 (1976)   [Sec.5.8.1]

[Kla 68]   Klaiber, B., in "1967 Boulder Lectures in Theoretical Physics" (Gor-
            don and Breach, 1968), Vol.XA, p.141   [Secs.2.5.1, 2.5.2, 2.5.3]

[Kob 63]   Kobayashi, S. and Nomizu, K., "Foundations of Differential Geome-
            try" (Interscience Publishers, 1963)   [App.A.1]

[Kog 74]   Kogut, J. and Susskind, L., Phys. Rev. **D9**, 3501 (1974)   [Sec.4.4.1]

[Kon 76]   Konoue, M. and Nakanishi, N., Prog. Theor. Phys. **56**, 972 (1976)
                                                           [Secs.2.4.4, 2.4.5]

[Kuc 81]   Kuchař, K., in "Quantum Gravity 2, A Second Oxford Symposium"
            ed. by Isham, C.J. et al. (Oxford Univ. Press, 1981), p.329 [Sec.5.1]

[Kug 78a]   Kugo,T. and Ojima,I., Phys. Lett. **73B**, 459 (1978)
                                                   [Secs.3.3.1, 3.4.1, 3.4.3, 4.1.4]

[Kug 78b]   Kugo, T. and Ojima, I., Nucl. Phys. **B144**, 234 (1978)
                                                   [Secs.5.3.2, 5.5.1, 5.8.1]

[Kug 78c]   Kugo,T. and Ojima,I., Prog. Theor. Phys. **60**, 1869 (1978)
                                                    [Secs.3.3.1, 3.4.3, 4.1.4]

[Kug 79a]   Kugo, T. and Ojima, I., Prog. Theor. Phys. **61**, 294 (1979)
                                                    [Secs.4.2, 4.2.1, 4.4.1]

[Kug 79b]   Kugo, T. and Ojima, I., Prog. Theor. Phys. **61**, 644 (1979)
                                                    [Secs.4.2, 4.2.2]

[Kug 79c]   Kugo, T., Phys. Lett. **83B**, 93 (1979)   [Sec.4.4.3]

[Kug 79d]   Kugo, T. and Ojima, I., Prog. Theor. Phys. Suppl. No.**66**, 1 (1979)
                                        [Secs.3.4.1, 3.4.3, 4.1.3, 4.1.4, 4.4.1, 4.4.2]

[Kug 80]   Kugo, T. and Uehara, S., Prog. Theor. Phys. **64**, 1395 (1980)
                                                    [Secs.4.1.2, 4.3.2]

[Kug unp.a] Kugo, T., private communication (1979)   [Sec.3.5.1]

[Kug unp.b] Kugo, T. and Ojima, I., preprint KUNS-486 (1979)
                                                    [Secs.4.1.3, 4.4.1, 4.4.2]

[Kug unp.c] Kugo, T., Soryushiron-Kenkyu (mimeographed circular in Japanese),
            71, 49 (1985)   [Add.1.A]

[Kul 70]   Kulish, P.P. and Faddeev, L.D., Teor. Mat. Fiz. **4**, 153 (1970) [transl.
            Theor. Math. Phys. **4**, 745 (1971)]   [Add.2.A]

[Lan 55]   Landau, L., in "Niels Bohr and the Development of Physics" ed. by
            Pauli, W. (McGraw-Hill, 1955), p.52   [Sec.5.8.2]

[Lang 86]   Langacker, P., in "Proc. of the 1985 Intern. Symp. on Lepton and
            Photon Interactions at High Energies" (Kyoto, 1986), p.186
                                                    [Sec.4.4.4]

[Lau 67]   Lautrup, B., Mat. Fys. Medd. Dan. Vid. Selsk. **35**, No.11 (1967)
                                                    [Sec.2.3]

[Lee 72]   Lee, B.W., Phys. Rev. **D5**, 823 (1972)   [Sec.2.4.3]

[Leh 55]   Lehmann, H., Symanzik, K. and Zimmermann, W., Nuovo Cim. **1**,
            205 (1955)   [Secs.1.4.2, 1.4.4, 4.2.2, App.A.4]

[Leh 57]   Lehmann, H., Symanzik, K. and Zimmermann,W., Nuovo Cim. **6**,
            319 (1957)   [Sec.1.4.4]

[Lei 75]   Leibbrandt, G., Rev. Mod. Phys. **47**, 849 (1975)   [Sec.5.8.2]

[Low 71]   Lowenstein, J.H. and Swieca, J.A., Ann. Phys. **68**, 172 (1971)
                                                    [Secs.2.5.4, 2.5.5]

[Luk 67]   Lukierski, J., Nuovo Cim. Suppl. **5**, 739 (1967)   [Sec.2.3]

[Mai 72]   Maison, H.D., Nuovo Cim. **11A**, 389 (1972)   [App.*A*.3]

[Man 75]   Mandelstam, S., Phys. Rev. **D11**, 3026 (1975)   [Sec.2.5.3]

[Mas unp]   Maskawa, T., private communication (1979)   [Sec.4.4.2]

[Mat 74]   Mathews, P.M., Seetharaman, M. and Simon, M.T., Phys. Rev. **D9**,
            1700 (1974)   [Sec.2.2.2]

[Mit 78]   Mitra, P., Prog. Theor. Phys. **60**, 892 (1978)   [Sec.2.5.3]

[Miy 87]   Miyake, S. and Shizuya, K., Phys. Rev. **D36**, 3781 (1987) [Sec.2.5.4]

[Miy 88]   Miyake, S. and Shizuya, K., Phys. Rev. **D37**, 2282 (1988) [Sec.2.5.4]

[Miy 89]   Miyake, S. and Shizuya, K., Mod. Phys. Lett. **A4**, 775 (1989)
                                                                    [Sec.2.5.4]

[Mor 83]   Morchio, G. and Strocchi, F., Nucl. Phys. **B211**, 471 (1983)
                                                                    [Add.2.*A*]

[Nak 66]   Nakanishi, N., Prog. Theor. Phys. **35**, 1111 (1966)   [Sec.2.3]

[Nak 67]   Nakanishi, N., Prog. Theor. Phys. **38**, 881 (1967)   [Sec.2.3]

[Nak 69]   Nakanishi, N., Prog. Theor. Phys. Suppl. No.**43**, 1 (1969)
                                                                    [Sec.4.1.3]

[Nak 72a]   Nakanishi, N., Phys. Rev. **D5**, 1324 (1972)   [Sec.2.4.2]

[Nak 72b]   Nakanishi, N., Prog. Theor. Phys. Suppl. No.**51**, 1 (1972)
                                                                 [Secs.2.3, 2.4.2]

[Nak 73a]   Nakanishi, N., Prog. Theor. Phys. **49**, 640 (1973)
                                              [Secs.2.4.3, 2.4.4, 4.2.1, 4.3.4]

[Nak 73b]   Nakanishi, N., Prog. Theor. Phys. **50**, 1388 (1973)   [Sec.2.4.5]

[Nak 74a]   Nakanishi, N., Prog. Theor. Phys. **51**, 952 (1974)   [App.*A*.5]

[Nak 74b]   Nakanishi, N., Prog. Theor. Phys. **52**, 1929 (1974)   [Sec.4.2.2]

[Nak 75a]   Nakanishi, N., "Quantum Field Theory" (Baifukan, 1975), Chapters
            3 and 6 (in Japanese)   [Secs.2.3, 2.3.7, 2.3.8]

[Nak 75b]   Nakanishi, N., Prog. Theor. Phys. **54**, 840 (1975)
                                                              [Secs.2.4.3, 2.5.4]

[Nak 77a]   Nakanishi, N., Prog. Theor. Phys. **57**, 269 (1977)

                                                    [Secs.2.5.1, 2.5.2]

[Nak 77b]   Nakanishi, N., Prog. Theor. Phys. **57**, 580 (1977)   [Secs.2.5.3, 2.5.4]

[Nak 77c]   Nakanishi, N., Prog. Theor. Phys. **57**, 1025 (1977) [Secs.2.5.3, 2.5.4]

[Nak 77d]   Nakanishi, N., Prog. Theor. Phys. **57**, 1079 (1977)   [Sec.2.5.6]

[Nak 77e]   Nakanishi, N., Prog. Theor. Phys. **58**, 1007 (1977) [Secs.2.5.2, 2.5.3]

[Nak 77f]   Nakanishi, N., Prog. Theor. Phys. **58**, 1580 (1977)   [Sec.2.5.5]

[Nak 77g]   Nakanishi, N., Prog. Theor. Phys. **58**, 1927 (1977)   [Sec.2.5.4]

[Nak 78a]   Nakanishi, N. and Ojima, I., Prog. Theor. Phys. **59**, 242 (1978)

                                                    [App.*A*.4]

[Nak 78b]   Nakanishi, N., Prog. Theor. Phys. **59**, 972 (1978). [I]

                                            [Secs.5.3.1, 5.3.2, 5.5.1, 5.8.1]

[Nak 78c]   Nakanishi, N., Prog. Theor. Phys. **59**, 2175 (1978)   [Sec.5.3.2]

[Nak 78d]   Nakanishi, N., Prog. Theor. Phys. **60**, 284 (1978)   [Sec.3.3.2]

[Nak 78e]   Nakanishi, N., Prog. Theor. Phys. **60**, 1190 (1978) [II]

                                                [Secs.5.4, 5.4.3, 5.4.4]

[Nak 78f]   Nakanishi, N., Prog. Theor. Phys. **60**, 1890 (1978) [III]   [Sec.5.5.1]

[Nak 79a]   Nakanishi, N., Prog. Theor. Phys. **61**, 1536 (1979) [IV]   [Sec.5.6.2]

[Nak 79b]   Nakanishi, N. and Ojima, I., Phys. Rev. Lett. **43**, 91 (1979)

                                                    [Sec.5.5.4]

[Nak 79c]   Nakanishi, N., Prog. Theor. Phys. **62**, 779 (1979) [V]

                                                [Secs.5.7.2, 5.7.3, 5.8.1]

[Nak 79d]   Nakanishi, N., Prog. Theor. Phys. **62**, 1101 (1979) [VI]   [Sec.5.7.4]

[Nak 79e]   Nakanishi, N., Prog. Theor. Phys. **62**, 1385 (1979) [VII]

                                                [Secs.5.7.6, 5.8.1]

[Nak 79f]   Nakanishi, N., Prog. Theor. Phys. **62**, 1396 (1979) [Secs.4.1.2, 4.3.2]

[Nak 80a]   Nakanishi, N., Zeits. f. Physik C, Particles and Fields **4**, 17 (1980)

                                                    [Sec.2.5.2]

[Nak 80b]   Nakanishi, N., Prog. Theor. Phys. **63**, 656 (1980) [VIII]

                                                [Secs.5.6.1, 5.7.4]

416   REFERENCES

[Nak 80c]   Nakanishi, N., Prog. Theor. Phys. **63**, 1823 (1980) [Secs.5.8.2, 5.8.3]

[Nak 80d]   Nakanishi, N., Prog. Theor. Phys. **63**, 2078 (1980) [IX]
                                          [Secs.2.4.3, 5.5.2, 5.8.1, 5.9.3]

[Nak 80e]   Nakanishi, N., Prog. Theor. Phys. **64**, 639 (1980) [X]   [Sec.5.5.2]

[Nak 80f]   Nakanishi, N. and Ojima, I., Zeits. f. Physik C, Particles and Fields
             6, 155 (1980)   [Secs.3.5.2, 5.7.5]

[Nak 81a]   Nakanishi, N. and Ojima, I., Prog. Theor. Phys. **65**, 728 (1981) [XI]
                                                              [Sec.5.5.4]

[Nak 81b]   Nakanishi, N. and Ojima, I., Prog. Theor. Phys. **65**, 1041 (1981) [XII]
                                                              [Sec.5.9.1]

[Nak 81c]   Nakanishi, N. and Yamagishi, K., Prog. Theor. Phys. **65**, 1719 (1981)
             [XIII]   [Sec.5.5.4]

[Nak 81d]   Nakanishi, N., Prog. Theor. Phys. **66**, 1843 (1981) [XIV]
                                                              [Secs.5.5.3, 5.7.6]

[Nak 82]   Nakanishi, N., Prog. Theor. Phys. **68**, 947 (1982) [XV]
                                                              [Secs.5.6.1, 5.6.2]

[Nak 83a]   Nakanishi, N., Prog. Theor. Phys. **69**, 1617 (1983) [XVI]
                                                              [Secs.5.6.3, 5.6.4]

[Nak 83b]   Nakanishi, N., Prog. Theor. Phys. **70**, 551 (1983) [XVII]
                                                              [Secs.5.6.5, 5.6.6]

[Nak 83c]   Nakanishi, N., Phys. Lett. **131B**, 381 (1983)   [Sec.2.2.1]

[Nak 83d]   Nakanishi, N., Publ. RIMS, Kyoto Univ. **19**, 1095 (1983)   [Sec.5.1]

[Nak 84a]   Nakanishi, N. and Ojima, I., Prog. Theor. Phys. **71**, 1359 (1984)
                                                              [Sec.4.4.3]

[Nak 84b]   Nakanishi, N. and Ojima, I., Prog. Theor. Phys. **72**, 1197(1984)
                                                              [Sec.4.4.3]

[Nak 84c]   Nakanishi, N., Prog. Theor. Phys. **72**, 1233 (1984) [XVIII]
                                                              [Sec.5.6.3]

[Nak 85]   Nakanishi, N., Prog. Theor. Phys. **73**, 1016 (1985)   [Sec.5.9.2]

[Nak 87]   Nakanishi, N., Prog. Theor. Phys. **77**, 1533 (1987)   [Sec.5.9.4]

[Nakw 84a]   Nakawaki, Y., Prog. Theor. Phys. **72**, 134 (1984)   [Sec.2.5.4]

[Nakw 84b] Nakawaki, Y., Prog. Theor. Phys. **72**, 152 (1984)   [Sec.2.5.4]

[Nam 61]   Nambu, Y. and Jona-Lasinio, G., Phys. Rev. **122**, 345 (1961)
                                                           [Secs.1.5, 1.5.3]

[Nes 81]   Nester, J.M., Phys. Lett. **83A**, 241 (1981)   [Sec.5.2.3]

[Nis 73]   Nishijima, K., Prog. Theor. Phys. **50**, 1381 (1973)   [Sec.2.4.3]

[Nis 78]   Nishijima, K. and Okawa, M., Prog. Theor. Phys. **60**, 272 (1978)
                                                           [Secs.5.3.2, 5.5.1]

[Nis 82]   Nishijima, K., Phys. Lett. **116B**, 295 (1982)   [Sec.4.4.2]

[Nis 84]   Nishijima, K., Prog. Theor. Phys. **72**, 1214 (1984) [Secs.3.5.2, 4.4.2]

[Nis 85a]   Nishijima, K., Prog. Theor. Phys. **73**, 536 (1985)   [Sec.3.5.2]

[Nis 85b]   Nishijima, K., Prog. Theor. Phys. **74**, 889 (1985)   [Sec.4.4.2]

[Nis 86]   Nishijima, K., Prog. Theor. Phys. **75**, 1221 (1986)   [Sec.4.4.2]

[Nis 87]   Nishijima, K., "Field Theory" (Kinikuniya, 1987), p.376 (in
            Japanese)   [Sec.3.4.3]

[Oeh 80a]   Oehme, R. and Zimmermann, W., Phys. Rev. **D21**, 471(1980)
                                                               [Sec.4.4.2]

[Oeh 80b]   Oehme, R. and Zimmermann, W., Phys. Rev. **D21**, 1661(1980)
                                                               [Sec.4.4.2]

[Oji 78]   Ojima, I., Nucl. Phys. **B143**, 340 (1978)   [Secs.3.4.3, 4.3.1, 4.3.2]

[Oji 79]   Ojima, I. and Hata, H., Zeits. f. Physik C, Particles and Fields **1**,
            405 (1979)   [Sec.4.3.3, Add.2.*A*]

[Oji 80a]   Ojima, I., Prog. Theor. Phys. **64**, 625 (1980)   [Secs.3.4.1, 3.5.1]

[Oji 80b]   Ojima, I., Zeits. f. Physik C, Particles and Fields, **5**, 227 (1980)
                                                               [Sec.4.3.4]

[Oji 82]   Ojima, I., Zeits. f. Physik C, Particles and Fields, **13**, 173 (1982)
                                                    [Secs.3.3.2, 3.5.2, 4.1.4]

[Orz 70]   Orzalesi, C.A., Rev. Mod. Phys. **42**, 381 (1970)   [App.*A*.3]

[Pag 81]   Page, D.N. and Geilker, C.D., Phys. Rev. Lett. **47**, 979 (1981)
                                                                 [Sec.5.1]

[Pas 82]   Pasti, P. and Tonin, M., Nuovo Cim. **69B**, 97 (1982)   [Sec.5.9.3]

[Pro 36]   Proca, J., J. Phys. Rad. **7**, 347 (1936)   [Sec.2.4.1]

[Qui 81]   Quirós, M., de Urries, F.J., Hoyos, J., Mazón, M.L. and Rodriguez,
              E., J. Math. Phys. **22**, 1767 (1981)   [Sec.3.5.1]

[Ree 61]   Reeh, H. and Schlieder, S., Nuovo Cim. **22**, 1051 (1961)
                                            [Secs.4.3.2, 4.3.3, App.*A*.2]

[Ree 68]   Reeh, H., Forts. der Physik **16**, 687 (1968)   [Sec.1.5.1, App.*A*.3]

[Rob 62]   Robinson, D.W., Helv. Phys. Acta **35**, 403 (1962)
                                            [Secs.1.4.2, 3.1, App.*A*.4]

[Rue 62]   Ruelle, D., Helv. Phys. Acta **35**, 147 (1962)
                                            [Sec.1.4.3, App.*A*.2, App.*A*.4]

[Sat unp]   Sato, K., preprint EPHOU 87 APR003 (1987)   [Sec.5.9.2]

[Sch 51]   Schwinger, J., Phys. Rev. **82**, 664 (1951)   [Sec.2.5.4]

[Sch 59]   Schwinger, J., Phys. Rev. Lett. **3**, 296 (1959)   [Sec.1.4.1]

[Sch 62a]   Schwinger, J., Phys. Rev. **125**, 397 (1962)   [Sec.4.4.1]

[Sch 62b]   Schwinger, J., Phys. Rev. **128**, 2425 (1962)   [Secs.2.5.4, 4.4.1]

[Sch 63]   Schwinger, J., Phys. Rev. **130**, 402 (1963)   [Sec.2.3]

[Schl 69]   Schlieder, S., Comm. Math. Phys. **13**, 216 (1969)   [Sec.4.3.4]

[Scho 79]   Schoen, R.M. and Yau, S.-T., Comm. Math. Phys. **65**, 45 (1979)
                                                              [Sec.5.2.3]

[Scho 82]   Schoen, R.M. and Yau, S.-T., Phys. Rev. Lett. **48**, 369 (1982)
                                                              [Sec.5.2.3]

[Schr 63]   Schroer, B., Forts. der Physik **17**, 1 (1963)   [Sec.2.5.6]

[Sen 67]   Sen, R.N. and Umezawa, H., Nuovo Cim. **50**, 53 (1967)
                                                    [Sec.1.5.1, App.*A*.4]

[Ser 65]   Serre, J.-P., "Lie Algebras and Lie Groups" (W.A.Benjamin, Inc.,
              1965)   [Sec.3.2.1]

[Sha 74]   Sharp, D.H. and Wightman, A.S., "Local Currents and their Appli-
              cations", (North-Holland, 1974) p.(ix)   [Sec.4.3.4]

[Sla 72]   Slavnov, A., Teor. Mat. Fiz. **10**, 153 (1972) [transl. Theor. Math.
              Phys. **10**, 99 (1972)]   [Secs.3.1, 3.4.4]

[Smi 74]   Smit, J., Phys. Rev. **D10**, 2473 (1974)   [Sec.4.4.1]

[Sol 56]   Solov'ev, L.D., Doklady Akad. Nauk. USSR **110**, 203 (1956) [transl.
           Sov. Phys. Doklady **1**, 536 (1956)]   [Add.2.*A*]

[Ste 77]   Stelle, K.S., Phys. Rev. **D16**, 953 (1977)   [Sec.5.3.1]

[Str 67]   Strocchi, F., Phys. Rev. **162**, 1429 (1967)   [Sec.2.2.1]

[Str 74]   Strocchi, F. and Wightman, A.S., J. Math. Phys. **15**, 2198 (1974)
                                                            [Secs.4.3.1, 4.3.4]

[Str 76]   Strocchi, F., Phys. Lett. **62B**, 60 (1976)   [Secs.4.3.2, 4.3.4]

[Str 77]   Strocchi, F., Comm. Math. Phys. **56**, 57 (1977)
                                               [Sec.4.3.4, App.*A*.2, App.*A*.3]

[Str 78]   Strocchi, F., Phys. Rev. **D17**, 2010 (1978)   [Secs.4.3.2, 4.3.4]

[Stre 64] Streater, R.F. and Wightman, A.S., "PCT, Spin and Statistics and
           All That" (Benjamin, 1964; revised ed. 1978)   [App.*A*.2]

[Stu 38]   Stueckelberg, E.C.G., Helv. Phys. Acta **11**, 299 (1938)   [Sec.2.4.1]

[Sug 77]   Sugamoto, A., Phys. Rev. **D16**, 1065 (1977)   [Sec.4.4.2]

[Sul 77]   Sullivan, D., Publ. IHES **47**, 269 (1977) [Secs.4.1.2, 4.3.2, App.*A*.1]

[Suz 82a] Suzuki, T. and Tameike, S., Prog. Theor. Phys. **67**, 920 (1982)
                                                                    [Sec.2.5.2]

[Suz 82b] Suzuki, T. and Tameike, S., Prog. Theor. Phys. **67**, 935 (1982)
                                                                    [Sec.2.5.2]

[SuzI 85] Suzuki, I., Prog. Theor. Phys. **73**, 197 (1985)   [Sec.2.4.5]

[Swi 67]   Swieca, J.A., Nuovo Cim. **52A**, 242 (1967)   [Sec.4.4.1]

[Tan 76]   Tanaka, A. and Yamamoto, K., Prog. Theor. Phys. **56**, 241 (1976)
                                                                    [Sec.2.5.3]

[Tay 71]   Taylor, J.C., Nucl. Phys. **B33**, 436 (1971)   [Secs.3.1, 3.4.4]

[Thi 58]   Thirring, W., Ann. Phys. **3**, 91 (1958)   [Sec.2.5.3]

[Thi 64]   Thirring, W.E. and Wess, J.E., Ann. Phys. **27**, 331 (1964) [Sec.2.5.5]

[Thie 79] Thierry-Mieg, J. and Ne'eman, Y., Ann. Phys. **123**, 247 (1979)
                                                                    [Sec.3.5.1]

[tHo 71]   't Hooft, G., Nucl. Phys. **B35**, 167 (1971)   [Secs.4.1.5, 4.2.1]

420   REFERENCES

[tHo 72]   't Hooft, G. and Veltman, M., Nucl. Phys. **B50**, 318 (1972)
                                                                [Sec.3.4.4]

[tHo 74]   't Hooft, G. and Veltman, M., Ann. Inst. Henri Poincaré **20**, 69
            (1974)   [Sec.5.8.2]

[Ton 86]   Tonomura, A., Osakabe, N., Matsuda, T., Kawasaki, T., Endo, J.,
            Yano, S. and Yamada, H., Phys. Rev. Lett. **56**, 792 (1986) [Sec.2.1.2]

[Tow 77]   Townsend, P.K. and Nieuwenhuizen, P.van, Nucl. Phys. **B120**, 301
            (1977)   [Sec.5.3.1]

[Tye 75]   Tye, S.-H.H., Tomboulis, E. and Poggio, E.C., Phys. Rev. **D11**, 2839
            (1975)   [Sec.4.4.1]

[Tyu unp]   Tyutin, I.V., Lebedev preprint FIAN No.39 (1975)   [Sec.3.1]

[Ume 65]   Umezawa, H., Nuovo Cim. **40**, 450 (1965)   [Sec.1.5.1]

[Uti 59]   Utiyama, R., Prog. Theor. Phys. Suppl. No.9, 19 (1959)   [Sec.2.3]

[Völ 74]   Völkel, A.H., Rep. on Math. Phys. **6**, 313 (1974)   [App.*A*.3]

[Völ 77]   Völkel, A.H., "Fields, Particles and Currents", Lecture Notes in
            Physics No.66 (Springer-Verlag, 1977)   [Apps.*A*.2, *A*.3]

[Wei 64]   Weinberg, S., Phys. Rev. **135**, 1049 (1964)   [Sec.4.4.4]

[Wei 72]   Weinberg, S., "Gravitation and Cosmology: Principles and Applica-
            tions of the General Theory of Relativity" (John Wiley and Sons,
            1972)   [Sec.4.3.2]

[Wei 73]   Weinberg, S., Phys. Rev. Lett. **31**, 494 (1973)   [Sec.4.4.2]

[Wes 83]   Wess, J. and Bagger, J., "Supersymmetry and Supergravity" (Prince-
            ton Univ. Press, 1983)   [Sec.5.9.4]

[Wes 74]   Wess, J. and Zumino, B., Nucl. Phys. **B70**, 39 (1974)   [Sec.5.9.4]

[Wig 56]   Wightman, A.S., Phys. Rev. **101**, 860 (1956)   [App.*A*.2]

[Wig 67]   Wightman, A.S., in "1964 Cargèse Lectures in Theoretical Physics"
            (Gordon and Breach, 1967), p.171   [Sec.2.5.2, App.*A*.4]

[Wit 81]   Witten, E., Comm. Math. Phys. **80**, 381 (1981)   [Sec.5.2.3]

[Wit 88]   Witten, E., Nucl. Phys. **B311**, 46 (1988)   [Sec.5.8.4]

[Yan 50]   Yang, C.N. and Feldman, D., Phys. Rev. **79**, 972 (1950)
                                                        [Sec.1.4.2, App.*A*.4]

[Yen 61]   Yennie, D.R., Frautschi, S.C. and Suura, H., Ann. Phys. **13**, 379
            (1961)   [Add.2.*A*]

[Yng 77]   Yngvason, J., Rep. on Math. Phys. **12**, 57 (1977)   [App.*A*.2]

[Yok 68]   Yokoyama, K., Prog. Theor. Phys. **40**, 160 (1968)   [Sec.2.3]

[Yok 74a]   Yokoyama, K., Prog. Theor. Phys. **51**, 1956 (1974)   [Sec.2.3.8]

[Yok 74b]   Yokoyama, K., Prog. Theor. Phys. **52**, 1669 (1974)   [Sec.2.4.2]

[Zim 58]   Zimmermann, W., Nuovo Cim. **10**, 597 (1958)   [App.*A*.4]

[Zum 84]   Zumino, B., in "Relativity, Groups and Topology II, Les Houches
            1983", ed. by DeWitt, B.S. and Stora, R. (North-Holland, 1984),
            p.1291   [Sec.4.3.2]

[Zwa 76]   Zwanziger, D., Phys. Rev. **D14**, 2570 (1976)   [Add.2.*A*]

INDEX